D1617211

ACS SYMPOSIUM SERIES 492

Polymer Latexes

Preparation, Characterization, and Applications

Eric S. Daniels, EDITOR
Lehigh University

E. David Sudol, EDITOR
Lehigh University

Mohamed S. El-Aasser, EDITOR
Lehigh University

Developed from a symposium sponsored
by the Division of Polymeric Materials: Science and Engineering
and the Division of Colloid and Surface Chemistry
at the 201st National Meeting
of the American Chemical Society,
Atlanta, Georgia,
April 14–19, 1991

American Chemical Society, Washington, DC 1992

Library of Congress Cataloging-in-Publication Data

Polymer latexes: preparation, characterization, and applications / Eric S. Daniels, editor, E. David Sudol, editor, Mohamed S. El-Aasser, editor.

p. cm.—(ACS symposium series, 0097–6156; 492)

"Developed from a symposium sponsored by the Division of Polymeric Materials: Science and Engineering and the Division of Colloid and Surface Chemistry at the 201st National Meeting of the American Chemical Society, Atlanta, Georgia, April 14–19, 1991."

Includes bibliographical references and indexes.

ISBN 0–8412–2305–X

1. Latex, Synthetic—Congresses. 2. Polymerization—Congresses.

I. Daniels, Eric S., 1955– . II. Sudol, E. David, 1952– . III. El-Aasser, Mohamed S. IV. American Chemical Society. Division of Polymeric Materials: Science and Engineering. V. American Chemical Society. Division of Colloid and Surface Chemistry. VI. American Chemical Society. Meeting (201st: 1991: Atlanta, Ga.) VII. Series.

TS1925.P629 1992
678′.5—dc20

92–8683
CIP

The paper used in this publication meets the minimum requirements of American National Standard for Information Sciences—Permanence of Paper for Printed Library Materials, ANSI Z39.48–1984. ♾

PRINTED IN THE UNITED STATES OF AMERICA

ACS Symposium Series

M. Joan Comstock, *Series Editor*

1992 ACS Books Advisory Board

Foreword

THE ACS SYMPOSIUM SERIES was founded in 1974 to provide a medium for publishing symposia quickly in book form. The format of the Series parallels that of the continuing ADVANCES IN CHEMISTRY SERIES except that, in order to save time, the papers are not typeset, but are reproduced as they are submitted by the authors in camera-ready form. Papers are reviewed under the supervision of the editors with the assistance of the Advisory Board and are selected to maintain the integrity of the symposia. Both reviews and reports of research are acceptable, because symposia may embrace both types of presentation. However, verbatim reproductions of previously published papers are not accepted.

Contents

INDEXES

Preface

THE FIELD OF EMULSION POLYMERS and latex technology continues to develop at a rapid pace. The driving force for this development is both commercial and environmental; latexes are water-based polymers in the form of colloidal particles dispersed in an aqueous medium. Millions of pounds of latexes are manufactured today for use as commodity and specialty polymers in a wide variety of applications, such as synthetic rubber, coatings, paints, adhesives, caulks, sealants, binders for nonwoven fabrics, high-impact thermoplastic composites, latex foam, carpet backing, and as additives in construction materials such as cement and concrete. Latexes are finding new and numerous applications in the biomedical and biotechnology fields, such as in diagnostic tests for AIDS, immunoassays, biological cell-labeling, and drug-delivery systems.

These applications and future developments in the latex field are based on advances in our understanding of the mechanism and kinetics of emulsion polymerization, the process by which latex is manufactured. Equally important are the relationships between process parameters and the colloidal, surface, and morphological properties of the latex particles themselves, as well as the properties of the end products in which the latex particles are employed.

The symposium on which this book is based focused on three aspects of emulsion polymerization and latex technology: the mechanisms of particle nucleation (the most important and most controversial step in emulsion polymerization) and particle growth; microstructure and particle morphology; and recent advances in applications such as immunodiagnostics, encapsulation of pigments, and catalysis.

Two leading scientists in the field were invited to review the state of the art of the mechanisms of particle nucleation and growth. Finn K. Hansen (University of Oslo, Norway) contributed a personal view of where the theories of nucleation in emulsion polymerization currently stand and where further advances are required. In the next chapter, Robert G. Gilbert (University of Sydney, Australia) and his colleagues first examine data that have been used to prove or disprove micellar nucleation; they then provide further information and arguments for a combined nucleation mechanism.

Of the following chapters, six contributions relate to recent progress in the area of particle formation and growth. Advances in the study and characterization of polymer microstructure and particle morphology are discussed in 11 chapters. Seven chapters deal with diverse applications of and extensions from polymer latexes in such areas as catalysis and immunodiagnostics.

The best excuse for organizing a symposium and the best way of ensuring its success in terms of participation is to honor a colleague. This symposium was organized to honor Robert M. Fitch and his contributions to the field of emulsion polymerization. His fame in this field is due to, among other contributions, the quantification of the homogeneous mechanism for particle formation in emulsion polymerization. The symposium, which succeeded in attracting numerous scientists from academia and industry, resulted in lively discussions about the current status and outstanding issues in the field. This volume should prove useful to engineers and scientists who are actively involved in the fields of emulsion polymerization and latex technology.

We thank all of the authors for their prompt responses to our requests. In addition, we thank the many reviewers, whose names will remain anonymous, for putting up with our intrusion on their time and for making valuable suggestions to the authors.

ERIC S. DANIELS
E. DAVID SUDOL
MOHAMED S. EL-AASSER
Lehigh University
Bethlehem, PA 18015

January 24, 1992

Chapter 1

Overview of Emulsion Polymerization

Stepping Toward Prediction

E. David Sudol[1], Eric S. Daniels[1], and Mohamed S. El-Aasser[1,2]

[1]Emulsion Polymers Institute and [2]Department of Chemical Engineering, Lehigh University, Bethlehem, PA 18015

Since the first conceptual and mathematical descriptions of emulsion polymerization were advanced over forty years ago, new processes have necessitated new and modified theories for the nucleation and growth of polymer particles in the presence and absence of micelles and submicron monomer droplets. The advances in our understanding have been considerable and yet numerous uncertainties remain; *prediction* continues to be the ultimate goal.

The use of multiple monomers, introduced through a number of often complex process variations, greatly complicates the development of these predictive models. An understanding of the development and control of particle microstructure and morphology during emulsion polymerization is an intense area of investigation, driven by the commercial need and application of polymer particles possessing narrowly defined properties. Advances in our abilities to prepare and characterize these particles continue to evolve.

The word *chaos* has been used to describe physical behavior which is macroscopically unpredictable in detail despite knowledge of the physics of individual events which are a subset of this behavior (*1*). The oft cited example of this is the motion of a spherical pendulum which when set in motion rapidly reaches a state in which its exact movements are no longer predictable (i.e., its motion deviates from predictions). In some ways, emulsion polymerization can be viewed as *chaotic*. The fate of a single species, such as a monomer molecule, is not predictable. At best, some estimate of its probability of reacting and becoming a part of a polymer particle can be made. But which particle, where in the particle, at what position in a polymer chain, these cannot be predicted.

Although *chaos* is not used to describe emulsion polymerization, it has frequently been referred to as an *art*. And those who *practice the art*, recognize the difficulty in consistently reproducing a batch of latex in terms of desired properties such as stability, particle size and distribution, mechanical properties in application, etc. The origins of irreproducibilities most often derive from a lack of a complete understanding of a process and the variables which affect this process. In emulsion polymerization, the process of creating latex particles is by far considered the greatest source of irreproducibilities observed. Typically, nucleation is described as being chiefly dependent on the

0097–6156/92/0492–0001$06.00/0

emulsifier (its concentration and properties) and the initiator (its concentration or the rate of radical generation). The monomer (type(s) and concentration(s)), added electrolyte (type and concentration), and reaction temperature are also key variables in the nucleation process. Quantification of the time evolution of the number of latex particles as a function of these variables has been advanced by many authors striving to transform the *art* into a *science*, and perhaps in a sense, to impose *order* on *chaos*. These efforts have been generally successful in modeling the behavior of specific systems under specific conditions but not without some adjustment of mechanistic parameters which are at best uncertain. This uncertainty has led others to invest effort in the clarification of important rate parameters governing the kinetics of emulsion polymerization (*2, 3*). These include the initiator decomposition rate constant (k_d), the monomer propagation rate constant (k_p), the first- (ρ) and second-order (k_a) rate coefficients for radical entry into latex particles, the desorption rate coefficient from the polymer particles (k), and the termination rate constants in the particle (k_t) and aqueous phases (k_{tw}). Activity in this area continues as an essential element to improved mechanistic model development. However, true predictability has yet to be achieved.

One means of exerting some control over an emulsion polymerization is to introduce a preformed seed. Although this may circumvent the more complex nucleation phase of the reaction, provided that conditions are such that no particles are generated or lost, sufficient uncertainties exist which make the predictability of further reactions difficult. Even in the simplest case where the seed and second stage polymer are the same, serious questions persist regarding the nature of free radical entry and the radical distribution within the polymer particles. Radicals originating in the aqueous phase (by decomposition of an initiator, typically $K_2S_2O_8$) can suffer a variety of fates, principally propagation and termination in the aqueous phase or in the polymer particles. The mechanism by which a radical arrives and propagates in a particle has been described in turn by collisional, diffusive, colloidal, and propagative mechanisms. The latter has been advanced most recently with support from a large body of experimental work (*4*). Unlike the other mechanisms, it claims no dependence of entry on the particle diameter.

On the other hand, the particle size may affect the resulting "morphology" through its effect on the radial distribution of radicals within the particles. This concept has been advanced, rejected, readvanced, and so on, through experimental and theoretical developments. What was once considered "simple" may not be so simple. Although this subject has been controversial in homopolymerizations of compatible monomer/polymer systems, morphology development in copolymerization of incompatible polymers has long been recognized and efforts at controlling morphology have recently become focussed and intense. Guidelines continue to evolve for the preparation of homogeneous copolymers (with a single T_g) or particles possessing core-shell structures, these representing the morphological limits of these latex particles. In practice, particles with morphologies ranging between these extremes are often found and even multiple morphologies can be found in a single latex.

Although the papers included in this volume cover a broad range of research activities related to polymer colloids, much of the activity is reflected in the preceeding areas of nucleation and particle morphology. (A bibliography of books published on the subject of *Emulsion Polymerization* over the past twenty years is given at the end of this chapter.)

Nucleation

Harkins (*5*) and Smith & Ewart (*6*) began more than forty years ago what would become a long series of incremental steps (some large & some small) towards the understanding of the complex mechanisms of particle formation and growth in a process simply referred to as *emulsion polymerization*. The research is ongoing and published work reflects the difficulties (and successes) in advancing the science; both forward and backward steps have been taken, each contributing to the body of knowledge arrayed at understanding the phenomena of this complex heterogeneous process. And yet the original process has broadened considerably.

Much attention was initially focussed on the emulsion copolymerization of styrene and butadiene through the U.S. Government Synthetic Rubber Program. Relatively high emulsifier ("soap" SF Flakes) concentrations were employed in the "Mutual Recipe" to produce large numbers of small latex particles. Emulsion polymerizations were carried out with these relatively water insoluble monomers and at emulsifier concentrations well above the critical micelle concentration (cmc). It seemed quite logical to conclude that the micelles acted as loci for the formation of growing polymer particles and convincing arguments were put forth by Smith and Ewart to support the Harkins conceptualization.

Variations introduced to the process of emulsion polymerization brought new information which could not be so readily described by early concepts. Polymerizations producing colloidal polymer particles were being successfully carried out with emulsifier concentrations well below their cmc and even without any added emulsifier. And the "emulsifiers" themselves have become more complex in the sense that combinations of a bewildering variety of ionic, nonionic, and polymeric stabilizers are being employed not only to nucleate and stabilize particles but also to control surface tensions and rheological properties. More water soluble monomers have become commonly employed alone and in combination; these polymerizations have not been found to readily fit into the predictions of the theory.

Efforts to modify, expand, and rewrite the theories for nucleation and growth in emulsion polymerization have proceeded with increasing sophistication and success (see Chapter 2, this volume, which reviews much of the history and developments leading to our current understanding). Particle formation is considered to take place in the aqueous phase, surfactant micelles, and monomer droplets. An illustration of the early stages of emulsion polymerization is presented in Figure 1 which depicts three scales of observation: the macroscopic, microscopic, and submicroscopic. The unaided eye typically notices few (if any) remarkable changes during the polymerization; an opaque, white emulsion produces an opaque, white latex (subtle changes are observable to the experienced practitioner, however). An optical microscope can be used to observe grander changes as the broad distribution of monomer droplets, created by the shear of the impeller, begin to shrink and disappear being replaced by a background of fine specks, the latex particles. It is on the submicroscopic scale, however, that most of what is of greatest interest occurs and yet this scale is the most difficult to observe. The dynamic processes occuring on this level determine the formation and stabilization of latex particles, whether in the continuous phase, in micelles, or in monomer droplets. Much of the theory of this process has been based on indirect observation, evidence, and speculation.

A mathematical description of nucleation in these loci is encompassed in what has become known as HUFT theory, representing the contributions of Hansen & Ugelstad (*7*) and Fitch & Tsai (*8*). The relative contribution of each nucleation site in a particular system is governed by such factors as the solubility of the monomer in the continuous phase, the number of micelles, and the size and number of monomer droplets. Although this theory is comprehensive, voids in our understanding persist. The dynamics of aqueous phase radical growth and subsequent ad(b)sorption/desorption in micelles, droplets, and particles are not yet resolved. Questions continue to be raised regarding the nature of the radicals (particularly oligomeric) and their interaction with these somewhat different species and additionally the free surfactant. These lie at the heart of the complexities in further understanding nucleation in emulsion polymerization. (Chapters 2, 3, and 4 of this volume address these issues and raise further questions.)

Nucleation in monomer droplets becomes important when their size is reduced to within the same order of magnitude as the latex particles. (Chapters 6 and 7 of this

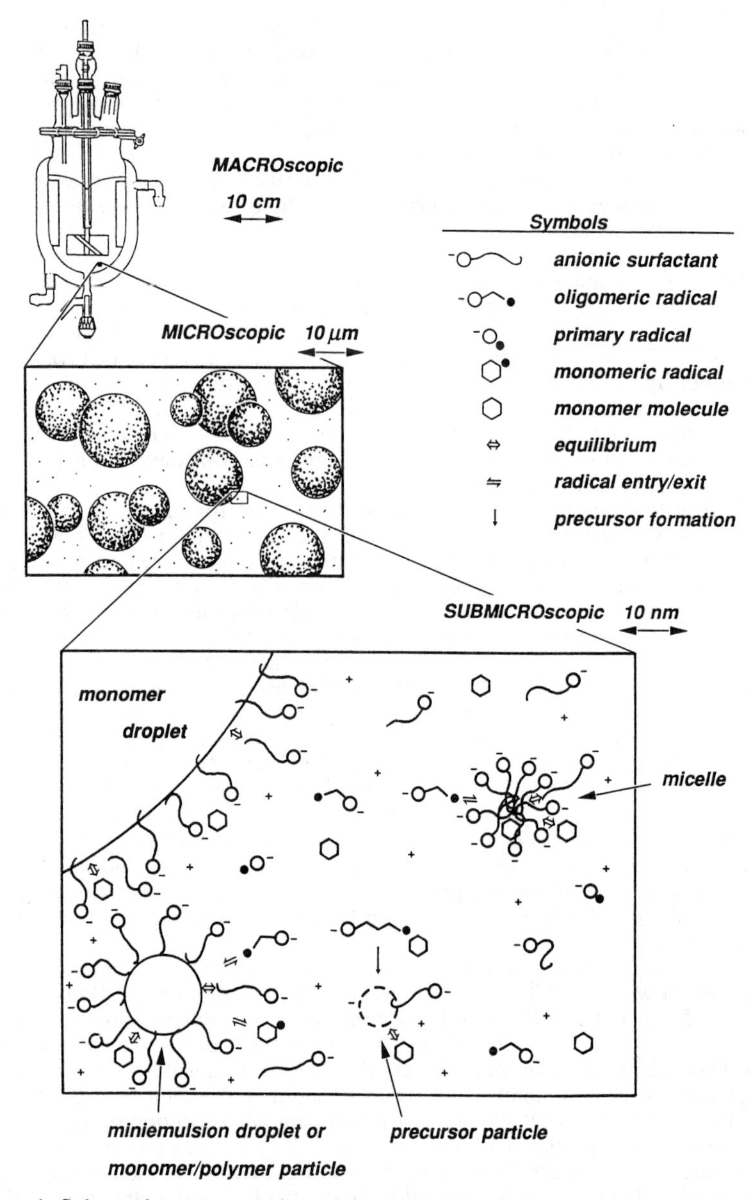

Figure 1. Schematic representation of the early stages of emulsion polymerization illustrating three scales of observation: macroscopic, microscopic, and submicroscopic.

volume describe polymerization in miniemulsions and microemulsions, respectively.) This can be accomplished by addition of relatively large amounts of "co-surfactant" as in microemulsions or energy in the form of shear, as in miniemulsions (which require some "co-surfactant" for long term stability). Efforts at modelling the polymerization of these systems have indicated that radical entry and growth (sometimes defined as "entry") in these droplets is "slow" in the sense that the efficiency of radicals effecting polymerization in droplets resulting in polymer particles is low compared to polymer particles. Similar findings are reported for micelles (Chapter 2, this volume). Clearly, our understanding of this process is incomplete and requires more work.

The immediate goal of modelling efforts is generally to gain some understanding of a process through typically "excellent" fits of existing data. However, the long range goal is *prediction*. One would like to specify initial conditions, and obtain the evolution of the properties of interest. Better yet would be the converse, to specify the end point and obtain the optimum means of reaching it. Although our current understanding of emulsion polymerization can provide "guidelines" for preparing various latexes, *prediction* is not a reality.

As mentioned earlier, emulsion polymerization is still considered by many to be an *art*. Why is this? One major reason is that the success of a process often seems to rely on the experience and skills of the person performing the work and often these cannot be passed on to succeeding individuals. A more understandable reason lies in the large effects that small variations in certain variables may have on the process and products, such as the presence of impurities. To illustrate the problem, we present some results obtained for the emulsion polymerization of styrene using sodium lauryl sulfate (SLS) as emulsifier and potassium persulfate as initiator. This perhaps is one of the most studied systems in this field and as such is often considered "well-understood". Figure 2 presents the conversion-time behaviors of four emulsion polymerization experiments conducted at 70°C (*9*); the variables are the emulsifier concentration and the impeller rpm. Both above and below the cmc (≈7mM), the agitation appears to influence the polymerization kinetics, but the results are not consistent and an explanation is not obvious. The first question that you immediately ask yourself is "what is the reproducibility of these experiments?" And this is often the crux of the problem. It really means "are there unexpected variables which are affecting the process?" The reader should also be asking: how were these reactions conducted...in what kind of reactor...with what kind of impeller...were there baffles...how much headspace...how good was the temperature control...were the ingredients purified and were they from the same lot of chemicals... how was the conversion measured...was the final particle size consistent with the kinetic results...was there coagulum and so on. Once these questions are satisfactorily addressed and provided there are no apparent experimental errors, then questions are asked regarding the nature of the process and how the deliberately chosen variables should or could affect this process.

For the sake of argument let's consider that there is nothing evident regarding the manner in which the experiments were conducted that allows us to disregard the data. It is well known that agitation is important in emulsion polymerization but is seldom elaborated upon. The standard assumption is that the agitation should be sufficient to produce monomer droplets in the micron to tens of microns size range. The conditions that fulfill this are dependent on a number of factors such as the reactor size and configuration, the impeller size, configuration and placement, and the nature of the fluid. Guidelines exist for producing various types of mixing but empirical studies are required to achieve the desired results (turbulent mixing is an example of a *chaotic* process). In the cases presented above, the emulsions (prior to initiation) appeared uniformly white, however, some pooling of monomer in the vortex region of the one-liter, baffled reactor was noted at the lower rpm (pitched blade impeller). This kind of difference is generally sufficient to suspect that there could be a difference in the polymerization results and typically such results as obtained for the lower rpm are discarded as experimentally irreproducible and "undesirable".

Nucleation models do not contain expressions explicitly accounting for the degree of agitation of a system and yet the effect exists and must be addressed by the experimentalist. It can influence the mass transfer of monomers, the availability of surfactant, the introduction of "impurities" through the vapor/liquid interface (*10*), the stability of the latex particles, and the size of monomer droplets. All of these can affect the rate of polymerization and thus the product quality such as the particle size and distribution.

Morphology

Commercially driven products and processes typically preceed fundamental understanding; emulsion polymers and emulsion polymerization are no exception to this rule. The control of composition and structure formation in latex particles prepared with two or most often more monomers is of great commercial interest as reflected in 1986, when a group representing both industry and academia met to discuss Future Directions in Polymer Colloids (*11*) having as one of their primary issues "Emulsion Copolymerization and Particle Morphology". In their *Position Paper*, they reviewed the "State of the Art" and proposed goals for research over the coming decade. These recommendations encompassed virtually all aspects related to the process and the products thereby reflecting the state of the transition between *art* and *science*.

An important goal of this research was also stated explicitly to be *prediction*. However, superimposing the copolymerization and its complexities on particle nucleation and growth adds yet another order of magnitude difficulty onto predictability. Added to the uncertainties outlined previously regarding particle formation are those associated with the presence of multiple monomers: their reactivity ratios and their distributions between the phases; polymer compatibility, phase separation, and morphology development; and the polymerization process. In the five years that have passed since these "Future Directions" were defined, significant progress has been made in a number of the areas.

Thermodynamics has become increasingly important in describing phenomena in emulsion polymerization. Initially, thermodynamic equilibrium was applied to describe the swellability of latex particles in the presence of monomer droplets (*12*) in homopolymerization systems. Successful application of this concept however required use of parameters subject to some uncertainty (the interaction parameters χ_{ij} and the ratios of the equivalent number of molecular segments between components i and j, m_{ij}) and these most often have to be determined experimentally. Nonetheless, this approach was subsequently extended to copolymerizations with particular emphasis on the determination of the monomer concentrations in the particle and aqueous phases (i.e., for monomers with significant water solubilities) (*13*). This was an important contribution since it allowed the use of *true* reactivity ratios rather than the *apparent* ones obtained from emulsion polymerization kinetics and recognized the potential importance and differences in polymers produced in the aqueous phase. These developments were particularly appropriate for systems and processes in which homogeneous copolymers were produced or at least where no appreciable phase separation occurred. Generally, this implies semi-continuous "starved" polymerization conditions where despite disparities in reactivities of the monomers a relatively uniform copolymer is produced by imposing low concentrations of the more reactive species. This approach was also applied with relative success to less ideal systems (e.g., batch polymerization) where copolymer composition drift was measured experimentally.

Monomer partitioning in miniemulsion systems ((*14*) and Chapter 6, this volume) has also been modeled based on the same thermodynamic approach, but in this case the role of the "co-surfactant" in limiting monomer transport from droplets to polymer particles is one of the key differences between this polymerization and the conventional process.

Phase separation and morphology development have more recently become an area for application of these thermodynamic concepts to emulsion polymerization. Generally, interest has been directed towards preparing one of two morphologies: homogeneous (i.e., no phase separation) or core-shell. Although these appear physically distinct, there may be in reality some point of similarity. As mentioned earlier, several authors have concluded, based on theoretical modelling of emulsion polymerization, that core-shell type morphologies can evolve simply based on a gradient of free radicals in the polymer particles even if the polymer being produced is completely compatible with the previous polymer (such as a seed) (*15, 16*). This gradient results from the anchoring of free radical oligomers entering from the aqueous phase due to the presence of the hydrophilic initiator endgroups (i.e., $-SO_4^-$). Controversial experimental evidence for this had been presented in the early 70's (*17, 18*).

Core-shell latex particles are typically prepared by semicontinuous emulsion polymerization where a second stage monomer (or monomers) is fed to a reactor containing a preformed seed. This seed may consist of a rubbery polymer (or copolymer) and the second stage polymer may be a rigid polymer (or copolymer) as used in impact modifiers (e.g., ABS). Although there is some advantage in these polymers being incompatible, this is often the source of the difficulty in obtaining true core-shell latexes; incompatibility at the polymer/polymer interface can lead to a host of undesired morphologies described as half-moon (or acorns), sandwich, raspberry-like, inverted core-shell, and so on (see Chapter 15, this volume, for further elaboration on these morphologies). The thermodynamic approach has been applied to gain some more fundamental understanding of the formation of these morphologies (*19, 20*) . Polymer/polymer interfacial tension as well as the polymer/aqueous phase interfacial tensions have been shown to play crucial roles in determining these morphologies. Interfacial tensions are determined not only by the type of polymer, but the initiator and surfactant used in the preparations. In general, core-shell morphologies are not favored; half-moon type morphologies are often thermodynamically preferred. And it is possible to obtain more than a single morphology in a given sample as illustrated in Figure 3. Both half-moon and sandwich type morphologies can be seen in this polystyrene/poly(ethyl methacrylate) latex (*21*). How can this be explained? One possibility is that true heterogeneities exist among these particles that lead to variations in the morphology (i.e., their polymerization histories may differ suffiently to bring about small but significant differences in properties leading to these variations). However, it is more likely that the "slowness" of the equilibrium process is responsible. Kinetic considerations are also crucially important as a determinant of particle morphology. The slow diffusivity of polymer chains can delay the attainment of equilibrium. This, of course, complicates even further the "predictability" of particle morphology; the interplay of the polymerization kinetics and polymer diffusion kinetics must be addressed to attain a true understanding of these processes.

Chemical or physical bonding of one polymer phase to another is often used to counter thermodynamic inevitabilities. Grafting between core and shell polymers and the use of compatibilizing agents aid in preserving desired morphology. Crosslinking in the form of simultaneous interpenetrating polymer networks can also be used to physically prevent phase separation (22) or in contrast it can be applied to effect phase separation of even similar polymers (*23*).

Our ability to prepare these and more complex particle morphologies will continue to require empirically oriented research but with ever increasing support from the many incremental steps being made in our understanding of structure development in these polymer latexes.

Characterization. The study of microstructure development in emulsion copolymers has proceeded in a number of directions. Direct examination by electron microscopy is popular not only for determining particle size and distributions but also particle morphology. This, however, is not usually a simple matter of examining a dried film in the

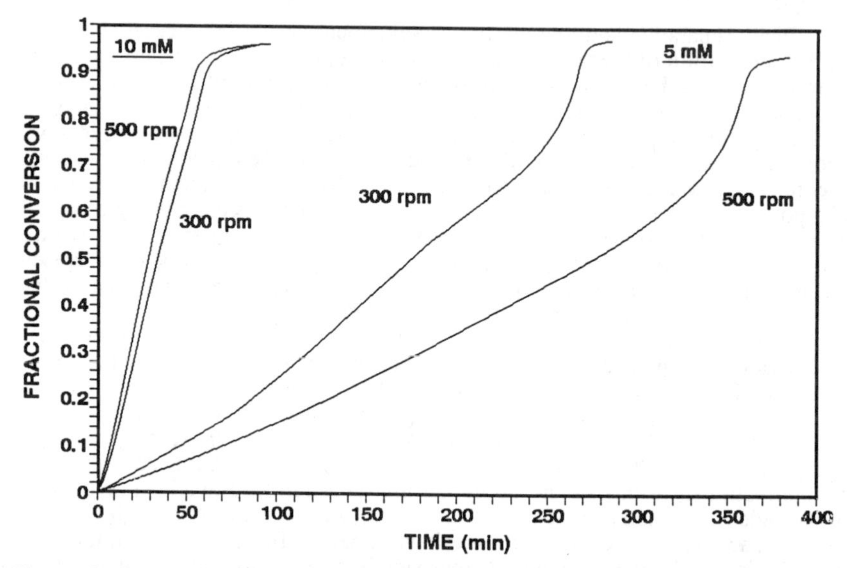

Figure 2. Conversion histories for the emulsion polymerization of styrene using two emulsifier concentrations (5 and 10 mM SLS) and two agitation speeds (300 and 500 rpm); reaction temperature = 70°C (*9*).

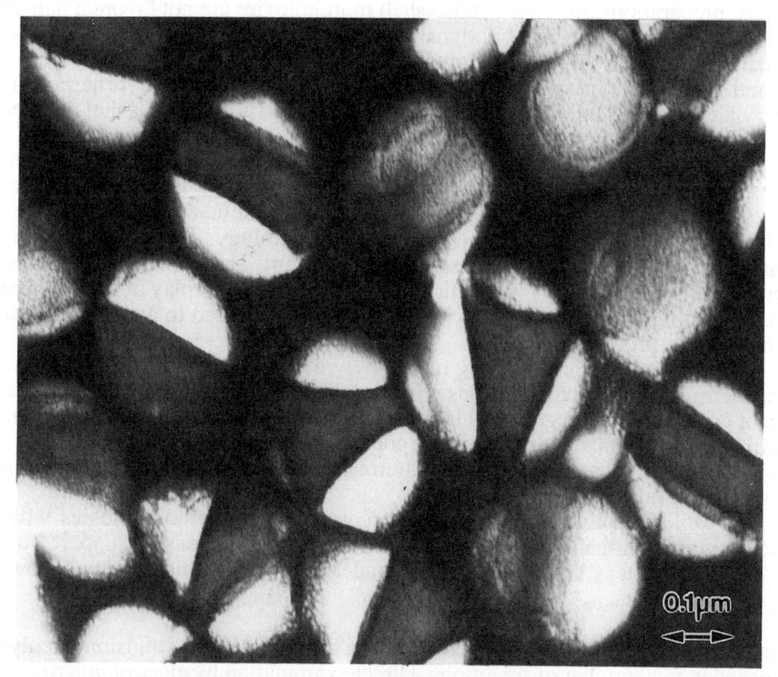

Figure 3. Transmission electron micrograph of composite latex particles of polystyrene (dark areas within particles) and poly(ethyl methacrylate) (light areas) stained with RuO_4 and phosphotungstic acid (*21*).

electron microscope. Techniques such as selective staining, cold stage, freeze fracture, and cryomicrotomy are often required and sometimes these are not yet sufficiently developed to yield a conclusive picture of the polymer structure.

Other indirect methods are being applied separately and in combination. Glass transitions measured via DSC or DMS are used as an indication of the evolution of structure development with conversion in copolymers in which composition drifts are evident (Chapter 10, this volume). Likewise the variation in the surface properties of composite latexes can be used as an indication of the evolution of the polymer composition; this has been accomplished by determination of the saturation adsorption areas of surfactants which vary depending on the nature of the polymer (Chapters 10 and 11, this volume).

The use of analytical instrumentation continues to evolve in characterizing copolymer latexes. Spectroscopy (NMR, FTIR, ESCA) and chromatography (GC, GPC, HPLC, SEC, TLC, ITP, CHDF) have been and are being used to great advantage. Copolymer composition and sequence distributions, branching (Chapter 12, this volume) and grafting, molecular weights, etc., are sought through application of these techniques.

Applications. Invention and high volume application of polymer latexes have long been the industrial domain where the *art* of emulsion polymerization flourishes. A modest representation in the scientific literature is balanced by the abundant representation in the patent literature underscoring what is traditionally viewed as the difference between basic and applied research. And although basic research often lags the applied, the reverse can also be true.

High volume applications of copolymer latexes include synthetic rubber, paints, adhesives, paper coatings, carpet backing, toughened plastics, and so on. These latexes are prepared either for the polymer (i.e., taking advantage of the high reaction rates and molecular weights), the ability to form films upon drying, or the combined colloidal and polymeric properties. For example, butadiene-styrene latexes were initially developed for the polymer as used in synthetic rubber. High molecular weight water soluble polymers (for use in waste water treatment, enhanced oil recovery, paper making, etc.) are recovered from the latexes prepared by inverse emulsion polymerization and more recently the development of high molecular weight oil soluble copolymers as fuel additives have been and continue to be investigated (Chapter 14, this volume).

Latexes used in applications requiring film formation are usually relatively soft (T_g near or below room temperature). Styrene-butadiene is again used extensively here. Also, vinyl acetate has proven to be a particularly popular monomer for homopolymerization and copolymerization with other monomers such as butyl acrylate (Chapters 8, 9, 10, 11, and 13, this volume) and ethylene. Many other acrylate ester copolymer latexes have also found extensive use in coatings.

Relatively rigid latex particles which do not lose their identity can be found in applications such as opacifying agents (e.g., hollow particles, Chapters 17 and 18, this volume) and impact modifiers (e.g., ABS, Chapter 16, this volume). Low volume monodisperse latexes, long the tool of basic research, find use as model colloids (Chapter 20, this volume), calibration standards, and in immunodiagnostic tests (Chapter 21, this volume). These are only a few of the growing list of uses for polymers prepared through application of emulsion polymerization technology.

Literature Cited

1. Gleick, J., *Chaos, Making a New Science*, Penguin Books, New York, 1987.
2. Gilbert, R.G.; Napper, D.H., *J. Macromol. Sci. Rev., Macromol. Chem. Phys.*, **1983**, *C23(1)*, p 127.
3. Asua, J.M.; Adams, M.E.; Sudol, E.D., *J. Appl. Polym. Sci.*, **1990**, *39*, p 1183.

4. Maxwell, I.A.; Morrison, B.R.; Napper, D.H.; Gilbert, R.G., *Macromolecules*, **1991**, *24*, p 1629.
5. Harkins, W.D., *J. Am. Chem. Soc.*, **1947**, *69*, p 1428.
6. Smith, W.V.; Ewart, R.H., *J. Chem. Phys.*, **1948**, *16*, p 592.
7. Hansen, F.K.; Ugelstad, J., in *Emulsion Polymerization*, Piirma, I., ed., Academic Press, New York, 1982, p 51.
8. Fitch, R.M.; Tsai, C.H., in *Polymer Colloids*, Fitch, R.M., ed., Plenum Press, New York, 1971, p 73.
9. Varela de la Rosa, L.; Sudol, E.D.; Klein, A., Emulsion Polymers Institute, Lehigh University, unpublished data.
10. Nomura, M.; Harada, M.; Eguchi, W.; Nagata, S., *J. Appl. Polym. Sci.*, **1972**, *16*, p 835.
11. *Future Directions in Polymer Colloids*, El-Aasser, M.S., and Fitch, R.M., eds., NATO ASI Series: Applied Science - No. 138, Nijhoff, Dordrecht, 1987, p 3.
12. Morton, M.; Kaizerman, S.; Altier, M.W., *J. Colloid Sci.*, **1954**, *9*, p 300.
13. Guillot, J., *Acta Polymer.*, **1981**, *32(10)*, p 593.
14. Delgado, J.; El-Aasser, M.S.; Silebi, C.A.; Vanderhoff, J.W., *J. Polym. Sci.: Part A: Polym. Chem.*, **1989**, *27*, p 193.
15. Chern, C.S.; Poehlein, G.W., *J. Polym. Sci.: Part A: Polym. Chem.*, **1987**, *25*, p 617.
16. de la Cal, J.C.; Urzay, R.; Zamora, A.; Forcada, J.; Asua, J.M., *J. Polym. Sci.: Part A: Polym. Chem.*, **1990**, *28*, p 1011.
17. Grancio, M.R.; Williams, D.J., *J.Polym. Sci., Part A-1*, **1970**, *8*, p 2617.
18. Keusch, P.; Price, J.; Williams, D.J., *J, Macromol. Chem., Chem.*, **1973**, *A7*, p 623.
19. Berg, J.; Sundberg, D.; Kronberg, B., *Polm. Mater. Sci. Engrg.*, **1986**, *54*, p 367.
20. Chen, Y.C.; Dimonie, V.; El-Aasser, M.S., *Macromolecules*, **1991**, *24*, p 3779.
21. Dimonie, V.L., Emulsion Polymers Institute, Lehigh University, unpublished data.
22. Sperling, L.H.; Chiu, T.W.; Hartman, C.P.; Thomas, D.A., *Intern. J. Polymeric. Mater.*, **1972**, *1*, p 331.
23. Sheu, H.R.; El-Aasser, M.S.; Vanderhoff, J.W., *J. Polym. Sci.: Part A: Polym. Chem.*, **1990**, *28*, p 629.

Bibliography

An Introduction to Polymer Colloids, Candau, F., and Ottewill, R.H., eds., Kluwer Academic, Dordrecht, 1990.

Scientific Methods for the Study of Polymer Colloids and Their Applications, Candau, F., and Ottewill, R.H., eds., NATO ASI Series C - No. 303, Kluwer Academic, Dordrecht, 1990.

Future Directions in Polymer Colloids, El-Aasser, M.S., and Fitch, R.M., eds., NATO ASI Series E: Applied Sciences - No. 138, Martinus Nijhoff, Dordrecht, 1987.

Science and Technology of Polymer Colloids (2 volumes), Poehlein, G.W., Ottewill, R.H., and Goodwin, J.W., eds., NATO ASI Series E: Applied Sciences - Nos. 67 and 68, Martinus Nijhoff, Dordrecht, 1983.

Emulsion Polymerization, Piirma, I., ed., Academic Press, New York, 1982.

Emulsion Polymers and Emulsion Polymerization, Bassett, D.R. and Hamielec, A.E., eds., ACS Symposium Series No. 165, Washington, D.C., 1981.

Emulsion Polymerization of Vinyl Acetate, El-Aasser, M.S. and Vanderhoff, J.W., eds., Applied Science, London, 1981.

Polymer Colloids II, Fitch, R.M., ed., Plenum, New York, 1980.

Emulsion Polymerization, Piirma, I. and Gardon, J.L., eds., ACS Symposium Series No. 24, Washington, D.C., 1976.

Emulsion Polymerization, Blackley, D.C., Applied Science, London, 1975.

RECEIVED January 24, 1992

Chapter 2

Is There Life Beyond Micelles?

Mechanisms of Latex Particle Nucleation

F. K. Hansen

Department of Chemistry, University of Oslo, P.O. Box 1033 Blindern, 0315 Oslo 3, Norway

The theory for particle nucleation in emulsion polymerization has been generally described by the Smith-Ewart/Roe and HUFT (Hansen, Ugelstad, Fitch & Tsai) theories. The HUFT theory recognizes 3 different loci for particle nucleation; homogeneous, micellar and droplet nucleation. More recently several new investigations have been published, considerably inpoving the amount and quality of experimental data; in addition, the stability of new particles above the CMC has been questioned. Some of the results seem to confirm the theory at its present stage, while other data call for novel theoretical considerations. From the HUFT theory, it can be calculated that micelles have a high probability for nucleation. However, the view of the micelles as dynamic species, and the dynamic interaction between micelles, monomer and oligomer chains could give better understanding of the physical processes involved. This view might result in a more flexible and general theory. In this paper the state-of-the-art of particle formation theory will be reviewed and the new ideas will be discussed.

The theories for particle nucleation in emulsion polymerization up to 1980 have been extensively described by this author earlier (*1*) and will not be repeated here. In this paper I will give a personal view of the present state-of-the-art of particle nucleation, especially on the basis of the HUFT theory, and including later work. In this background the main historic data and arguments must be repeated. The part of nucleation theory that concerns nucleation in monomer droplets will not be treated in this article. At the end of the paper I will discuss some propositions that may bring further understanding to the field and act as a basis for further experimental and theoretical ventures.

0097–6156/92/0492–0012$06.00/0

The Smith-Ewart theory

The attempt to quantitatively understand the latex particle nucleation phenomenon starts with the Smith-Ewart theory (2) based on Harkins' micellar theory. The fundamental assumption of this theory is that free radicals generated by initiator in the aqueous phase are absorbed in surfactant micelles at a rate proportional to the surface area of the micelle. The rate of absorption is either assumed constant (upper limit) or decreasing with time because of competition with new particles (lower limit). Nucleation stops when all surfactant has been consumed by adsorption onto the new particles. The particle number is not proportional to the number of micelles (or total surfactant area) because particles grow at a constant rate during the nucleation period, thereby decreasing the number of micelles at a higher rate than that of pure nucleation. The Smith-Ewart expression is usually presented as:

$$N = k\,(\rho_i/\mu\,)^{0.4}(a_s S)^{0.6} \qquad (1)$$

where ρ_i is the rate of radical generation, $\mu = dv/dt$ is the (constant) rate of particle volume growth, a_s the specific area of the surfactant, and S the surfactant concentration. The constant k is 0.53 for the upper limit, and 0.37 for the lower limit.

This theory has been extensively discussed by many workers, and it seemed (and still seems) to fit well for monomers of low water solubility (e.g., styrene) (*3*). The theory was developed with the early emulsion polymerization systems in mind (e.g., in the artificial rubber industry) where such monomers were often utilized, together with surfactants (e.g., soaps) of low critical micelle concentration (CMC). Before long it became apparent, however, that the Smith-Ewart theory has several drawbacks. The most important objections are:

a) Particles are formed even if no micelles are present.
b) Estimated particle numbers are double that found by experiment.
c) More water soluble monomers do not fit the theory.
d) A maximum rate in interval I is predicted, but never observed.

The Roe theory

Objection (a) is considered by Roe (*4*), where the Smith-Ewart expression is redeveloped on a homogeneous nucleation assumption. Roe's argument is also that in a mixture of nonionic and anionic micelles, the final particle number is not directly related to the NUMBER of micelles. Homogeneous nucleation replaces micellar absorption by the argument that the water-soluble free radicals cannot absorb into the hydrophobic interior of a micelle, but will start particle growth in the aqueous phase. The growing radicals will then adsorb surfactant and thus become new stable particles. In the Roe theory Smith and Ewart's parameters and derivations are replaced with exact copies based on homogeneous grounds. For instance the CMC is replaced by a critical stabilization concentration ("CSC"). The volume growth, etc. are assumed the same. By no surprise, Roe ended up with the Smith-Ewart expression (1), proving that agreement with this expression does not have to imply micellar nucleation. It does however, imply a rather abrupt stop in nucleation at the "CSC". Roe does not make any estimates about the magnitude and reason for this "CSC", but he proposes to use "some sort of adsorption isotherm" below the CMC. The Roe theory also does not consider the other arguments.

The Fitch-Tsai theory

Because argument (a) is quite obvious, there seems to be general agreement that a homogeneous mechanism must be active **below** the CMC, but the exact physical model and mathematical treatment have been the subject of discussions. The Fitch and Tsai derivation was the first serious attempt to produce a quantitative model for homogeneous nucleation (*5-7*). It is based on the Priest idea (*8*) of propagation in the water phase and self-nucleation when the chains reach a critical degree of polymerization (c.d.o.p.), jcr. The assumption of self-nucleation by means of a c.d.o.p. is different from ordinary nucleation theory where it is the critical (supersaturation) concentration that comes into play. The Fitch theory considers that the final particle number is determined by competition between oligomer precipitation and absorption in already formed particles,

$$dN/dt = \rho_i - \rho_A \qquad (2)$$

where ρ_i is the rate of radical generation (corrected for aqueous phase termination) and ρ_A is the rate of radical absorption. This rate is derived from geometrical considerations about the average length, L, the radicals can travel before self-nucleation. If the radical collides with an existing particle before it can travel this distance, it will not produce a new particle. ρ_A is expressed by

$$\rho_A = \pi \rho_i L N r_p^2 \qquad (3)$$

where r_p is the particle radius. This type of absorption model is called the "collision theory" because it does not consider any concentration gradient outside the particle surface and thereby treats the radicals as points that collide with the particles. The absorption rate is then determined by pure geometric factors, and is proportional to the particle surface $A_p = Nr_p^2$. Fitch and Tsai could express the distance L by means of the c.d.o.p., jcr, through Einstein's relationship and the particle radius as a function of time by means of non-steady-state kinetics. They ended up with an analytical, although somewhat complicated, expression for the particle formation rate:

$$dN/dt = \rho_i[1-(\pi N)^{1/3} \{[3\bar{v}k_p M_w/(4k_{tw}\phi_{pp}N_A)] \ln[\cosh(\rho_i k_{tw})^{1/2}t]\}^{2/3}]L \qquad (4)$$

This theory was tested on methyl methacrylate systems and called for a very low, but in itself not unreasonable, initiator efficiency. However, the "collision theory" for absorption (Nr_p^2) has been criticized on more formal physical grounds. Fitch also found later from seed experiments that the absorption rate was proportional to Nr ("diffusion theory", see below) (*7*). Proportionality with Nr_p^2, is not physically impossible, however, but requires special conditions (see below). Another important observation of Fitch and coworkers was that at submicellar surfactant concentrations, the particle number did not increase monotonically with time, but rather went through a maximum. This phenomenon, which is called limited coagulation (limited because the coagulation stops when the surface charge and potential has reached a level where the particles are stable) is believed to be the common mechanism for establishing the final particle number in low surfactant and surfactant free systems. At that time, no quantitative theory for this had been presented.

The Hansen and Ugelstad theory

The treatment by Hansen and Ugelstad (*9-12*) was partly based on the Fitch theory as concerns the oligomer propagation and precipitation. The basic idea is threefold:

(1) The aqueous phase kinetics is fully developed to take into consideration propagation, termination and absorption of oligomeric radicals.
(2) The absorption (capture) of radicals in particles is a reversible diffusion process that is dependent on oligomer solubility, reaction rate in particles (propagation, termination) and surface potential.
(3) The new particles - "primary particles" - are not stable unless they are fully covered by emulsifier, and will undergo limited coagulation.

(1) The radical kinetics in the aqueous phase are represented by one rate expression for each radical type.

For initiator radicals:

$$dR_i/dt = \rho_i - k_{pi}R_iM_w - k_{twi}R_iR_w - R_ik_{api}N_p \tag{5}$$

For oligomer radicals of d.o.p. =j :

$$dR_j/dt = k_pR_{j-1}M_w - k_{pi}R_jM_w - k_{twi}R_jR_w - R_jk_{apj}N_p \tag{6}$$

The rate of nucleation is given by:

$$dN/dt = k_pR_{jcr-1}M_w \tag{7}$$

which is the propagation rate of the chain with d.o.p. 1 less the critical. These equations have been elaborated further to include many details (*1*) but are in principle simple. They may be solved either numerically or by a steady state assumption to different degrees of accuracy. A relatively simple solution for the nucleation rate can be obtained under certain conditions,

$$dN/dt = \rho_i(1 + k_{tw}R_w/k_pM_w + \langle k_a\rangle N/k_pM_w)^{1-jcr} \tag{8}$$

with the total oligomer radical concentration in the aqueous phase given by

$$R_w = \{[(\langle k_a \rangle N)^2 + 4\rho_i k_{tw}]^{1/2} - \langle k_a \rangle N\}/2k_{tw} \quad (9)$$

(2) The constants k_{aj} are the 2nd order rate constants of absorption (capture) in the particles and are represented in Equation (8) and (9) by an average value, $\langle k_a \rangle$. The use of an average value is necessary in order to obtain Equation (9), but is physically not very correct. Use of these equations with reasonable values for the constants, also tends to generate far too few particles, especially compared to "stable almost CMC" conditions. The reason for this is that the absorption rate as expressed by the "diffusion theory" is too high. The Hansen and Ugelstad work therefore also included a more detailed reevaluation of the absorption rate constants. By considering the particle geometry and concentration gradients both outside and inside the particles, a new expression for the absorption rate constant was developed. Thus k_{aj} could be expressed as

$$k_{aj} = 4\pi\, r_p D_{wj} F_j \quad (10)$$

where the factor F, the "efficiency factor" represents the lumped effect of both reversible diffusion and electrostatic repulsion. This factor is given by the expression

$$1/F = (D_w/aD_p)/(X \coth X - 1) + W' \quad (11)$$

where $$X = r_p(k/D_p)^{1/2} \quad (12)$$

and $$k = k_p M_p + n k_{tp}/v_p \quad (13)$$

where $$n = 0 \text{ or } 1$$

(The other constants are given in the symbol list.) The physical and mathematical significance of the F-expression is easier understood by observing the limiting cases

(i) $F = 1/W'$ => $k_a = 4\pi D_w r_p/W'$ (14)

(ii) $F = ar_p(kD_p)^{1/2}/D_w$ => $k_a = 4\pi r_p^2(kD_p)^{1/2}$ (15)

(iii) $F = ar_p^2k/3D_w$ => $k_a = av_pk$ (16)

Case (i) corresponds to irreversible diffusion, and it is this that is called the "diffusion theory". In this case there may be some effect of electrostatic repulsion, represented by W' (≥1). In case (ii) the rate is controlled by diffusion and reaction **inside** the particles (very low diffusion constant D_p), while in case (iii) there is a dynamic **equilibrium distribution** of radicals between the particles and the aqueous phase. Both case (ii) and (iii) will lead to much lower values of k_a than case (i). One important consequence of this expression is the significance of k, which is the total rate constant (1st order) for reaction of a radical inside a particle by propagation or termination. The expression also includes the termination rate through the second term where n is the number of radicals already in the particle. If k=0 then also F=0 and there will be no **net** radical absorption. This expression does not include the consequences of a (static) equilibrium distribution of radicals between the two phases, but in most nucleation cases, this will be negligible. If there is a very dense surface layer involved, the diffusion equations may be modified according to an alternative equation. Also if radicals are merely **adsorbed**, the equations will be different. In Figure 1 is shown some calculations of F in the cases n=0 and n=1, and it is seen that most cases where F<<1 corresponds to case (iii) (slope 2 in log F vs. log r_p). In the case where n=1 and r_p<10 nm the slope becomes -1; at the same time the value of F is much higher than when n=0. This corresponds to case (iii) when k becomes dominated by termination inside the particles, i.e., the 2nd term in Equation (13) is dominating. The expression thus predicts that small particles that contain 1 radical should absorb radicals at a much higher rate than particles and micelles that do not contain a radical. This can explain objection (c) against the Smith-Ewart theory, because this will lower the radial number at the end of Interval I. It is also seen from the figures that the absorption rate is very dependent on the parameter **a**, which is the (equilibrium) distribution constant between particles and water.

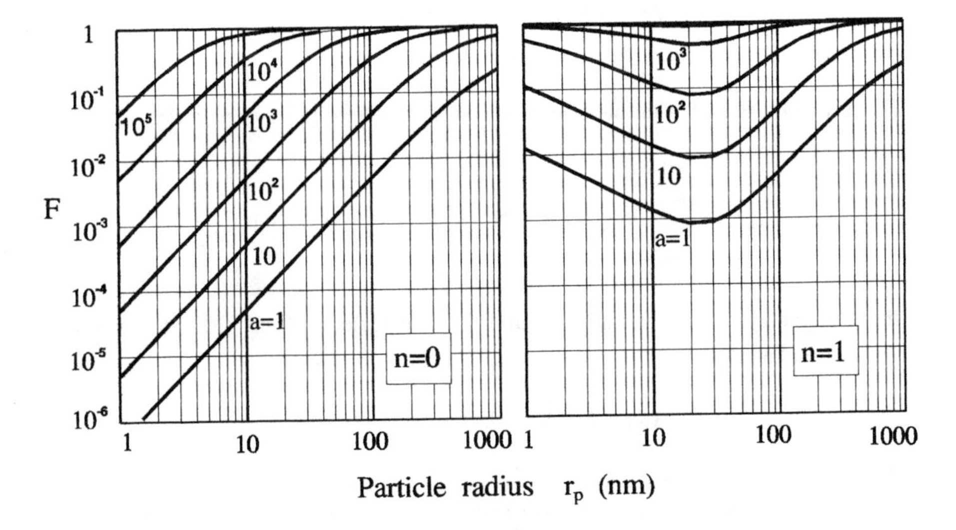

Figure 1. Absorption efficiency, F, as a function of particle size for n=0 and n=1 and different values of the ratio, **a**, between the solubility of radicals in oil phase and water phase. Other constants: k_p=300 dm^3/mol s, M_p=5 mol/dm^3 , D_w=1×10^{-7} dm^2/s , D_p=5×10^{-8} dm^2/s k_{tp}=7×10^7 dm^3/mol s , W'=1.0.

Hansen and Ugelstad also made a simplified model for styrene that takes the consequence of the low absorption of shorter chains into consideration by assuming that only the radicals 1 unit below the critical chain length are absorbed. Equation (8) is then greatly simplified, and one may use the factor F to calculate the absorption rate of these radicals. This model was used to explain results obtained above and around the CMC (styrene) where all particles are stable. A disadvantage with this model is the "adjustable parameter" problem. In the Hansen and Ugelstad work there was too little experimental data to fully check the theory. Also the number of computer simulations were limited (only for styrene). The j_{cr} used by Hansen and Ugelstad was 5, as this value best fitted the kinetics and particle number data, and also because chains of this size were found by others in GPC analysis of water and latex early in Interval I (27).

(3) The theory for limited coagulation uses Smoluchowski coagulation kinetics and the DLVO theory to calculate W-ratios. The fundamental equations that describe the complete system is a combination of the aqueous kinetic equations (5)-(6) and the coagulation equations given below:

The rate of nucleation, Equation (7), is now:

$$dN_1/dt = kR_{jcr-1} - N_1 \sum_{p=1}^{\infty} k_{fp1} - k_{a1}N_1R_w \tag{17}$$

The formation rate of other particles that consist of s=p+q primary particles is:

$$dN_s/dt = (1/2) \sum_{p=1,q=s-p}^{p=s-1} k_{fpq}N_pN_q - N_s \sum_{q=1}^{\infty} k_{fsq}N_q + [k_{a(s-1)}N_{s-1} + k_{as}]\, R_w \tag{18}$$

The coagulation rate constants k_f are those calculated from the Smoluchowski coagulation kinetics:

$$k_{fpq} = 4\pi(D_p + D_q)(r_p + r_q)/W_{pq} \tag{19}$$

The system becomes very complex if coagulation is extensive and all coagulate sizes must be accounted for. The Hansen and Ugelstad work described the coagulation equations and solution method, but calculations were very limited because of the complexity. Also good theoretical expressions for W between unequally sized particles were not available at that time.

The HUFT theory

The combination of the ideas and derivations of Fitch and Tsai and Hansen and Ugelstad has eventually been adopted as the "HUFT" theory. As most theories these ideas have some basic physical relevance, but much work remained to be done, both experimentally and computationally to test the theory. Some of this later work is described below.

Further Fitch work

Fitch and Watson (*13*) and Fitch et al. (*14*) set out to investigate the limited coagulation process. They performed coagulation experiments with MMA, using photo initiation of homogeneous solutions and light scattering detection. Fitch and Watson used flash initiation and investigated the subsequent coagulation process. They clearly showed that coagulation takes place below the CMC and they could calculate the stability ratio as a function of surfactant (SDS) concentration. They found that W was a linear function of SDS concentration below 9×10^{-4} mol/l (ca. 10% of the CMC).

Fitch et al. used continuous illumination and observed the more complicated effect of simultaneous particle formation and coagulation. To describe the system quantitively they utilized the full kinetic description, Equations (5)-(7), and the reversible diffusion equation (11) to calculate particle numbers and subsequent Rayleigh ratios. Their calculations showed that diffusion must be reversible, i.e. that $F<1$, and they used Equation (16). The predicted increased absorption rate in $n=1$ - particles, however, did not fit. They also found that $W=1$ at zero emulsifier concentration and that $W=\infty$ at and above the CMC, meaning that all particles are stable. Their main problem was the unknown relationship between the distribution constant, **a**, and the d.o.p. j. They tried two methods for calculating this relationship, the Flory-Huggins approximation and an empirical exponential expression given by

$$\mathbf{a}_j = f\,[\exp(0.2(j-1)] \tag{20}$$

This expression was adjusted such that the longest chain ($j=53$) had $\mathbf{a}=10^6$. There is a question if not a shorter chain length with the same value of a could have fitted. They also did not consider chain transfer to monomer.

The Australian contribution

The Australian group, Feeney, Lichti, Gilbert and Napper (FLGN) have initiated and continued work on particle nucleation during the 80's. Especially they have contributed with new experimental work, and this has been followed up partly by new theoretical ideas. Traditionally the comparison between theory and experiment with respect to particle nucleation is done by comparing (final) particle numbers and/or the rate of polymerization. FLGN argue that several other parameters provide additional and more sensitive information about the nucleation mechanism. Such parameters are the particle size distribution, molecular weight distribution (also in the aqueous phase), and the rate parameters for absorption (entry) and desorption (exit). By measuring the rate constants explicitly, they are aiming to avoid the "curve fitting" dilemmas that are inherently present in the theoretical calculations cited above. They measured the particle size distribution as a function of time (*15,16*), and from the observation that these distributions are positively skewed, they concluded that the particle formation rate must be an increasing (or at least not decreasing) function of time, and that this may only be explained by a limited coagulation mechanism (they call this coagulative nucleation). That such a mechanism is active below the CMC comes as no big surprise, while it seems contradictory to other experimental and theoretical work that this should also be a governing mechanism **above** the CMC, especially for monomers such as styrene that adsorbs surfactants well, and emulsifiers such as SDS that form gaseous/liquid expanded layers when used alone and therefore have very fast adsorption/desorption kinetics.

The theoretical calculations of FLGN are performed both for systems with SDS and in emulsifier-free systems. In the work with SDS (*16*) they used a Smoluchowski coagulation model (similar to that of Hansen and Ugelstad, above). These calculations are simplified in several areas, both aqueous phase reactions and radical absorption kinetics are considered in a very simplified way (capture and exit are considered as independent and adjustable parameters), and also the DLVO equations seem too simplified.

In their work with emulsifier-free systems (*17*) they use the HUFT theory for primary particle generation and the Baruch/Matijevic method for calculation of stability ratios for unequal particles. This provides good simulation of limited coagulation in these systems (styrene), but extrapolation to other systems may demand a more detailed treatment of oligomer absorption and water phase termination. An important contribution in this work is the method of numerical solution of the particle generation and coagulation equations, Equations (5),(6),(17)-(19). The number of simultaneous differential equations may become quite large and their stability is not always good. Feeney et al. solved this problem by writing the equations in a continuum version (exchanging summation with integration) and then expressing these as finite difference equations with a convenient grain size. By this method it is now possible to do the limited coagulation simulations with satisfactory computer time and accuracy.

Recently this group also has performed calculations on emulsifier containing systems (*18*) at submicellar concentrations using an adsorption isotherm for the surfactant coverage as proposed by Roe and others. They show that this gives good agreement between theory and experiment. They also have used the simplified HUFT model for styrene, but have arrived at a lower value of j_{cr}=2 or 3. (*19*). For such a low value to fit experimentally, the aqueous phase termination constant, k_{tw}, must be 1-2 orders of magnitude higher than earlier expected from low conversion bulk type kinetics (it must be in the diffusion controlled regime), leading to a quite low initiator efficiency. They have good argument for their claims, but definite confirmation still seems to be lacking.

The role of micelles

The role of the micelles in nucleation is still being discussed. The main argument for the micellar theory is that micelles are so numerous above the CMC that they **must** be the locus for nucleation. The arguments against this theory are mentioned above, but it is important to notice that arguments against the Smith-Ewart theory are not necessarily contradictory to micellar nucleation. For instance, does the possibility for limited coagulation above the CMC not necessarily exclude the possibility of micelles as nucleation loci. Monomers such as styrene **do** show a very steep change in particle numbers around the CMC, while more water soluble monomers (MMA, VAc, VC) do not show such behavior. Seed experiments show an even steeper change at the CMC for styrene. Such experiments also show changes around the CMC for more water soluble monomers such as MMA. Hansen and Ugelstad explained this with competitive absorption of radicals between seed particles and micelles. Because micelles are so numerous, they can effectively compete with seed particles for radicals, while just below the CMC, absorption in seed particles effectively hinders homogeneous nucleation. These changes may also, however, at least qualitatively be explained by competitive coagulation of primary particles.

The fact (c) that the Smith-Ewart theory fits badly more water soluble monomers is partly explained by homogeneous nucleation (below CMC). The high particle numbers and slopes of almost 1 in logN/logC plots do not exclude the micellar mechanism above the CMC. Both Nomura et al. (*20,21*) and Hansen and Ugelstad (*22*) have shown that if chain transfer to monomer is high, the particle numbers increase because desorbed monomer radicals also may form new particles. The total rate of particle nucleation (in micelles) can then be written as

$$dN/dt = (\rho_i + k_d\ \bar{n})/\{1 + [(N/\delta N_M)(r_p/r_M)^x\} \qquad (21)$$

Here x represents the exponent of r in the absorption equation, Equation (14)-(16) and δ is the efficiency of radical absorption in micelles relative to particles (of the same size). It is shown that the Smith- Ewart expression (1) in this case is transformed to

$$N \propto S^z \rho_i^{1-z} \qquad (22)$$

where $0.6<z<1$. When there is no desorption, $z=0.6$ and absorption and reabsorption dominates over initiation, $z=1$. The picture becomes more complex if water phase termination is important, when the first term in Equation (21) must be exchanged with $\rho_i + k_d - k_{tw}R_w^2$. A complete quantitative explanation for the more water soluble monomers (VC and Vac) requires that δ is less than 1. Degradative chain transfer and/or the effect of oligomer absorption (see below) can explain this.

More recent results. Several other more recent investigations have resulted in arguments both for and against micellar nucleation. Huo et al. (*23*) found that an oil soluble inhibitor causes a large increase in particle numbers. Nomura et al. (*24*) found higher styrene levels in MMA/ST copolymers in the presence of micelles. Barton et al. (*25*) showed that a hydrophobic spin label did not effect the rate or particle number, but Pearson et al. (*26*) found an increase in particle numbers above CMC when using 9-vinyl anthracene inhibitor. All of these results may be explained by a micellar nucleation mechanism, except the results of Barton et al. However, if the spin label has a solubility in the same order of magnitude as styrene or higher, it may be absorbed and desorbed at a high rate, and may not have any influence on the particle number. Also the results of Pearson et al. may be explained by either mechanism.

The micellar process. The theories that assume homogeneous nucleation above the CMC do not give good physical reasons why nucleation should **not** take place in micelles, even if there is limited coagulation. Pure homogeneous nucleation would seem to require some mechanism that hinders absorption/nucleation in micelles. It therefore seems necessary to investigate the micellar processes in more detail. Especially the variation between different types of emulsifiers and the length of the oligomer chain relative to the micelle size have not been considered in detail. The different factors may be divided between the effects of surfactant type and the effects of monomer (oligomer) type. The variation between surfactants results in differences in micellar size and lifetime, differences between the diffusion constant into micelles and differences between where in the micelle the monomer is solubilized (in the interior or in the palisade layer). The type of monomer that makes up the oligomer causes differences in solubility of the oligomer in the micelle, differences in the diffusion constant of the oligomer, in the size of the oligomer relative to the micelle and in surface activity (possibility for mixed micelle formation). This becomes even more complicated with copolymers. A quantitative evaluation of all these factors may be called upon, but will be beyond the scope of this paper. I will, however, discuss the competition between diffusion and propagation in micelles relative to the aqueous phase.

The absorption rate of radicals into micelles may be generally described by HUFT theory as given above. The limiting equation for most values of a gives $F = \mathbf{a} r_p^2 k / 3D_w << 1$ as in Equation (16) (see Figure 1, n=0). This means that the radical absorption rate is equal to the radical reaction rate inside the micelles, and that we have an equilibrium distribution of all radicals between micelles and the aqueous phase in the same way as for monomer. The absorption/reaction rate is

$$\rho_m = R_w \mathbf{a}\, k_p M_m V_m \tag{23}$$

where the index m means micelles and M represents monomer. The propagation rate in the aqueous phase is

$$\rho_w = R_w k_p M_w \tag{24}$$

assuming that the propagation constant is the same in water and micelles. The ratio between propagation in the micelles and the aqueous phase is then

$$\rho_m/\rho_w = \mathbf{a}\ \mathbf{a}_M V_m \tag{25}$$

where $\mathbf{a}_M$ is the distribution constant for the monomer. In Table I below this ratio is calculated for some common monomers. The values for $\mathbf{a}_M$ in the first column are those used by Nomura. In the second column are values for ρ_m/ρ_w for monomer radicals (with $\mathbf{a}=\mathbf{a}_M$) and in the third column corresponding values for a fixed value of $\mathbf{a}=10^6$.

Table I. Values for ρ_m/ρ_w for Different Monomer and Oligomer Radicals. Micelle volume V_m=0.01 (-)

Monomer	a_M	$a=a_M$	$a=10^6$
Styrene	1300	17000	1.3×10^7
BMA	830	6900	8.3×10^6
MMA	48	23	4.8×10^5
VC	35	12.3	3.5×10^5
Vac	28	7.8	2.8×10^5

At least for the 3 lower monomers, chain transfer to monomer is an important process, and the values for $\mathbf{a=a_M}$ have importance for the absorption/propagation of these radicals. For the higher oligomers, the value of a increases steadily with the d.o.p. (j). A value of $\mathbf{a=10^6}$ has been chosen to represent the oligomers with the critical d.o.p. (j_{cr}) (Fitch et al. (*14*)) so that this value will give approximately irreversible absorption also in micelles (from Figure 1). However, this value is rather arbitrarily chosen, and the basis for more theoretically founded values should be further explored. Maxwell et al. (*19*) have made an attempt to estimate j_{cr} from the free energy of bringing the hydrocarbon groups out of solution, and obtained rather low values for j_{cr} that is also consistent with their data on radical entry. Their basis for the choice of critical solubilities can also, however, be debated.

In spite of the uncertainties in j_{cr} and critical solubilities, the main features of micellar nucleation may be predicted from the considerations above. Because of the higher solubility of all radicals (except for the low chain length ionic oligomers) most of them will be distributed in the micellar phase. But because the number of micelles is very high (10^{19}-10^{21} dm^{-3}) relative to the steady state concentration of free radicals (10^{15}-10^{17} dm^{-3}), very few of the micelles will contain a radical. Even if most of the radicals will be present inside a micelle, they will not stay inside the **same** micelle for a long time, but will diffuse in and out of micelles, adding most of their monomer units inside some micelle. This means that **micelles will act as yielding an apparent increase in aqueous phase monomer concentration.** Therefore, the apparent rate of propagation of oligomers should be much higher above than below the CMC. The micelles will only be the locus for **nucleation**, however, if the radicals **stay** in the micelle. From the HUFT theory, the rate of nucleation is given as the rate of propagation of the chains of one unit less than the critical chain length, meaning that the ratio between micellar and aqueous phase nucleation also is given by the ratio ρ_m/ρ_w for j_{cr}.

The ratio between absorption in seed or newly formed particles and micellar nucleation will be more complex with more water soluble monomers because the different oligomers may absorb into particles. The "efficiency factor" δ for absorption of radicals in micelles relative to particles (of the same size) will go down. This is in accord with experiment (S: δ=0.6 , VAc: $\delta=1.6\times10^{-4}$ (22)). It is the micellar volume and not the size, number or surface area that enters the absorption equation and this gives the necessary low micelle absorption efficiencies (x=3 in Equation (21)). The **nucleation rate** will also be dependent on aqueous phase kinetics and competition with absorption in new particles. The specific surface and stabilizing properties of the emulsifier must come into play when the final **number** of particles is to be calculated. Therefore, the Smith-Ewart **type** of equation (i.e. emulsifier dependence) will still be correct for some monomers, as long as absorption into micelles is the dominating process during nucleation. The results of Dunn and Al-Shahib (*28*), who found that the same molar concentration of micelles gave the same particle number, may be explained by this type of behavior.

For the more water soluble monomers, interaction between the surfactant and the oligomer may be more complex. If the oligomer is long, its interaction with micelles may be more gradual; micelles may form around the oligomer as easily as absorption of oligomers in micelles because the diffusion constant of surfactant becomes comparable to that of the oligomer. Also oligomers may associate with monomer in the aqueous solution, leading to true "precursor" particles. These phenomena, together with a lower absorption efficiency in new particles because of the higher water solubility and lower $\bar{n}$ (that both give smaller particles and lower absorption efficiency) may be the reason we do not see any abrupt change in the particle number at the CMC for these monomers.

Limited coagulation above the CMC. The conclusions above do not exclude the possibility for limited coagulation above the CMC, but it may seem difficult to believe that completely surfactant covered particles are sufficiently unstable to coagulate on this time scale. Calculations indicating unstability have been peformed by the Australian group, but is still under debate; also the results of Fitch et al. seem to contradict this. Dunn (*29*) has proposed the possibility for non-equilibrium adsorption because of the high propagation rate. This, however requires a very low diffusion constant for the emulsifier on the order $<10^{-8}$ cm^2/s, while surfactants like SDS have quite high diffusion constants (*30*). However, this effect may be possible for nonionic emulsifiers that form much larger micelles with longer relaxation times. Also mixed emulsifier systems containing fatty alcohols come into this category. Both micelles and larger particles seem to be stable against coagulation, otherwise instability of surfactant covered primary particles would require some sort of minimum in the repulsive potential. Such calculations would be interesting to perform.

Conclusion.

From these reflections it may be concluded that both theory and experimental evidence have come quite far since the Smith-Ewart theory, but there is still a considerable amount of interesting and unsolved problems to investigate in latex particle nucleation.

Symbols

a, a_j Ratio (equilibrium constant) between the solubility of radicals (general and of d.o.p.=j) in oil phase and water phase.

a_M	Solubility ratio for the monomer.
a_S	Specific surface area of the emulsifier.
D_p	Diffusion constant of radicals in particles.
D_p, D_q	Diffusion constant of particle coagulate containing p or q primary particle units.
D_w, D_{wj}	Diffusion constant of radicals (general and of d.o.p=j) in water.
f	A constant
F, F_j	Absorption efficiency of radicals.
j	The number of monomer units in an oligomer = degree of polymerization.
j_{cr}	The critical degree of polymerization where an oligomer precipitates to form a primary particle.
k	Reaction constant (1st order) for radicals in particles or micelles.
k_a, k_{aj}	Absorption constant (2nd order) for radicals in particles or micelles.
k_d	Desorption constant (2nd order) for radicals from particles.
k_{fpq}	Rate constant for coagulation between particles containing p and q primary particles.
k_p	Propagation constant for radicals with monomer.
k_{pi}	Propagation constant of initiator radicals.
k_{tp}	Termination constant (2nd order) in particles.
k_{tw}	Termination constant (2nd order) in water.
k_{twi}	Termination constant (2nd order) of initiator radicals in water.
$\langle k_a \rangle$	Average absorption constant for all oligomer types.
L	The average diffusion distance of radicals from initiation to self-nucleation.
M_m	The concentration of monomer in micelles.
M_w	The concentration of monomer in water.
n	The number of propagating radicals in a particle.
$\bar{n}$	The average number of radicals per particle.
N	The total number of particles per unit volume of water.
N_1	The number of primary particles per unit volume of water.
N_A	Avogadro's number.
N_M	The number of micelles per unit volume of water.
$N_{s(p,q)}$	The number of particles formed by coagulation of s, p or q number of primary particles.
r_M	The average radius of micelles.
r_p	The radius of a polymer particle.
R_i	The number (concentration) of initiator radicals in water.
R_j	The number (concentration) of growing j-mer oligomers in water.
R_w	The total number (concentration) of growing oligomers in water.
S	The total micellar concentration of surfactant (emulsifier) in water.
t	Time.
v, v_p	Particle volume.

V_m The total volume of micelles per unit volume of water.

W′ The electrostatic stability ratio between a unit charge (radical) and a particle or micelle.

W_{pq} Fuchs' stability ratio between p- and q-type particles.

x The order (exponent) of the absorption rate with respect to particle size.

X Dimensionless variable defined by Equation (12).

z Order (exponent) of particle number with respect to surfactant concentration.

δ Efficiency of radical absorption in micelles relative to particles of the same size.

μ The constant rate of volume growth of particles.

ρ_A Net absorption rate of radicals in particles or micelles.

ρ_i The rate of initiator splitting.

ρ_m The rate of absorption/reaction in micelles.

ρ_w The rate of reaction in water

ϕ_{pp} Volume fraction of polymer in particles.

Literature cited

1. Hansen, F.K., Ugelstad, J. In *Emulsion Polymerization*; Piirma I., Ed.; Acad.Press; 1982, pp 51-92
2. Smith, W.V., Ewart, R.H., *J.Chem.Phys.* **1948** *16*, 592
3. Gerrens, H., *Fortschr. Hochpolym.-Forsch.* **1959** *1*, 234
4. Roe, C.P., *Ind.Eng.Chem.* **1968** *Sept.*, 20
5. Fitch, R.M., Tsai, C.H., *Polymer Colloids*; Fitch R.M., Ed.; Plenum press; 1971, p.73 and p.103
6. Fitch, R.M., *Br.Polymer J.* **1973** *5*, 467
7. Fitch, R.M., Shih L.-B., *Progr.Colloid Polymer Sci.* **1975** *56*, 1-11
8. Priest, W.J., J.Phys.Chem. **1952** *56*, 1077
9. Hansen, F.K., Ugelstad, J., *J.Polymer Sci., Polymer Chem.Ed.* **1978** *16*, 1953
10. Hansen, F.K., Ugelstad, J., *J.Polymer Sci., Polymer Chem.Ed.* **1979** *17*, 3033
11. Hansen, F.K., Ugelstad, J., *J.Polymer Sci., Polymer Chem.Ed.* **1979** *17*, 3047
12. Hansen, F.K., Ugelstad, J., *J.Polymer Sci., Polymer Chem.Ed.* **1979** *17*,3069
13. Fitch, R.M., Watson, R.C., *J.Colloid Interface Sci.* **1979** *68*, 14
14. Fitch, R.M., Palmgren, T.H., Aoyagi, T., Zuikov, A., *Angew.Makromol.Chem.* **1984** *123/124*, 261
15. Lichti, G., Gilbert, R., Napper D.H., *J.Polymer Sci., Polymer Chem.Ed.* **1983** *21*, 269
16. Feeney, P.J., Napper, D.H., Gilbert, R., *Macromolecules* **1984** *17*, 2520
17. Feeney, P.J., Napper, D.H., Gilbert, T., *Macromolecules* **1987** *20*, 2922
18. Richards, J.R., Congalidis, J.P., Gilbert, R.G., *J.Appl.Polymer Sci.* **1989** *38*, 2727
19. Maxwell, I.A., Morrison, B.R., Napper, D.H., Gilbert R., *Makromolecules* **1991** *24*, 1629
20. Nomura, M., Harada, M.,Euguchi, S., Nagata, S., *ACS Symp.Ser.* **1991** *24*, 104
21. Nomura, M., Private communication

22. Hansen, F.H., Ugelstad, J., *Makromol.Chem.* **1979** *180*, 2423
23. Huo, B.P., Campbell, J.D., Penlidis, A., Macgregor, J.F., Hamielec, A.E., *J.Appl.Polym.Sci.* **1987** *35*, 2009
24. Nomura, M., Satpathy, U.S., Kouno, Y., Fujita, K., *J.Polym.Sci, Polym.Letts.* **1988** *26*, 385
25. Barton, J., Juranicova, V., Hlouskova, Z., *Makromol.Chem* **1988** *189*, 510
26. Pearson, L.T., Louis, P.E.J., Gilbert, R., Napper, D.H., *J. Polym. Sci., Polym. Chem. Ed.*, submitted
27. Goodall, A.R, Wilkinson, M.C., Hearn J., Prog. *Colloid Interface Sci.* **1975** *53*, 327
28. Dunn, A.S., Al-Shahib, W., *Proc.Plast.Rubber Inst., Internat.conf. Polymer Latex*, London **1978**, Paper 2
29. Dunn, A.S., *This book*
30. Wennerström, H., Lindman, B., *Physics Reports* **1979** *52*, No.1, pp.1-86

RECEIVED January 6, 1992

Chapter 3

Testing Nucleation Models for Emulsion-Polymerization Systems

Bradley R. Morrison, Ian A. Maxwell, Robert G. Gilbert, and Donald H. Napper

School of Chemistry, Sydney University, New South Wales 2006, Australia

Particle formation in emulsion polymerization below the cmc is now accepted to take place by homogeneous-coagulative nucleation, but there is debate whether this or micellar entry is operative above the cmc. Literature data are shown to be unable to refute either model above the cmc. The mechanism for free-radical entry into latex particles (Maxwell, I.A.; Morrison, B.R.; Napper, D.H.; Gilbert, R.G., *Macromolecules*, **1991**, *24*, 1629) suggests that observed particle numbers above the cmc are only obtained if homogeneously nucleated free radicals spend time growing in micelles; this and other considerations suggest that above the cmc nucleation is by a combination of micellar and homogeneous mechanisms. New experimental results on the time evolution of a bimodal particle size distribution support the assumptions used to draw this conclusion. This suggests that both mechanisms are simultaneously operative above the cmc. The methodology can also predict conditions for onset of secondary nucleation below the cmc.

The question of the mechanism for particle formation in emulsion polymerizations is anything but a settled question. The pioneering work of Harkins [*1*,*2*], and of Smith and Ewart [*3*], and most contemporary textbooks, assume that particle nucleation occurs entirely through the entry of free radicals into micelles ("micellar entry"). However, there is considerable evidence to support a rival theory: homogeneous nucleation, which was first quantified by Fitch [*4*], and has become known as "HUFT" (Hansen-Ugelstad-Fitch-Tsai) theory [*5*]. Indeed, it is now generally accepted that homogeneous nucleation must be operative at low (or zero) surfactant concentrations, where there are no micelles present.

At surfactant concentrations near and above the critical micelle concentration (cmc), we shall show that, while data are readily found which are *consistent* with either model, it has so far proved impossible to find data which are generally accepted as being

0097–6156/92/0492–0028$06.00/0

able to *refute* [*6*] one or other model. Take for example one well-known effect which is often taken as furnishing definitive evidence for micellar entry: the large change observed in particle number, N_c, as a function of surfactant concentration [S], as [S] goes through the cmc (Figure 1). However, this turnover can also be successfully predicted by homogeneous-coagulative nucleation theory [*8,9*], an extension of HUFT theory which takes detailed account of coagulation of newly formed, or precursor, particles. In homogeneous-coagulative nucleation, this turnover arises from the adsorption isotherm of the surfactant on precursor particles. As shown in Figure 1, this model can reproduce the sigmoidal shape of N_c([S]) which has been observed experimentally, as shown in the data for styrene in this figure (note incidentally that extensive N_c([S]) curves have been reported in the literature [*10*], but those results were presented as smooth curves without the original data points, and so it is very hard to judge if effects shown therein, such as the apparent continuous slow rise in N_c as [S] increased well above the cmc, were indeed present or artifactual; Figure 1 shows the considerable scatter in actual experimental data). The coagulative-nucleation model takes no account of micelles except insofar as they serve as surfactant reservoirs.

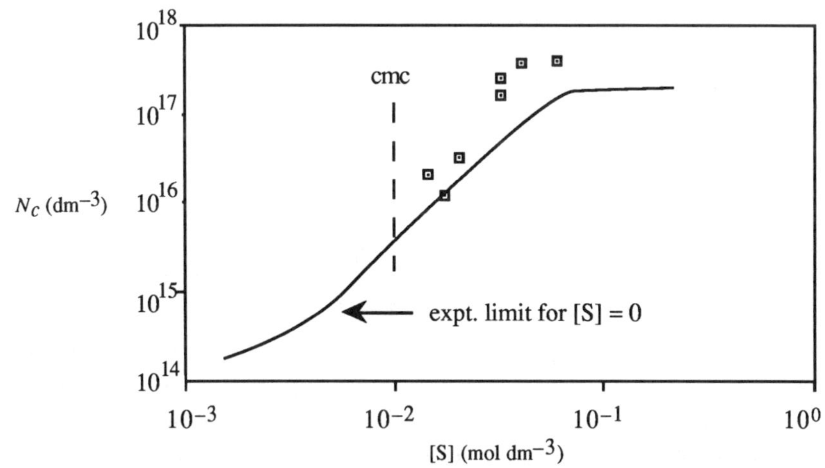

Figure 1. Particle number (N_c) as function of surfactant concentration [S], for styrene emulsion polymerization at 50°C, with Aerosol MA as surfactant, and 1.3×10^{-2} mol dm^{-3} $K_2S_2O_8$ initiator. Points: experiment (P. Hidi, R.G.Gilbert and D.H. Napper, unpublished data). Line: calculated from homogeneous-coagulative nucleation model. Appproximate cmc also shown. Experimental zero-surfactant limit at 70 °C are also indicated [*7*]; no [S]=0 data are available for 50 °C, but N_c for [S]=0 should not be strongly temperature-dependent on this scale.

In this paper, we show how new information on the mechanism for entry of free radicals into latex particles can be used to elucidate nucleation mechanisms. It will be shown that recent experimental and theoretical results can combine to provide evidence that micelles are in fact intimately involved in the nucleation process above the cmc.

The coagulative-nucleation extension of the HUFT model has been shown [*8,9*] to fit a very large range of data (including all types of data previously interpreted as unambiguous evidence for micellar entry). However, at least for data where [S] is near or exceeds the cmc, we show that such data are also consistent with micellar entry. We first consider, in a somewhat different context, an example of how data can be used to refute a mechanism; this example will in fact later prove crucial in understanding nucleation above the cmc.

An Example of Refutation: Initiator Efficiencies

In the growth of latex particles after nucleation has ceased, a major determinant of the kinetics is the process of free radical entry into a pre-existing particle. Until recently, there were no reliable *a priori* models for this entry process, and various semi-empirical treatments were used perforce (e.g., [*8,9*]). We first consider some postulated rate-determining steps, show how these can be refuted, and then summarize a new model [*11*] that successfully reproduces all extant entry data.

Two postulates that had been put forward for the rate-determining step in entry were (a) the displacement of a surfactant molecule from the latex surface, or (b) a colloidal barrier arising from double-layer effects. These postulates were tested by examining the variation of the entry rate coefficient ρ with (a) surfactant coverage and (b) ionic strength [*12*]. These studies showed that ρ *showed no variation* with either o these quantities, whereas either postulated mechanism would predict a significant change.

These studies led to an entry model [*11*] which fits these and indeed all extant data. This model is as follows. Initiator-derived free radicals propagate in the aqueous phase until they reach a critical degree of polymerization z such that they become surface-active. These z-mers enter instantly, and no smaller oligomeric free radical species can enter. Aqueous-phase termination of all smaller species may also occur. Thus one has (for persulfate initiator):

$$S_2O_8^{2-} \rightarrow 2\,{}^\bullet SO_4^-(k_d);\ {}^\bullet SO_4^- + M \rightarrow {}^\bullet MSO_4^-\ (k_p^0);$$
$${}^\bullet M_iSO_4^- + M \rightarrow {}^\bullet M_{i+1}SO_4^-\ (k_p) \qquad (1)$$

$${}^\bullet M_zSO_4^- + \text{particle} \rightarrow \text{entry}\ (k_e);\ {}^\bullet M_iSO_4^- + T^\bullet \rightarrow \text{termination}\ (k_t^{aq}),\ i<z \qquad (2)$$

where $T^\bullet$ is any aqueous-phase free radical. When account is taken of the experimental result that k_p^0 and k_t^{aq} are both very large (essentially diffusion-controlled), and if k_e is assumed sufficiently large so as not to be rate-determining, this reaction scheme is found to fit, both qualitatively and quantitatively, a wide variety of data on the entry rate coefficient in *seeded* emulsion polymerizations (e.g., [*12,13*]). Note that the value of z is typically 2 for an insoluble monomer such as styrene.

The steady-state solution of the kinetic equations for (1) and (2) yields the following expression for ρ:

$$\rho = \frac{2k_d[I]}{N_c}\left(\frac{2\sqrt{k_d[I]\,k_t^{aq}}}{k_p[M_{aq}]} + 1\right)^{1-z} \qquad (3)$$

($[I]$ = initiator concentration and $[M_{aq}]$ = aqueous phase monomer concentration). This is of course the familiar HUFT expression of Fitch and co-workers, albeit in a rather different context.

Can Extant Data Refute Either Nucleation Mechanism Above the cmc?

We now consider whether there are any available types of data which can *qualitatively* refute either homogeneous-coagulative nucleation or micellar entry above the cmc, in the same way as the surfactant coverage and ionic strength dependences (or rather, lack of them) were able to refute two postulated mechanisms for entry.

Dependence of Particle Number on Surfactant Concentration. As exemplified in Figure 1 and in the preceding discussion, the dependence of N_c on [S] cannot be used to refute homogeneous-coagulative nucleation, and different types of data must be examined for such purposes.

Early-time Particle Size Distribution. It has been claimed [*14*] that early-time particle size distribution data, illustrated in Figure 2 [*15*], can refute micellar entry. The reason that such data were supposed to refute this model is the following. The results of Figure 2 show that, just after nucleation has ceased, most particles have small volumes; hence most particles must have been formed late in the nucleation period. However, formation of particles from micelles would be expected to show the *opposite* trend: viz., the rate of particle formation should *decrease* in time, as micelles disappear; this inference therefore appears to refute micellar entry. On the other hand, the observed behavior is characteristic of coagulation kinetics, where formation of colloidally stable latex particles is through coagulative growth, which is autoaccelerating. However, this apparent refutation of micellar entry is not definitive, since it has been pointed out by us [*13*] that the observed results are also consistent with the supposition that very small particles grow slowly because of low monomer concentration (as predicted by the Morton equation [*16*]), so that growth by propagation in a particle newly formed by entry into a micelle would also be autoaccelerating, as the equilibrium monomer concentration increases with increasing radius. Hence data such as Figure 2 cannot refute micellar entry. The same argument can be used to show that the formation of Liesegang rings in unstirred emulsion polymerization systems [*17*], which must also be due to an autoaccelerating process, cannot be used to refute micellar entry.

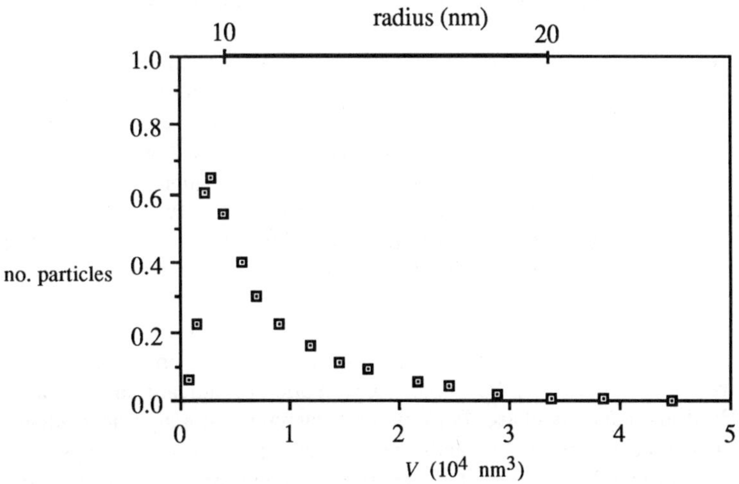

Figure 2. Early-time particle size distribution (as number of particles with unswollen volume V), for styrene emulsion polymerization with SDS surfactant.

Early-time Molecular Weight Distribution. It has been claimed [*13,18*] that the early-time molecular weight distribution (MWD) observed in emulsion polymerization can be used to refute micellar entry. This is illustrated in Figure 3, redrawn from Whang *et al.* [*18*]. This shows the *number* MWD $S(M)$; obviously the number of chains with a given molecular weight is easier to interpret mechanistically than is the weight of chains with that molecular weight, and hence it is $S(M)$ which is the best distribution function to employ for phenomenological understanding. It must be realized that a GPC trace is approximately the *weight* MWD, $W(M) = MS(M)$, and thus the GPC trace must be deconvoluted to yield $S(M)$. If the main chain-stopping event is transfer, then $S(M)$ is proportional to $\exp(-k_{tr}M/k_pM_0)$, where M_0 is the molecular weight of monomer. Figure 3 indeed shows the expected straight-line region in a plot of $\ln S(M)$ at higher M, and the slope of this transfer-dominated regime (broken line) gives $k_{tr}/k_p = 7\times10^{-5}$. This compares well with the literature value [*19*] of 6×10^{-5}, suggesting that the techniques involved are correct. However, Figure 3 shows a *maximum* at low M. This entails an increase (at $M\approx 9\times10^5$) in the number of chains above the transfer limit, and then a fall below this limit for $M< 5\times10^5$, and hence cannot be ascribed to an artifact arising from baseline subtraction, etc. The most likely explanation of the increase above the transfer-dominated limit is some form of termination by combination, wherein two macroradicals (each of degree of polymerization of the order of 5×10^5) combine to form a non-growing chain whose degree of polymerization is the sum of the two component macroradicals: shorter chains become a longer one. However, the styrene system of Figure 3 is one where the average number of free radicals per particle, $\bar{n}$, does not exceed $1/2$, and hence where combination between two macroradicals inside a single particle is kinetically insignificant. These data were obtained at low conversion, and so it is likely that the kinetics affecting particle nucleation play a major role in the observed properties. Coagulation between two precursor particles, each of which contains growing macroradicals (of relatively low molecular weight, since the precursor particles are very small) would lead to combination between the growing macroradicals in the separate particles. This would thus convert lower molecular weight chains to higher ones: i.e., it would produce a maximum in $S(M)$. Hence the observed maximum can be readily explained if there is significant coagulation occurring during the nucleation process. There however appears to be no explanation for the data that is consistent with micellar entry (in the absence of coagulation), apparently refuting that mechanism. However, the maximum is at low M, and it is conceivable (although not especially likely) that this maximum could be an artifact of the GPC deconvolution. Hence this apparent refutation must await further work on accurate GPC deconvolution to obtain $S(M)$ before the refutation can be said to be rigorous.

Free-radical Traps. Another possible source of information about nucleation mechanism is available from the effects on particle formation of radical traps which are almost insoluble in the aqueous phase. For example, carrying out a styrene emulsion polymerization in the presence of 9-vinyl anthracene (VAn), above the cmc, brings about a large increase in particle number (due to the creation of many smaller particles) and a much slower rate of polymerization [*20*]. This might appear to furnish a refutation of homogeneous-coagulative nucleation, since one might suppose that the VAn resides only in micelles or particles, and would therefore severely retard micellar entry and hence prolong the nucleation period, leading to the observed effects. However, the solubility of VAn in water, although extremely low, is still sufficient to retard the growth of homogeneous nucleation in the aqueous phase, and thus the observations are consistent with either mechanism [*20*].

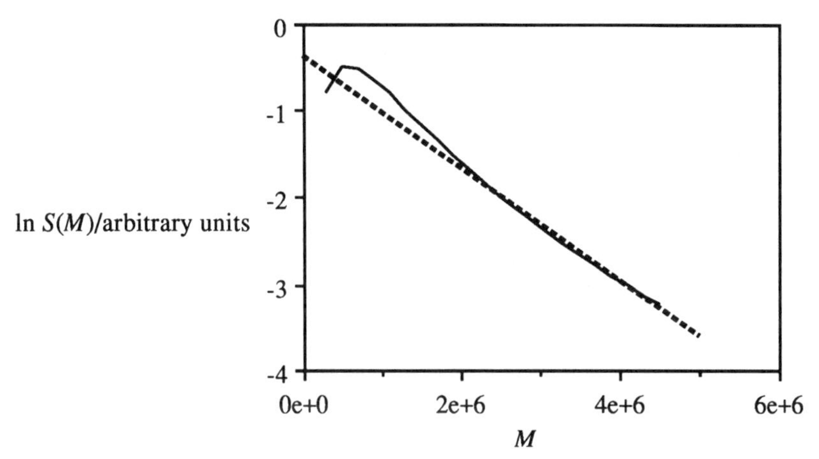

Figure 3. Number MWD for styrene emulsion polymerization with oleate as surfactant, taken at 8.8% conversion (just after the cessation of formation of new particles). Dashed line shows transfer-dominated limit.

Oscillations in CSTR Systems. It is sometimes supposed that the observation of oscillatory behavior in continuous stirred-tank reactors for monomers such as vinyl acetate furnishes a refutation of homogeneous-coagulative nucleation, since this behavior is easily explained by appearance and disappearance of micelles [*21,22*]. However, in principle this effect is also consistent with homogeneous-coagulative nucleation, since the coagulative event is autocatalytic, which is a necessary (but not sufficient!) condition for the feedback (overshoot) which can give rise to oscillatory behavior in a kinetic scheme. Despite this assertion, it must be stated that at present no simulations based solely on the homogeneous-coagulative mechanism have succeeded in reproducing the observed oscillatory behavior.

Effects of Inert Diluent and of Indifferent Electrolyte. The presence of an indifferent electrolyte (e.g., NaCl) and of an inert diluent (e.g., ethyl benzene) each has a qualitatively significant effect on the nucleation rate and particle size distribution in an (unseeded) emulsion polymerization system [*23*]; this is illustrated in Figure 4. One sees that NaCl causes a smaller polydispersity, while ethyl benzene results in a much increased polydispersity and the presence of many more smaller particles. Moreover, it is observed that NaCl decreases the nucleation period, while ethyl benzene lengthens it. These results are completely consistent with homogeneous-coagulative nucleation: the NaCl causes increased coagulative growth, and hence decreases the nucleation time and hence also decreases the polydispersity. The presence of an inert diluent slows the rate of propagative growth and hence the time at which particles become sufficiently large to become colloidally stable. That is, the polydispersity is higher with diluent because the first-born mature particles have a much longer period in which to grow (relative to a system without diluent) than those formed at the end of the nucleation period. However, these results can also be made consistent with micellar entry. The rationalization for the results with inert diluent are the same as for the homogeneous-coagulative nucleation model, while the effects of indifferent electrolyte can be rationalized by the changes that this induces in micelles.

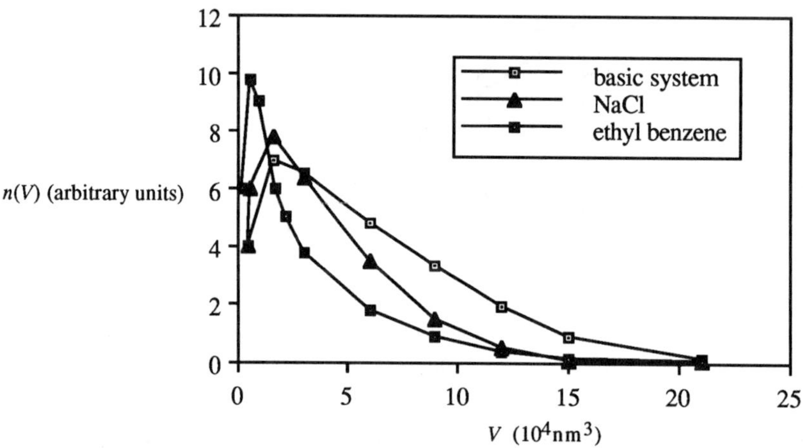

Figure 4. Particle size as function of unswollen volume V for styrene emulsion polymerization systems at 50°C, initiator 2.6×10^{-4} mol dm^{-3} persulfate, 1.06×10^{-2} mol dm^{-3} borax and 3.1×10^{-2} mol dm^{-3} sodium dodecyl sulfate, and for the same system with the addition of 8.3×10^{-2} mol dm^{-3} NaCl, and for the same system with the replacement 25% of the styrene with ethyl benzene.

Secondary Nucleation. One set of experiments which originally appeared to provide strong support for micellar entry was reported by Hansen and Ugelstad [*24*] who compared the dependence of N_c on [S] in seeded and unseeded systems. They observed that below the cmc, the number of new particles in seeded systems was much less than for unseeded ones, while both were the same above the cmc; this suggests immediately that micelles are involved in the nucleation process. Such data can however be interpreted by homogeneous-coagulative nucleation, as follows [*14*]. Below the cmc, the rate of formation of latex particles from primary particles is relatively slow, and the extent of coagulation relatively large. In these circumstances, the seed particles are able to compete kinetically with the coagulation process for the primary particles. Above the cmc, so much surfactant is present that the relative importance of the coagulation of primary particles in new particle formation is reduced. Once again, these data can neither refute nor confirm the validity of either mechanism.

Because of the clear lack of any truly discriminatory experiments, we explore here a new approach to nucleation, based on information which has come out of our new model for free radical entry [*11*].

Extending entry model to account for homogeneous nucleation.

The entry model of equations 1-3 can be extended to take into account homogeneous nucleation (in the absence of coagulation), as follows. In the absence of (new) particle formation, the entry model assumes that the actual event of an aqueous-phase free radical penetrating the particle and propagating therein is so fast as not to be rate-determining. We now wish to consider what happens when particles are being formed, and for this reason must consider the competition for the fate of an aqueous-phase free radical between *either* penetrating (entering) a pre-existing particle *or* forming a new one. The actual penetration event may under these circumstances not be sufficiently fast to avoid the alternative fate, and therefore we must specifically include the rate coefficient for this penetration. The extension is therefore to let nucleation (which is presumed to involve the collapse of a free radical whose degree of polymerization, x, exceeds the value for entry, z) compete with entry. That is, when the possibility of (new) particle formation is now considered, we must allow for oligomers of degree of polymerization *exceeding* z surviving in the aqueous phase and forming new particles. Thus one has the following extension to (1) and (2):

$$^{\bullet}M_zSO_4^- + M \rightarrow \ldots \rightarrow {}^{\bullet}M_xSO_4^- \ (k_p);\ {}^{\bullet}M_iSO_4^- + T^{\bullet} \rightarrow \text{termination } (k_t^{aq}),\ i<x \quad (4)$$

$$^{\bullet}M_xSO_4^- \rightarrow \text{new particle} \quad (5)$$

In calculations using this extended scheme (which is still based on the fundamental HUFT ideas), the second-order rate coefficient for the penetration step, k_e, is assumed tc take the diffusion-controlled limit:

$$k_e = 4\pi D\, r_s \quad (6)$$

where D is the diffusion coefficient of an x-mer in water and r_s is the swollen radius of the latex particle. D is calculated from the Stokes-Einstein relation

$$D = \frac{k_B T}{6\pi r_i \eta} \quad (7)$$

where η is the viscosity of the aqueous medium and r_i the radius of an i-meric oligomer. We take the latter from the LeBas volume expression

$$r_i = 1.18 \left(\frac{3}{4\pi} V_i\right)^{1/3} \quad (8)$$

where the volume V_i is taken as iV_1, where V_1 is the volume of a monomeric unit. Equation 6 is consistent with the original entry model [*11*], since the entry (penetration) rate calculated from this equation is indeed so fast, in the presence of a sufficient population of pre-existing particles, as not to be rate-determining. Solution of the steady-state kinetics for this extended scheme (a trivial extension of the method which leads to equation 3) can be easily carried out numerically. The rate of nucleation decreases as the rate of entry of newly-formed free radicals into pre-formed particles exceeds the rate of these radicals forming new particles, and eventually no new particles will be formed when the entry rate sufficiently exceeds the rate of new particle formation. This then yields a value of the rate of x-mer formation, which is the *maximum* rate of formation of new particles. Of course, now that we specifically allow for the competition between entry and new particle formation, the number of radicals captured by the particles will be an increasing function of the number of pre-existing particles.

Some results of calculations using this model are shown in Figure 5 for the maximum rate of new particle formation as a function of particle number for a seeded emulsion polymerization of styrene at 50°C, with 10^{-3} mol dm^{-3} persulfate initiator. These calculations used $z = 2$ (as found previously, and justified on thermodynamic grounds [*11*]) and $x = 4$, the value of the insoluble limit from the Krafft temperature of similar surfactant species [*11*]. The values of other rate coefficients were those found previously to fit all extant data for this system [*11*]. It is especially important to note that this includes $k_p^0 = 10^9$ dm^3 mol^{-1} s^{-1} and $k_t^{aq} = 7\times10^9$ dm^3 mol^{-1} s^{-1}: i.e., values in the diffusion limit.

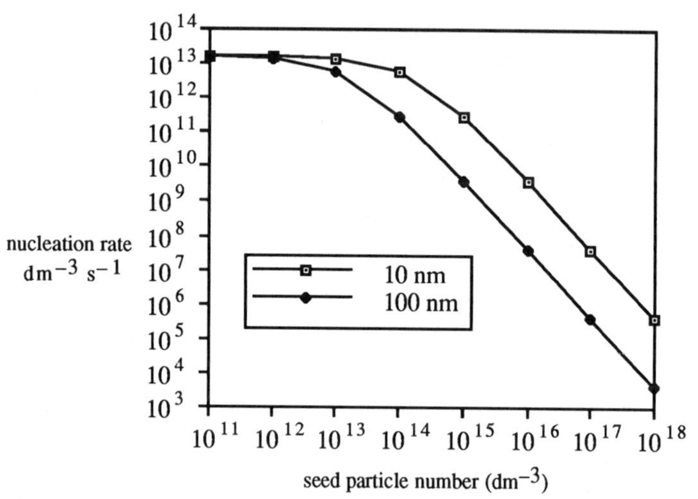

Figure 5. Calculated rate of x-mer formation (which is maximum possible rate of particle formation) as function of concentration of seed particles of different radii, for styrene at 50°C.

Figure 5 shows a drastic decrease in the nucleation rate for $N_c \gtrsim 10^{15}$ dm^{-3}. This is also reflected in the initiator efficiency calculated from the same model, shown in Figure 6. It is important to note here that above 10^{14}–10^{15} particles dm^{-3}, the initiator efficiency is independent of the particle concentration, as implicit in equation 3 and in accord with data on seeded systems [*11*]. The results of Figure 5 imply that, in the presence of a seed, the maximum rate for the formation of new particles never exceeds 10^{12} particles per dm^3 per second when the particle number is 10^{15} dm^{-3} (these conclusions are not qualitatively dependent on the values of parameters such as x). Now, typical nucleation times are 10^3 s, and hence the number of new particles formed under such conditions cannot exceed (say) 10^{15} dm^{-3}. However, as is typified by Figure 1, particle numbers greatly exceed this value for [S] above the cmc. The results of Figure 5 imply that to form enough particles to be observable (say, 0.1% of existing particles) would, according to homogeneous nucleation theory, take at least 10^4 s in the presence of pre-existing particles of radius 10 nm at typical N_c: however, this time is very much longer than experimentally observed.

Hence the homogeneous nucleation model (using parameters found from the new entry theory that fit a huge range of types of data) implies that, at a comparatively low particle concentration, all new free radicals are snapped up by pre-existing particles before they can propagate more to form new particles. This occurs at an N_c such that the time for a free radical diffusing between pre-existing particles is less than the times for propagation or aqueous-phase termination, as exemplified in Figure 7. This conclusion is clearly *incorrect*: it implies that particle concentrations exceeding 10^{15} dm^{-3} cannot form in typical systems! This, it is emphasized, is based on the maximum possible rate of particle formation (since it ignores coagulation). This implies that that there must be **another mechanism** operating: not exclusive of homogeneous-coagulative nucleation, but presumably in addition to it. Thus this new information appears finally able to refute the hypothesis that homogeneous-coagulative nucleation is the **sole** mechanism responsible for particle formation above the cmc.

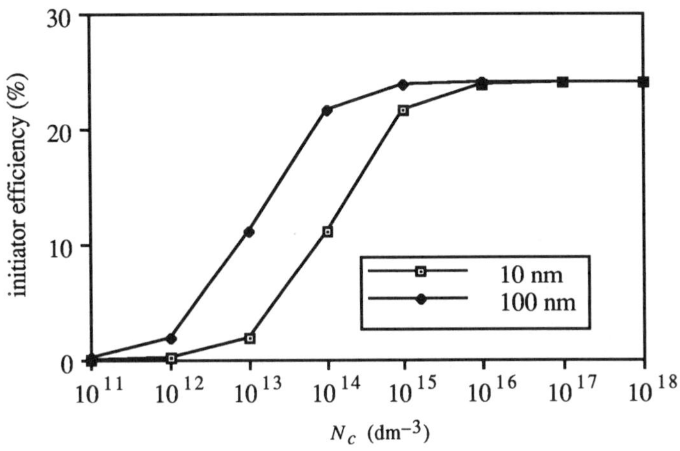

Figure 6. Initiator efficiency as function of particle number for the same system as in Figure 5.

The simplest additional mechanism is the popular one of direct involvement of micelles in the nucleation process above the cmc. That is, the problem (that the system could never form enough particles if homogeneous nucleation were the only mechanism operative) could be overcome if the oligomeric radicals formed in the aqueous phase were to grow inside micelles (which are entered in competition with pre-existing particles). This micellar growth could arise from oligomers of degree of polymerization z or higher (i.e., any species which is sufficiently surface-active or insoluble in the aqueous phase). Inside micelles, the oligomeric free radicals are able to propagate faster (because of the higher monomer concentration therein), and, more importantly, are protected (at least in part) from the very high rate of aqueous-phase termination that would otherwise usually be their fate. It is important to realize in this context that (because free energy is minimized when the surfactant head-groups in a micelle maximize their contact with the aqueous phase) *micelles* are stable to coagulation. However, it is likely that precursor particles, when first formed, are of marginal colloidal stability. Since there are no experimental data which can be said to provide reliable information on this question for such small particles (although the means for such measurments are becoming available [25]), theoretical estimates provide the only current means of making inferences. Table I shows the electrostatic barrier (E_m) to coagulation of two styrene particles; these E_m values were calculated using the expressions of Healy and co-workers [26,27]. The sizes and surface potentials used in these calculations cover typical ranges for precursors (a micelle has a surface potential of about 140 mV, and the surface charge density of a particle of micellar size containing significant amounts of polymer is likely to be less than this, since a micelle contains the maximum possible amount of surfactant for a given size). The barriers to homo-coagulation in Table I are seen to be less than $15k_BT$ for smaller particles; moreover, hetero-coagulation (i.e., between different sized particles) involves a much smaller barrier, because of the effect of asymmetry in the double-layer curvature. Since a latex particle is stable if $E_m/k_BT \geq 15$ (when the lifetime to coagulation is of the order of hours for typical particle concentrations), it is likely that precursors will undergo significant, although limited, coagulation. The E_m results of Table I suggest that homogeneously-nucleated free radicals which enter micelles will have ample time to propagate significantly before any coagulation could occur, and this growth will be sufficient to form new particles at a rate corresponding to that observed experimentally. Hence the nucleation mechanism above the cmc probably involves *both* homogeneous-coagulative nucleation and growth in micelles.

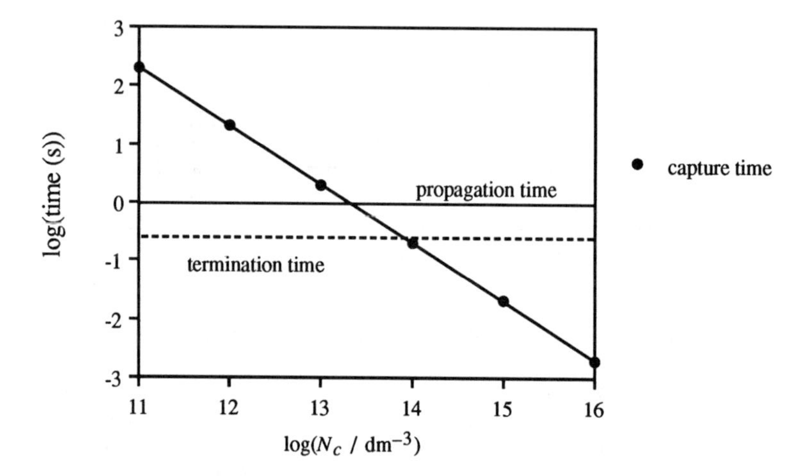

Figure 7. Comparison of times of propagation, termination and capture by pre-existing particles for the system of Figure 5.

A most important by-product of the present development is that the methodology leading to Figures 5-7 can be used to predict the **onset of secondary nucleation** in systems where only homogeneous nucleation is operative, viz., below the cmc; this is regarded as an especially useful outcome of the present calculations. That is, below the cmc, when there is no possibility of micellar involvement, the onset of secondary nucleation can be identified as the conditions when the capture time of x-mers exceeds either their propagation or termination time.

Table I. Calculated barrier to coagulation, as E_m/k_BT, for homo-coagulation of two particles of radius r with various surface potentials, using the expressions of Healy and co-workers [26,27]. Inverse of double-layer thickness κ for these calculations taken as 4.73×10^8 m^{-1}

surface potential (mV):	75	100	150
radius (nm)			
3	6.8	13	29.6
4	9.0	17	39.5

Competitive growth

The discussion of oligomer capture given above depended critically on the assumption that the rate coefficient for capture of oligomers by pre-existing particles is given by the Smoluchowski equation, equation 6: i.e., that there was no barrier to entry of initiator-derived oligomeric free radicals into the particles. Now, the successful entry model on which our present development is based can give no information on this point, since it merely assumes that this process is sufficiently fast so as not to be rate-determining.

However, if one were to assume that there were a small barrier so as to reduce k_e by (say) a few orders of magnitude (while still being sufficiently rapid as not to be rate-determining in normal entry), then the above conclusions would be invalid. It is therefore essential to test the validity of equation 6. Such a test can be carried out using the technique of competitive growth, wherein the particle size distributions of a latex made up of two different-sized (and monodisperse) particles is measured. This technique was pioneered by Vanderhoff and co-workers [*28*], with some further development of quantitative theory culminating in work by Ugelstad and co-workers [*29*]. However, this early work did not take into account more recent discoveries on the aqueous-phase kinetics of the desorbed free radicals, and did not cover on particle sizes sufficiently small to extrapolate with confidence to the very small particles involved in nucleation kinetics. For these reasons, we have carried out an extensive series of such experiments and concomitant theory, some of whose results are shown in Figure 8. Such data can be used to obtain information on the radial dependence of k_e: i.e., the radial dependence predicted by equation 6 can be directly tested.

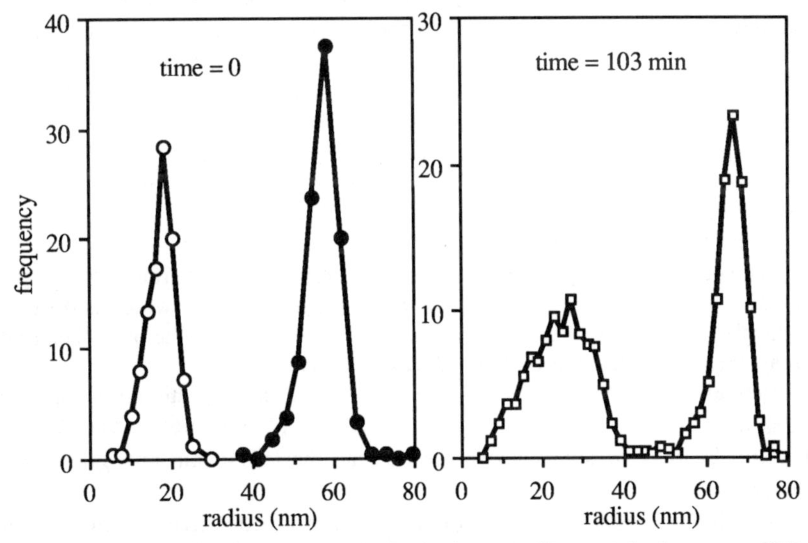

Figure 8. Time evolution of particle size distribution in competitive growth of styrene at 50°C. Particle numbers for small and large particles were 2.3 and 2.6×10^{16} dm^{-3} respectively; initiator was 10^{-4} mol dm^{-3} persulfate. Experiments were all in Interval II.

The data interpretation must take explicit account of the exit (desorption) of free radicals, and of the fate of these exited free radicals. This is because exit is very significant kinetically for particles of these sizes, and this has a strong effect on the particle growth kinetics and hence on the value of the entry rate coefficient deduced from kinetic and particle size data. There are two extreme cases as to the fate of exited free radicals [*30*]: complete aqueous-phase termination (in which case the exited free radicals never re-enter), and complete re-entry. Unfortunately, present data are insufficient to distinguish between either mechanism (or anything intermediate). For example, Figure 9 shows values for the fitted first-order exit rate coefficient k determined by the standard chemically-initiated technique [*31*], as a function of swollen particle surface area; in this figure, results for all except 18 nm particles are from [*31*]; data for 18 nm radius are from present work. It is apparent that k is inversely proportional to surface area, a result consistent with a first-order loss process wherein all free radicals are terminated in the aqueous phase [31,32]. However, the same data can also be successfully interpreted by complete re-entry, wherein the loss becomes kinetically second order, but with a rate coefficient still proportional to particle area.

Because these experiments do not give definitive information on the fate of exited free radicals, we are forced to interpret competitive growth data exemplified by Figure 8 using both hypotheses.

(1) *Complete aqueous-phase termination.* The simplest interpretation is that there is complete termination of exited free radicals in the aqueous phase in a chemically initiated system. This gives the following model for the growth kinetics of a bimodal (or rather, doubly monodisperse) particle size distribution, with two particle sizes a and b:

$$\frac{d\bar{n}_a}{dt} = \rho_a(1 - 2\bar{n}_a) - k_a\bar{n}_a \tag{9}$$

where $\bar{n}_a$ is the average number of free radicals in particles of size a, etc.; a similar equation holds for particles of size b. The dependence of the exit rate coefficient on particle size, and hence the values of k_a and k_b, can be found directly from the data of Figure 9 (noting that the data of Figure 8 refer to unswollen radius, and those of Figure 9 to swollen radius); the values of $\bar{n}_a$ and $\bar{n}_b$ are found directly from the measured change in unswollen particle radius and the known [31] propagation rate coefficient.

(2) *Complete re-entry.* The interpretation using the assumption of complete re-entry is more complex. It is necessary to take account of the following events:

(a) transfer and propagation inside each particle;

(b) the escape of a transferred free radical with rate coefficient k_{dm}, before it propagates; we use k_{dm}, referring to desorptionof a *monomeric* free radical, which is different from the phenomenological exit rate coefficient k which refers to the overall exit rate from particles containing radicals of any degree of polymerization;

(c) re-entry of all exited free radicals, whose aqueous-phase concentration is denoted E;

(d) propagation of a monomeric free radical inside the particle, noting that this rate coefficient, k_p^1, may be significantly greater than the long-chain propagation rate coefficient k_p;

(e) as implicit in the complete re-entry assumption for this model, termination involving aqueous-phase *exited* free radicals is ignored; indeed, it can be shown that the rate of this process is considerably less than that for re-entry;

(f) instantaneous termination however takes place when any aqueous-phase free radical enters a particle already containing a free radical;

(g) it is necessary to distinguish between particles of size a containing a *monomeric* free radical (whose relative population we denote N_{a01}), and those containing a *propagated* free radical (whose population is denoted N_{a1}), as well as particles containing no free radicals (N_{a0}); one has the normalization $N_{a1}+N_{a01}+N_{a0} = 1 \approx N_{a1}+N_{a0}$.

The resulting equations (where the subscript a refers to particles of size a, and the corresponding equations for b have been omitted for brevity) are:

$$\frac{dN_{a01}}{dt} = k_{tr}C_mN_{a1} + k_{ea}EN_{a0} - k_p^1C_mN_{a01} - k_{dma}N_{a01} \tag{10}$$

$$\frac{dN_{a1}}{dt} = \rho_aN_{a0} - \rho_aN_{a1} - k_{tr}C_mN_{a1} + k_p^1C_mN_{a01} - k_{ea}EN_{a1} \tag{11}$$

$$\frac{dE}{dt} = k_{dma}N_{a01}N^a + k_{dmb}N_{b01}N^b - k_{ea}EN^a - k_{eb}EN^b \tag{12}$$

Here N^a and N^b the (molar) concentrations of particles of size a and b; C_m is assumed independent of particle size (an assumption whose validity we proved experimentally by direct measurement of this quantity), and it is assumed that entry of an initiator-derived free radical (ρ_a and ρ_b) does not result in exit (i.e., escape can only occur after transfer). Applying the steady-state assumption in equations 10 and 12 (justified because the individual gain and loss rates involved are large compared to the overall polymerization rate), one has:

$$(k_p^1 C_m + k_{dma})N_{a01} - (k_{tr}C_m N_{a1} + k_{ea}EN_{a0}) = 0 \tag{13}$$

$$(k_p^1 C_m + k_{dmb})N_{b01} - (k_{tr}C_m N_{b1} + k_{eb}EN_{b0}) = 0 \tag{14}$$

$$(k_{ea}N^a + k_{eb}N^b)E - (k_{dma}N_{a01}N^a + k_{dmb}N_{b01}N^b) = 0 \tag{15}$$

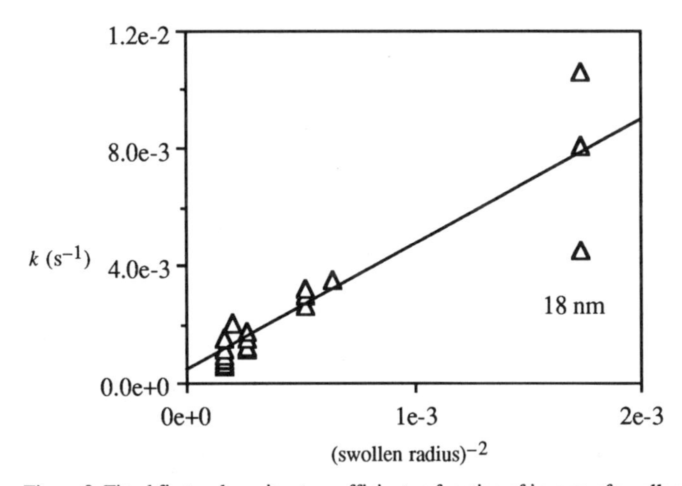

Figure 9. Fitted first-order exit rate coefficient as function of inverse of swollen particle area.

Equations 13-15 constitute three linear simultaneous equations which can be solved, given the values of $N_{a1} = \bar{n}_a$ and $N_{b1} = \bar{n}_b$, and k_{ea} and k_{eb}, and noting that the escape and entry rate coefficients are related by microscopic reversibility:

$$\frac{k_e}{k_{dm}} = \frac{4}{3}\pi\, r_s^3\, N_A \frac{C_m}{C_{aq}} \tag{16}$$

where C_{aq} is the aqueous-phase solubility of a monomeric unit and N_A is Avogadro's constant. We can make the steady-state approximation in equation 11 to obtain:

$$\rho_a = \frac{k_p^1 C_m N_{a01} - \bar{n}_a(k_{tr}C_m + k_{ea}E)}{2\bar{n}_a - 1} \tag{17}$$

$$\rho_b = \frac{k_p^1 C_m N_{b01} - \bar{n}_b(k_{tr}C_m + k_{eb}E)}{2\bar{n}_b - 1} \tag{18}$$

where N_{a01}, N_{b01} and E can be found by solution of equations 13-15. The value of k_e (i.e., the rate coefficient for re-entry of exited radicals) is assumed to be given by equation 6; that is, we assume the validity of the Smoluchowski equation for re-entry of exited free radicals, which is not unreasonable as it is hard to envisage any barrier to entry for these uncharged species. Equations 17 and 18 then yield values of the required entry rate coefficients for radicals derived from the initiator, ρ_a and ρ_b, given the *experimental* values of $\bar{n}_a$ and $\bar{n}_b$ from the observed particle size distributions.

The only unknown parameter in the use of equations 17 and 18 to find these quantities from the experimental particle size evolution is k_p^1. Since k_p^1 is not known, the data were processed assuming two values of this quantity: the macroradical limit (260 $dm^3\ mol^{-1}\ s^{-1}$ for styrene at 50°C [*31*]) and 1000 $dm^3\ mol^{-1}\ s^{-1}$; the higher value is more likely, since there is theoretical and experimental evidence that the propagation rate coefficient for a monomeric free radical is significantly greater than for a large macroradical. Results for the processing are shown in Figure 10. It can be seen that both models can fit the data, with the fit for the complete-re-entry version being good for both assumed values of k_p^1. In the same figure, we show the ratios expected if the entry rate coefficient were to scale as the ratio of the radii (as expected for the Smoluchowski model) or as the square of the radius.

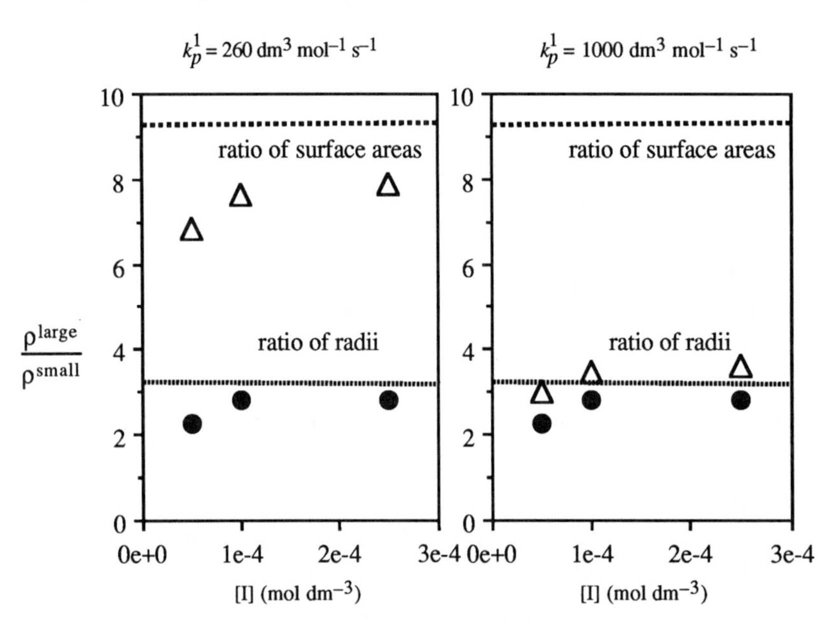

Figure 10. Ratios of entry rate coefficients for different sized particles, from data of Figure 8 and similar experiments with different initiator concentrations, assuming either complete aqueous-phase termination or complete re-entry of exited free radicals. Δ = complete re-entry assumption; • = complete aqueous-phase termination assumption.

It will be recalled that the objective of these experiments is to test the validity of the Smoluchowski equation (upon which our demonstration of the importance of micelles in nucleation above the cmc was based). The complete aqueous-phase termination assumption, if true, shows that the data are completely consistent with the radius dependence predicted by this model. On the other hand, this dependence is only consistent with the complete re-entry assumption if $k_p^1 \gg k_p$. Since such is expected to be the case, the data are all-in-all supportive of our model assumptions. However, this conclusion cannot be said to be definite until a direct measurement of k_p^1 has been carried out, a project which we are currently pursuing.

Conclusions

The incorporation into standard HUFT nucleation theory of an accurate model for the rate of entry of free radicals into pre-existing latex particles strongly suggests that micelles are directly involved in the nucleation process at surfactant concentrations above the cmc. This must occur **in addition** to homogeneous nucleation and coagulation of precursor particles, which dominate nucleation at low surfactant concentrations. Thus it is reasonable to suppose that above the cmc, nucleation involves homogeneous-coagulative nucleation, with oligomeric free radicals entering micelles (in competition with existing particles) for (probably limited) protection against aqueous-phase termination (which is identical to coagulation for oligmers) so they can grow enough to become true particles. Accurate models of nucleation must therefore take this additional effect into account; fortunately, some such models (e.g., [*8,9*]) are sufficiently flexible that this can be readily implemented.

This mechanism is genuinely *different* from conventional micellar entry, which assumes that species such as $SO_4^{-\bullet}$ enter micelles directly. The present description **combines** homogeneous-coagulative and micellar descriptions, because the precursors (x- to z-mers) are still formed by homogeneous nucleation. Indeed, this *combined* mechanism may explain why no definitive evidence refuting either the micellar-entry or the homogeneous-coagulative mechanism could be found. However, these conclusions as to the involvement of micelles depend on reasonable but untested assumptions for interpreting the new data; only further experimental investigation can determine this.

Finally, the methodology developed here can also be used to predict the conditions for onset of secondary nucleation below the cmc.

Acknowledgements

The financial support of the Australian Research Grants Scheme, and generous provision of resources by the Sydney University Electron Microscope Unit, are gratefully acknowledged.

Literature Cited

(1) Harkins, W. D. *J. Polym. Sci.* **1950,** *5*, 217.
(2) Harkins, W. D. *J. Am. Chem. Soc.* **1947,** *69*, 1428.
(3) Smith, W. V.; Ewart, R. H. *J. Chem. Phys.* **1948,** *16*, 592.
(4) Fitch, R. M.; Tsai, C. H. In *Polymer Colloids*; R. M. Fitch, Ed.; Plenum: New York, 1971.
(5) Ugelstad, J.; Hansen, F. K. *Rubber Chem. Technol.* **1976,** *49*, 536.
(6) Casey, B. S.; Maxwell, I. A.; Morrison, B. R.; Gilbert, R. G. *Makromol. Chem., Macromol. Symp.* **1990,** *31*, 1.

(7) Goodwin, J. W.; Hearn, J.; Ho, C. C.; Ottewill, R. H. *Colloid Polym. Sci.* **1974**, *252*, 464.
(8) Feeney, P. J.; Napper, D. H.; Gilbert, R. G. *Macromolecules* **1987**, *20*, 2922.
(9) Richards, J. R.; Congalidis, J. P.; Gilbert, R. G. *J. Appl. Polym. Sci.* **1989**, *37*, 2727.
(10) Sütterlin, N. In *Polymer Colloids II*; R. M. Fitch, Ed.; Plenum: New York, 1980.
(11) Maxwell, I. A.; Morrison, B. R.; Napper, D. H.; Gilbert, R. G. *Macromolecules* **1991**, *24*, 1629.
(12) Adams, M. E.; Trau, M.; Gilbert, R. G.; Napper, D. H.; Sangster, D. F. *Aust. J. Chem.* **1988**, *41*, 1799.
(13) Napper, D. H.; Gilbert, R. G. In *Comprehensive Polymer Science*; G. A. Allen, J. C. Bevington and G. C. Eastwood, Ed.; Pergamon: Oxford, 1989; Vol. 4; pp 171.
(14) Lichti, G.; Gilbert, R. G.; Napper, D. H. *J. Polym. Sci. Polym. Chem. Edn.* **1983**, *21*, 269.
(15) Feeney, P. J.; Napper, D. H.; Gilbert, R. G. *J. Colloid Interface Sci.* **1987**, *118*, 493.
(16) Morton, M.; Kaizerman, S.; Altier, M. W. *J. Colloid Sci.* **1954**, *9*, 300.
(17) Feeney, P. J.; Gilbert, R. G.; Napper, D. H. *J. Colloid Interfacial Sci.* **1985**, *107*, 159.
(18) Whang, B. C. Y.; Ballard, M. J.; Napper, D. H.; Gilbert, R. G. *Aust. J. Chem.* **1991**, *44*, 1133.
(19) Brandrup, A.; Immergut, E. H. *Polymer Handbook;* 3rd ed.; Wiley Interscience: New York, 1989.
(20) Pearson, L. T.; Louis, P. E. J.; Gilbert, R. G.; Napper, D. H. *J. Polym. Sci., Polym. Chem. Ed.* **1991**, *29*, 515.
(21) Rawlings, J. B.; Ray, W. H. *Polym. Eng. Sci.* **1988**, *28*, 257.
(22) Rawlings, J. B.; Ray, W. H. *Polym. Eng. Sci.* **1988**, *28*, 237.
(23) Pascal, P.; Gilbert, R. G.; Napper, D. H. *manuscript in preparation.*
(24) Hansen, F. K.; Ugelstad, J. *J. Polym. Sci., Polym. Chem. Ed.* **1979**, *17*, 3047.
(25) Feeney, P. J.; Geissler, E.; Gilbert, R. G.; Napper, D. H. *J. Colloid Interface Sci.* **1988**, *121*, 508.
(26) Wiese, G. R.; Healy, T. W. *Trans. Faraday Soc.* **1970**, *66*, 490.
(27) Hogg, R.; Healy, T. W.; Furstenau, D. W. *Trans. Faraday Soc.* **1966**, *62*, 1638.
(28) Vanderhoff, J. W.; Vitkuske, J.F.; Bradford, E. B.; Alfrey, T. *J. Polym. Sci.* **1956**, *20*, 265.
(29) Ugelstad, J.; El-Aasser, M. S.; Vanderhoff, J. W. *J. Polym. Sci., Polym. Letters Ed.* **1973**, *11*, 503.
(30) Whang, B. C. Y.; Napper, D. H.; Ballard, M. J.; Gilbert, R. G.; Lichti, G. *J. Chem. Soc. Faraday Trans. 1* **1982**, *78*, 1117.
(31) Hawkett, B. S.; Napper, D. H.; Gilbert, R. G. *J. Chem. Soc. Faraday Trans. 1* **1980**, *76*, 1323.
(32) Gilbert, R. G.; Napper, D. H. *J. Macromol. Sci. - Rev. Macromol. Chem. Phys. C* **1983**, *23*, 127.

RECEIVED December 4, 1991

Chapter 4

Rate of Emulsifier Adsorption as a Factor in the Nucleation of Polymer Latex Particles

Alexander S. Dunn

Chemistry Department, University of Manchester Institute of Science and Technology, Manchester M60 1QD, United Kingdom

Although there is no question that the predominant mechanism for the nucleation of latex particles in the emulsion polymerization of the more water-soluble monomers such as methyl methacrylate and vinyl acetate is the oligomeric precipitation mechanism of Fitch and Roe and this is the only mechanism which can apply to styrene despite its low solubility in water in absence of micellar emulsifier, micellar nucleation has been generally believed to predominate in presence of micellar emulsifier. But micellar nucleation appears to be inconsistent with the evidence for the participation of a coalescive step in the nucleation of polystyrene latex particles. Recently, Richards, Congalidis, and Gilbert have been able to model the emulsion polymerization of styrene without the need to include micellar nucleation. However they assume a value of the equilibrium constant for emulsifier adsorption which is much smaller than is justified by experimental values of Zuikov and Vasilenko. It seems that the initial rate of increase of the surface area of a polystyrene latex particle nucleated from a micelle would exceed the rate of adsorption of emulsifiers (especially nonionics) so that the surface potential of small particles could periodically fall to a level permitting a significant rate of coalescence.

The amount of styrene which can be solubilized in micellar solutions of emulsifiers greatly exceeds the apparently insignificant amount of styrene (0.032% at 25 °C) in molecular solution in the aqueous phase so it was natural for Harkins (1) and other early workers in the field of emulsion polymerization to assume that the polymer latex particles (in which the bulk of the reaction occurs) were nucleated from monomer solubilized in emulsifier micelles. This assumption is inappropriate in the case of the more water-soluble monomers such as methyl methacrylate (solubility 1% at 60 °C) and vinyl acetate

0097-6156/92/0492-0045$06.00/0

(solubility 2.8% at 60 °C) for which the amount in molecular solution greatly exceeds any amount which may be solubilized. For these monomers, latex particles are certainly nucleated by the precipitation of oligomers in the aqueous phase (2) which then undergo limited coalescence and adsorb surfactant until their surface potential increases sufficiently to prevent further coalescence (3). Richards, Congalidis, and Gilbert (4,5) have recently succeeded in modelling the emulsion polymerization of styrene without the need to allow for the participation of micellar nucleation. However their model fails to predict the approximate 3/5 power dependence of particle number on emulsifier concentration above the critical micelle concentration (cmc) which is well established for styrene and, to get a good fit to their experimental results, they have to assume a value (50 dm^3 mol^{-1}) for the equilibrium constant for the adsorption of sodium dodecyl sulfate by polystyrene which seems to be seven times too low. When this value at 80 °C is combined with the experimental value at 25 °C [2400 dm^3 mol^{-1} (6)] the enthalpy of adsorption is calculated as - 60.5 kJ mol^{-1} which is twice the experimental value (- 30.1 kJ mol^{-1}) determined by Zuikov and Vasilenko (7) which makes the constant 350 dm^3 mol^{-1} at 80 °C given the 25 °C value.

There is no question that latex particles can be nucleated from monomer-swollen micelles. Although latex particles are formed in water saturated with styrene in the absence of emulsifier micelles, its less soluble homologue tert-butyl styrene [solubility 0.00053% at 25 °C (8)] does not polymerize in the presence of potassium persulfate at 60 °C until emulsifier is injected, at which point latex forms immediately (9).

Doubts have often been expressed as to whether a negatively charged sulfate radical could enter a negatively charged micelle. However, Fitch and Shih (10) calculated the energy barrier between a sulfate radical and a latex particle with a surface charge density of 0.65 μC cm^{-2} and found it to be of the order of kT which would not reduce the entry rate significantly. On the other hand, the net charge on a monomer-swollen sodium dodecyl sulfate micelle (after allowing for the effect of counter-ion binding in the Stern layer) is much larger (3.9 μC cm^{-2}) which would produce a repulsive energy of about 6 kT at the micelle surface (taken to include the bound counter-ions) which would be large enough to inhibit entry of sulfate radicals although probably not of oligomer radicals having several monomer residues in addition to a sulfate end-group. Nevertheless, no difference has ever been observed on the effect of persulfate initiation compared with initiation by neutral radicals from hydrogen peroxide. Also sulfate radicals are effective in nucleating tert-butyl styrene solubilized in micelles where any oligomer formed is evidently terminated in the aqueous phase before reaching a sufficient degree of polymerization to become insoluble and precipitate.

Convincing evidence has been presented (11,12) for the participation of a coagulative step in the nucleation of latex particles in the emulsion polymerization of styrene. Under the sort of conditions normally chosen for conventional emulsion polymerization sufficient emulsifier is adsorbed on the surface of the final latex particles to ensure that they are kinetically stable, or at any rate, only coalesce very slowly when the fully polymerized latex is maintained at a high temperature (13). Monomer-swollen micelles are thermodynamically stable. There does not appear to be any scope for a coagulative step if the surface of growing particles nucleated from monomer-swollen micelles remains saturated with adsorbed emulsifier so long as micelles remain in the system although a delay in the dispersal of micelles has often been suggested to explain some phenomena observed in continuous emulsion polymerization systems (14).

Rates of Adsorption of Surfactants

Work on the kinetics of surfactant adsorption is comparatively sparse and mostly relates to adsorption at the air/water interface at ambient temperature.

Static (e.g. capillary rise, Wilhelmy plate) and dynamic (e.g. oscillating jet, maximum bubble pressure) methods of measuring surface tension give concordant results when applied to pure liquids. However, discrepancies are found when these techniques are applied to surfactant solutions because the surfactant is adsorbed at the interface only at a finite rate when the interface expands. Addison (15) used the oscillating jet method to study the rate of adsorption of alkanols at freshly formed air/water interfaces and found that whereas with n-pentanol at 20 °C equilibrium was established within 0.08 s, n-octanol required about 0.2 s. Using the drop weight method King [quoted by Ward and Tordai (16)] found that more than 30 min was required to establish equilibrium with decanoic acid: the time required for other alkanoic acids increased with alkyl chain length. Klimenko et al. (17) found experimentally that the rate of adsorption of nonionic surfactants increased in proportion to their concentration in solution, but reached a maximum at the critical micelle concentrations. Ward and Tordai (16) deduced the relation:

$$\Gamma = 2c_o\sqrt{\frac{Dt}{\pi}}$$

for the initial adsorption of surfactant at a planar surface where Γ is the surface concentration, c_o the bulk concentration, and D the diffusion coefficient of the surfactant in the solution. Joos and Rillaerts (18) show that for a sphere expanding at a constant volume growth rate (like the dropping mercury cathode used in polarography) this expression becomes:

$$\Gamma = 2c_o\sqrt{\frac{3Dt}{7\pi}}$$

Thomas and Hall (19) used the oscillating jet method to determine diffusion coefficients as a function of concentration for 11 nonionic surfactants and found that the diffusion coefficients decreased by a factor of between 2 and 4 when the concentration was doubled. A 5.6×10^{-4} mol dm^{-3} solution 'Triton' X-100 (an iso-nonylphenol polyoxyethylene condensate) required about 0.2 s to reach equilibrium. Davies and Rideal (20) suggested that the rate of adsorption should be represented by the equation:

$$\frac{d\Theta}{dt} = k_a(1 - \Theta)c_o - k_d\Theta$$

where Θ is the fraction of the surface covered, c_o the initial concentration in solution, and k_a, k_d rate constants for adsorption and desorption. They show that the time required for adsorption equilibrium to be established is generally less than 0.4 s and that the very slow changes which some previous workers had observed in long-chain alkanoic acid solutions could be attributed to reaction with very low concentrations of polyvalent metal ions present in the water used. Miyamoto and Tagawa (21) have recently determined the rate constants for the adsorption of a range of sodium alkyl sulfates on nylon at 25 °C to be in the range 2-8 x 10^4 dm^3 mol^{-1} s^{-1}, finding them to have a uniform adsorption area of 0.72 nm^2 in a saturated monlayer with equilibrium constants 3.5-73 x 10^3 dm^3 mol^{-1}. A logarithmic spiral was used by van Voorst Vader et al. (22) to drive a pair of barriers apart on a Langmuir trough in such a way that the relative rate of increase of surface area:

$$\frac{1}{A}\frac{dA}{dt} = \frac{d \ln A}{dt}$$

was constant. Relative rate of dilatation up to 0.01 s^{-1} could be attained. At the highest relative rate, 0.012 s^{-1} with a dodecyl polyoxyethylene ether surfactant 1.9 x 10^{-5} mol dm^{-3} at 21 °C, a maximum rise of 10 mN m^{-1} in the surface tension was observed. By running a strip of polyethylene through the surface of a solution, Rillaerts and Joos (23) were able to extend the range of relative rates of dilatation up to 1 s^{-1}. They found the surface tension only reached its steady state value after 140 s with the relative rate of dilatation at 0.03 s^{-1} and sodium dodecyl sulfate at about half its critical micelle concentration. However enough surfactant to stabilize particles could be absorbed in a time much less than that required to saturate the surface. Feinerman (24) criticized the design of the apparatus used by Rillaerts and Joos: he used an apparatus in which the rate of flow of a surfactant solution could be varied to give relative rates of dilatation in the range 0 - 0.2 s^{-1}. In recent years a much improved design (25) of Langmuir trough has been marketed commercially for use in the preparation of Langmuir-Blodgett films; this enables the area of the film to be varied while it is confined within a constant perimeter. It is normally operated to keep the surface pressure constant but it could be used for studies of the effect of dilatation rate on surface tension although no such studies have yet been published. Klimenko et al. (17) used a turbidimetric method to measure the rates of adsorption of nonionic surfactants on silica gel; they found the rate to be proportional to concentration up to the cmc. This means that rates of adsorption of nonionic surfactants are generally much lower than those of ionic surfactants because their critical micelle concentrations are much lower (by a factor of 100 or thereby). Their molecular weights are also higher implying lower values for their diffusion coefficients.

Dilatation Rates of Growing Polymer Particles

It has always been assumed that if latex particles were nucleated from monomer-swollen micelles their surfaces would remain saturated with adsorbed emulsifier so long as emulsifier micelles remained in the system. Although the surface area of the particles continues to increase after all emulsifier has been adsorbed reducing the surface charge density and consequently the stability of electrostatically stabililized particles, the reduction is usually insufficient to permit particles to coagulate. Soap titration (26) can be used to determine the amount of additional surfactant required to saturate the surface of completely polymerized latex particles. Exceptionally, when unusually high emulsifier concentrations are used, micelles may remain at the end of the reaction when the surface of the finished latex will be covered with a saturated monolayer of adsorbed surfactant(27,28).

However, it appears that the relative rates of increase of the surface area of very small latex particles (i.e. particles newly nucleated from monomer-swollen micelles) are quite comparable with the relative rates of dilatation of surfaces in experiments in which rates of adsorption of surfactants have been measured so that it is possible that in the early stages of particle growth the concentration of adsorbed emulsifier may periodically fall to a value low enough to permit some slow coalescence of growing latex particles to occur although rates of emulsifier adsorption will generally be sufficient to ensure that resting particles (i.e. those not containg a radical at a particular instant) are stable. Growing particle could undergo coalescence at varying rates leading to a particle size distribution just as mutual termination between radicals of different lengths produces a distribution of molecular weights (with the simplification that the termination rate constant is not a function of radical size). If the stabilizer concentration is low enough (e.g. as in 'emulsifier-free' emulsion polymerization in which ionic end-groups derived from the initiator are solely responsible for the stabilization or when the initial emulsifier concentration is less than its cmc) no small particles are ever sufficiently stable to be exempt from involvement in coalescive processes which generally ultimately lead to a monodisperse particle size distribution under these conditions.

The volume growth rate of growing particles can be calculated from the appropriate value of the propagation rate constant of the monomer, k_p. Although the value of k_p for styrene determined by Matheson et al. (29) (126 dm^3 mol^{-1} s^{-1} at 50 °C) has long been favoured, it appears that this value is low because of the neglect of primary radical termination (30). This correction has been generally overlooked but it does bring Matheson's value into reasonable accord with recent determinations (31,32) which indicate that a value of k_p = 254 dm^3 mol^{-1} s^{-1} for styrene at 50 °C should be used (33).

The rate of polymerization within a growing latex particle R_{pp} is $k_p[M]_p$ where $[M]_p$ is the monomer concentration when a particle of the relevant size is swollen to equilibrium. This expression gives the number of monomer molecules added to a polymer radical in unit time: division by the Avogadro number L is required to obtain the amount of monomer polymerized in unit time. The volume growth rate $\mu = R_{pp}/[M]_p = k_p/L$. Hence taking k_p = 254 dm3 mol^{-1} s^{-1}, μ = 4.22 x 10^{-22} dm^3 s^{-1}. Publications on emulsion polymerization have rarely included values of the volume growth rate. Van der Hoff (34) using oil-soluble initiators with styrene at 50 °C finds μ = 1.36 x 10^{-22} dm^3 s^{-1} during Interval II although his values of R_{pp} and $[M]_p$ actually give μ' = 1.02 x 10^{-22} dm^3 s^{-1}. On the assumption that only half the latex particles would be actively polymerizing at any time in Interval II, $\mu = 2\mu'$ = 2.04 x 10^{-22} dm^3 s^{-1} which is in excellent agreement with the value (2.09 x 10^{-22} dm^3 s^{-1}) calculated from Matheson's value for k_p. If the new higher value of k_p is actually correct there are two factors (or a combination of them) which could explain the discrepancy (a) an underestimate of the average particle size leading to the value calculated for N, the number of latex particles per unit volume of the aqueous phase, being too large and/or (b) $\bar{n}$ the average number of radicals per polymer particle during Interval II being less than 0.5. This latter possibility is perhaps the more probable because the work did not include a perturbation experiment to see whether an increase of initiator concentration during the course of an experiment had any effect on the rate.

However this expression for μ does not take account of the contraction due to polymerisation and needs to be multiplied by the ratio of the densities of the monomer and polymer, σ_1/σ_2, at the temperature of polymerization (0.832 for styrene at 50 °C). This gives the volume growth rate of unswollen particles as might be observed by electron microscopy. Under polymerization conditions the particles are swollen with monomer to an extent which depends on particle size. If it is assumed that equilibrium swelling is maintained the volume fraction of monomer in the particles ϕ can be estimated from the curve calculated by Gardon (35) taking the value of the polymer-solvent interaction parameter χ = 0.4 with the interfacial tension as 5 mN m^{-1}. The volume growth rate of monomer-swollen latex particles then becomes:

$$\mu^* = \frac{k_p}{L}\frac{\sigma_1}{\sigma_2}\frac{\phi}{1-\phi}$$

as given by Gardon (35).

According to Almgren and Swarup (36) the radius of a sodium dodecyl sulfate micelle saturated with toluene is 2.11 nm which should be a typical value for an ionic emulsifier althoughsaturated micelles of lower homologues will be smaller and those of higher homologues larger. Saturated micelles of nonionic emulsifiers would be much larger.

Relative rates of dilatation for styrene latex particles at 50 °C are calculated in Table I for particles in the range of sizes likely to be encountered in emulsion polymerization. It is apparent that although the relative rate of dilatation falls rapidly with increase of particle size, the rates to be expected in the initial stages of the growth of a latex particle by polymerization of monomer solubilized in an emulsifier micelle (assuming that diffusion of monomer to the growing particle does not become rate limiting) are as large as any which have been employed in experimental studies of the rates of adsorption of surfactants at expanding interfaces.

Table I
Dependence on Particle Size of the Relative Rate of Dilatation of Polystyrene Latex Particles at 50 °C

Diameter/nm	4	5	6	10	20	60
ϕ	0.535	0.580	0.600	0.670	0.760	0.835
$\mu^*/nm^3\ s^{-1}$	404	484	526	712	1111	1176
$\frac{1}{A}\frac{dA}{dt}/\ s^{-1}$	12.1	8.8	6.5	3.2	1.7	1.4

These calculations provide an explanation of the observations we made some years ago (37) that when equal micellar concentrations of the homologous series of potassium carboxylates or sodium alkyl sulfates are used in the emulsion polymerization of styrene the same number of latex particles with the same average size is ultimately obtained although the duration of Interval I during which these particles are nucleated decreases with increase of alkyl chain length of the surfactant. Although a much higher concentration of the lower members of the series is present in molecular solution with presumably larger diffusion coefficients, the effect of the greater relative rate of dilatation when the smaller micelles begin to polymerize is evidently dominant permitting more extensive coalescence to occur before stable particles form. The final latices all had identical particle size distributions: it was conspicuous that the breadth of the particle size distributions was not related to the duration of Interval I.

Comparison of the Effects of Ionic and Nonionic Emulsifiers

It is well known that the average particle size of latices produced using a nonionic emulsifier alone are much larger than those produced using comparable concentrations of ionic emulsifiers although they are smaller than latices produced by 'emulsifier-free' emulsion polymerization. Moreover, such latices are generally monodisperse. Mixed emulsifiers containing low concentrations (below the cmc) of an ionic emulsifier in addition to the nonionic emulsifier may be used to reduce the average particle size obtained. Although the micelles (and mixed micelles) of nonionic emulsifiers are much larger than those of ionic emulsifiers (and consequently will have relatively low relative rates of dilatation when they polymerize) nonionic emulsifiers also have lower diffusion coefficients and much lower critical micelle concentrations than ionic emulsifiers. Consequently their rates of adsorption are evidently relatively low compared with the dilatation rate, permitting more extensive limited coalescence to occur to produce the larger average particle sizes observed. Klimenko et al. (17) found that the adsorption rate of the E-9 isononyl aryl polyoxyethylene ether ('Triton' X-100) was three times that of the E-23 dodecyl polyoxyethylene ether $C_{12}H_{25}O(C_2H_3O)_{23}H$ above their critical micelle concentrations. On this basis it may be predicted that if surfactant adsorption rates are a critical factor in determining latex particle sizes, and if micellar concentrations of these emulsifiers were chosen so as to produce equal areas of saturated monolayer on the polymer surface at the polymerization temperature, then the E-23 ether should produce a latex with a larger average particle size because its slower adsorption rate should permit more extensive limited coalescence of latex particle nuclei and that correspondingly the Interval II polymerization rate observed would be lower. It would however be essential to ensure that the 'Triton' X-100 used was not peroxidized.

Polymerization at high emulsifier/monomer ratios

Although the monomer/water ratio is not generally a critical factor in emulsion polymerization, Chatterjee, Banerjee, and Konar (36) observed that styrene polymerized in emulsion at a higher rate at a very low monomer/water ratio. This observation has been confirmed (Fig.1). Nomura and Harada (39,40) have found that a larger number of latex particles is formed at low monomer/water ratios at a constant

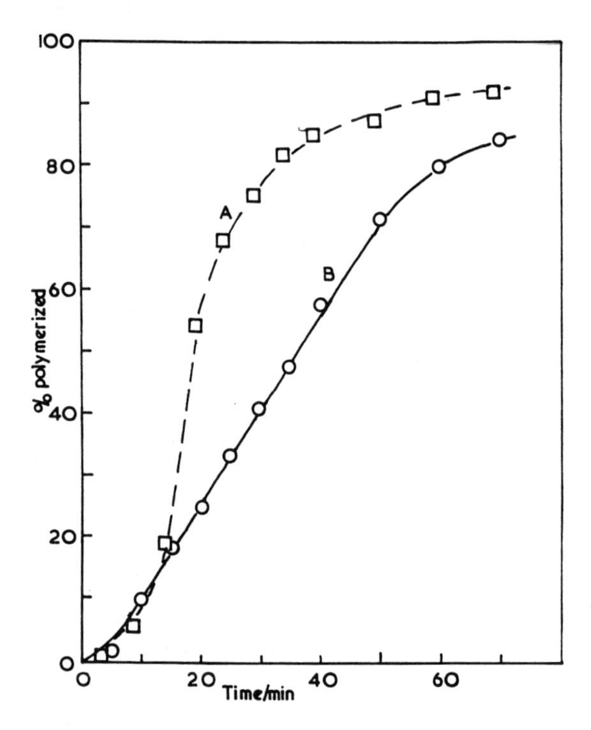

Figure 1. Emulsion polymerization of (A) 0.625% (B) 5% styrene at 50 °C with 0.1% potassium peroxydisulfate and 0.6% sodium dodecyl sulfate (41).

emulsifier concentration. Since the Interval II rate in the emulsion polymerization of styrene is directly proportional to the number of latex particles this explains the observations of a higher rate of polymerization. However, it is still necessary to find an explanation for the formation of larger numbers of latex particles at low monomer/water ratios. Reducing the amount of monomer used while keeping the emulsifier concentration in the aqueous phase constant means that the emulsifier/monomer ratio increases and this appears to be the factor responsible for the stabilization of a larger number of latex particles.

When the amount of monomer is reduced below the amount required to saturate the emulsifier micelles with solubilized monomer it is likely that the rate of polymerization in the latex particles will ultimately become diffusion controlled reducing the rate of dilatation so that the surface of the growing particles can be kept continually saturated with adsorbed emulsifier precluding all particle coalescence. Below the critical monomer/water ratio the number of latex particles should rise to a limit at a maximum of which there is indeed some indication in Nomura and Harada's results [cf. Fig.6 of (40)].

Conclusion

It has been generally assumed hitherto that the rate at which emulsifier is adsorbed on the expanding polymer/water interface of polymerizing latex particles would be sufficient to ensure that the surface remained covered with a saturated monolayer of adsorbed emulsifier so long as emulsifier micelles remained in the reaction mixture. However there is sufficient evidence to make it doubtful whether this assumption is justified during the period of particle nucleation in emulsion polymerizations. A critical experiment which might resolve the issue has been suggested.

References

1 Harkins,W.D. J.Amer.Chem.Soc. 1947, 69,1428.
2 Fitch,R.M.;Tsai,C.H. in Fitch,R.M. Ed. 'Polymer Colloids'; Plenum: New York, 1973; p 73.
3 Dunn,A.S.; Chong,L.C.-H. Br.Polym.J. 1970,2,49.
4 Congalidis,J.P.; Richards,J.R.;Gilbert,R.G. in 'Computer Applications in Polymer Science II'; Provder,T.Ed.; A.C.S. Symposium Series 404; American Chemical Society: Washington D.C. 1989.
5 Richards,J.R.; Congalidis,J.P.; Gilbert,R.G. J.Appl.Polym.Sci. 1989, 37, 2727.
6 Ahmed,S.M.; El-Aasser,M.S.; Micale,F.J.; Poehlein,G.W.; Vanderhoff,J.W. in 'Polymer Colloids II'; Fitch,R.M. Ed.; Plenum, New York, 1980, p 265.
7 Zuikov,A.V.; Vasilenko, A.I. Zhur.Fiz,Khim. 1983,57,1457. English translation: Russ.J.Phys.Chem. 1983,57,880.
8 Dow Technical Bulletin CX-2290 on tert-Butylstyrene Monomer.
9 Satpathy,U.S. unpublished work at UMIST.
10 Fitch,R.M.; Shih,L.-B. Prog.Coll.Sci. 1975,56, 1.
11 Feeney,P.J.; Napper,D.H.; Gilbert,R.G. Macromolecules 1984,17,2520.
12 Feeney,P.J.; Gilbert,R.G.; Napper,D.H. J.Coll.Interface Sci. 1985, 107, 159.

13 Dunn,A.S.: Hassan,S.A. Polym.Mat.Sci.Eng. 1986,54,439.
14 Brooks,B.W. Br.Polym.J. 1973,5,199.
15 Addison,C.C. J.Chem.Soc. 1944,252,477; 1945,98.
16 Ward,A.F.H.: Tordai,L. J.Chem.Phys. 1946,14,453.
17 Klimenko,N.A.; Permilovskaya,A.A.; Tryasoukova,A.A.; Koganovskii,A.M. Koll.Zh. 1975,37, 972: English translation, Coll.J.USSR, 1975,37,873.
18 Joos,P.;Rillaerts,E. J.Coll.Interface Sci. 1981,79,96.
19 Thomas,W.D.E.;Hall,D.J. in 'Surface Active Agents' (Proceedings of a Symposium at Nottingham, 1979) SCI, London,1979, p.107.
20 Davies,J.T.; Rideal,E.K. 'Interfacial Phenomena' Academic, New York, 2nd Edn. 1963, p.154.
21 Miyamoto,S.; Tigawa,M. Coll.Polym.Sci. 1988,266,1126.
22 van Voorst Vader, F.; Erkens,Th.F.; van der Tempel,M. Trans. Faraday Soc. 1964,60,1170.
23 Rillaerts,E.; Joos,P. J.Coll.Interface Sci. 1982,88,1.
24 Feinerman,V.B. Koll.Zh. 1986,48,758: English translation, Coll.J.USSR, 1986,48,640.
25 Petty,M.C.; Barlow,W.A. in 'Langmuir-Blodgett Films' Roberts,G., Ed.; Plenum, New York,1990,p.103.
26 Maron,S.H.; Elder,M.E.; Ulevitch,I.N. J.Coll.Sci. 1954,9,89.
27 Bakker,J. Philips Research Reports 1952,7,344.
28 Al-Shahib,W. Ph.D. Thesis, Manchester, 1977.
29 Matheson,M.S.;Auer,E.E.; Bevilacqua,E.B.; Hart,E.J. J.Amer.Chem. Soc. 1951,73,1700.
30 Henrici-Olivé,G.; Olivé,S. Makromol.Chem. 1960,37,71.
31 Mahabadi,H.K.; O'Driscoll,K.F. J.Macromol.Sci.Chem. 1977,A 11,967.
32 Lansdowne,S.W.; Gilbert,R.G.; Napper,D.H.; Sangster,D.F. J.Chem. Soc. Faraday Trans. 1980,76,1344.
33 Buback,M.; Garcia-Rubio,L.H.; Gilbert,R.G.; Napper,D.H.; Guillot,J. Hamielec,A.E.; Hill,D.; O'Driscoll,K.F.; Olaj,O.F.; Shen,J.; Solomon,D.; Moad,G.; Stickler,M.; Tirrell,M.; Winnik,M.A. J.Polym.Sci. C Polym.Lett. 1988,26, 293.
34 van der Hoff,B.M.E. J.Polym.Sci. 1960,44,241.
35 Gardon,J.L. J.Polym.Sci. A-1,1968,6,2859.
36 Almgren,M.; Swarup,S. J.Phys.Chem. 1982,86,4212.
37 Dunn,A.S.; Al-Shahib,W.A. in 'Polymer Colloids II' Fitch,R.M., Ed.; Plenum, New York, 1980, p.619.
38 Chatterjee,S.P.; Banerjee,N.; Konar,R.S. J.Polym.Sci. Polym.Chem. Edn. 1979,17,219.
39 Harada,M.; Nomura,M.; Eguchi,W.; Nagata,S. Kobunshi Kagaku, 1972, 29, 844.
40 Nomura,M.; Harada,M.in 'Emulsion Polymers and Emulsion Polymerization' Bassett,D.R.; Hamielec,A.E.,Eds., A.C.S. Symposium Series 165, American Chemical Society, Washington,D.C. 1981,p.121.
41 Farooqui,T. M.Sc.Thesis, Manchester, 1981.

RECEIVED December 4, 1991

Chapter 5

Kinetics and Mechanisms of Particle Formation and Growth in the Emulsion Polymerization Initiated by the Oil-Soluble Initiator, 2,2′-Azobisisobutyronitrile

M. Nomura, J. Ikoma, and K. Fujita

Department of Materials Science and Engineering, Fukui University, Fukui, Japan

This paper experimentally clarifies that in emulsion polymerizations initiated by oil-soluble initiator, 2,2'-azobisisobutyronitrile, polymer particles are generated from emulsifier micelles and that the polymerization proceeds mainly inside the polymer particles. Furthermore, it is demonstrated that the radicals which initiate the polymerization are those stemming from the initiator dissolved in the water phase. Based on these findings, a kinetic model for seeded emulsion polymerizations initiated by oil-soluble initiators is proposed and compared with experiment.

Oil-soluble initiators are known to initiate polymerization in emulsion and the kinetic behavior of this polymerization system is very similar to that of emulsion polymerizations initiated by water-soluble initiators despite the difference in the principal loci of radical production in both systems*(1)*. We have reported that the kinetic behavior of the emulsion polymerization of styrene (ST) initiated by the oil-soluble initiator, 2,2'-azobisisobutyronitrile (AIBN) with sodium lauryl sulfate (NaLS) as emulsifier is quite similar to that initiated by the water-soluble initiator, potassium persulfate ($K_2S_2O_8$) *(2)*. The reasons for this similarity, however, have not been elucidated. Recently, Asua et al. *(3)* have presented a model which predicts the average number of radicals per particle in emulsion polymerization initiated by oil-soluble initiators, and arrived at a conclusion that the initiator radicals generated from the initiator distributed inside the polymer particles initiate the polymerization in this system.

The aim of this paper is first to provide experimental evidence which supports the hypothesis that polymer particles are generated from emulsifier micelles and that the principal loci of the polymerization shifts at a comparatively early stage of polymerization from the monomer droplets to the resulting polymer particles, when the emulsion polymerization of styrene is initiated by the oil-soluble initiator, AIBN with NaLS as emulsifier.

Secondly, the results in this paper demonstrate that the polymerization inside the polymer particles is initiated not by the radicals produced from the initiator distributed in the polymer particles and the monomer droplets, but mainly by the radicals generated from the initiator dissolved in the water phase. Based on these findings, a kinetic model for seeded emulsion polymerization which accounts for the rates of polymerization in both the monomer droplets and the polymer particles will be

0097–6156/92/0492–0055$06.00/0

presented, introducing the theoretical approach reported previously *(4)*, and further, the validity of the proposed kinetic model will be demonstrated by comparing the model predictions with the experimental data reported in the previous article *(2)*. This study will, therefore, show why the kinetic behavior of emulsion polymerizations initiated by oil-soluble initiators are very similar to those initiated by water-soluble initiators.

Experimental

Polymerization Apparatus and Procedure. Commercial ST monomer was purified by distillation after first washing with 15% KOH aqueous solution to remove hydroquinone inhibitor and then with deionized water until KOH was not detected in the effluent. AIBN and NaLS of extra-pure grade were used without further purification as initiator and emulsifier, respectively. All polymerization experiments were conducted at 50°C with the same experimental apparatus and procedure as previously described *(5)*, except that high purity nitrogen gas (purity>99.995%) was used without further purification. In these experiments, the impeller speed was constant at 400 rpm. Monomer conversion was determined gravimetrically using methanol as a precipitant for the polymer. The number of polymer particles produced was determined by electron microscopy with the following expressions.

$$N_T = \frac{6\, M_0\, X_M}{\pi\, d_p^3\, \rho_p} \tag{1}$$

$$d_p^3 = \frac{\sum n_i\, d_{pi}^3}{\sum n_i} \tag{2}$$

where N_T is the number of polymer particles per cm^3-water, M_0 the amount of monomer initially charged per cm^3-water, X_M the monomer conversion, ρ_p the density of polymer, and d_p the volume average diameter of the polymer particles.

Water-Solubility of AIBN. The water-solubility of AIBN and its partition coefficient between the monomer and water phases were determined as follows. An appropriate amount of AIBN was dissolved in styrene monomer. A portion of this solution was allowed to contact distilled water for more than 30 min in a flask maintained at 50°C with the use of a thermostatted water bath. Moderate stirring was applied using a magnetic stirring bar to quicken the transport of AIBN molecules from the monomer droplets to the water phase. The mixture in the flask was then completely separated into a monomer layer and a water phase with a centrifuge and a small sample taken from each separated phase was subjected to measurement of the concentration of AIBN by high-performance liquid chromatography (HPLC). AIBN was separated on a TOSOH TSK-GEL ODS-80TM column by using a mixture of water-methanol (50:50 by volume) as the eluent. The concentration of AIBN was measured by a UV detector at a wavelength of 346 nm. Figure 1 is a plot of the measured equilibrium concentration of AIBN in the monomer phase, $[I]_d$, versus that in the water phase, $[I]_w$, at 50°C. When $[I]_d$ is less than about 40 g/liter, the partition of AIBN between the monomer and water phases can be correlated by a linear relation.

$$[I]_d = \lambda\, [I]_w \qquad (\lambda = 115 \text{ at } 50°\text{C}) \tag{3}$$

The partition coefficient of AIBN between the polymer particle and water phases, λ', is usually different from λ. However, it is reasonable to consider in this case that $\lambda=\lambda'$, so that we regard, as a first approximation, also that $\lambda'=\lambda=115$ at 50°C.

Since the thermal decomposition rate constant for AIBN is comparatively small and also the volume change of the total organic phase due to polymerization is not so large in this system, we can safely regard that the concentration of AIBN in the monomer droplets is approximately equal to the initial value, $[I]_0$, which is given by dividing I_0, the initially charged weight of AIBN, by the total volume of the monomer and seed polymer particles initially charged.

$$[I]_d = [I]_p = [I]_0 \tag{4}$$

Considering the monomer partition in the case of emulsion copolymerization *(6)*, the following relationships appear to hold and hence, to be applicable in obtaining more precise values for $[I]_p$, $[I]_d$, and $[I]_w$.

$$\frac{[I]_d}{[M]_d} = \frac{[I]_p}{[M]_p}, \quad V_d[I]_d + V_p[I]_p + V_w[I]_w = I_o\exp(-k_d t) \tag{5}$$

where $[M]_d$ is the monomer concentration in the monomer droplets, V_d is the total volume of the monomer droplets contained per cm^3-water, k_d the thermal decomposition rate constant for AIBN, and t the reaction time. The suffixes, d, p, and w denote the properties associated with the monomer droplet, the polymer particle, and the water phase, respectively. Furthermore, the total volume of organic phase in the reactor, (V_d+V_p), is approximately constant and equal to the total volume of the monomer and seed polymer particles initially charged, because the densities of the monomer and the resulting polymer are not very different.

Thermal Decomposition Rate Constant for AIBN. AIBN initiator thermally decomposes to produce a pair of radicals in the water, monomer droplet and polymer particle phases, respectively. As mentioned above, however, only radical production in the water phase and, in some cases, also in the polymer particles is important and their rates are expressed, respectively, as

$$r_{iw} = 2k_{dw}f_w[I]_w \tag{6}$$

$$r_{ip} = 2k_{dp}f_p[I]_p v_p N_T \tag{7}$$

where r_i and f are the rate of radical production and the initiator efficiency in each phase and v_p is the average volume of a polymer particle.
The equality of these parameters among the phases is also a reasonable assumption.

$$k_{dw}=k_{dp}=k_d, \quad f_d=f_p=f \tag{8}$$

In order to determine the thermal decomposition rate constant for AIBN, bulk polymerizations of styrene were carried out at 50°C with varying initial concentrations of AIBN, $[I]_0$. The value of k_d was determined from the measured values of the rate of polymerization, R_p, calculated from the slope of the conversion versus time curves and the number average molecular weight of the resulting polymers, $\overline{P}_n$, measured by gel permeation chromatography (GPC) with the expression:

$$\frac{1}{\overline{P}_n} = k_d f \frac{[I]_0}{R_p} + \frac{k_{mf}}{k_p} \tag{9}$$

According to Equation 9, the reciprocal of the value of $\overline{P}_n$ observed experimentally was plotted against the corresponding value of $[I]_0/R_p$ in Figure 2. All the experimental points fall on a straight line whose slope is $k_d f$ and the intercept with the ordinate is k_{mf}/k_p, the chain transfer constant to monomer. Thus, the following values were determined:

$$k_d f = 8.4 \times 10^{-7} \text{ 1/sec}, \quad k_{mf}/k_p = 9.2 \times 10^{-4} \tag{10}$$

Experimental Results and Discussion

Polymerization Loci. Figure 3 shows the effect of the initial emulsifier concentration, S_0, on the progress of the unseeded emulsion polymerization of styrene carried out with the initial initiator and monomer concentrations fixed at I_0=36.1 g/dm^3-monomer and M_0=0.2 g/cm^3-water, respectively *(2)*. The number of polymer particles, N_T, reached a constant value very early and is plotted against S_0 in Figure 4, along with the rate of polymerization, R_p, calculated from the slope of the linear portion (30 to 50% X_M) of the monomer conversion versus time curves shown in Figure 3. The number of polymer particles formed increases abruptly in the vicinity of S_0=2.0 g/dm^3-water. This particle nucleation behavior is quite similar to that observed in the case of the emulsion polymerization of styrene initiated by the water-soluble initiator, $K_2S_2O_8$ with NaLS as emulsifier. These data are also shown in Figure 4. Considering that the CMC of NaLS in the absence of electrolyte is about 8×10^{-3}mole/dm^3-water (2.3 g/dm^3-water), these experimental results appear to imply that the polymer particles are generated from the emulsifier micelles independently of the kind of initiator used.

As an example, the polymerization was started under the reaction conditions shown in Figure 5. 15 min after the start of the polymerization, about 20 g sample of the reaction mixture was withdrawn from the sampling cock attached at the bottom of the reaction vessel and a portion of this sample was poured into excess methanol to precipitate the polymer. The precipitated polymer was separated by filtration with a glass crucible and dried in a vacuum oven. Then, the total monomer conversion was gravimetrically determined to be 4.3%. The solid line A shown in Figure 5 indicates the molecular weight distribution (MWD) determined by GPC of the total polymer thus collected. The rest of the sample was, on the other hand, subjected to separation of the mixture with a centrifuge into a monomer layer and a serum containing polymer particles. The polymers contained in each separated phase were precipitated and collected by the same procedure as mentioned above. The broken line B shows the GPC chromatogram of the polymer collected from the monomer layer, the dashed line C indicating that of the polymer collected from the serum. The MWD shown by line B agreed perfectly with that of the polymer obtained in the very beginning of the bulk polymerization of styrene carried out under the same temperature and the same concentration of AIBN initiator. Therefore, we believe that in this emulsion polymerization system, polymerization takes place simultaneously in both the monomer droplets and the polymer particles. Suspension polymerization which follows homogeneous bulk kinetics occurs in the monomer droplets. On the other hand, Figure 6 shows how the shape of the MWD changes with the progress of polymerization. The polymerization in the monomer droplets is important in the beginning, but the polymerization in the polymer particles becomes dominant with the progress of the polymerization.

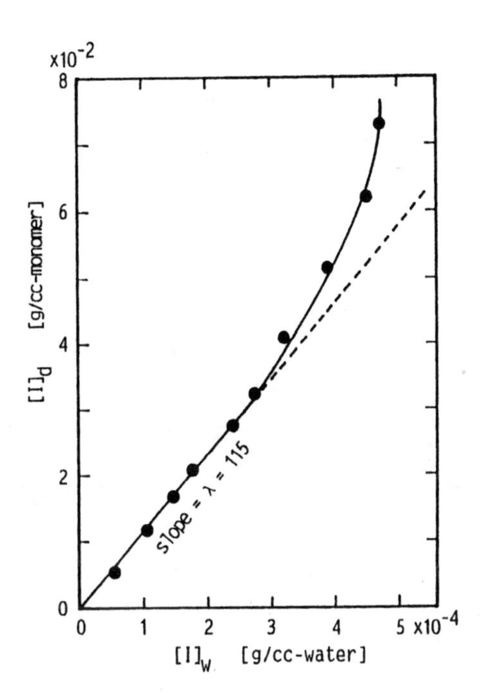

Figure 1. Determination of the partition coefficient of AIBN, λ, between the monomer droplets and the water phase.

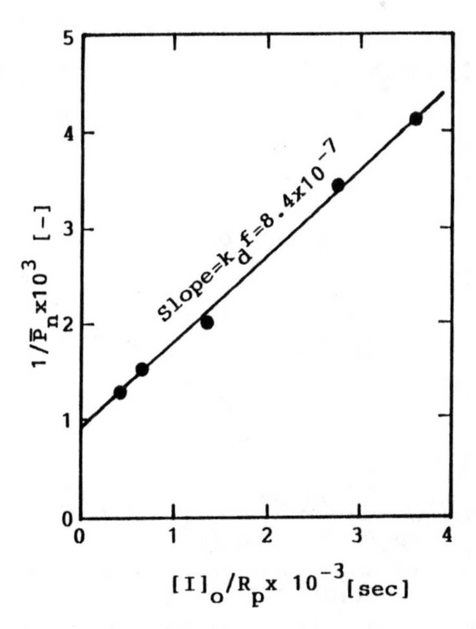

Figure 2. Determination of the thermal decomposition rate constant of AIBN.

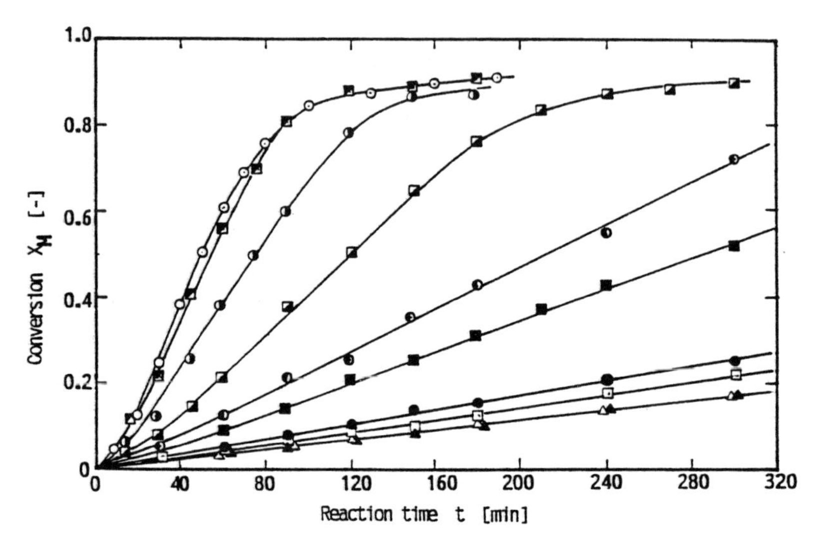

Figure 3. Effect of initial emulsifier concentration on the progress of polymerization (I_0=36.1 g/dm^3-water, M_0=0.2 g/cm^3-water, S_0= ○ 25.0, ◩ 20.0, ◑ 12.5, ◪ 6.25, ◐ 3.13, ■ 2.50, ● 2.0, □ 1.25, ▲ 0.8, ▲ 0.5, △ 0 g/dm^3-water), (Reproduced with permission from ref. 2. Copyright 1991 Wiley.)

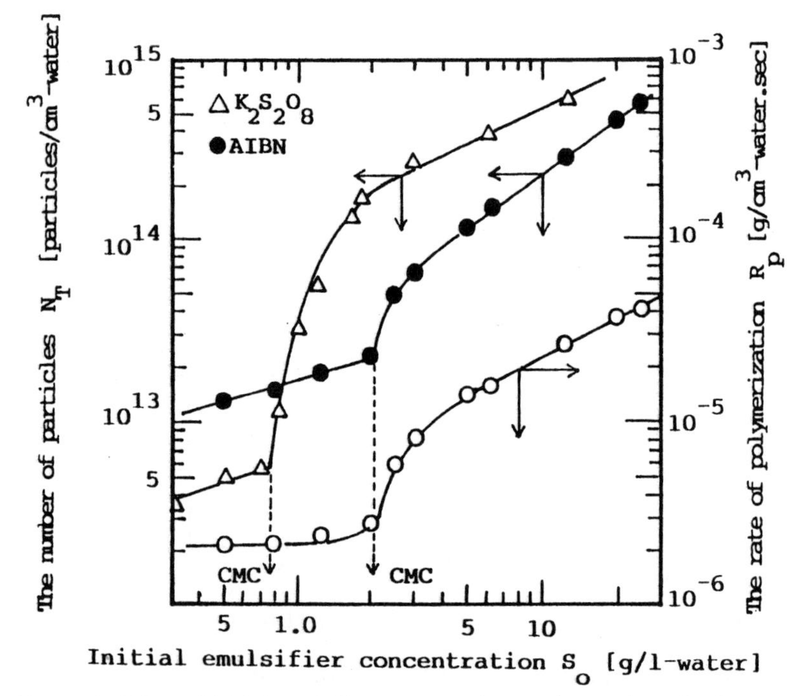

Figure 4. Effect of initial emulsifier concentration on the number of polymer particles produced, N_T, and the rate of polymerization, R_p.

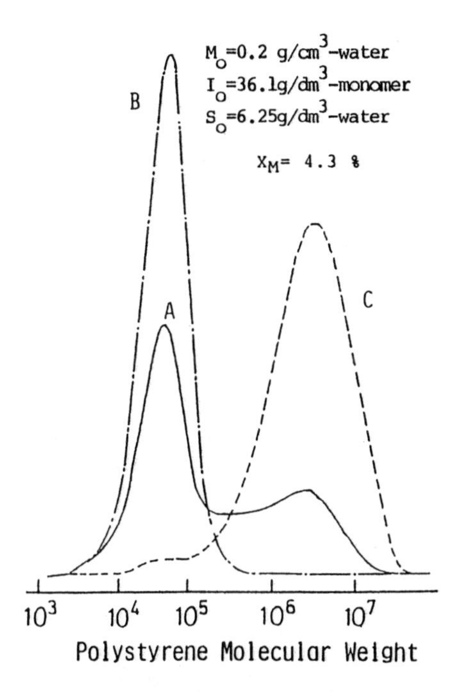

Figure 5. The molecular weight distributions (MWD's) of the polymer produced in the monomer droplets and in the polymer particles (line A: the MWD of the total polymer, line B: the MWD of the polymer produced in the monomer droplets, line C: the MWD of the polymer produced in the polymer particles).

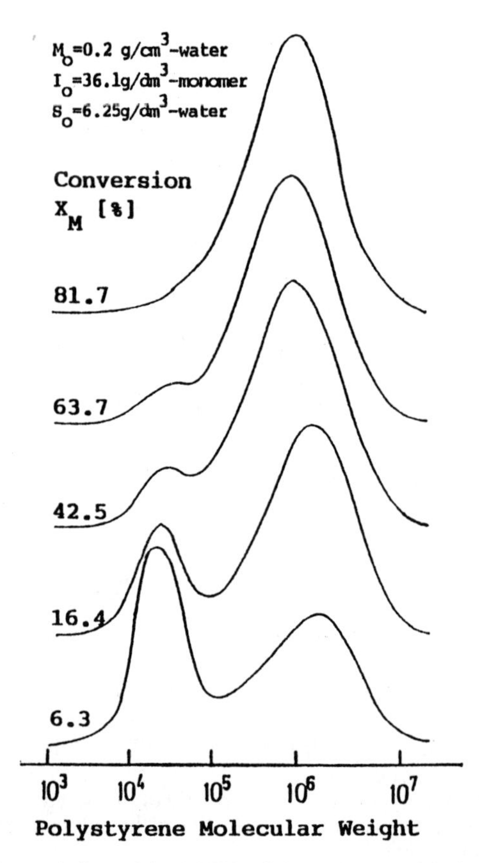

Figure 6. The variation of the MWD of the polymer with the monomer conversion.

The amount of polymer produced in each reaction locus can be determined first by separating, for example, the chromatogram A with two peaks shown in Figure 5 into two individual chromatograms and then, by calculating the area bounded by each chromatogram and the base line. Let P_p and P_d be the amounts of polymer produced in the polymer particles and in the monomer droplets, respectively. The experimental value of P_d thus determined was plotted against the monomer conversion, X_M, in Figure 7. The solid lines indicate the values calculated by the kinetic model for the seeded emulsion polymerization proposed later. The data in Figure 7 show that the polymerization in the monomer droplets is dominant only in the very beginning and becomes negligible above ca. 30% conversion. This is because most of the monomer in the monomer droplets, which also contains a small amount of polymer, has already been transferred into the polymer particles at ca. 30% conversion. Thus, the rate of polymerization calculated from the slope between 30 and 50% conversion of the conversion versus time curve, R_p, can be safely regarded to show the rate of polymerization in the polymer particles, r_p, which is expressed by Equation 11 given later. Hence, the average number of radicals per particle, $\bar{n}$, can be calculated by applying the observed values of r_p and $[M]_p$ to Equation 11 presented later.

The Average Number of Radicals per Particle. The average number of radicals per particle is one of the important parameters which determine the average volumetric growth rate per particle. The solid lines in Figure 8 show the theoretical values of $\bar{n}$ which were computed by applying the method previously proposed for the case of $K=\alpha_w/\alpha_p=r_{iw}/r_{ip}=0.05$ *(4)*. The experimental values of $\bar{n}$ (closed circles) obtained by applying the procedure mentioned above to the experimental data shown in Figure 3 are also plotted against $\alpha_w(=r_{iw}v_p/k_{tp}N_T)$ in Figure 8. The value of r_{iw} was calculated by Equation 6. For the concentration of AIBN in the monomer droplets, $[I]_d$, which is necessary for estimating the value of $[I]_w$ using the correlation shown in Figure 1, the initial value $[I]_0$, defined by Equation 4, was employed, because the change in the total volume of the organic phase in the reaction system would not be so large, as already presumed, even with the density change in the polymer particles and the monomer droplets occurring during the polymerization. The experimental data shown by open circles, on the other hand, indicate those found in the emulsion polymerization of styrene initiated by the water-soluble initiator, $K_2S_2O_8$ *(7)*. The dotted lines show the theoretical values of $\bar{n}$ corresponding to the case for water-soluble initiators($K=\infty$). From this figure, it is clear that the solid lines are perfectly superimposed upon the dotted lines in the range where the values of α_w and m are both less than about 10^{-2}. This means that in this range, any radical production inside the polymer particles does not contribute to an increase in the average number of radicals per particle. This can also be taken as a theoretical evidence that the polymerization in the polymer particles proceeds mainly by the radicals stemming from the initiator dissolved in the water phase, not by the radicals produced from the initiator distributed in the polymer particles as long as the rate of radical desorption from the polymer particles is not so high (that is, in the range where the values of m and α_w are both less than about 10^{-2} when $K=5\times10^{-2}$) *(4)*. The fact that the experimental values of $\bar{n}$ obtained with AIBN (closed circles) are in complete agreement with those obtained with $K_2S_2O_8$ (open circles) can be regarded as a decisive experimental evidence for supporting the validity of the conclusions mentioned above. Furthermore, this result also supports the validity of the assumption that the radicals produced in the monomer droplets do not affect the radical balance in the water phase because only a negligible amount of them can desorb into the water phase owing to a small surface area per unit volume of the monomer droplets, so that almost all of them are terminated inside the monomer

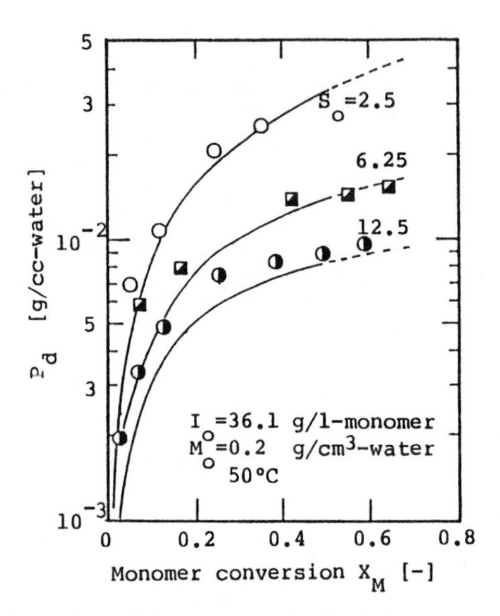

Figure 7. The effect of emulsifier concentration initially charged on the amount of polymer produced in the monomer droplets, P_d versus monomer conversions (solid line: model predictions by Equation 13).

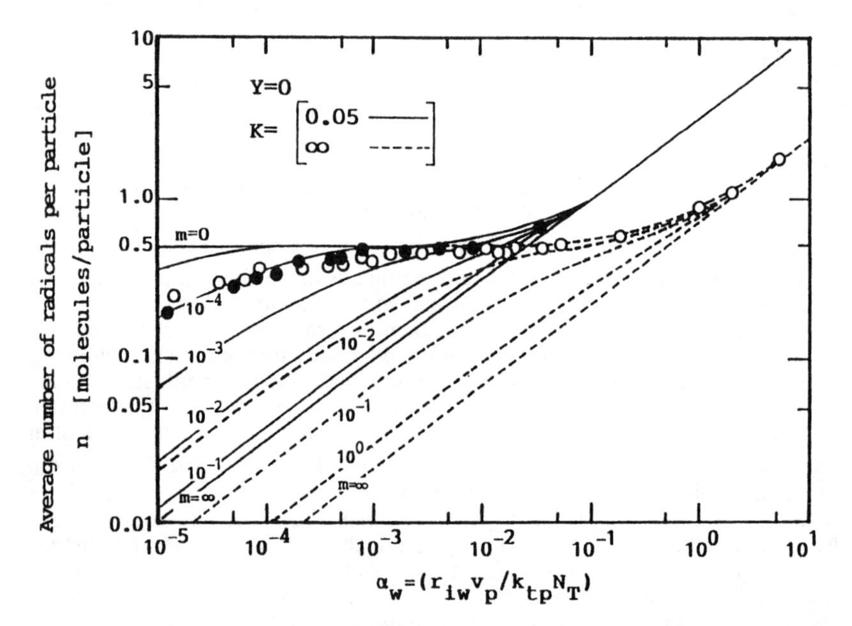

Figure 8. A plot of $\bar{n}$ versus α_w (Experimental:●=AIBN, ○=$K_2S_2O_8$; Theoretical value: dotted line=$\bar{n}$ for water-soluble initiators with K=∞; solid line =$\bar{n}$ for oil-soluble initiators with K=0.05).

droplets. The radicals produced inside the polymer particles will not contribute to the polymerization inside the polymer particles because the volume of a polymer particle is so small that instantaneous radical termination takes place as soon as a radical pair is born inside the polymer particles.

Based on the analysis given above, we propose a kinetic model for seeded emulsion polymerization which can predict the rates of polymerization in both the monomer droplets and the polymer particles.

A Seeded Emulsion Polymerization Model.
(1) The polymerization rate in the polymer particles is given by:

$$r_p = \frac{dP_p}{dt} = (k_p[M]_p\bar{n}N_T)\cdot\left(\frac{M_g}{N_A}\right) \tag{11}$$

where P_p is the weight of polymer produced inside the polymer particles per cm^3-water, k_p the propagation rate constant, M_g the molecular weight of styrene, N_A Avogadro's number.

The average number of radicals per particle, $\bar{n}$ can be calculated by the method previously proposed *(4)*, or by the following approximate expression *(6)*:

$$\bar{n} = \frac{1}{2}\left[\left\{\left(\alpha_w + \frac{\alpha_w}{m}\right)^2 + 2\left(\alpha_w + \frac{\alpha_w}{m}\right)\right\}^{1/2} - \left(\alpha_w + \frac{\alpha_w}{m}\right)\right] + \left(\frac{1}{4} + \frac{\alpha_w}{4}\right)^{1/2} - \frac{1}{2} \tag{12}$$

(2) The polymerization rate in the monomer droplets is:

$$r_d = \frac{dP_d}{dt} = \left(k_p[M]_d[R^*]_d v_d N_d\right)\cdot\left(\frac{M_g}{N_A}\right) \tag{13}$$

where P_d is the weight of polymer produced inside the monomer droplets per cm^3-water. v_d is the average volume of a monomer droplet and N_d the number of monomer droplets per cm^3-water. The product v_dN_d is expressed by Equation 19 shown later. $[R^*]_d$ is the concentration of radicals in the monomer droplets, which is determined exclusively by the events occurring inside the monomer droplets, that is, the production and mutual termination of radicals, if the exchange of radicals between the monomer droplets and the water phase can be neglected as mentioned above. This means that polymerization in the monomer droplets follows suspension polymerization kinetics. Therefore, $[R^*]_d$ is given by

$$[R^*]_d = (k_d f[I]_d/k_{td})^{1/2} \tag{14}$$

where k_{td} is the mutual termination rate constant for the radicals in the monomer droplets.

The total monomer conversion, X_M, is defined by

$$X_M = (P_p + P_d)/M_0 \tag{15}$$

(3) Monomer concentrations in the polymer particles and the monomer droplets are treated as follows. The chemical potential of monomer in a given phase i in the form of spheres with diameter d_i is well known to be expressed by

$$\frac{\Delta\mu_i}{RT} = \ln(1-\phi_i) + \phi_i + \chi\phi_i^2 + \frac{4\bar{v}_m\gamma}{d_iRT} \tag{16}$$

where ϕ_i is the volume fraction of polymer, χ the Flory-Huggins interaction parameter, $\bar{v}_m$ the partial molar volume of monomer, γ the interfacial tension, R the gas constant, and T the absolute temperature.

When the monomer in the monomer droplets and the polymer particles are in thermodynamical equilibrium, the chemical potential of the monomer in every phase is equal, so that the following equation holds:

$$\ln(1-\phi_p) + \phi_p + \chi\phi_p^2 + \frac{4\bar{v}_m\gamma}{d_pRT} = \ln(1-\phi_d) + \phi_d + \chi\phi_d^2 + \frac{4\bar{v}_m\gamma}{d_dRT} \tag{17}$$

If we assume the volume additivity of polymer and monomer, the average volumes of a polymer particle, v_p, and a monomer droplet, v_d, and their diameters can be expressed as follows:

$$v_p = \left(\frac{P_p}{\rho_p} + \frac{M_p}{\rho_m}\right)/N_T \tag{18}$$

$$d_p = \left(\frac{6v_p}{\pi}\right)^{1/3} \tag{18'}$$

$$v_d = \left(\frac{P_d}{\rho_p} + \frac{M_d}{\rho_m}\right)/N_d \tag{19}$$

$$d_d = \left(\frac{6v_d}{\pi}\right)^{1/3} \tag{19'}$$

where d_d is the average diameter of monomer droplets, M_p and M_d the weight of monomer contained in the polymer particles and in the monomer droplets per cm^3-water, respectively, and ρ_p and ρ_m the densities of polymer and monomer, respectively.

The expressions which define ϕ_p and ϕ_d are given by

$$\phi_p = \frac{P_p/\rho_p}{M_p/\rho_m + P_p/\rho_p} \tag{20}$$

$$\phi_d = \frac{P_d/\rho_p}{M_d/\rho_m + P_d/\rho_p} \tag{21}$$

Furthermore, the following mass balance equation must be satisfied:

$$(M_p + P_p) + (M_d + P_d) = M_0 + P_0 \tag{22}$$

where P_0 is the weight of the seed polymer particles initially charged per cm^3-water.

By solving the set of simultaneous equations from Equation 17 to Equation 22 for ϕ_p and ϕ_d and inserting them into Equations 23 and 24 given below, we can determine $[M]_p$ and $[M]_d$.

$$[M]_p = \frac{(1-\phi_p)}{\bar{v}_m} \quad (23)$$

$$[M]_d = \frac{(1-\phi_d)}{\bar{v}_m} \quad (24)$$

Figure 9 shows a typical comparison between the observed and predicted monomer concentrations. In this calculation, the interfacial energy term given by $4\bar{v}_m\gamma/d_dRT$ in the right-hand side of Equation 17 was neglected because the average diameter of the monomer droplets would be at least an order of magnitude larger than that of the polymer particles. Figure 9 demonstrates that the approach used in this study gives reasonable estimates for the monomer concentrations in both the monomer droplets and the polymer particles. The value of γ employed in this calculation was established so that the predicted monomer concentrations would agree with those observed. The value of γ thus determined is shown in Table I. The literature value of 0.48 was employed as the value of χ. All the numerical values used in this calculation are listed in Table I.

Comparison of Experimental Results with Model Predictions. The solid lines in Figure 7 show the predicted values of P_d obtained by numerically solving the set of differential equations mentioned above. In this calculation, seeded emulsion polymerization was assumed and the experimental N_T values shown in Figure 4 were employed as the number of seed polymer particles. The kinetic model proposed in this study for the seeded emulsion polymerization of styrene initiated by AIBN can predict very well the progress of polymerization in the monomer droplets.

Figure 10 shows a comparison between the predicted and observed total monomer conversion versus time histories when the number of seed polymer particles was widely varied. The model predictions given by the solid lines indicate a good agreement with the experimental conversion versus time data shown in Figure 3, where the corresponding experimental value of N_T shown in Figure 4 was applied to Equation 11 as the number of seed polymer particles.

Figure 11 shows, on the other hand, the effect of the initial initiator concentration on the conversion-time profiles, although the number of polymer particles changes. Fairly good agreement can also be seen between the predicted and observed results.

CONCLUSIONS

We can conclude from this study that in the emulsion polymerization of styrene initiated by the oil-soluble initiator AIBN:

(1) The polymerization takes place in both the monomer droplets and the polymer particles. However, the polymerization inside the monomer droplets is important only in the very beginning and the amount of polymer produced inside the monomer droplets is only several percent of that produced in the polymer particles;

(2) The radicals produced from the initiator dissolved in the water phase enter the polymer particles and participate in the polymerization there. The radicals produced

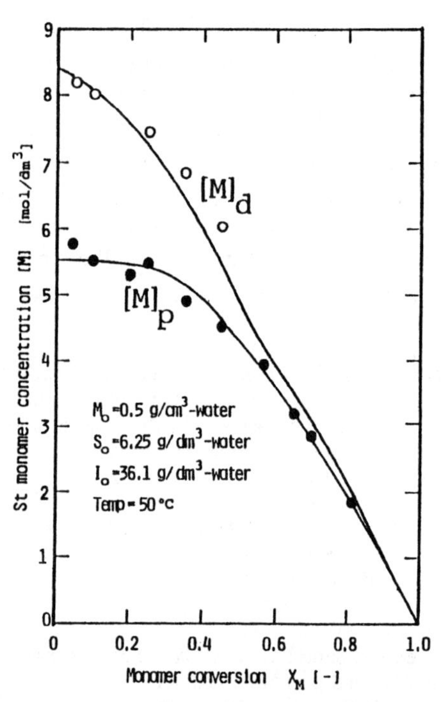

Figure 9. Comparison between the observed and predicted values of the monomer concentrations in the monomer droplets, $[M]_d$ and the polymer particles, $[M]_p$ (O =$[M]_d$, ● = $[M]_p$; solid lines=model predictions).

Table I. Numerical Constants Used (50° C)

Constant	Value	Units
k_p	212	$dm^3/mol\cdot sec$
k_d	1.4×10^{-6}	1/sec
f	0.6	--
D_w	1.2×10^{-5}	cm^2/sec
m_d	1300	--
k_{mf} / k_p	9.2×10^{-4}	--
δ	0.2	--
χ	0.48	--
$\bar{v}_m$	118.4	cm^3/mol
γ	11.3	dyne/cm
ρ_m	0.88	g/cm^3-monomer
ρ_p	1.05	g/cm^3-polymer
λ	115	--

$k_t = 2.0 \times 10^7 \{\exp(-0.94w - 3.88w^2 + 0.494w^3)\}^2$

w: weight fraction of polymer

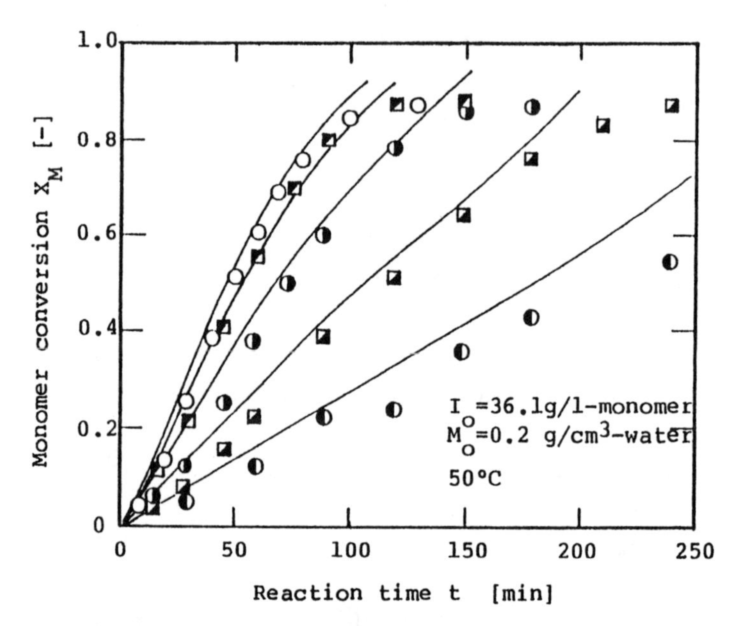

Figure 10. Comparison between the observed and predicted conversion versus time curves when the number of polymer particles, N_T, was varied (○ =5.8x10^{14}, ◪ =4.6x10^{14}, ◑ =2.5x10^{14}, ◩ =1.5x10^{14}, ◐ =6.5x10^{13} particles/cm^3-water).

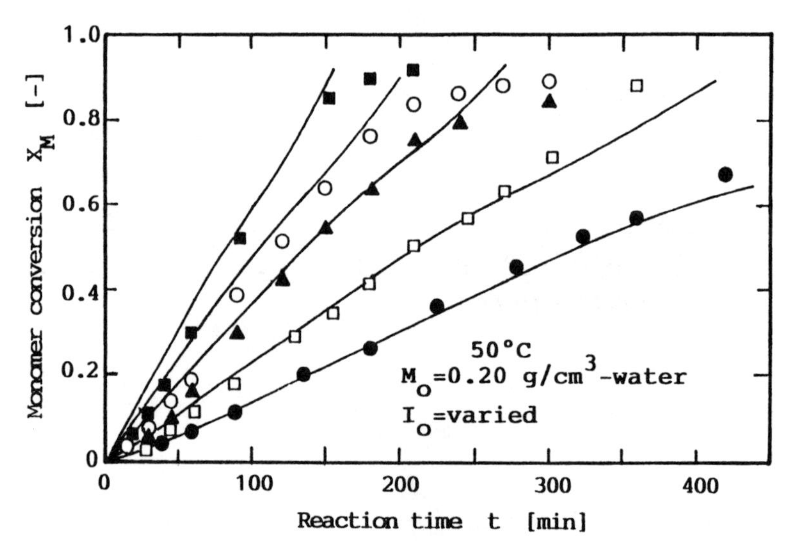

Figure 11. Comparison between the predicted and observed conversion versus time curves when the initial initiator concentration was varied (■= 72.2, ○=36.1, ▲=18.0, □=5.4, ●=1.8 g/dm^3-water).

inside the polymer particles, on the other hand, play almost no role in the polymerization in the polymer particles because the volume of a polymer particle is so small that they terminate as soon as a pair of radicals is produced by thermal decomposition of an initiator molecule;
(3) The radicals produced inside the monomer droplets make almost no contribution to a change in the water-phase radical concentration because the rate of radical exchange (by entry and exit) between the monomer droplets and the water phase is negligible due to the small surface area per volume of the monomer droplets, compared with both the rate of radical production in the water phase and the rate of radical exchange between the polymer particles and the water phase. Therefore, it can be safely assumed that inside the monomer droplets, the rates of radical production and termination are equal to each other and hence, the polymerization proceeds according to suspension polymerization kinetics, almost independently of the events occurring outside the monomer droplets;
(4) The kinetic model for seeded emulsion polymerization proposed in this study based on the conclusions given above shows good estimates for the rates of polymerization in both the monomer droplets and the polymer particles. This kinetic model can, therefore, be applied to any seeded emulsion polymerization system if the water-solubility and the thermal decomposition rate constant of the oil-soluble initiators used are known.

Furthermore, the proposed kinetic model and the knowledge obtained in this study will be of great use for clarifying the kinetics and mechanisms of miniemulsion, microemulsion, and dispersion polymerizations and other emulsion polymerization systems initiated by AIBN or different oil-soluble initiators.

Literature Cited

1) Capek, I.; Barton, J.; Karpatyova, A. *Makromol. Chem.* **1987**, *188*, p.703 .
2) Nomura, M.; Yamada, A.; Fujita, K.; Sugimoto, A.; Ikoma, J.; Fujita, K. *J. Polym. Sci.* **1991**, *29*, p.987.
3) Asua, J.M.; Rodriguez, V.S.; Sudol, E.D.; El-Aasser, M.S. *J. Polym. Sci., Part A: Polym. Chem.* **1989**, *27*, p.3569.
4) Nomura, M.; Fujita, K. *Makromol. Chem., Rapid Commun.*, **1989**, *10*, p.581.
5) Nomura, M.; Sasaki, S.; Harada, M.; Eguchi, W. *J. Appl. Polym. Sci.* **1978**, *22*, p.1043.
6) Nomura, M.; Fujita, K. *Makromol. Chem. Suppl.* **1985**, *10/11*, p.24.
7) Nomura, M. *Kobunshi* **1987**, *36*, p.680.

RECEIVED December 4, 1991

Chapter 6

Miniemulsion Polymerization

P. L. Tang[1,3], E. David Sudol[1], M. E. Adams[1,4], C. A. Silebi[1,2], and Mohamed S. El-Aasser[1,2]

[1]Emulsion Polymers Institute and [2]Department of Chemical Engineering, Lehigh University, Bethlehem, PA 18015

The creation of polystyrene latex particles through the nucleation of comparably sized styrene monomer droplets was investigated through experiment and mathematical modelling. Important preparative variables controlling the formation and stability of miniemulsion droplets were studied to reconcile disparate information gained in prior works. Conditions for the preparation of the smallest and most uniform droplets were optimized through inference from the subsequent polymerization kinetics and final particle size distributions. The effect of the presence of cetyl alcohol on the entry (and exit) of radicals was studied through seeded experiments in an attempt to explain the relatively slow nucleation observed in the polymerization of the systems prepared with this co-surfactant. A previously developed mathematical model was extended to include the thermodynamics describing the partition of components between the phases during the course of the polymerization.

Miniemulsions are relatively stable submicron (50 to 500 nm) dispersions of oil (such as monomer) in water prepared by shearing a system containing oil, water, surfactant, and additionally, a "co-surfactant". The principle behind the making of stable miniemulsions is the introduction of a low molecular weight and relatively water-insoluble compound (the "co-surfactant") inside the monomer droplets to substantially retard the diffusion of monomer out of the monomer droplets. For the cetyl alcohol system, the enhanced stability is also attributed to the formation of "intermolecular complexes" at the oil/water interface. These complexes would be liquid condensed and electrically charged creating a low interfacial tension and high resistance to droplet coalescence (*1 - 4*).

[3]Current address: Polymer Products Department, E. I. du Pont de Nemours and Company, 4200 Campground Road, Louisville, KY 40216

[4]Current address: General Electric Corporate Research and Development Center, Building K-1/CEB 126, P.O. Box 8, Schenectady, NY 12301

0097–6156/92/0492–0072$07.75/0

Higuchi and Misra (*5*) used diffusion theory to show that the diffusion of the more water-soluble compound in the droplets is governed by the diffusion of the water-insoluble one. Initially, when the monomer is broken up into droplets, some monomer would diffuse from the smaller droplets to the larger ones and thus the concentration of the water-insoluble compound would be increased in the smaller droplets. This would continue until the concentration of the water-insoluble compound is high enough to offset the chemical potential of the smaller droplets and significantly retard the diffusion process. Some miniemulsions can be prepared that are stable upon standing for months.

In conventional emulsion polymerization, the principal locus of particle nucleation is the aqueous phase or the monomer swollen micelles depending on the degree of water solubility of the monomers and the amount of surfactant used; lower water solubility monomer and higher amounts of surfactant would favor particle nucleation in monomer swollen micelles (*6*). Monomer droplets are only considered to act as monomer reservoirs supplying monomer to the growing polymer particles. However, in miniemulsion polymerization, the small size of the monomer droplets enables them to become the principal locus of particle nucleation. Thus, different conditions used in their preparation produce miniemulsions with different average droplet sizes and size distributions which in turn exhibit different polymerization kinetics and final particle size distributions.

A common feature of miniemulsion polymerizations, using cetyl alcohol as co-surfactant, is the relatively slow nucleation stage of the reaction as indicated by the long rise to a rate maximum in the experimental conversion-time data (*7 - 9*). This phenomenon has been attributed to a reduced radical absorption rate possibly caused by the interfacial layer formed by the mixed emulsifier at the droplet-water interface (*8*). It has been postulated that a reduced radical absorption rate into monomer droplets is due to the fact that the entering oligomer must first displace a surfactant molecule from the surface of the particle and the surfactant/monomer-rich surfaces have a lower free energy than surfactant/polymer-rich ones, thus the entry into the former should be lower (*7, 10*). Another possible cause for a reduced radical absorption rate is the lowering of the monomer concentration in the aqueous phase due to the presence of cetyl alcohol (by acting as a swelling promoter for the monomer droplets) thus slowing the polymer radical growth in the aqueous phase, thereby affecting the relative rates of aqueous phase radical termination and absorption into the droplets. This has not been established conclusively, however, and more direct evidence is needed. Thus, the effect of cetyl alcohol on the entry and exit of radicals in emulsion polymerization was studied (*11*).

The effect of co-surfactant on the entry of radicals into the monomer droplets can be examined indirectly by the fitting of mathematical models of miniemulsion polymerization to experimental data. The first mathematical model for the miniemulsion polymerization of styrene was developed by Chamberlain et al. (*7*) This model was based on the mathematical treatment by Hawkett et al. (*12*) seeded emulsion polymerization. It was assumed that polymer particles were formed only upon the entry of free radicals into the monomer droplets and each entry event resulted in a polymer particle. Chamberlain et al.'s work showed that the rate coefficient for radical entry (ρ) into the monomer droplets has to be an order of magnitude lower than the value determined for the seeded emulsion polymerization of styrene in order to fit the experimental data obtained for miniemulsion polymerizations using dodecanol as co-surfactant. An extension of this work was carried out by considering the possibility of desorption of a radical once it enters a droplet, to compare the relative effects of cetyl alcohol and hexadecane on the entry of radicals into the monomer droplets (*11*).

Effect of Preparative Variables on the Kinetics and Particle Size Distributions in Miniemulsion Polymerization

As reported in the literature, miniemulsions for subsequent polymerization have been prepared with a number of coincident variations which make conclusive comparisons concerning systems and methods impossible. For example, Choi (*8*) studied the polymerization kinetics of styrene miniemulsions prepared with cetyl alcohol (CA) and sodium lauryl sulfate (SLS) using a Microfluidizer M-110 (Microfluidics Corp., Newton, MA) shear device. Delgado (*13*) copolymerized vinyl acetate and butyl acrylate in miniemulsions prepared with hexadecane (HD) and sodium hexadecyl sulfate (SHS) using an Omni mixer (Ivan Sorvall, Inc.) as the dispersing device, and Rodriguez (*14*) copolymerized methyl methacrylate and styrene from miniemulsions prepared with HD/SLS and CA/SLS, respectively, using a sonifier (Branson Sonic Power Co.) to provide the necessary shear to create the fine droplets. Subsequently, because of these differences, a systematic study was initiated to tie together the disparate information gained in these studies.

Effect of Homogenization. Three shear devices were applied under typical operating conditions to the preparation of styrene miniemulsions using a single recipe. These were: 1) the Microfluidizer, operated at 7000 psig with the A10 orifice and ten passes of the fluid; 2) the sonifier with a 1/2" horn, operated for 60 seconds at 50% duty and power level 7 on a fluid volume of approximately 50 cc; and 3) the Omni mixer, operated for 20 minutes at 16,000 rpm. The HD/SLS (20 mM/5 mM) mixed emulsifier system was chosen for the preparation of 20 wt% styrene miniemulsions. Polymerizations were conducted at 70°C with the kinetics monitored by dilatometry. The results are shown in Figure 1. The polymerization in run M12 (Microfluidizer) was much faster than that of run M13 (sonifier) which in turn was faster than that in run M21 (Omni mixer). These results agree with the final particle sizes as measured by transmission electron microscopy (TEM) (Figure 2); the faster polymerization rate produced the smaller particle size (i.e., larger number of particles). Figure 2 shows that the PSD of the latex produced in run M12 was narrower than those of runs M13 and M21 (which each contained considerable amounts of larger particles) which is consistent with the claim that the Microfluidizer produces a finer and more uniform distribution than can be achieved by other means. The Microfluidizer was therefore chosen as the device for preparing all subsequent miniemulsions.

Effect of the Degree of Aging of the Emulsion. When a co-surfactant is not used in the preparation of miniemulsions, the emulsion degrades faster than when a co-surfactant is employed. The results in Figure 3 were obtained by polymerization of miniemulsions prepared using the Microfluidizer but without co-surfactant (referred to below as "conventional") and after aging under varying conditions. These emulsions were relatively unstable without the co-surfactant, the droplets becoming larger with time. This is likely to be due to their instability against diffusive degradation (also referred to as Ostwald ripening) as well as coagulation when the size differences between droplets are large. The emulsion in run C8 was initiated just after degassing, whereas C7 was stirred at 70°C for one hour, and C6 was left unstirred for two hours at room temperature before initiation. The results in Figure 3 show that the polymerization rate decreased with increased aging of the emulsion resulting in fewer and larger droplets.

A similar study was also carried out with miniemulsions prepared with cetyl alcohol (15 mM) and SLS (5 mM). The results are shown in Figure 4. The miniemulsion in run M16 was initiated after one hour (unstirred) at room temperature and the miniemulsion in run M17 was initiated after four hours (unstirred) at room temperature. The results show that the conversion-time behavior is the same for one

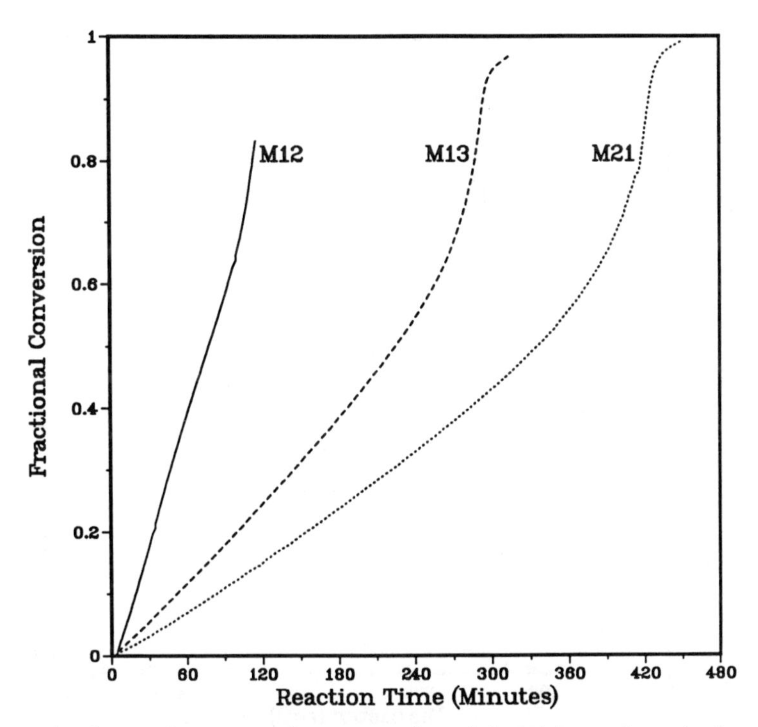

Figure 1: Conversion versus time curves for miniemulsion polymerizations of styrene: M12 (prepared with the Microfluidizer); M13 (sonifier); and M21 (Omni mixer); all contained 5 mM SLS/20 mM HD and 20 wt% styrene; 70°C.

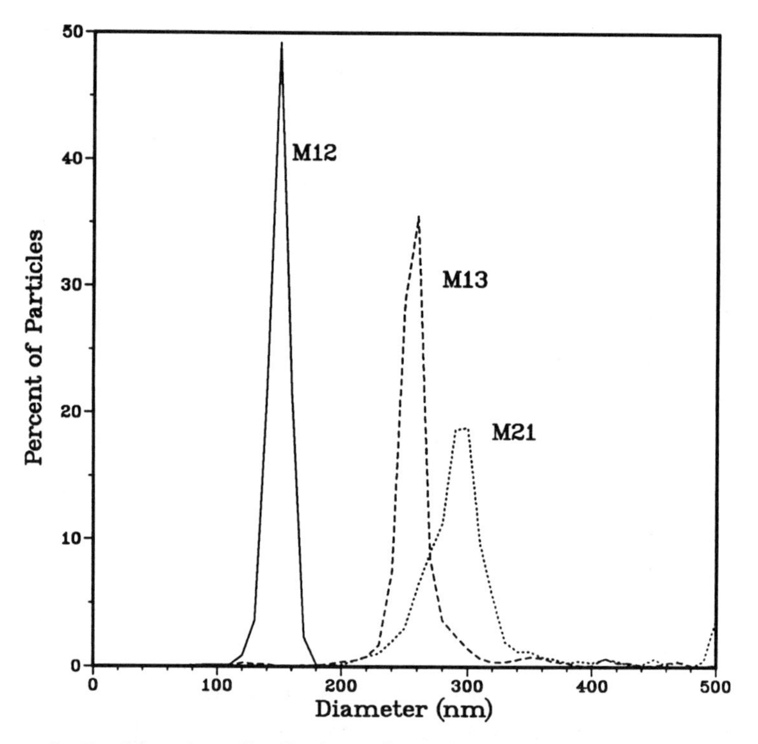

Figure 2: Particle size distributions for miniemulsion polymerizations of styrene: M12 (Microfluidizer); M13 (sonifier); and M21 (Omni mixer); all contained 5 mM SLS/20 mM HD and 20 wt% styrene.

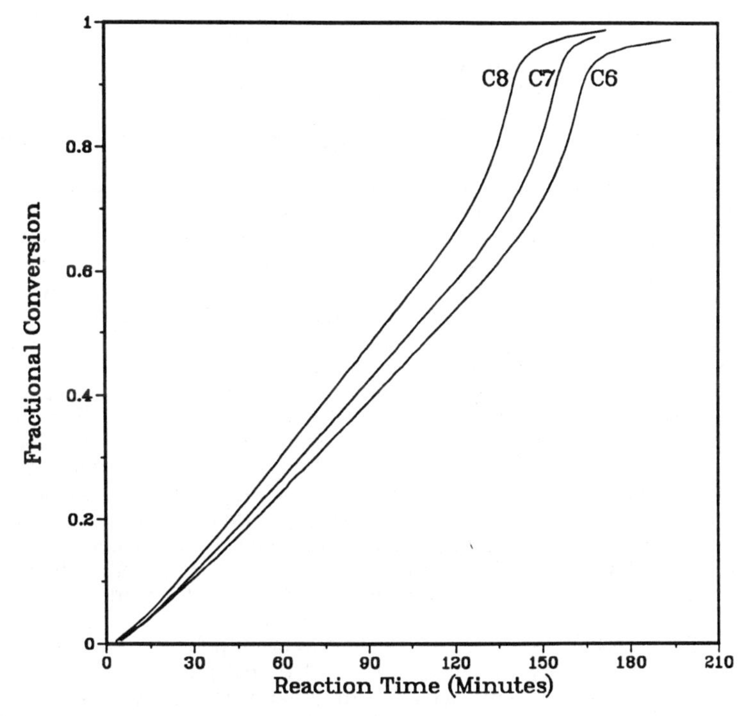

Figure 3: Conversion versus time curves for miniemulsion polymerizations of styrene prepared without co-surfactant: C8 (Microfluidizer, 5 mM SLS and no aging time); C7 (Microfluidizer, 5 mM SLS and one hour at 70°C); and C6 (Microfluidizer, 5 mM SLS and two hours at 25°C); 70°C. (Reproduced with permission from reference 9. Copyright 1991 Wiley.)

or four hours of aging. This indicates the monomer droplets are not changing in size between these aging times and that cetyl alcohol is responsible for the increased stability of the monomer droplets.

Effect of Cetyl Alcohol and Sodium Lauryl Sulfate. In order to carry out the "conventional" emulsion polymerizations (i.e., without a co-surfactant) in the dilatometer, the emulsions were first homogenized with the Microfluidizer. True conventional emulsion polymerizations could not be carried out in the dilatometer because of monomer separation in the capillary during the polymerization. To obtain these kinetics, the polymerizations were carried out in a reaction calorimeter (RC1, Mettler Instrument Corp.) for comparison and the results are shown in Figure 5. The dotted curves are the results obtained without homogenization whereas the solid curves are those resulting from homogenization of the emulsion (dilatometer). The other variable is the concentration of sodium lauryl sulfate. The results show that above the CMC (7.4 mM, as determined by measuring the conductivity of SLS solutions of varying concentration at room temperature) the run with no homogenization reacted slightly faster than the one with homogenization; but below the CMC, the run with no homogenization was much slower. Above the CMC, the run with no homogenization would be expected to have more nucleation sites (presumably micelles) as compared to the one with homogenization. Based on the final particle size distribution of the latter (C2 in Figure 7), it seems plausible that two competing mechanisms of nucleation were present in this system, namely, micellar and droplet. However, below the CMC nucleation was reduced in the non-homogenized system (homogeneous nucleation) compared to that in which fine droplets were present as produced by the Microfluidizer.

As seen in the preceeding results, it is possible to create and polymerize miniemulsions without the addition of a co-surfactant. However, these emulsions are not considered to have long term stability (as addressed later). The addition of the co-surfactant, cetyl alcohol, brings increased stability and smaller droplet sizes. The polymerization kinetics and resulting particle size distributions of miniemulsions prepared with SLS/CA (1/3) at levels of 10 mM and 5 mM SLS, are given in Figures 6 and 7. For comparison, the preceeding results obtained by homogenizing without CA are included. Below the CMC, smaller particles and a faster overall polymerization were obtained with the miniemulsion prepared with CA (M10) compared to the one without CA (C8). The opposite was seen when the added SLS was above its CMC (M6 and C2). As mentioned above, the high rate of polymerization and smaller particle size were primarily attributed to a micellar mechanism of nucleation in this system with some particles being nucleated from monomer droplets. When CA was used as cosurfactant, finer and stabler droplets were produced thereby reducing significantly the presence of micelles and their competition as nucleation sites.

Effect of Co-surfactant on the Polymerization Kinetics. Polymerization kinetics for various emulsions prepared with 5 mM SLS are compared in Figure 8. This includes results obtained with hexadecane as co-surfactant. The use of either co-surfactant produced faster rates than without, however there are some subtle differences. In particular, in the early stages of the polymerizations (as magnified in Figure 9), the initial polymerization rate of the miniemulsion prepared with CA (M10) was lower than the comparable case with HD (M12) (the final average particle sizes of these latexes were 147 nm and 145 nm, respectively). However, this rate increased to nearly the same as that in the HD system by 5% conversion. This phenomenon has been previously attributed to slow nucleation of particles in these systems resulting from a reduced entry of radicals into monomer droplets stabilized by the SLS/CA mixed-emulsifiers. No direct evidence has been obtained to support this supposition, however.

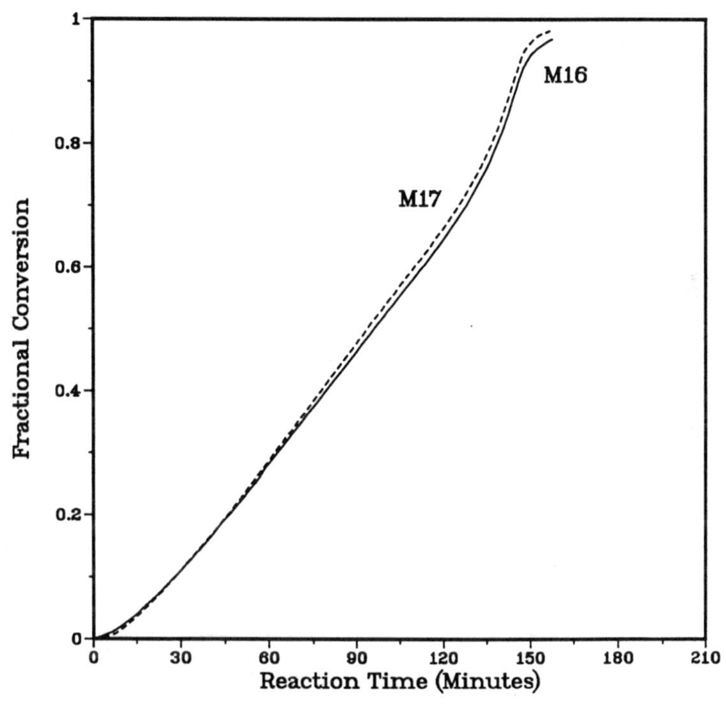

Figure 4: Conversion versus time curves for miniemulsion polymerizations of styrene: M16 (Microfluidizer, 5 mM SLS, 15 mM CA, and one hour of aging); and M17 (Microfluidizer, 5 mM SLS, 15 mM CA, and four hours of aging); 70°C.

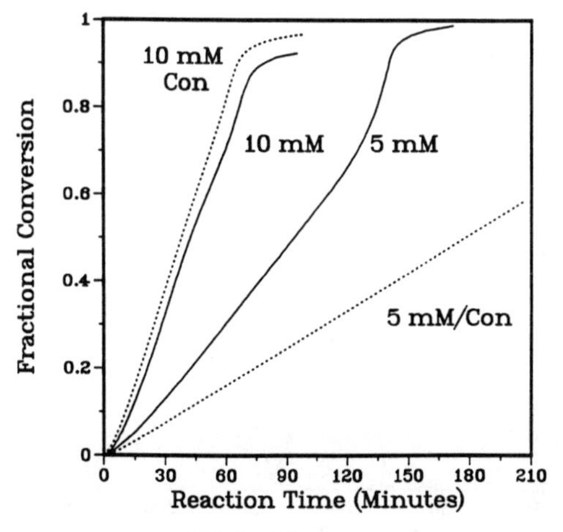

Figure 5: Conversion versus time curves for conventional polymerizations of styrene with 5 mM and 10 mM SLS; dotted lines: no homogenization of the emulsion before polymerization in the RC1 reactor, and solid lines: homogenization of the emulsion with the Microfluidizer before polymerization in the dilatometer; 70°C.

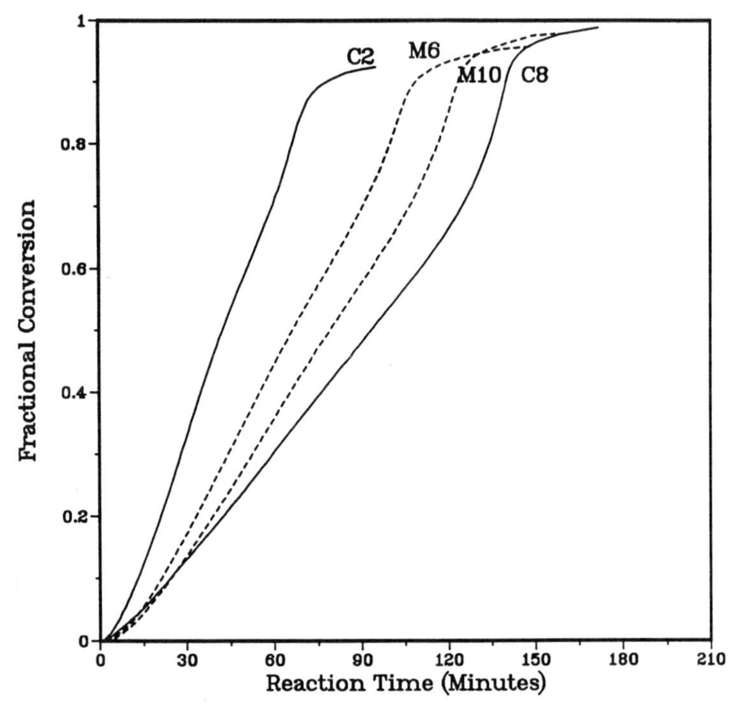

Figure 6: Conversion versus time curves for polymerizations of styrene: C2 (Microfluidizer, 10 mM SLS) and C8 (Microfluidizer, 5 mM SLS); and miniemulsion polymerizations of styrene: M6 (Microfluidizer, 10 mM SLS/30 mM CA) and M10 (Microfluidizer, 5 mM SLS/15 mM CA); 70°C. (Reproduced with permission from reference 9. Copyright 1991 Wiley.)

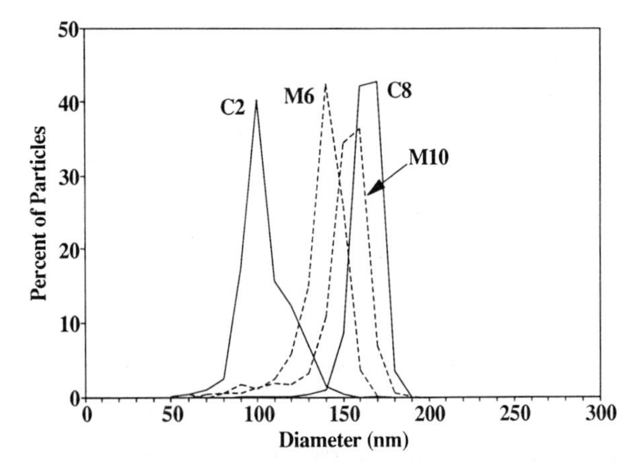

Figure 7: Particle size distributions for latexes prepared by emulsion polymerizations of styrene: C2 (Microfluidizer, 10 mM SLS); C8 (Microfluidizer, 5 mM SLS); M6 (Microfluidizer, 10 mM SLS/30 mM CA); and M10 (Microfluidizer, 5 mM SLS/15 mM CA).

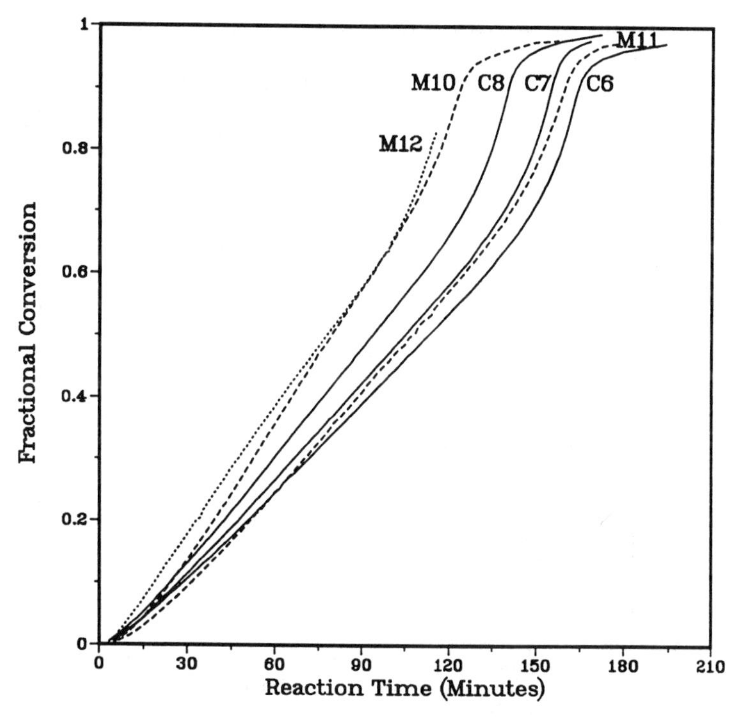

Figure 8: Conversion versus time curves for emulsion polymerizations of styrene: C6, C7, and C8, "conventional" cases; M10 and M11, miniemulsions with cetyl alcohol; and M12, miniemulsion with hexadecane; 70°C. (Reproduced with permission from reference 9. Copyright 1991 Wiley.)

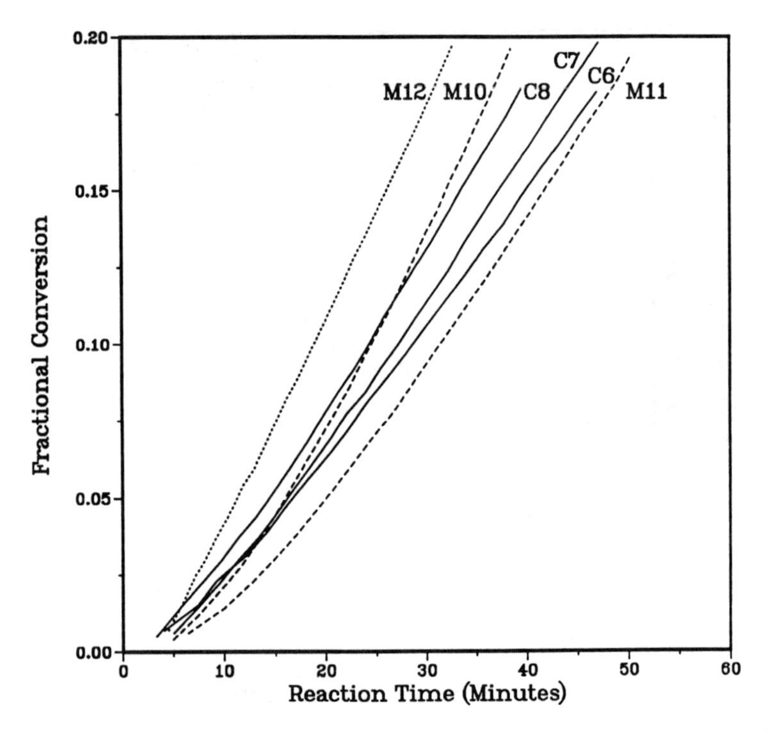

Figure 9: Magnified conversion versus time curves for emulsion polymerizations of styrene: C6, C7 and C8, "conventional" cases; M10 and M11, miniemulsions with cetyl alcohol; and M12, miniemulsion with hexadecane. (Reproduced with permission from reference 9. Copyright 1991 Wiley.)

Effect of Cetyl Alcohol on the Entry and Exit of Radicals in Emulsion Polymerization

Seeded emulsion polymerizations of styrene in polystyrene were carried out to study the effect of cetyl alcohol (CA) on the entry and exit of radicals in miniemulsion polymerization (*11*). A monodisperse seed (92 nm diameter, LS 1039E, Dow Chemical Co.) was used to ensure a constant particle concentration throughout the reactions, as required for employing parameter estimation methods to determine the entry and exit rate coefficients. Cetyl alcohol and sodium lauryl sulfate were added (as a gel phase) to the system to mimic the oil/water interphase of the miniemulsion droplets. The absence of the CA in the aqueous phase was checked by gas chromatography. The amount of cetyl alcohol, the initial concentration, the seed size, and the monomer/polymer swelling ratio were varied.

A reproducibility study was carried out and the results are shown in Figure 10. These indicate first, that the reproducibility is good and second, that the polymerization rate in a seeded system increases when cetyl alcohol is present. These results appear to contradict the idea that cetyl alcohol hinders the entry of radicals into the monomer-swollen particles by forming a complex at the oil-water interface, thereby reducing the polymerization rate. An increased polymerization rate with the presence of cetyl alcohol is also seen at all other levels of initiator in the concentration range of 0.013 mM to 1.33 mM (based on the aqueous phase), as shown by the results presented in Figure 11.

The data in Figure 11 for the three highest initiator concentrations were used to estimate the second order entry rate coefficient for radicals into the particles (k_a^p) and the exit rate coefficient of radicals from the particles (k) by the method developed by Asua et al. (*15*). The data for each curve used in this analysis included all points from zero conversion to 30 - 40% conversion. In this range, $\bar{n}$ was always less than 0.5. The results obtained for the set of experiments containing cetyl alcohol were $k_a^p = 1.4 \times 10^6 dm^3 mol^{-1} s^{-1}$ and $k = 6.7 \times 10^{-3}\ s^{-1}$, while those obtained for the set of experiments without cetyl alcohol were $k_a^p = 6.2 \times 10^6 dm^3 mol^{-1} s^{-1}$ and $k = 17.0 \times 10^{-3}\ s^{-1}$. These results suggest that both the entry and exit of radicals are slowed by the presence of cetyl alcohol with exit being the more dominant effect. It should be noted, however, that only three conversion-time results were used for each analysis which may introduce some uncertainty into the results.

Further experiments were sought to corroborate these findings. In the preceeding experiments, radical desorption was expected to be important because of the small particle size of the seed (≈ 100 nm). Experiments employing a larger particle size seed can be utilized to minimize the kinetic importance of radical desorption and provide more definitive information concerning the effect of cetyl alcohol on the entry and exit of radicals in seeded systems. Comparisons of the kinetics of experiments with cetyl alcohol and those without were also carried out using a 357 nm diameter seed to study the effect of cetyl alcohol on radical absorption. Final particle size distributions were also determined and the results showed that no particles were formed or lost during the polymerizations. The conversion-time results shown in Figure 12 indicate that the polymerization rate was the same (at a given initiator concentration) regardless of whether cetyl alcohol was present in the system or not. The average number of radicals per particle versus time for the set of experiments performed at an initiator concentration of 1.33 mM is shown in Figure 13 (the propagation rate constant, $k_p = 258\ dm^3 mol^{-1} s^{-1}$ at 50°C). The effect of exit on the kinetics is considered to be negligible with $\bar{n}$ much greater than 0.5.

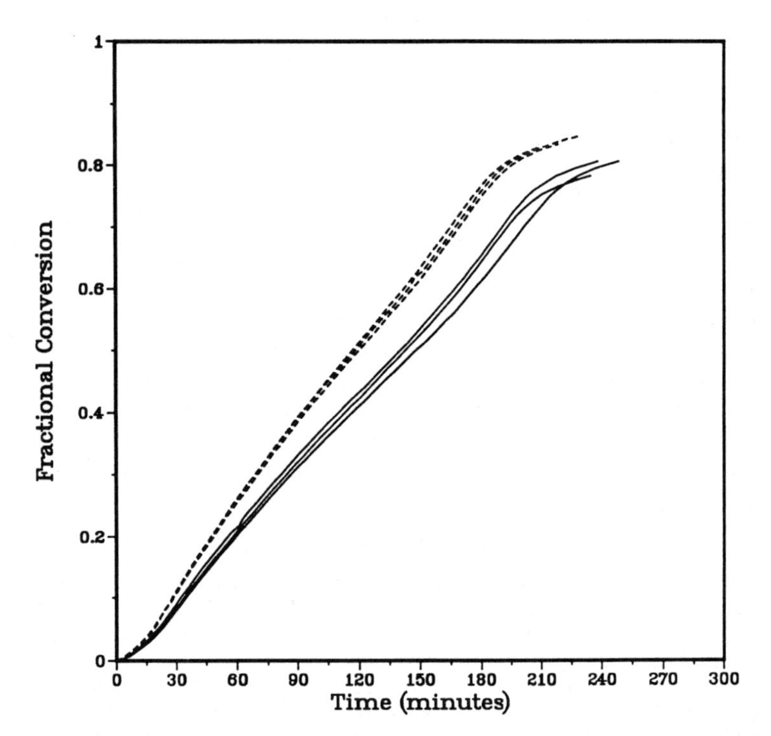

Figure 10: Repeatability of conversion versus time curves for seeded emulsion polymerizations of polystyrene/styrene; initiator concentration = 0.66 mM, SLS concentration = 10 mM, and seed diameter = 92 nm; solid curves obtained with no CA, dashed curves with 30 mM CA; 50°C.

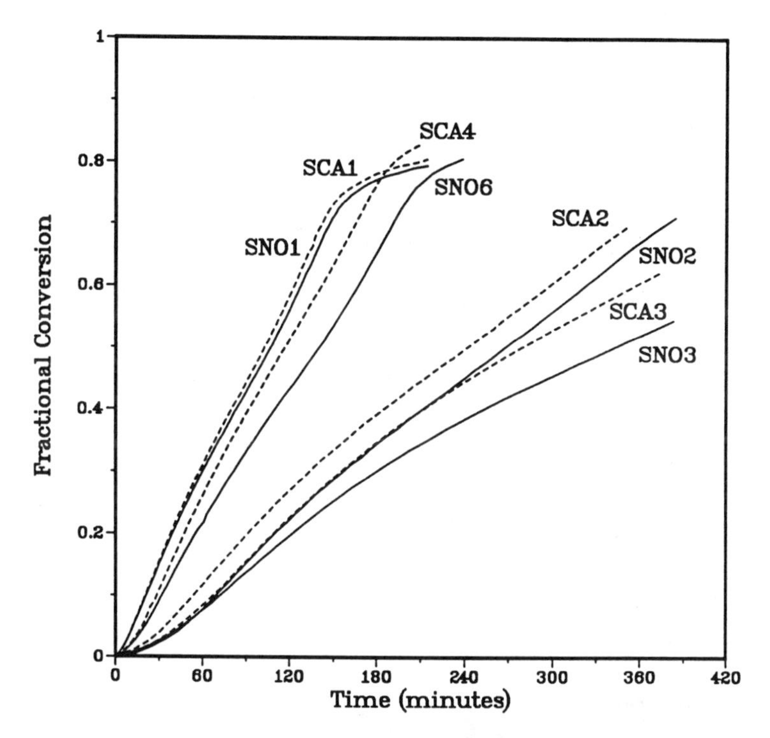

Figure 11: Conversion versus time curves for seeded emulsion polymerizations of polystyrene/styrene: SCA1 (1.33 mM $K_2S_2O_8$ and 10 mM CA); SNO1 (1.33 mM $K_2S_2O_8$ and no CA); SCA4 (0.66 mM $K_2S_2O_8$ and 30 mM CA); SNO6 (0.66 mM $K_2S_2O_8$ and no CA); SCA2 (0.13 mM $K_2S_2O_8$ and 10 mM CA); SNO2(0.13 mM $K_2S_2O_8$ and no CA); SCA3 (0.013 mM $K_2S_2O_8$ and 30 mM CA); and SNO3 (0.013 mM $K_2S_2O_8$ and no CA); SLS concentration = 10 mM, seed diameter = 92 nm; dashed lines obtained with CA and solid lines without CA; 50°C.

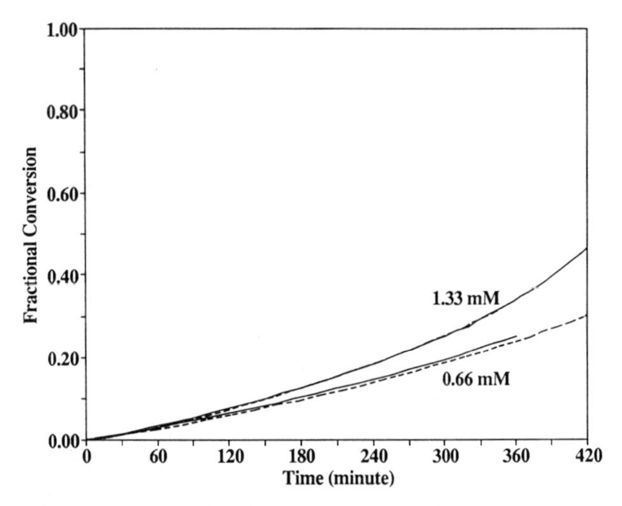

Figure 12: Conversion versus time curves for seeded emulsion polymerizations of polystyrene/styrene obtained using a 357 nm diameter seed; 10 mM SLS; 1.33 mM and 0.66 mM $K_2S_2O_8$; dashed lines obtained with CA, solid lines without CA; 50°C.

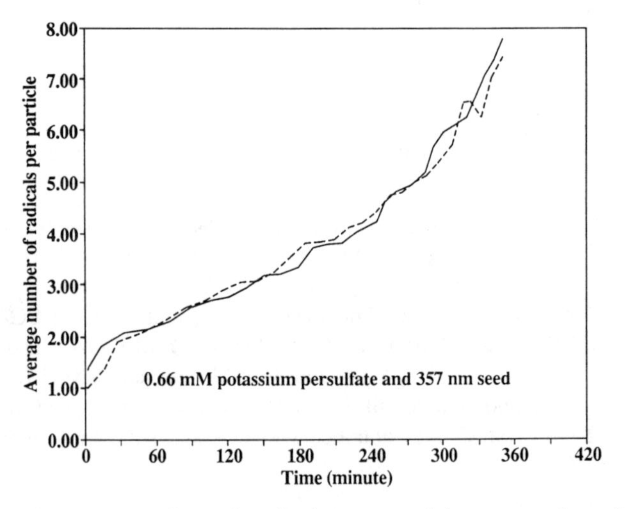

Figure 13: Average number of radicals per particle versus time for seeded emulsion polymerizations of polystyrene/styrene obtained using a 357 nm diameter seed, 10 mM SLS, and 1.33 mM $K_2S_2O_8$; dashed line obtained with CA, solid line without CA.

These results would indicate that cetyl alcohol does not effect the entry of radicals and therefore, may only effect the exit of radicals from the particles for the small size seed. These results only partially support the preceeding findings (i.e., exit is reduced with the presence of cetyl alcohol). Exit could be affected if cetyl alcohol acts as a chain transfer agent and in its radical form it is reactive and less likely to desorb than a single unit monomer radical (radical exit in emulsion polymerization generally refers to a single monomeric unit (*16 - 18*).

Experiments were carried out to determine the chain transfer constant for cetyl alcohol by carrying out bulk polymerizations of styrene at 50°C with different cetyl alcohol concentrations (*19*). AIBN was used as initiator and the conversion was held to less than five percent. The number average molecular weight was determined by gel permeation chromatography from which the number-average degree of polymerization, X_n, was obtained. The slope of $1/X_n$ versus the ratio of cetyl alcohol concentration to styrene concentration represents the value of the chain transfer constant for cetyl alcohol. The results of these experiments are shown in Figure 14 from which the dimensionless chain transfer constant for cetyl alcohol, C_s, was determined to be 11×10^{-4}. This is more than one order of magnitude larger than the chain transfer constant to styrene monomer at 50°C ($0.35 - 0.78 \times 10^{-4}$) (*20*). From this information, a rough estimate of the relative probabilities of chain transfer to CA versus chain transfer to styrene can be made from the ratio $k_{tr,CA}[CA]/k_{tr,S}[S]$. By assuming that all of the CA is present inside the polymer particles (upper limit) this ratio is approximately 4 (monomer/polymer = 1.35, [CA] = 30 mM on the aqueous phase or 1.17 M in the particle phase, and C_s(styrene) $\approx 0.6 \times 10^{-4}$). Thus, chain transfer to CA is more likely and would increase with decreasing monomer concentration (increasing conversion).

The results from the above experiments suggest that when a co-surfactant (such as cetyl alcohol) is used in miniemulsion polymerization, it may affect the exit of radicals from the swollen polymer particles if it has a higher chain transfer constant than that of the monomer and a lower water solubility. These act to reduce the radical exit rate and increase the rate of polymerization.

More importantly, the apparent contradiction between the seeded and unseeded (miniemulsion) polymerizations seems to suggest that miniemulsion droplets differ significantly from the polymer particles that they become. This difference might be attributed to differences in their respective entry rate coefficients, this being due to the differences in their physico-chemical nature. However, the entry rate coefficient for monomer droplets has not been determined directly due to the inherent difficulties of controlling the process, i.e., fixing the number of droplets entered during the reaction.

An attempt was made to further investigate the entry rate of radicals into the monomer droplets by carrying out seeded polymerizations with various monomer/polymer swelling ratios. The concentration of particles (based on the aqueous phase) was the same for all experiments. The variables were the amount of styrene (or swelling ratio) and the initiator concentration. Monomer/polymer swelling ratios (by weight) of up to 50 to 1 were achieved. The final particle size for run SCA11 (monomer/polymer swelling ratio of 50 to 1) was determined by TEM to be 296 nm diameter with a coefficient of variation of 8% (9% for the 92 nm diameter seed). No significant amount of new particles were formed during the polymerization. The conversion-time data are shown in Figure 15. Conversion is plotted in terms of the grams of polymer formed per liter of aqueous phase to aid in the comparison of the results. Runs SCA9, SCA8, SCA7, SCA11, and SCA13 have monomer/polymer swelling ratios (by weight) of 1.36, 10, 15, 50, and 50 to 1, respectively. All runs used an initial initiator concentration of 1.33 mM (potassium persulfate) except for run SCA11 which used twice this amount (2.66 mM). For the experiments at the same intiator concentration, the rate of polymerization (indicated by the slope of the conversion-time curve) increased with increasing swelling ratio up to a ratio of 15:1.

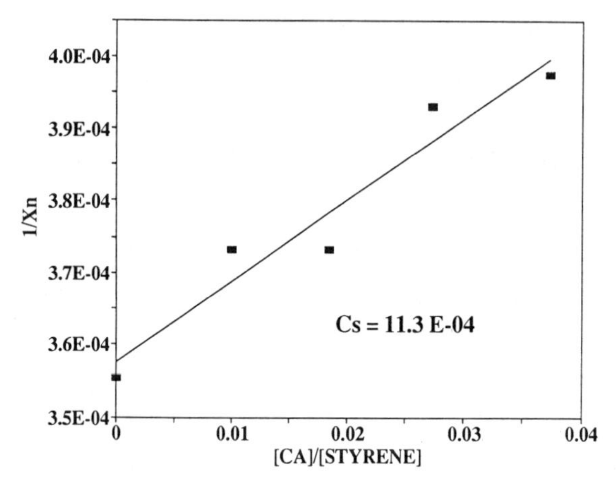

Figure 14: Ratio of cetyl alcohol concentration to styrene concentration versus one over the number-average degree of polymerization using AIBN at 50°C.

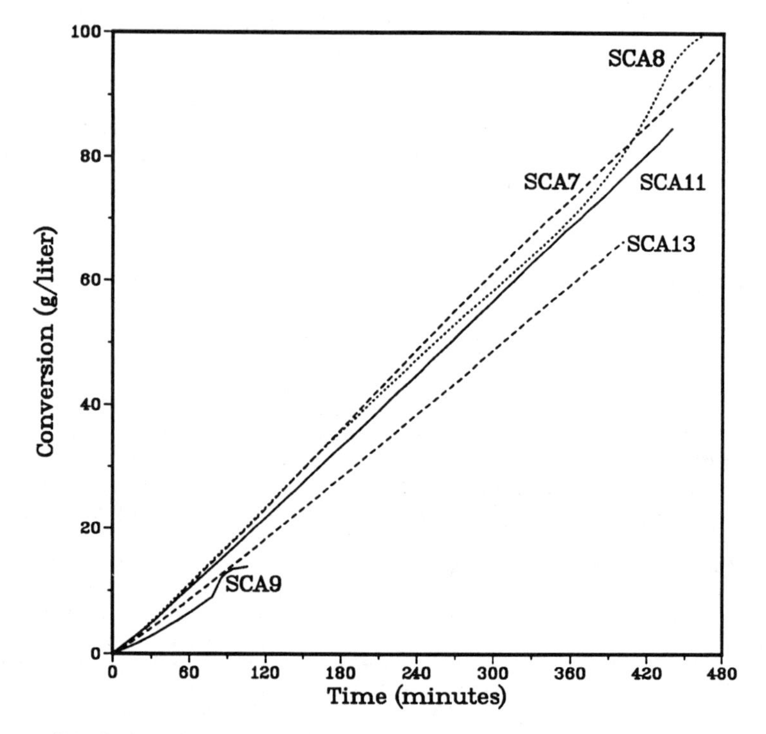

Figure 15: Conversion versus time for seeded emulsion polymerizations of polystyrene/styrene at different monomer to polymer swelling ratios; SCA9: 1.36 to 1 ratio and 1.33 mM [I]; SCA8: 10 to 1 ratio and 1.33 mM [I]; SCA7: 15 to 1 ratio and 1.33 mM [I]; SCA13: 50 to 1 ratio and 1.33 mM [I]; and SCA11: 50 to 1 ratio and 2.66 mM [I]; 50°C.

In these experiments, the rate of polymerization is proportional to the product of the concentration of monomer in the particles and the average number of radicals per particle. The lower polymerization rate in run SCA9 was due to the lower initial monomer concentration and a smaller value of $\bar{n}$ caused by a greater desorption (smaller particle size) and perhaps reduced absorption of radicals.

It is interesting to note that when the swelling ratio increased from 15:1 (SCA7) to 50:1 (SCA13), the rate of polymerization decreased; even doubling the amount of initiator (SCA11) produced a polymerization rate which was still lower than that with the 15:1 swelling ratio. This is because the average number of radicals per particle decreased when the swelling ratio increased from 15:1 to 50:1 (Figure 16). This is not caused by radical desorption, since the higher swelling ratio should result in a larger particle size and thus a lower desorption rate. The internal viscosity in these cases does not affect the diffusion of monomeric radicals due to the high concentration of monomer in the particles (*21*). In other words, the low average number of radicals per particle for the high swelling experiment is due to a slower entry of radicals. These experiments, therefore, suggest that when the swelling ratio is high enough that the polymer particles exhibit a behavior closer to that of the monomer droplets, that is they have a reduced entry rate.

The parameter estimation method (*15*) was used to estimate the second order entry rate coefficient for radicals into the particles (k_a^p) and the exit rate coefficient for radicals from the particles (k) for runs SCA7 and SCA13 using the first 120 minutes of data where $\bar{n} < 0.5$ (Figure 16). For run SCA7, the following values were obtained: $k_a^p = 2.2 \times 10^6$ dm^3 mol^{-1} s^{-1} and $k = 4.9 \times 10^{-4}$ s^{-1}; and for run SCA13, $k_a^p = 1.4 \times 10^6$ dm^3 mol^{-1} s^{-1} and $k = 2.0 \times 10^{-4}$ s^{-1}. These results indicate that the lower swelling ratio has both higher exit and entry rate coefficients. The differences between the values of k_a^p for runs SCA7 and SCA13 is about 40%, and these values are close to that found by Asua et al. (22). Still, at this high swelling ratio, the polymer concentration in the particles is about two percent, and it was suggested that the reduced entry rate of radicals into droplets may be better seen at less than one percent conversion (9). Although these experiments cannot determine the absolute value of the entry rate coefficient for radicals into the monomer droplets due to the presence of the polymer (2%) in the particles, it suggests that the entry rate coefficient for radicals into the monomer droplets is much smaller than that of the particles.

Mathematical Modelling of the Effect of Co-surfactant

A mathematical model which takes into account the nucleation of monomer droplets and the equilibrium swelling thermodynamics was developed to study the effect of cetyl alcohol and hexadecane on the entry of radicals into the monomer droplets (*11*).

Modeling Framework. The rate of change of fractional conversion in an emulsion polymerization is expressed as:

$$\frac{dx}{dt} = \frac{k_p C_m^p \bar{n} N_p}{M_m^o N_A} \tag{1}$$

where k_p is the rate coefficient for radical propagation, C_m^p the monomer concentration in the polymer particles, $\bar{n}$ the average number of radicals per particle, N_p the number of particles per unit volume of the continuous phase, M_m^o the initial moles of monomer per unit volume of the continuous phase, and N_A Avogadro's number.

The population balance for the number of particles containing n radicals was expressed by Smith and Ewart (*23*) as:

$$\frac{dN_n}{dt}=k_a^p C_R^w (N_{n-1}-N_n)+k[(n+1)N_{n+1}-nN_n]$$
$$+c[(n+2)(n+1)N_{n+2}-n(n-1)N_n] \quad (2)$$

where k_a^p is the second-order entry rate coefficient for radicals into the polymer particles, C_R^w the concentration of radicals in the aqueous phase in molecular units, k the desorption rate coefficient for radicals from the polymer particles, and c the pseudo-first-order bimolecular termination rate coefficient for free radicals in a particle.

Monomer droplets can disappear by nucleation, coagulation with droplets and particles, and diffusion to larger droplets and growing particles. However, during nucleation the latter two mechanisms are assumed to be neglegible, and thus the rate of monomer droplet disappearance can be written as:

$$\frac{dN_d}{dt}=-k_a^d C_R^w N_d \quad (3)$$

where N_d is the number of monomer droplets and k_a^d the second-order rate coefficient for radical entry into monomer droplets. The rate of change of the number of particles containing one radical is expressed as:

$$\frac{dN_1}{dt}=k_a^p C_R^w (N_0-N_1)+k(2N_2-N_1)+k_a^d C_R^w N_d \quad (4)$$

The rate of change of the concentration of radicals in the aqueous phase can be written in terms of the generation of primary radicals by initiator decomposition, desorption of radicals from the particles, absorption of radicals by the particles and droplets, and termination of radicals in the aqueous phase:

$$\frac{dC_R^w}{dt} = 2fk_I C_I+\frac{k_d \phi_w N_1}{N_A}+\frac{2k_d \phi_w N_2}{N_A}-\frac{(k_a^d N_d+k_a^p N_p)C_R^w \phi_w}{N_A}-2k_{tw}(C_R^w)^2 \quad (5)$$

where f is the initiator efficiency factor, k_I the rate constant for initiator decomposition, C_I the initiator concentration, ϕ_w the volume fraction of water in the aqueous phase, and k_{tw} the radical termination rate constant in the aqueous phase.

The equilibrium concentration of each component in each phase can be determined from the equilibrium swelling thermodynamics which is based on the idea that equilibrium conditions are attained when the partial molar free energies of mixing of monomer in all phases are the same. The equilibrium conditions can be expressed as:

Monomer droplets—aqueous phase

$$\left(\frac{\overline{\Delta G}}{RT}\right)_i^d=\left(\frac{\overline{\Delta G}}{RT}\right)_i^a \quad (6)$$

Polymer particles—aqueous phase

$$\left(\frac{\overline{\Delta G}}{RT}\right)_i^p = \left(\frac{\overline{\Delta G}}{RT}\right)_i^a \tag{7}$$

where $\overline{\Delta G}$ is the change in the Gibbs free energy, R the gas constant, T the temperature; a stands for aqueous phase, d for droplets, p for particles, and i for a component.

The partial molar free energy of mixing of a component i in phase q is given by the Flory-Huggins lattice theory of polymer solutions (*24*), with the addition of an interfacial energy term for spherical phases (*25*), and was expressed by Ugelstad et al. (*26*) as:

$$\left(\frac{\overline{\Delta G}}{RT}\right)_i^q = \ln \phi_{i,q} + \sum_{j=1, j\neq i}^{n} (1-m_{ij})\phi_{j,q} + \sum_{j=1, j\neq i}^{n} \chi_{ij}\phi_{j,q}^2$$

$$+ \sum_{j=1, j\neq i}^{n-1} \sum_{k=j+1, \neq i}^{n} \phi_{j,q}\phi_{k,q}(\chi_{ij} + \chi_{ik} - \chi_{jk}m_{ij}) + \frac{2\gamma\overline{V}_i}{rRT} \tag{8}$$

where $\phi_{i,q}$ is the volume fraction of component i in phase q, m_{ij} the ratio of the equivalent number of molecular segments between i and j (usually expressed as the ratio of molar volumes of i and j), χ_{ij} the Flory-Huggins interaction parameter, γ the interfacial tension, $\overline{V}_i$ the molar volume of component i, and r the radius of the phase. Thus the partial molar free energy of monomer in each phase can be expressed as:

i. monomer droplets:

$$\left(\frac{\overline{\Delta G}}{RT}\right)_m^d = \ln \phi_{m,d} + (1-m_{mc})\phi_{c,d} + \chi_{mc}\phi_{c,d}^2 + \frac{2\gamma_d\overline{V}_m}{r_d RT} \tag{9}$$

ii. aqueous phase:

$$\left(\frac{\overline{\Delta G}}{RT}\right)_m^a = \ln \phi_{m,w} + (1-m_{mw})\phi_{w,a} + \chi_{mw}\phi_{w,a}^2 \tag{10}$$

iii. monomer-polymer particles:

$$\left(\frac{\overline{\Delta G}}{RT}\right)_m^p = \ln \phi_{m,p} + (1-m_{mc})\phi_{c,p} + (1-m_{mp})\phi_{p,p} + \chi_{mc}\phi_{c,p}^2 +$$

$$\chi_{mp}\phi_{p,p}^2 + \phi_{c,p}\phi_{p,p}(\chi_{mc} + \chi_{mp} - \chi_{cp}m_{mc}) + \frac{2\gamma_p\overline{V}_m}{r_p RT} \tag{11}$$

The material balances for each phase are:

Material balance for the droplets

$$\phi_{m,d} + \phi_{c,d} = 1 \tag{12}$$

Material balance for the aqueous phase

$$\phi_{m,a}+\phi_{w,a}=1 \tag{13}$$

Material balance for the polymer particles

$$\phi_{m,p}+\phi_{c,p}+\phi_{p,p}=1 \tag{14}$$

The material balances for the components are:

Material balance for monomer

$$M_m^o\overline{V}_m(1-x)=\phi_{m,p}V_p+\phi_{m,d}V_d+\phi_{m,a}V_a \tag{15}$$

Material balance for co–surfactant

$$M_c^o\overline{V}_c=\phi_{c,d}V_c+\phi_{c,p}V_p \tag{16}$$

Material balance for water

$$M_w^o\overline{V}_w=\phi_{w,a}V_a \tag{17}$$

Material balance for polymer

$$M_m^o\overline{V}_m x=\phi_{p,p}V_p \tag{18}$$

where M_i^o is the initial moles of component i, x the fractional conversion, and V_q the volume of phase q.

The solution to the above set of equations will provide the volume fraction of every component in the three phases and the average radius of droplets and particles as a function of conversion. The equilibrium concentrations can be expressed as:

$$C_{ei,q}=\phi_{i,q}/\overline{V}_i \tag{19}$$

Results and Discussion. The model was applied to study miniemulsion polymerization using different co-surfactants (cetyl alcohol and hexadecane), particularly the results of runs M10 and M12 presented earlier. Run M10 is the miniemulsion polymerization of styrene using 5 mM SLS/15 mM CA and 2.66 mM potassium persulfate and run M12 is the miniemulsion polymerization of styrene using 5 mM SLS/20 mM HD and 2.66 mM potassium persulfate. The variables investigated through the modeling were the initial monomer droplet diameter and the value of the entry rate coefficient for radicals into the monomer droplets. An entry rate coefficient for radicals into the polymer particles of 1.25 x 10^6 dm^3 mol^{-1} s^{-1} was used. This value has been found to fit a wide variety of data obtained for seeded emulsion polymerizations of styrene (22).

The results of the model predictions for the styrene miniemulsion polymerization of run M10 are shown in Figure 17. These indicate that the model predicts the experimental data well when an initial monomer droplet diameter of 132 nm and an entry rate coefficient for radicals into the droplets of 0.087 x 10^6 dm^3 mol^{-1}

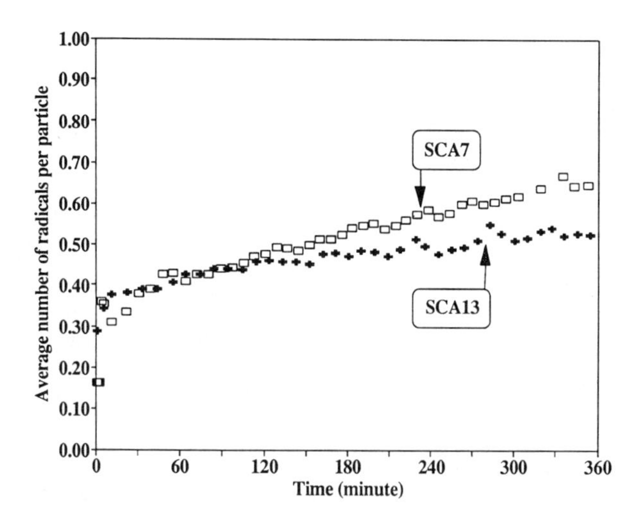

Figure 16: Average number of radicals per particle versus time for seeded emulsion polymerizations of polystyrene/styrene at different monomer to polymer swelling ratios; SCA7: 15 to 1 ratio and 1.33 mM [I]; and SCA13: 50 to 1 ratio and 1.33 mM [I].

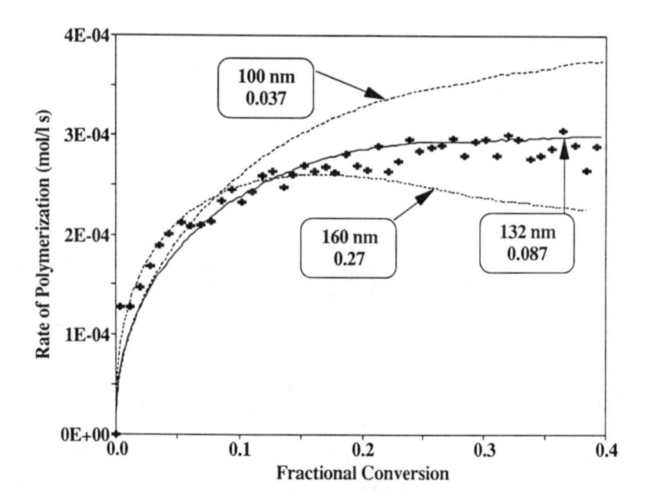

Figure 17: Comparison of simulated and experimental rate of polymerization versus fractional conversion for the styrene miniemulsion polymerization using 5 mM SLS/15 mM CA and 2.66 mM potassium persulfate; lines: simulations at different initial monomer droplet diameters and entry rate coefficients for radicals into the droplets (# x $10^6 dm^3 mol^{-1} s^{-1}$), symbols: experimental data from run M10.

s^{-1} were used. The results for two other initial monomer droplet diameters (100 nm and 160 nm) are also shown and the model is unable to fit the data for any value of the entry rate coefficient for radicals into the droplets.

Since the final particle sizes for runs M10 and M12 were found to be similar, an initial average droplet size of 130 nm was used in an attempt to model the experimental results of run M12 obtained using HD as coemulsifier. The model was unable to produce a good fit of the data for this experiment; the results are shown in Figure 18. When the same value of the entry rate coefficient for radicals into the monomer droplets was used, as found in the preceeding results (0.087 x 10^6 dm^3 mol^{-1} s^{-1}), the model was only able to fit the experimental polymerization rate at higher conversions ($x > 0.15$), failing to predict the experimental data at lower conversions ($x < 0.15$). When a higher value of the entry rate coefficient for radicals into the monomer droplets (0.37 x 10^6 dm^3 mol^{-1} s^{-1}) was used, the model was only able to fit the experimental data at lower conversions ($x < 0.03$). These results suggest that the entry rate coefficient for radicals into monomer droplets containing hexadecane should be higher than for those prepared using cetyl alcohol.

A further attempt was made to explain the experimental data for the styrene miniemulsion polymerization using hexadecane as cosurfactant (run M12). In the preceeding it was assumed that each entry of a radical into a monomer droplet resulted in the creation of a polymer particle. This, however, may not be the case. A significant fraction of radicals may desorb before any propagation occurs, thus preserving the existance of the droplet. Thus a particle is only nucleated from a monomer droplet when a radical enters and begins to propagate. By assuming the effect of coagulation between droplets and particles at low conversion is negligible, the rate of monomer droplet disappearance can be written as:

$$\frac{dN_{d,0}}{dt} = -k_a^d C_R^w N_{d,0} + k^d N_{d,1} \tag{20}$$

where $N_{d,0}$ is the number of monomer droplets containing no radicals, $N_{d,1}$ the number of monomer droplets containing one radical, and k^d the exit rate coefficient for a radical from the monomer droplets. Thus, the balance for the number of droplets containing one radical can be expressed as:

$$\frac{dN_{d,1}}{dt} = k_a^d C_R^w N_{d,0} - k^d N_{d,1} \tag{21}$$

A distinction is made between the desorption from droplets and particles; desorption can only take place from particles by a transfer reaction from a growing radical (i.e., absorption of radicals by particles is assumed to be irreversible). The rate of change of the number of particles containing one radical is written as:

$$\frac{dN_1}{dt} = k_a^p C_R^w (N_0 - N_1) + k(2N_2 - N_1) + k_a^d C_R^w N_{d,0} - k^d N_{d,1} \tag{22}$$

The entry rate coefficient for radicals into the monomer droplets was assumed to be the same as that of the particles (1.25 x 10^6 dm^3 mol^{-1} s^{-1}). An average initial droplet size of 130 nm and an exit rate coefficient for radicals from the droplets containing one radical of 2.8 x 10^{-3} s^{-1} were used. Figure 19 indicates that the modified model is able to predict the experimental data of run M12. This model was also applied to the case of miniemulsion polymerization using cetyl alcohol as cosurfactant (run M10) and the results are shown in Figure 20. The entry rate coefficient for radicals into the particles of 1.25 x 10^6 $dm^{-3}mol^{-1}s^{-1}$ and the average initial droplet size of 132 nm were used in the simulation. The entry rate coefficient for radicals into the droplets and the exit rate coefficient for radicals from the

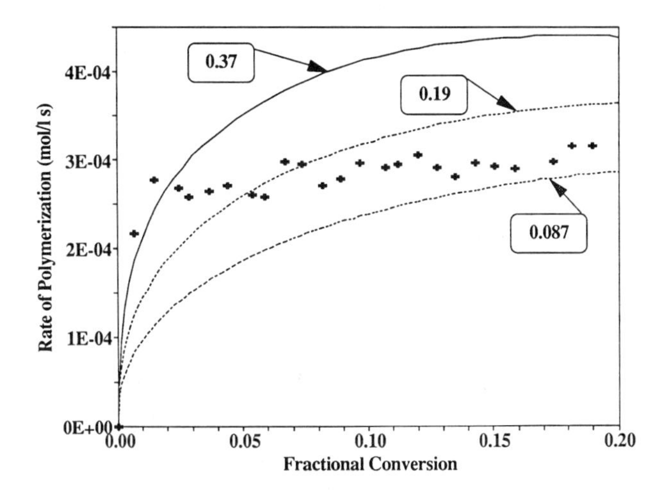

Figure 18: Comparison of simulated and experimental rates of polymerization versus fractional conversion for the styrene miniemulsion polymerization using 5 mM SLS/20 mM HD and 2.66 mM potassium persulfate; lines: simulations with 130 nm initial monomer droplet diameter and different entry rate coefficients for radicals into the droplets (# x $10^6 dm^3 mol^{-1} s^{-1}$), symbols: experimental data from run M12.

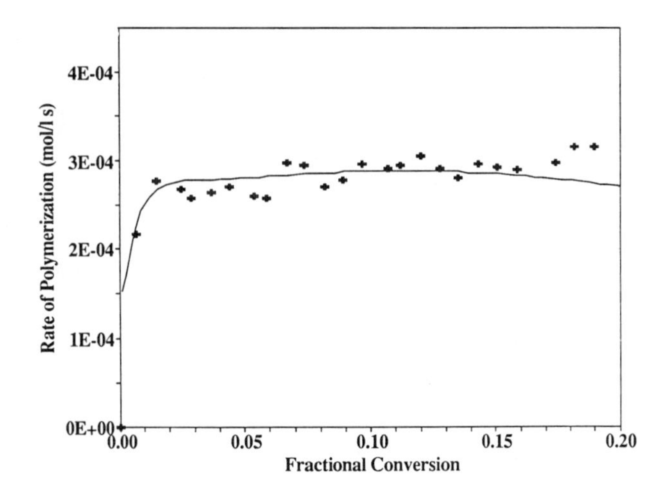

Figure 19: Comparison of simulated and experimental rates of polymerization versus fractional conversion for the styrene miniemulsion polymerization using 5 mM SLS/20 mM HD and 2.66 mM potassium persulfate; lines: simulations with 130 nm initial monomer droplet diameter and the exit rate coefficient for droplets containing one monomeric radical of 2.8 x 10^{-3} s^{-1}, symbols: experimental data from run M12.

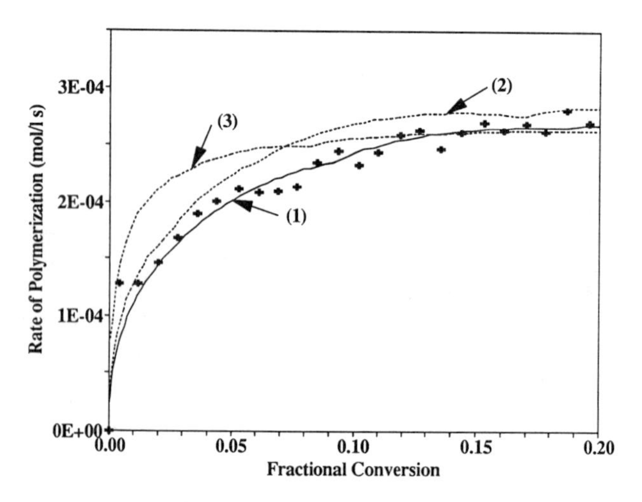

Figure 20: Comparison of simulated and experimental rates of polymerization versus fractional conversion for the styrene miniemulsion polymerization using 5 mM SLS/15 mM CA and 2.66 mM potassium persulfate; lines: simulations with different entry rate coefficients for radicals into the monomer droplets and exit rate coefficients for radicals from the droplets: (1) $k_a^d = 0.17 \times 10^6 dm^3mol^{-1}s^{-1}$ and $k^d = 0.4 \times 10^{-3} s^{-1}$; (2) $k_a^d = 0.25 \times 10^6 dm^3mol^{-1}s^{-1}$ and $k^d = 0.5 \times 10^{-3} s^{-1}$; and (3) $k_a^d = 0.62 \times 10^6 dm^3mol^{-1}s^{-1}$ and $k^d = 2.0 \times 10^{-3} s^{-1}$, symbols: experimental data from run M10.

monomer droplets containing one radical were adjusted to fit the shape of the polymerization rate-conversion data of run M10. The best fit to the experimental data was obtained when the entry rate coefficient for radicals into the droplets was 0.17 x 10^6 $dm^3mol^{-1}s^{-1}$ and the exit rate coefficient for radicals from the droplets containing one radical was 0.4 x $10^{-3}s^{-1}$. These values are both about an order of magnitude smaller than those used for the particles. These results again show that miniemulsions prepared with CA and HD differ significantly in their polymerization characteristics and that CA may affect both the entry and exit of radicals as found in the seeded polymerization studies.

Summary and Conclusions

A comparative study was carried out to evaluate the effect of the variation in parameters important to the preparation of styrene miniemulsions on the polymerization kinetics and the resulting particle size distributions. The variables studied were the type of co-surfactant and the means of homogenization. The results indicated that, under the reported conditions, the Microfluidizer provided a greater and more uniform shear than the sonifier and the Omni mixer, thus resulting in smaller particles and more uniform particle size distributions. When the emulsion without a co-surfactant was allowed to age, the polymerization rate was found to decrease with aging time as a result of the increased monomer droplet size. The miniemulsions prepared with cetyl alcohol were found to be unchanged within four hours of aging at room temperature. The results also showed that when the concentration of SLS was above its CMC, the overall rate of the conventional polymerization was slightly faster than that of the miniemulsion system. However, these results were reversed when the concentration of SLS was below its CMC. The results also suggested that particle nucleation is slowed by the presence of cetyl alcohol on the surface of the monomer droplets.

The effect of cetyl alcohol on the entry and exit rate coefficients was studied using seeded emulsion polymerization. It was found that cetyl alcohol increases the polymerization rate in a seeded system by acting as a chain transfer agent competing with monomer and decreasing the exit of radicals. These results also suggest that when a miniemulsion droplet becomes a polymer particle, the polymerization behavior is similar to conventional polymer particles. Experiments with monomer/polymer swelling ratios of up to 50:1 were carried out to study the entry rate coefficient of radicals into the monomer droplets. The results indicated that when the swelling ratio increased from 15:1 to 50:1, the rate of polymerization decreased. This suggested that the entry rate coefficient for radicals into the monomer droplets containing cetyl alcohol was smaller than that of the particles.

A mathematical model which takes into account the nucleation of monomer droplets and the equilibrium swelling thermodynamics was applied to study the effect of co-surfactant on the entry of radicals into the monomer droplets. The results showed that the entry rate coefficient for radicals into the monomer droplets is about an order of magnitude smaller than that of the particles when cetyl alcohol was used as co-surfactant, and this smaller value contributes to the slow nucleation rate in miniemulsion polymerization using cetyl alcohol as a cosurfactant. When hexadecane was used as co-surfactant, the entry rate for radicals into the monomer droplets was found to be similar to that of the particles.

Acknowledgments

Financial support for this program from the National Science Foundation under Grant No. CTS-8907600 is greatly appreciated. Also partial support for PLT from the United States Department of Education Graduate Fellow Program in Chemical Engineering is acknowledged.

Literature Cited

1. Choi, Y.T., *Formation and Stabilization of Miniemulsions and Latexes*, Ph.D. Dissertation, Lehigh University, 1986.
2. Cockbain, E.G. and McRoberts, T.S., *J. Colloid & Interface Sci.*, **1953**, *8*, p 440.
3. Blackley, D.C. and Lawrence, A.C.S., *Discuss. Faraday Soc.*, **1954**, *28*, p 268.
4. Pithayanukul, P. and Pipel, N., *J. Colloid & Interface Sci.*, **1982**, *89*, p 494.
5. Higuchi, W.I. and Misra, J., *J. Phar. Sci.*, **1962**, *51*, p 459.
6. Vanderhoff, J.W., *J. Polym. Sci., Polym. Symp.*, **1985**, *72*, p 161.
7. Chamberlain, B.J., Napper, D.H., Gilbert, R.G., *J. Chem. Soc., Faraday Trans. I*, **1982**, *78*, p 591.
8. Choi, Y.T., El-Aasser, M.S., Sudol. E.D. and Vanderhoff, J.W., *J. Polym. Sci., Polym. Chem. Ed.*, **1985**, *23*, p 2973.
9. Tang, P.L., Sudol, E.D., Silebi, C.A. and El-Aasser, M.S., *J. Appl. Polym. Sci.*, **1991**, *43*, p 1059.
10. Wood, D.F., Whang, B.C.Y., Napper, D.H., Gilbert, R.G. and Lichti, G., *J. Polym. Sci., Polym. Chem. Ed.*, **1983**, *21*, p 985.
11. Tang, P.L., *Kinetic Factors in Miniemulsion Polymerization*, Ph.D. Dissertation, Lehigh University, 1991.
12. Hawkett, B.S., Gilbert, R.G. and Napper, D.H., *J. Chem. Soc., Faraday Trans. I*, **1975**, *71*, p 2288.
13. Delgado, J., *Miniemulsion Copolymerization of Vinyl Acetate and n-Butyl Acrylate*, Ph.D. Dissertation, Lehigh University, 1986.
14. Rodriguez, V.S., El-Aasser, M.S., Asua, J.M. and Silebi, C.A., *J. Polym. Sci., Polym. Chem. Ed.*, **1989**, *27*, p 3659.
15. Asua, J.M., Adams, M.E. and Sudol, E.D., *J. Appl. Polym. Sci.*, **1990**, *39*, p 1183.
16. Asua, J.M., Sudol, E.D. and El-Aasser, M.S., *J. Polym. Sci., Polym. Chem. Ed.*, **1989**, *27*, p 3903.
17. Hansen, F.K. and Ugelstad, J., *Makromol. Chem.*, **1979**, *180*, p 2423.
18. Nomura, M., in *Emulsion Polymerization*, Piirma, I., Ed., Academic Press, New York, 1982 p 191.
19. Collins, E.A., Bares, J. and Billmeyer Jr., F.W., *Experiments in Polymer Science*, Wiley-Interscience, New York, 1973, p 333.
20. *Polymer Handbook*, 2nd Ed., Brandrup, J. and Immergut, H., Ed., Wiley-Interscience, New York, 1975.
21. von Meerwall, E.D., Amis, E.J. and Ferry, J.D., *Macromolecules*, **1985**, *18*, p 260.
22. Asua, J.M. and de la Cal, J.C., *J. Appl. Polym. Sci.*, **1991**, *42*, p 1869.
23. Smith, W.V. and Ewart, R.W., *J. Chem. Phys.*, **1948**, *16*, p 592.
24. Flory, P.J., *Principles of Polymer Chemistry*, Cornell University Press, New York, 1953.
25. Morton, M., Kaizerman, S. and Altier, M.W., *J. Colloid Sci.*, **1954**, *9*, p 300.
26. Ugelstad, J. Mork, P.C., Mfutakamba, H.R., Soleimany, E., Nordhuus, I., Nustad, K., Schmid, R., Berge, A., Ellingsen, T. and Aune, O., in *Science and Technology of Polymer Colloids*, Poehlein, G.W., Ottewill, R.H. and Goodwin, J.W., Eds., NATO ASI Ser, Vol. I, 1983, p 51.

RECEIVED December 4, 1991

Chapter 7

Kinetics and Mechanism of Styrene Microemulsion Polymerization

J. S. Guo[1,4], E. David Sudol[1], J. W. Vanderhoff[1,2], and Mohamed S. El-Aasser[1,3]

[1]Emulsion Polymers Institute, [2]Department of Chemistry, and [3]Department of Chemical Engineering, Lehigh University, Bethlehem, PA 18015

The polymerization of styrene oil-in-water microemulsions was carried out using water-soluble potassium persulfate initiator at 70°C. The latexes were stable, bluish, and less translucent than the original microemulsions. The polymerization rates, measured by dilatometry, increased to a maximum and then decreased. The maximum polymerization rate and number of particles varied with the 0.47 and 0.40 powers of the potassium persulfate concentration. The small latex particle diameters (20 to 30 nm) and high polymer molecular weights (ca. 2×10^6) showed that on average each latex particle was comprised of only 2 to 3 polystyrene molecules. The microemulsion polymerization of styrene exhibited continuous nucleation. A mechanism based on nucleation in microemulsion droplets was proposed to explain the experimental findings, although other nucleation mechanisms were not totally ruled out. The competition in capturing radicals from the aqueous phase by microemulsion droplets and polymer particles was suggested to account for the decrease in the particle nucleation rate and the increase in the number of polymer chains per particle with increasing conversion. The low $\bar{n}$ (less than 0.5) of the microemulsion system was attributed to the exit of monomeric radicals.

Emulsion polymerization is an important process which produces a wide variety of polymer colloids or latexes for industrial applications such as synthetic rubber, coatings, paints, adhesives, flocculants, binders for non-woven fabrics, and engineering plastics. Latexes are also used in numerous biomedical applications such as diagnostic tests, immunoassays, biological cell-labelling, and drug delivery systems.

Recently, microemulsion polymerization has attracted considerable interest and attention due to the versatile nature of microemulsions. They are isotropic, optically transparent or translucent, and thermodynamically stable. Polymerizations have been carried out in oil-in-water (o/w), water-in-oil (w/o), and bicontinuous microemulsion systems.

Atik and Thomas (*1*) polymerized styrene o/w microemulsions, consisting of cetyltrimethylammonium bromide, hexanol, styrene, and water, using azobisisobutyro-

[4]Current address: Monsanto Chemical Company, Springfield, MA 01151

0097–6156/92/0492–0099$06.00/0

nitrile (AIBN) initiator or a Cs gamma-ray source. They reported that latexes of narrow size distribution with average diameters of 20 and 35 nm were obtained. Johnson and Gulari (2) polymerized dilute styrene o/w microemulsions, consisting of sodium dodecyl sulfate (SDS), pentanol, styrene, and water, using potassium persulfate or AIBN as initiator and characterized the size of microemulsion droplets and latex particles by photon correlation spectroscopy. Bimodal latex particle size distributions were obtained and these were attributed to the presence of two competing mechanisms for initiation. Jayakrishnan and Shah (*3*) studied the polymerization of styrene and methyl methacrylate microemulsions with AIBN or benzoyl peroxide as initiator. The rate of polymerization was found to be somewhat lower than for conventional emulsion polymerizations under identical conditions. The microemulsions were found to become turbid and unstable during the polymerization. Gratzel et al. (*4*) reported the photoinduced polymerization in o/w microemulsions using a surface active initiator, cetyltrimethylammonium persulfate. The monomers used in this study were styrene, divinylbenzene, acrylamide, methyl methacrylate, and acrolein. The polymerizations were initiated with ultraviolet light or visible light, with either eosin Y or the ruthenium bipyridyl complex as the photosensitizer. Latexes of large particle size, up to 11.5 μm, were produced, but electron microscopy indicated that these were formed by agglomeration of 10 to 30 nm particles. The photoinitiated polymerization of styrene o/w microemulsions was reported by Kuo et al. (*5*), who used dibenzylketone as initiator. High molecular weights (M_w = 1 - $5x10^5$) and narrow latex particle size distributions (D_n = 30 - 60 nm) were obtained. The principal locus for initiation of polymerization was the microemulsion droplets. It was also found that the final particle size increased with the increasing initiator concentration. Kuo et al. postulated that the particle growth in microemulsion polymerization was a combination of coagulation of the particles and monomer diffusion. Although these studies of o/w microemulsion polymerization were detailed and extensive, they did not give a definitive mechanism for particle nucleation and growth.

In this paper, the kinetics and particle nucleation process in styrene oil-in-water microemulsion polymerizations are investigated. The possible nucleation and growth mechanisms are discussed.

Experimental

The styrene monomer (Polysciences) was washed with 10% aqueous sodium hydroxide to remove the inhibitor, then washed with distilled-deionized water to remove residual base. After drying with anhydrous sodium sulfate, the styrene was vacuum distilled under nitrogen at 20 torr and 40°C. The purified styrene monomer was stored at -10°C until use. The 1-pentanol (Fisher Scientific, certified) was used as received without further purification. The potassium persulfate (KPS, Fisher Scientific) was recrystallized twice from water. The crystals were dried at room temperature in a vacuum oven and stored at -5°C. The sodium dodecyl sulfate (SDS, Henkel) was used as received. Its critical micelle concentration was determined by surface tension to be 5.4 mM. Distilled-deionized (DDI) water was used in all experiments.

The microemulsion polymerization recipe comprised 82.25% water, 9.05% sodium dodecyl sulfate, 3.85% 1-pentanol, and 4.85% styrene by weight. The rates of polymerization at 70°C were measured dilatometrically in a 25 ml Erlenmeyer flask equipped with a 45-cm long 1-mm ID capillary. The microemulsions containing initiator ([KPS] = 0.14-0.69 mM) were degassed, loaded into the dilatometer, and polymerized in a thermostated water bath. This polymerization procedure was previously described in more detail (*6*). The latex particle size distributions were determined using a Philips 400 transmission electron microscope with phosphotungstic acid negative-staining. After polymerization, the latex was poured into methanol; the precipitated polymer was filtered, washed with methanol and water, and dried. The polymer molecular weight distributions were determined using a Waters Model 440 gel permeation chromatograph with tetrahydrofuran as the eluant solvent.

A series of polymerizations using 0.27 mM potassium persulfate (based on water) was carried out at 70°C to predetermined conversions in a dilatometer. The conversion at which the polymerization was stopped by hydroquinone was determined by the change in the capillary height. This was double checked by gas chromatography (Hewlett-Packard 5890). The particle size distributions were determined as a function of conversion using a Phillips 400 Transmission Electron Microscope (TEM). To obtain a representative particle size distribution, at least 1200 particles were counted. The particle diameters were measured using a Zeiss MOP-3 Analyzer.

Results and Discussion

The latexes prepared using the water-soluble potassium persulfate initiator were stable, bluish, and less translucent than the original microemulsions. This change in appearance was attributed to the larger size and the higher refractive index of the dispersed phase. The refractive index is 1.594 for polystyrene and 1.546 for styrene at 20°C.

Effect of Potassium Persulfate Concentration. The effect of water-soluble initiator concentration on the polymerization kinetics of styrene microemulsions was studied using 0.14 to 0.69 mM potassium persulfate (based on water). Figures 1 and 2 show respectively the conversion versus time and rate of polymerization versus conversion curves at 70°C for the different initiator concentrations. The polymerization rate increased with increasing initiator concentration, and the polymerization rate versus conversion curves showed only two intervals; neither showed a constant rate or a gel effect. These findings are quite different from the kinetics of conventional emulsion polymerization, which is characterized by three intervals and sometimes a gel effect. However, the absence of a constant rate region was also observed for miniemulsion polymerizations of styrene (*7*).

Interval I of the polymerization rate-conversion curves was characterized by an increase to a maximum, Interval II, by a decrease from this maximum. The maximum polymerization rate was reached at 20 to 25% conversion, in contrast to the 2 to 15% usually observed in conventional emulsion polymerization, and the 30% found for miniemulson polymerization. In conventional emulsion polymerization, Interval I is defined as the particle nucleation stage. For miniemulsion polymerization, Chamberlain et al. (*8*) also interpreted Interval I as the particle nucleation stage. The polymerization rate thus increased with increasing numbers of particles nucleated during Interval I. The long nucleation stage for miniemulsion polymerizations compared to conventional emulsion polymerizations has been attributed to a slower rate of radical entry into the miniemulsion droplets, which in turn has been attributed to a higher concentration of adsorbed emulsifier on the miniemulsion droplet surface or adsorbed mixed emulsifier liquid crystals on the droplet surface, or the larger size of miniemulsion droplets compared to micelles (*7*).

In microemulsion polymerization, if we consider that Interval I is the nucleation stage and that nucleation occurs in the microemulsion droplets, the long nucleation period can be attributed to the adsorbed surfactant/co-surfactant complex on the microemulsion droplet surface, which hinders the entry of radicals and thus gives a low radical capture rate. Interval I ends when all of the microemulsion droplets have disappeared, either by becoming polymer particles or by being consumed by diffusion of their monomer to the growing polymer particles. In Interval II, the polymerization rate decreased because of the decrease of monomer concentration in the monomer-swollen polymer particles. The absence of a gel effect in this microemulsion polymerization is attributed to the small size of the latex particles. The entry of a second radical causes instantaneous termination with the radical growing inside the particle.

However, we must mention at this point that the interpretation of Interval I as the nucleation stage for styrene microemulsion polymerizations would be misleading. As we know, the rate of polymerization is determined by the number of particles, the monomer

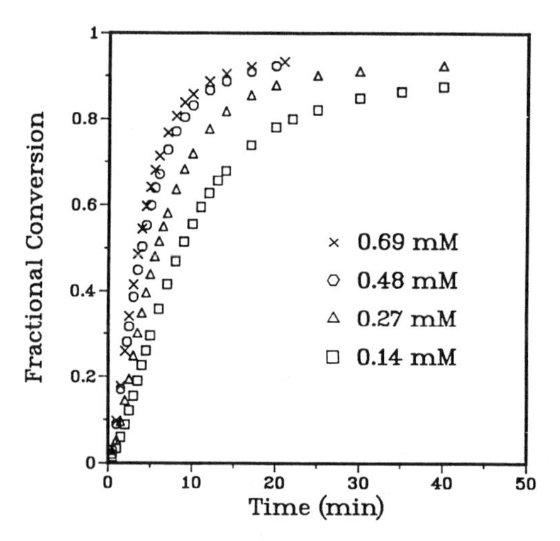

Figure 1: Conversion versus time curves for styrene microemulsion polymerizations with different KPS initiator concentrations at 70°C.

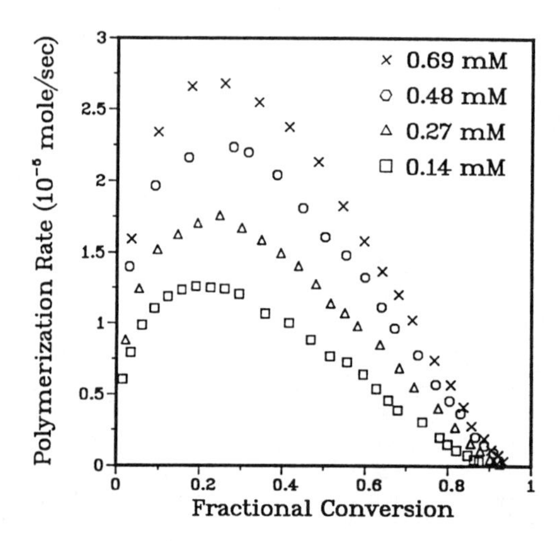

Figure 2: Polymerization rate versus conversion curves for styrene microemulsion polymerizations with different KPS initiator concentrations at 70°C.

concentration inside the particles, and the average number of radicals per particle. Therefore, a maximum in the polymerization rate does not necessarily mean that the maximum number of polymer particles is attained at that moment. Indeed, it will be shown in the next section that continuous nucleation was found for styrene microemulsion polymerizations and nucleation of polymer particles in the microemulsion droplets (or monomer swollen mixed emulsifier micelles) is the most probable mechanism.

Figures 3 and 4 show that the maximum polymerization rate and final number of polymer particles (calculated from the volume-average diameter D_v) varied according to the 0.47 and 0.40 powers, respectively, of the potassium persulfate concentration. These dependences are similar to the 0.40 power predicted for Smith-Ewart case 2 kinetics. However, the mechanism and kinetics of microemulsion polymerization are different from those of conventional emulsion polymerization. By considering that nucleation occurs in the microemulsion droplets, the fraction of the droplets that become particles is determined by the rate of radical generation. An increase in initiator concentration would increase the radical flux to the microemulsion droplets, which in turn would increase the fraction of droplets converted to polymer particles.

The final latex particle size distributions obtained by polymerization of styrene microemulsions were broad with particles in the range of 5 to 40 nm. These sizes are larger than the corresponding microemulsion droplet size of 4.6 nm diameter (*9*). The larger particle sizes of the latex particles are attributed to the fact that not all of the microemulsion droplets capture radicals and become polymer particles during the polymerization. This can be rationalized by comparing the microemulsion droplet number, 1.4×10^{18}/ml water, with the radical generation rate of 8.5×10^{14} radicals/ml/min for 0.27 mM KPS at 70°C. This polymerization was found to be completed in 30 minutes; therefore, the total number of radicals generated was only enough to nucleate 2% of the microemulsion droplets if the droplets at most capture one radical. Indeed, by dividing the total number of particles by the number of droplets, only 0.5% of the droplets became polymer particles.

Table I shows the final particle sizes and molecular weights produced using the different initiator concentrations. The particle size decreased and the molecular weight remained relatively constant as the initiator concentration was increased. Again, the decrease in the particle size with increased initiator concentration was due to the increasing radical flux to the microemulsion droplets, which led to the formation of more particles. The weight-average molecular weight (M_w) of the polymer produced was high (ca. 2×10^6), and a slight dependence of M_w on the KPS concentration (-0.07 power) was obtained. This is quite different from the -0.6 power dependence predicted by Smith-Ewart theory, which is based on bimolecular termination inside the particles. From the latex particle sizes and the polymer molecular weights, each latex particle was estimated to be comprised of only 2 to 3 polymer molecules. Similarly, in the inverse microemulsion polymerization of acrylamide, Candau et al. (*10*) obtained polyacrylamide of very high molecular weight (up to 10^7) and concluded that each final latex particle consisted of one single polymer chain. Again, this differs from conventional emulsion polymerization, in which a polymer particle consists of hundreds or thousands of polymer chains.

Particle Nucleation in Microemulsion Polymerization. Figure 5 shows representative TEM micrographs for the microemulsion latexes produced by polymerization to different conversions using 0.27 mM potassium persulfate initiator. The particle size distribution was narrow at the beginning of the polymerization (2% conversion) and became broader with increasing conversion (Figure 6). Table II shows that the average polymer particle size (unswollen) increased with conversion and leveled off after 77% conversion. Based on the volume-average particle diameter (D_v) and degree of conversion, the number of polymer particles (N_p) as a function of conversion was calculated and is depicted in Figure 7. This figure shows a continuous increase in the particle number with increasing conversion. In other words, the nucleation of particles continued throughout the course of the polymerization.

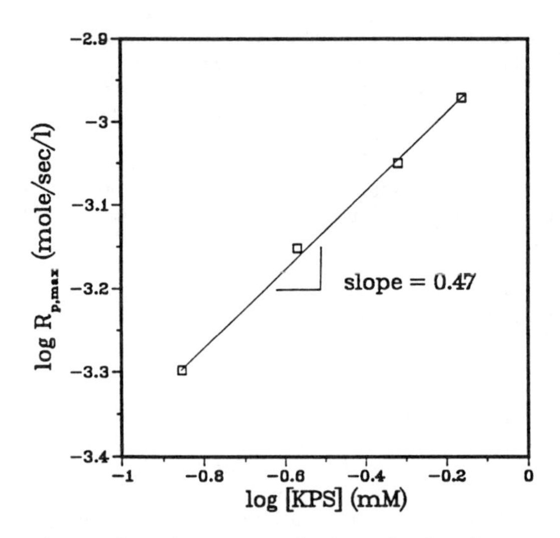

Figure 3: Dependence of maximum rate of polymerization, $R_{p,max}$, on KPS initiator concentration for styrene microemulsion polymerizations.

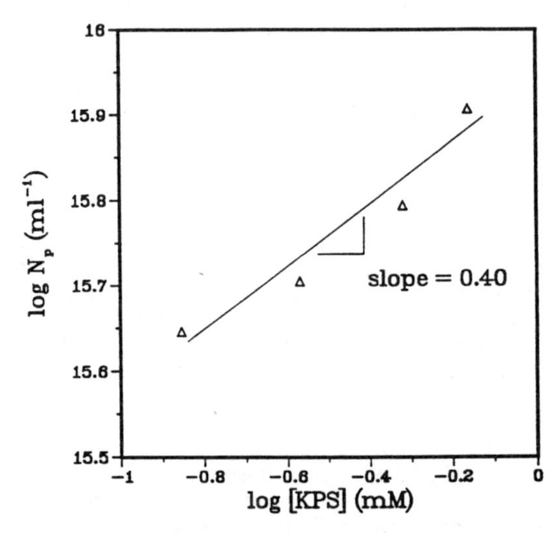

Figure 4: Dependence of final particle number, N_p, on KPS initiator concentration for styrene microemulsion polymerizations.

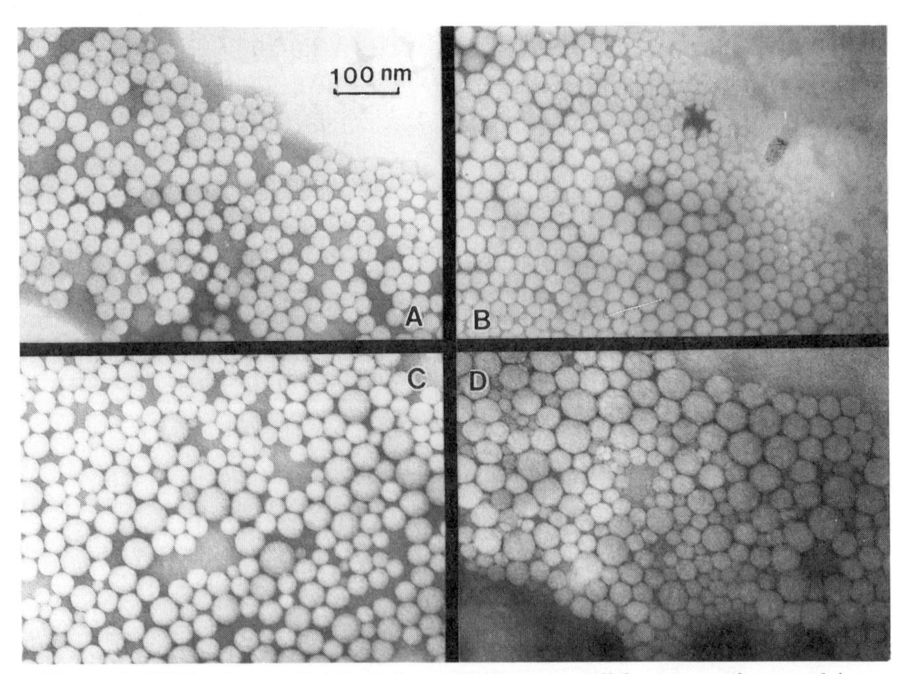

Figure 5: TEM micrographs for polymerized styrene oil-in-water microemulsion latexes at different conversions with 0.27 mM KPS at 70°C (A:2% conversion, B: 13% conversion, C: 43% conversion, D: 77% conversion).

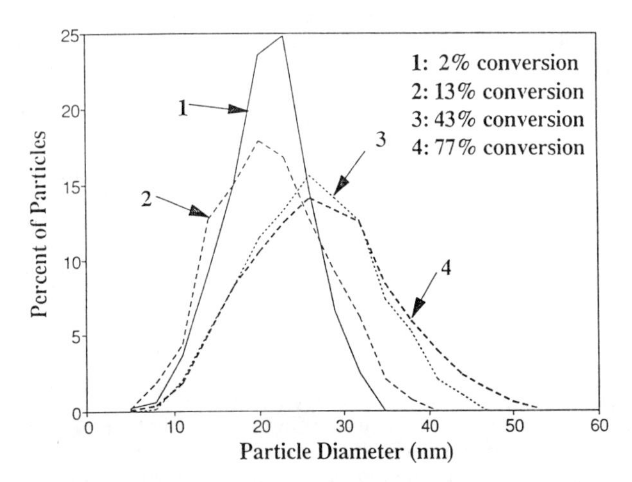

Figure 6: Particle size distributions for polymerized styrene oil-in-water microemulsion latexes at different conversions with 0.27 mM KPS at 70°C.

Table I. Resulting particle sizes and molecular weights for styrene microemulsion polymerizations using different KPS concentrations at 70°C

[KPS] (mM)	D_v	M_w	n_p
0.14	27.0	$1.96x10^6$	3.3
0.27	25.8	$1.94x10^6$	2.9
0.48	24.1	$1.76x10^6$	2.6
0.69	22.1	$1.80x10^6$	2.0

* D_v = volume-average diameter; M_w = weight-average molecular weight; n_p = average number of polymer chains inside each latex particle.

Table II. Average particle size of polymerized styrene microemulsion latexes as a function of conversion with 0.27 mM KPS at 70°C

% conversion	D_n nm	D_v nm	D_w nm	PDI	n_p
2	17.0	18.0	18.5	1.09	1.0
13	17.4	19.1	20.0	1.15	1.14
24	19.2	21.0	21.9	1.14	1.52
34	20.6	22.2	25.1	1.22	1.79
43	21.8	23.7	26.2	1.20	2.18
51	21.8	23.5	27.3	1.25	2.13
68	22.3	24.5	27.9	1.25	2.41
77	22.5	25.7	28.8	1.28	2.80
85	24.0	25.8	30.4	1.27	2.90
90	23.3	25.7	30.3	1.30	2.90

* D_n = number-average diameter; D_v = volume-average diameter; D_w = weight-average diameter; PDI = D_w/D_n; n_p = number of polymer chains per particle, calculated based on a polymer molecular weight of $2x10^6$.

In conventional styrene emulsion polymerization, nucleation is generally considered to occur in monomer-swollen micelles. The disappearance of the micelles (or an insufficient number of micelles) marks the end of the particle nucleation stage (Interval I) and the beginning of Interval II. Owing to the similarity between microemulsion droplets and monomer-swollen micelles, the most probable locus of nucleation was considered to be the microemulsion droplets for a styrene oil-in-water microemulsion polymerization. The other possible loci of nucleation will be discussed later.

For the polymerization of styrene oil-in-water microemulsions, the surfactant concentration always exceeds the critical micelle concentration (CMC) owing to the high concentration of surfactant in the system, which means that mixed micelles exist throughout the polymerization. The reason why we refer to these as mixed micelles instead of microemulsion droplets is due to the finding (*11*) that the oil cores of the microemulsion droplets disappear early in the polymerization (4% conversion). What remains are SDS micelles with 1-pentanol and a small amount of styrene in the palisade layer (i.e., mixed micelles). Based on the sizes of microemulsion droplets and polymer particles, the surface area of the final polymer particles is approximately 10% of that of the initial microemulsion droplets. Thus, at the end of the polymerization, the surface

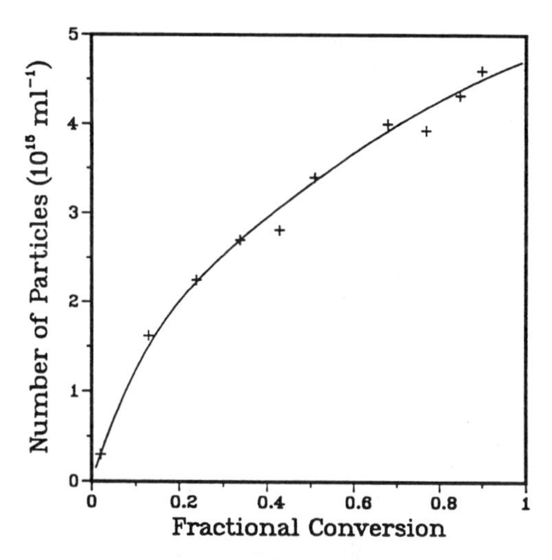

Figure 7: Number of polymer particles versus fractional conversion for styrene microemulsion polymerization with 0.27 mM KPS at 70°C.

area of the remaining mixed micelles is still eight times larger than that of the polymer particles. Therefore, the continuous nucleation of polymer particles throughout the polymerization should not be surprising. In Figure 5, it can be seen that small particles appear in every stage of the polymerization (which accounts for the newly formed particles), and the particles continue growing with conversion.

For the sake of clarity, the particle nucleation process in a styrene oil-in-water microemulsion polymerization is depicted in Figure 8. At the beginning of the polymerization, the radicals generated in the aqueous phase enter some of the microemulsion droplets and start the polymerization. Once polystyrene is formed inside the droplets (now particles), the system components redistribute to maintain equilibrium. As stated above, it was found that the oil cores of microemulsion droplets disappear at 4% conversion (*11*). The newly formed polymer particles grow via the supply of monomer from the microemulsion droplets (or mixed micelles after 4% conversion) and the inactive polymer particles either by diffusion or collision. Owing to the very small size of the polymer particles, there is at most one radical in each particle at a given time. The entry of a second radical causes instantaneous termination of the first radical. However, it was found (as discussed in detail in the final part of this paper) that most growing polymer chains inside the particles undergo chain transfer to monomer before a second radical enters. This monomeric radical readily diffuses out of the particle (desorbs).

With increasing conversion, the number and total surface area of the mixed micelles become smaller while the surface area of the polymer particles increases. Thus, the polymer particles increasingly compete with the mixed micelles in capturing radicals, although the mixed micelles still capture more radicals than the polymer particles. This leads to the continuous generation of new particles through the capture of radicals by the mixed micelles and allows the existing polymer particles to continue to grow with increasing conversion. The capture of radicals by the existing polymer particles was indicated by the increase in the number of polymer chains per particle (Table II) and the decrease in the polymer particle generation rate (Figure 7) with conversion. It is also interesting to note that, early in the polymerization (2% conversion), only one polymer chain per particle was obtained, which is consistent with the proposed mechanism. In Figure 7, the particle generation rate at the beginning of the polymerization of a styrene microemulsion was about the same as the initial primary radical production rate from initiator decomposition, 8.6×10^{14} $ml^{-1}min^{-1}$ (calculated from the decomposition rate constant of KPS = 4.4×10^{-5} sec^{-1} at 70°C). The initiator decomposition rate changed only slightly during the polymerization owing to the short period of the reaction. The decrease in the particle nucleation rate with increasing conversion was mainly due to the competition for radicals by the polymer particles. At the end of the polymerization, only polymer particles and SDS/1-pentanol mixed micelles remain. Owing to the very low monomer concentration at high conversion, any particles nucleated must be small, perhaps too small to be resolved by TEM. It is also possible that these are too small to be stable and are captured by the larger particles.

A similar particle nucleation process has also been reported for the inverse microemulsion polymerization of acrylamide (*12*). However, there are some notable differences. Only one polyacrylamide chain was found inside each particle at the end of the polymerization, which suggests that, once nucleated, particles did not capture more radicals (at least not more than one additional radical). The number of particles was also found to increase proportionally with conversion. These differences may be due to the much higher surfactant concentration (17% sodium bis(2-ethylhexyl) sulfosuccinate, Aerosol OT) used in their system compared to ours (9% SDS), which provides more surface area for formation of the microemulsion droplets. The other possible reason is that the acrylamide was observed to polymerize completely in only 2 to 3 minutes. In such a short time, the number of radicals generated was small. Therefore, the number of radicals captured by the polymer particles would be much smaller than in our system and hence it was unlikely a particle was entered more than once. Termination by chain transfer to monomer, toluene, or Aerosol OT, followed by desorption of the new radical species from the particles was also suggested (*13*).

In our system, the possibility of particle-particle flocculation appears to be ruled out with the finding of one polymer chain per particle at 2% conversion. The increase in the number of polymer chains per particle with conversion was attributed to the capture of additional radicals by the polymer particles; however, limited flocculation at a later stage can also lead to the same result. Therefore, we cannot completely rule out this possibility.

The contribution of homogeneous nucleation to the generation of polymer particles in microemulsion polymerization is considered to be small because: 1) the amount of styrene in the aqueous phase is lower than its solubility due to the presence of 1-pentanol (*9*); and 2) the microemulsion droplets provide a large surface area for capture of radicals. However, it is recognized that if radical capture by microemulsion droplets was retarded by the condensed interfacial layer then the oligomeric radicals may have the opportunity to grow and precipitate in the aqueous phase, and be stabilized by emulsifier desorbed from the microemulsion droplets, thus creating particles.

The Average Number of Radicals per Particle in Microemulsion Polymerization. The rate of polymerization, R_p, in a conventional emulsion polymerization can be expressed as follows:

$$R_p = \frac{k_p \bar{n} N_p [M]_p}{N_A} \quad (1)$$

where k_p is the propagation rate constant, $[M]_p$ is the concentration of monomer in the particle, $\bar{n}$ is the average number of free radicals per particle, and N_A is Avogadro's number. In order to calculate $\bar{n}$, all the other parameters in equation (1) must be known. In this case, the variations of R_p with conversion for four different initiator concentrations were calculated from dilatometric measurements and are shown in Figure 2. The particle number (N_p) versus conversion was determined only for the reaction employing 0.27 mM potassium persulfate (Figure 7). Since the final particle number was found to be proportional to the 0.4 power of the initiator concentration (Figure 4), this relationship was utilized to establish the N_p versus conversion curves for other initiator concentrations. The values of $[M]_p$ were determined as a function of conversion (see Figure 9) from the thermodynamic simulation based on the fact that, at equilibrium, the chemical potential of each component in every phase will be equal. The equations describing this are given in detail elsewhere (*11*). Therefore, the values of $\bar{n}$ were calculated and are depicted in Figure 10 for the different initiator concentrations. It can be seen that $\bar{n}$ decreased with increasing conversion and was always less than the 0.5. Increasing the initiator concentration gave higher values of $\bar{n}$. Before developing an explanation for the $\bar{n}$ data, we must have some understanding of the entry and exit of radicals in this system.

It can be seen in Table I that the molecular weight of the polymer, approximately $2x10^6$, was almost independent of the initiator concentration (M_w was proportional to $[KPS]^{-0.07}$), which is quite different from the -0.6 power dependence on water-soluble initiator concentration predicted by the Smith-Ewart theory. The polymer molecular weight is determined by the polymerization rate and the mode of termination. If we consider that chain termination was due mainly to chain transfer to styrene, based on the molecular weight data (M_w), the chain transfer constant to styrene can be calculated to be $0.5x10^{-4}$, which is within the range $0.3\text{-}0.6x10^{-4}$ found in the literature (*14*). This provides evidence that termination by chain transfer to monomer is dominant. This also explains the lack of dependence of M_w on initiator concentration. The lack of bimolecular termination can be explained as follows. The polymer particles compete with the microemulsion droplets in capturing radicals. However, owing to the much larger surface area provided by microemulsion droplets (or mixed micelles), the radical flux to the polymer particles is small. Thus, the time required for each polymer particle to capture a

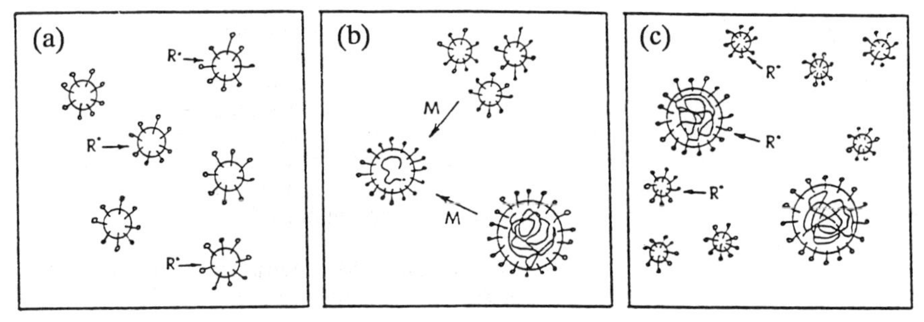

R•—free radical M—monomer

(a) Nucleation in Microemulsion Droplets
(b) Monomer Transport
(c) Competition in Radical Capture

Figure 8: Polymerization mechanism for styrene oil-in-water microemulsions.

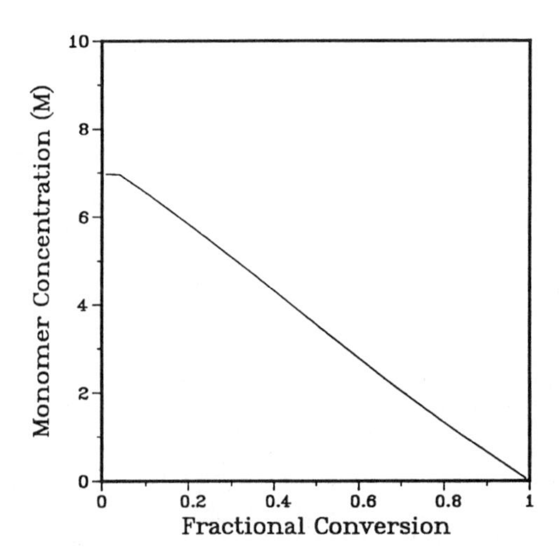

Figure 9: Simulation of styrene concentration in the polymer particles versus conversion for styrene microemulsion polymerization.

second radical to terminate the growing polymer chain is greater than that of chain transfer to monomer. For example, for the polymerization using 0.27 mM KPS at 70°C, the radical generation rate was 1.4×10^{13} radicals/ml/s, R_p was 7×10^{-4} mole/l/s, N_p was 2.3×10^{15} ml^{-1}, and $[M]_p$ was 5.5 M at 25% conversion. Since not all particles are growing, the growth rate of a polymer chain inside a particle ($R_p/\bar{n}N_p$) was calculated to be 1810 molecules/growing chain/s. Thus, it took 11 sec ($=X_n/1810$) for the polymer chain to grow to a molecular weight of 2×10^{-6} ($X_n=1.92 \times 10^4$) and then undergo chain transfer to monomer. The time scale for a chain transfer to styrene event, calculated from the literature value of the chain transfer constant (*14*), $C_M=0.3\text{-}0.6 \times 10^{-4}$, is in the range of 9 to 18 sec. The time needed for each particle to capture a radical can also be calculated. For the case where all radicals are captured by polymer particles, N_p divided by the radical production rate gives the time between successive entries of radicals, which is 164 sec. In fact, the time scale of radical capture by polymer particles should be much longer than 164 sec because of the competition for radicals by the mixed micelles. By comparing the time scales of radical capture by particles (»164 sec) and chain transfer to monomer (11 sec), it is apparent that before the entry of a second radical into the polymer particles, the polymer chain is terminated by chain transfer to monomer instead of bimolecular termination.

The monomer radicals generated by chain transfer can diffuse out of the polymer particles. It has been shown (*15*) that the exit rate of radicals increases with decreasing latex particle size. Therefore, in the polymerization of styrene microemulsions, the radical desorption rate from the particles is expected to be fast.

It is well known that radicals desorbed from polymer particles can have a major influence on the kinetics of emulsion polymerization. These radicals may re-enter the microemulsion droplets, the mixed micelles swollen with monomer, or the monomer swollen polymer particles, or propagate and terminate in the aqueous phase. It has been proposed that in seeded emulsion polymerizations of styrene, the desorbed monomer radicals terminate in the aqueous phase with primary radicals generated by decomposition of the initiator (*15*). For acrylate and methacrylate (*16*), or styrene with an organic solvent (*17*), these radicals re-enter the particles. It is shown in Figure 11 that the most probable fate of a desorbed monomeric radicals is to re-enter the particles or droplets in microemulsion polymerization. The lower curve represents the total number of polymer particles generated. The middle curve represents the total number of radicals generated by initiator decomposition. The upper curve represents the total number of polymer chains present in the polymer particles, which was calculated from the total number of polymer particles, N_p, and the average number of polymer chains per particle, n_p. It is interesting that the total number of polymer chains was greater than the total number of radicals generated from the initiator. This indicates that monomer radicals produced by chain transfer to monomer inside the particles do indeed desorb from the particles and then re-enter the particles or droplets to form new polymer chains.

Conclusions

Polymerization of styrene oil-in-water microemulsions using water-soluble potassium persulfate initiator gave stable latexes, which were bluish and less translucent than the original microemulsions. The kinetics of polymerization were different from those of conventional emulsion polymerization or miniemulsion polymerization. The polymerization rate versus conversion curves showed only two intervals, with neither a constant rate region nor a gel effect. The final latexes obtained were characterized by high molecular weights (2×10^6) and small average particle sizes (20 to 30 nm). Each latex particle contained only an average of two to three polystyrene molecules. The maximum rate of polymerization and the number of particles varied with the 0.47 and 0.40 powers of the potassium persulfate concentration.

The particle nucleation process in microemulsion polymerization was found to be continuous throughout the course of the polymerization. The termination of chain growth

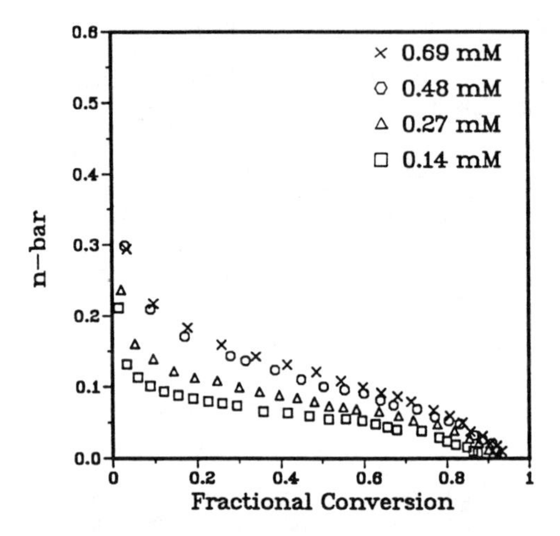

Figure 10: Average number of radicals per particle, $\bar{n}$, versus conversion with different KPS initiator concentrations for styrene microemulsion polymerizations at 70°C.

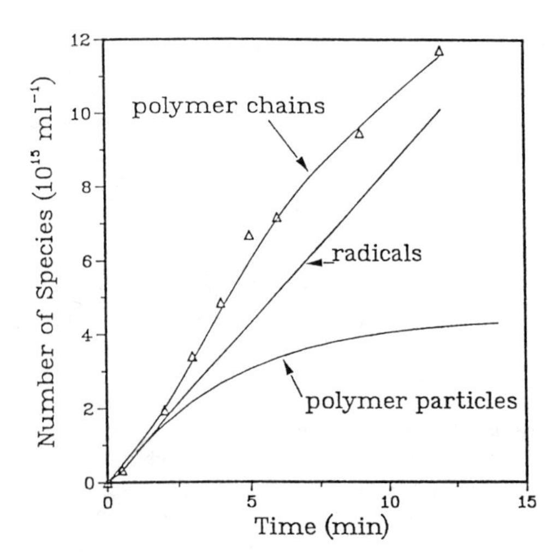

Figure 11: Comparison of the numbers of different species in the styrene microemulsion polymerization with 0.27 mM KPS at 70°C, where the number of radicals was calculated from k_I=4.4x10^{-5} sec^{-1}.

in polymer particles was attributed to the chain transfer reaction to monomer. The rate of desorption of monomer radicals from the small particles of this system is fast, which accounts for the low values of $\bar{n}$. The most probable fate of the desorbed radicals was re-entry into the particles or droplets. The average number of radicals per particle was found to decrease with increasing conversion and was always less than 0.5. Nucleation was postulated to occur in the microemulsion droplets (or mixed micelles after 4% conversion).

Literature Cited

1. Atik, S.S.; Thomas, J.K., *J. Am. Chem. Soc.*, **1981**, *103*, p 4279.
2. Johnson, P.L.; Gulari, E., *J. Polym. Sci., Polym. Chem. Ed.*, **1984**, *22*, p 3967.
3. Jayakrishnan, A.; Shah, D.O., *J. Polym. Sci., Polym. Lett. Ed.*, **1984**, *22*, p 31.
4. Gratzel, C.K.; Jirousek, M.; Gratzel, M., *Langmuir*, **1986**, *2*, p 292.
5. Kuo, P.L.; Turro, N.J.; Tseng, C.M.; El-Aasser, M.S.; Vanderhoff, J.W., *Macromolecules*, **1987**, *20*, p 1216.
6. Guo, J.S.; El-Aasser, M.S.; Vanderhoff, J.W., *J. Polym. Sci., Polym. Chem. Ed.*, **1989**, *27*, p 691.
7. Choi, Y.Y.; El-Aasser, M.S.; Sudol, E.D.; Vanderhoff, J.W., *J. Polym. Sci., Polym. Chem. Ed.*, **1985**, *23*, p 2973.
8. Chamberlain, B.J.; Napper, D.H.; Gilbert, R.G., *J. Chem. Soc., Faraday Trans. I*, **1982**, *78*, p 591.
9. Guo, M.S. El-Aasser, E.D. Sudol, H.J. Yue, and J.W. Vanderhoff, *J. Colloid Interface Sci.*, **140**, 175 (1990).
10. Candau, F.; Leong, Y.S.; Fitch, R.M., *J. Polym. Sci., Polym. Chem. Ed.*, **1985**, *23*, p 193.
11. Guo, J.S.; Sudol, E.D.; Vanderhoff J.W.; El-Aasser, M.S., *J. Polym. Sci. Polym. Chem. Ed.*, accepted.
12. Carver, M.T.; Hirsch, E.; Wittmann, J.C.; Fitch, R.M.; Candau, F., *J. Phys. Chem.*, **1989**, *93*, p 4867.
13. Carver, M.T.; Dreyer, U.; Knoesel, R.; Candau, F.; Fitch, R.M., *J. Polym. Sci., Polym. Chem. Ed.*, **1989**, *27*, p 2161.
14. Odian, G., *Principles of Polymerization*, 2nd Ed., Wiley-Interscience, New York, 1981.
15. Whang, B.C.Y.; Napper, D.H.; Ballard, M.J.; Gilbert, R.G., *J. Chem. Soc. Faraday Trans. I*, **1982**, *78*, p 1117.
16. Ballard, M.J.; Napper, D.H.; Gilbert, R.G., *J. Polym. Sci., Polym. Chem. Ed.*, **1984**, *22*, p 3225.
17. Lichti, G.; Sangster, D.F.; Whang, B.C.Y.; Napper, D.H.; Gilbert, R.G., *J. Chem. Soc. Faraday Trans. I*, **1984**, *80*, p 2911.

RECEIVED December 4, 1991

Chapter 8

Production of Vinyl Acetate–Butyl Acrylate Copolymer Latexes of Narrow Particle Size Distribution

Part 1, Effect of Reaction Variables

Gerald A. Vandezande and Alfred Rudin

Institute for Polymer Research, Department of Chemistry, University of Waterloo, Waterloo, Ontario N2L 3G1, Canada

There are many effects of reaction variables on the particle size and size distribution of vinyl acetate / butyl acrylate copolymer latexes. A monomer starved feed method was used to produce narrow particle size distribution (PSD) latexes of homogeneous composition and the effects of process parameters on particle size and size distribution of the latexes were examined. Increases in the ionic strength of the aqueous phase, using Na_2CO_3 buffer, and $(NH_4)_2S_2O_8$ initiator, increased the particle size and widened the particle size distribution of the latex. There appears to be a linear relationship between latex surface area and ionic strength of the aqueous phase in the ionic strength region studied. The batch addition of a fraction of the total pre-emulsion at the beginning of the reaction decreased the particle size and widened the size distribution of the particles. The method of initiator addition affects the conversion of the latex when a fixed amount of initiator is used. It was determined that the highest conversion was obtained by feeding the initiator, which is held at room temperature in an aqueous solution, concurrently with the monomer.

Most published studies of vinyl acetate polymerization have been investigations of the kinetics of batch homopolymerizations. The present work involves the semi-continuous emulsion copolymerization of a vinyl acetate / butyl acrylate (85/15 w/w) copolymer. This is the composition of the polymer used to make most latex paints in North America. Our focus has been on control of polymerization variables to produce latexes which had narrow PSD's and which had solids contents in the range of 40 - 60 (w/w) percent. This is a first stage in a study of the effects of particle size and size distribution on the properties of vinyl acrylic water-based coatings. While the thrust of the present work is different from the kinetic studies that have been reported to date, this different experimental approach has provided insights into the mechanism of vinyl acetate / butyl acrylate emulsion copolymerization under practical conditions. Both the polymerization protocol and mechanistic inferences are reported in this and a

0097–6156/92/0492–0114$06.00/0

companion paper. For the sake of clarity this article commences with a brief overview of the current ideas of the emulsion polymerization of vinyl acetate.

General Mechanism of Particle Formation.

Knowledge of the mechanism of emulsion polymerization has increased significantly over the years. The classical mechanism of emulsion polymerization applies to polymerization of a water insoluble monomer. The process is considered to occur in three intervals (*1,2,3*); interval I where nucleation takes place in surfactant micelles to produce particles, interval II where growth of these polymer particles takes place by the migration of monomer from monomer droplets to the particles, with subsequent polymerization of the monomer inside the particles, and interval III where no monomer droplets are left and polymerization takes place inside the polymer particles until all of the monomer has reacted. More currently, the coagulative nucleation theory, accepts the three interval theory of the classical model but the mechanism of formation of particles during interval I and the subsequent mechanism of polymerization of the monomer during intervals II and III have been modified.

Interval I is very important in the production of monodisperse particles since it controls the particle number and the size distribution of the particles in the latex. A short length of time for the completion of interval I facilitates production of a monodisperse latex since all of the particles have essentially the same growth period (*4*). Long interval I times produce polydisperse latexes. Smith and Ewart, in 1948, developed a mathematical model for interval I that was based on aqueous phase initiation (*5*). The end of interval I was determined as the time when the area of coverage of surfactant equalled the surface area of the particles.

Harkins elaborated on the classical model to predict the number of particles in a latex by presenting data on the effect of addition of salts to an emulsion (*6*). He noted that salts produced a screening effect which reduced the repulsion between the ionic ends of the surfactant molecules. The ultimate effect of increasing the ionic strength was the production of fewer micelles with more surfactant molecules per micelle, and thus fewer latex particles.

Priest (*7*), and subsequently Fitch (*8*), challenged the model of Smith, Ewart and Harkins since the classical model assumes the existence of water insoluble monomers. In Fitch's model for water soluble monomers the number of primary particles equals the number of chains initiated in solution minus the number of chains which had coagulated with each other. The resulting mathematics give the following simplified equation:

$$dN = R_i - R_c \qquad (1)$$

where dN is the instantaneous rate of change in the number of particles, R_i is the effective rate of initiation, and R_c is the overall rate of oligomeric capture of radicals by particles.

According to Fitch's experiments a maximum in particle number should develop shortly after the start of the reaction. These particles, called primary particles, contain oligomeric radicals initiated by ionic initiators in the aqueous phase, which have grown above a critical chain length and collapsed upon themselves. Coalescence of the primary particles with oligomeric radicals and with other primary particles takes place until the surface charge of the resulting particles increases to a point where mature latex particles are formed which are stable to coalescence (*9,10,11,12*). The particles continue to grow by monomer adsorption and coagulation with oligomeric radicals. This coagulative nucleation theory has been elaborated by Hansen and Ugelstad (*13*), Goodwin et al. (*14*), Lichti et al. (*15,16*), and others (*17,18*).

The surface charge of the primary and mature latex particles is produced by the ionic end groups of oligomers which have migrated to the surface of the particle (*15,19*). These form a double charged layer around the outside of the latex particle (*13*). Any anionic surfactant present adds to the stability of the particle by adsorbing onto the latex particle and increasing the surface charge density of the double layer. This double layer repels nearby particles by electrostatic repulsion preventing extensive coalescence.

Mechanism of Particle Formation in Vinyl Acetate Latexes. Dunn and Chong have used the coagulative nucleation theory in elucidating the batch polymerization mechanism of vinyl acetate (*20*). They produce evidence that the particle number is controlled by two factors: initiation rate and the coagulation rate in interval I, as described by Fitch (*8*). Aqueous phase oligomers are generated by initiation of the monomer in the aqueous phase. Once these have reached a critical length and are no longer soluble in the aqueous phase they collapse upon themselves and form primary particles. As the particle number increases, the oligomers collapse onto existing particles rather than precipitate to form new particles.

The theoretical rate of coagulation of oligomers and primary particles was calculated by Dunn and Chong through the application of the Deryaguin-Landau-Verwey-Overbeek (DLVO) theory of colloid stability. Zollars (*21*) points out that the results of the DLVO theory are most easily expressed in terms of the ratio of the rate of coagulation hindered by an electrostatic energy barrier to the rate of coagulation which would be observed without this hindrance (Fuch's stability ratio W). Particle stability increases with increasing surface potential or particle diameter and decreases exponentially with increasing ionic strength of the suspending medium.

Hydrolysis of Vinyl Acetate/Butyl Acrylate Latexes. When persulfate initiators are used in the production of vinyl acetate / butyl acrylate copolymer latexes, acids are produced which lower the pH as the reaction proceeds. This is a result of reaction of sulfate radicals with water to produce:

$$SO_4^{-\cdot} + H_2O \dashrightarrow HSO_4^- + HO\cdot \tag{2}$$

which decreases the pH of the solution (*20,22*). It has been reported that the decomposition rate for persulfates to form sulfuric acid is independent of pH above 3 but the decomposition increases rapidly below a pH of 3 (*22,23*). The increased acidity catalyses the hydrolysis reactions of vinyl acetate and butyl acrylate. Hydrolysis of poly(vinyl acetate) occurs at the carbonyl carbon to produce poly(vinyl alcohol) and acetic acid, while poly(butyl acrylate) hydrolyzes to form poly(acrylic acid) and butanol (*24*).

An additional mechanism of hydrolysis, discussed by Litt and Chang (*25*), includes the induced decomposition of persulfate initiator. Further hydrolysis may occur through the sulfate terminated polymer chains (*26*). At low pH's they can hydrolyze to form hydroxyl groups. Oxidizing conditions in the latex during polymerization can also oxidize hydroxyls to carboxylic acids (*16*).

Hydrolysis increases the hydrophilicity of the latex and decreases the pH. The use of buffers, typically $NaHCO_3$ (*27*) or Na_2CO_3, during the polymerization produces a latex with a neutral pH. This ensures that a latex is obtained with a minimum of hydrolysis.

Reactivity Ratios of Vinyl Acetate and Butyl Acrylate. The reactivity ratios of vinyl acetate and butyl acrylate are approximately 0.04 and 5.50, respectively (*28*), and therefore in batch polymerizations the butyl acrylate is polymerized much faster than the vinyl acetate, with the harder, more hydrophilic vinyl acetate on the exterior of the particle and the softer, more hydrophobic, butyl acrylate in the interior. To circumvent this problem, monomer starved feed methods are used to produce such latexes. A premix of the two monomers is slowly fed to the reactor during the reaction so that the faster reacting monomer concentration is constant in the reactor and thus a homogeneous particle is produced.

Studies have been done by other authors on vinyl acetate / butyl acrylate latex polymerizations but the ideas presented here deviate from other studies in that they deal with the synthesis of monodisperse latexes of 85/15 weight percent vinyl acetate / butyl acrylate, produced via a semibatch technique. The production of monodisperse latexes allowed for the study of the effects of the concentration and addition modes of buffer, initiator, and monomer, on the particle size and polydispersity of vinyl acetate / butyl acrylate copolymer latexes. The effects of such important production variables have not been studied systematically before, to the best of our knowledge.

Experimental

Latex Preparation. The polymerizations were carried out in a 1 liter reactor fitted with an overhead condenser and a water jacketed stirrer. The stirring rate was kept constant at 150 rpm. The position of the stirrer blades was adjusted throughout the reaction so that the paddle portion of the blade was immersed in only the top 2 centimeters of the reactor contents. This method of stirring helped to ensure that monomer did not pool on the surface of the reaction mixture. The temperature of the reactor contents was maintained at 80°C by partially immersing the reactor in a thermostated water bath.

A typical recipe is shown in Table I, Experiment 101. All ingredient quantities are listed in grams. The reactor was charged with deionized water and other ingredients except initiator. These ingredients were then heated in the water bath and purged with nitrogen for 30 minutes.

The monomer feed was composed of vinyl acetate and butyl acrylate at 85/15 weight percent. The monomers were obtained from Aldrich and used without further purification (vinyl acetate, 99+%, inhibited with 3-5 ppm of hydroquinone, and butyl acrylate, 99+%, inhibited with 10-50 ppm MEHQ). The monomers were mixed and then slowly added to a stirred solution of surfactant (DS10, sodium dodecylbenzene sulfonate, 98%, CMC 0.4 g/L, from Alcolac Inc.) and deionized water, to obtain an oil in water emulsion (pre-emulsion). The pre-emulsion was fed to the reactor over a period of 2.5 hours at a constant rate using an FMI Lab Pump, model RP-SY. No phase separation of the pre-emulsion was noticed over the feed period.

The initiator (ammonium persulfate from Aldrich Chemical Company Inc., 98+%) was added in two stages. The first stage was in the form of a batch addition just prior to the beginning of the addition of the monomer feed. The second stage initiator was dissolved in deionized water to form a concentrated solution of initiator. This initiator solution was also made up just prior to the start of the reaction and was added to the reactor concurrently with the monomer. The initiator was added using a Sage Instruments Model 355 syringe pump at a constant rate.

Table I. Effect of Na_2CO_3 on the particle size and size distribution of latexes

EXPERIMENT #	101	102	103	104
REACTOR CHARGE				
Deionized Water (g)	225.0	225.0	225.0	225.0
Na_2CO_3 (g)	0.0	0.10	0.50	2.50
Ammonium persulfate (g)	0.35	0.35	0.35	0.35
MONOMER FEED				
Vinyl Acetate (g)	180.0	180.0	180.0	180.0
Butyl Acrylate (g)	32.0	32.0	32.0	32.0
DS10 (g)	0.50	0.50	0.50	0.50
Deionized Water (g)	71.0	71.0	71.0	71.0
INITIATOR FEED				
Deionized Water (g)	20.0	20.0	20.0	20.0
Ammonium persulfate (g)	1.00	1.00	1.00	1.00
SIZE AND DISTRIBUTION				
D_n(nm.)	307	426	563	---
Polydispersity	1.034	1.035	1.037	---
Standard Deviation (nm.)	31.7	44.3	59.4	---
Coef. of Variation (%)	10.3	10.4	10.6	---

After the addition of all the ingredients the reaction mixture was heated for an additional 60 min. at 80°C. This ensured that the reaction was complete, 99.75%. The reactor contents were cooled and then filtered through a 100 mesh screen to remove any residual coagulum.

Latex Particle Size and Size Distribution Measurements. The particle size and size distribution of the latexes was measured using an ICI-Joyce Loebl Disk Centrifuge (*29, 30*). The output was recorded by a Brookhaven Instruments DCP-1000 Data System. Water was used as the spin fluid and the density gradient was produced using methanol. The latexes were diluted to 0.25 to 0.5 weight percent in a mixture of 80/20 water/methanol by weight.

The density of the latex, 1.17 g/cm^3, was calculated using a weighted average of the density of the two homopolymers. The rotational speed of the disc was selected so that the particles passed the detector after 1 minute. For the longer analysis times, the spin fluid was put into the disc and 1 mL of dodecane was added to the top of the spin fluid. This inhibited the evaporation of methanol from the top of the spin fluid and thus prevented the breakdown of the spin fluid gradient, especially on humid days.

Monomer Concentration Determination. Monomer concentration was determined using a Hewlett Packard 5890 Series II Gas Chromatograph, and recorded on a Hewlett Packard HP 3396A Integrator. Pressures for the air, hydrogen, and helium were 35, 17, and 40 psi, respectively. The oven temperature was 80°C, and the column used was a 6 foot, 10% carbowax 20M-TPA, Chromsorb W - DMCS, 60-80 mesh column. The detector was a flame ionization detector.

Samples were drawn from the reactor during and at the end of specific reactions, by syringe, through a septum. The latex was diluted 1/10 with deionized water and was injected directly into the gas chromatograph. The injector port contained a removeable glass tube to facilitate cleaning.

A standard monomer concentration curve was obtained by purging a latex, which had previously been diluted with water, with nitrogen until virtually no monomer was detected. A known quantity of vinyl acetate and butyl acrylate (85/15 weight percent) was added to this latex. The latex was successively diluted with water to obtain standards at lower monomer concentrations.

pH Measurements of the Latex. The pH of selected latexes was measured during the reaction to determine the change of pH with time. This was accomplished by drawing small aliquots of latex from the reactor and using wide range pH paper (pH 1-12 from Anachemia) to test the pH.

The pH was also measured in the final product. The pH of these latexes were measured using a Corning Model 610A Expand Portable pH Meter.

Polymer Characterization using NMR. Fourier transform ^{13}C-nuclear magnetic resonance (NMR) spectra were obtained using a Bruker AC 300 NMR Spectrometer. The spectra were recorded at 20°C. The copolymer samples were dried at room temperature and dissolved in a 50 weight percent solution of deuterated benzene and tetrachloroethylene to a solids concentration of 10 percent. The spectra were referenced to deuterated benzene. A 90° pulse width was used, with a delay time of 12 seconds. Each sample was analyzed using approximately 2100 scans. The spectra were analyzed essentially by the method of Pichot, Llauro, and Pham (*28*). This method confirms that a homogeneous copolymer was obtained. Such a product is possible only if the reaction was monomer starve fed, because of the wide difference in reactivity ratios of vinyl acetate and butyl acrylate.

Results and Discussion

Initial Narrow PSD Latex. The initial narrow PSD recipe that was developed, listed in Table I Experiment 101, consisted of deionized water and initiator ($(NH_4)_2S_2O_8$) charged to the reactor before the addition of any other ingredients. The pre-emulsion of monomer, water, and surfactant (DS10) was added slowly over a period of 2.5 hrs. to maintain starve feed conditions. A pre-emulsion was prepared to facilitate adsorption of monomer into the water phase and thus increase the rate of reaction. The pre-emulsion contained very little surfactant in order to keep the surfactant concentration low in the reactor. The low concentration of surfactant in the reactor also helped to ensure a short nucleation period and aided in reducing the possibility of second generation of particles (*31,32*). The second stage initiator was added concurrently with the monomer as an aqueous solution.

Conversion vs. Time. The conversion of monomer to polymer during the reaction is shown in Figure 1. The first curve labelled "Conversion in reactor" is the conversion of monomer which has been added up to a particular time. The second curve labelled "Total conversion" is the fraction of the whole monomer

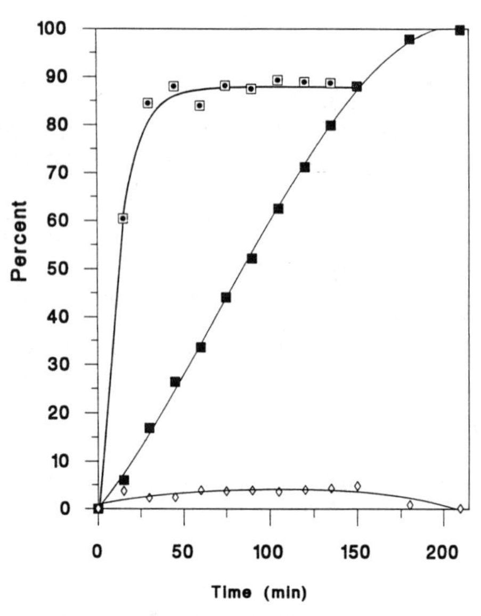

Figure 1. Conversion vs. time for experiment 101.

charge that has been converted to polymer at the given time. (Obviously, this increases with time since all the monomer is continually being fed into the reactor). The third curve labelled "Concentration of monomer" is the monomer concentration in the latex at any time. These results show that the reaction was starve fed since the conversion was at 90% throughout the reaction and therefore no drift in polymer composition occured. This was confirmed by NMR analysis. The reaction was completed at 99.75%.

Ionic Strength Effects.

The Effects of Na_2CO_3 on the Particle Size and Size Distribution of Latexes. A buffer, in this case Na_2CO_3, is needed in the production of vinyl acetate / butyl acrylate copolymer latexes to reduce hydrolysis, as mentioned earlier. In Table I the effects of increasing the Na_2CO_3 concentration on the particle size and distribution are presented (Experiments 101 to 104). As the Na_2CO_3 concentration was increased, the particle size increased until gross coagulation occurred at the highest ionic strength.

The increase in particle size with increases in the ionic strength of the aqueous phase was expected since the effect of changing the ionic strength on the particle size in vinyl acetate and other latexes, produced by a batch process, has been discussed (*4,6,7,14,20,21,24,33,34,35,36*). It was also expected that at high ionic strengths gross coagulation of the latex would occur (*21*).

At lower ionic strengths, coagulation of the primary particles apparently occurred during the nucleation period until the surface charge density produced by the ionic end groups of the collapsed oligomers had increased to the point where the particle became stable. Raising the ionic strength increased the coagulation rate during the nucleation period. Because of this, fewer, but larger particles were obtained.

The polydispersity (defined as D_w/D_n) appears to have remained constant with increases in Na_2CO_3 concentration. Suzuki, Ho, and Higuchi (*37*) state that when an electrical barrier is small, any initial distribution of particles in an emulsion or suspension becomes more polydisperse with time. Thus, when applied to latexes, an increase in polydispersity with increases in Na_2CO_3 concentration would be expected since increasing the ionic strength during the nucleation period decreases the electrical barrier to coagulation of charged particles. Also, in latex polymerizations, collisions between oligomeric radicals to produce primary particles are random and therefore it is expected that not all primary particles will contain the same number of oligomers. For the same reason, all latex particles are not expected to be comprised of the same number of primary particles. Thus an increase in coagulation events, such as that found in higher ionic strength systems, is expected to broaden the particle size distribution.

Under the conditions presented in this paper the nucleation period was apparently very short compared to the particle growth period, since rather narrow particle size distributions were obtained. During the growth period the larger particles need to accumulate more polymer to increase their radius by the same proportion as the smaller particles. This tendency allows the smaller particles to grow more rapidly than larger ones and thus the particle size should tend to become more uniform (*4*). For this reason the effects of ionic strength on the particle size distribution may not be as pronounced as those in longer nucleating systems.

It should also be noted that when the smaller particles are increased in size to that of the larger particles via seeded reactions, the resultant particles have a more narrow PSD (Table II, experimental conditions provided in the companion paper "Effects of Reaction Variables in the Production of Narrow

Particle Size Distribution Vinyl Acetate/Butyl Acrylate Copolymer Latexes(II) - Seeded Reactions"). Thus, when latexes are compared at equivalent particle sizes the polydispersity increases with increasing ionic strength of the aqueous phase.

When adding Na_2CO_3 the pH as well as the ionic strength of the latexes increases. It may be argued that the change in pH may contribute to the change in particle size and polydispersity. The polymer hydrolyzes at low pH's, and thus poly (vinyl alcohol), which functions as a stabilizer, may be produced and the particles are stabilized at smaller particle sizes. Oxidizing conditions in the latex during polymerization can also oxidize hydroxyl groups to carboxylic acids. This process may be offset by the hydrolysis of persulfate end groups. The pH of a series of latexes were measured during the reaction (Table III). Upon examination of the experiments it can be seen that at no time is the pH of the reaction mixture below a value of 5.5 when no Na_2CO_3 is present during the nucleation period (Table III and Experiment 105, Na_2CO_3 added at 30 minutes). Therefore, since the hydrolysis reaction is acid catalyzed (*24*), and the pH only dropped to a value of 5.5, it is assumed that no significant hydrolysis will have taken place during the nucleation period. This leads to the conclusion that the effect of changing particle size and distribution is strictly due to the ionic strength in the reaction mixture and not due to hydrolysis.

Determination of the End of the Nucleation Period. In Table IV the effects of changing the addition time of the Na_2CO_3 are presented. All experiments are identical to experiment 103 except for the addition time of Na_2CO_3. When the Na_2CO_3 was added at the beginning a narrow PSD product was obtained (Experiment 103). When the time of addition was delayed to 30 minutes after the start of the reaction, (Experiment 105) the particle size decreased and distribution increased. The decrease in particle size is attributed to the reduced ionic strength during the nucleation period which reduced the frequency of the initial coagulation events. The increase in the PSD may be due to coagulation of the particles which had not acquired sufficient sulfate ended polymer chains and surfactant to resist coagulation upon the addition of the Na_2CO_3.

Table II. Particle size and size distributions of seeded reactions with 0.07g DS10 in the pre-emulsion. Seeded polymerization from 206 nm latex particles (Latex 124)

Latex	Size (nm.)	Poly-dispersity (nm)	Standard Deviation (%)	Coefficient of Variation
124	206	1.035	21.7	10.5
125	305	1.033	30.6	10.0
126	454	1.029	43.5	9.6
127	690	1.025	61.4	8.9
128	1050	1.010	61.6	5.9

Table III. The pH of selected latexes during the reaction period

Experiment **Time (min)**	**102**	**103** **pH**	**104**	**105**
15	8.0	8.5	9.0	6.0
30	-	-	-	6.0
31	-	-	-	9.0
60	-	-	-	-
90	-	7.5	8.0	6.5
120	-	6.5	-	-
product	3.7	4.7	-	6.0

When the addition time of the Na_2CO_3 was delayed to 45 minutes after the start of the reaction (Experiment 106) the particle size did not change significantly as compared to Experiment 101. Thus, a significant portion of the particles may have acquired sufficient sulfate ended polymer chains and surfactant to resist coagulation upon the addition of the Na_2CO_3. The PSD of Experiment 106 is narrower than the distribution in Experiment 101. A possible explanation is that the smaller particles in the distribution were less resistant to coagulation than the larger ones since the concentration of sulfate ended polymer chains and surfactant on their surface is lower than that of larger particles. Goodwin and coworkers (*14*) have developed a simple model for persulfate-initiated, emulsifier free polystyrene latexes in which the surface charge densities of sulfate ions is seen to be directly proportional to the particle radius. There appears to be no reason why this useful model should not also apply here to vinyl acetate / butyl acrylate copolymerizations. Thus, it is possible that the smaller particles had coagulated upon the addition of Na_2CO_3, but the larger particles, being slightly more stable, may not have experienced any coagulation and thus a more narrow PSD was obtained.

It appears that under these reaction conditions, the nucleation period ends before 45 minutes into the reaction and even before 30 minutes since the particle sizes are virtually identical before and after the addition of Na_2CO_3 (Experiments 105 and 106).

The observation of the reaction mixture in the reactor during the initial stages of the reaction in Experiment 101 can give further insight into the length of interval I. Initially there was only water and initiator in the reactor. As the pre-emulsion and initiator was pumped into the reactor the contents of the reactor turned cloudy. At approximately 7 minutes after the start of the addition of the pre-emulsion, the mixture turned to a clear, slightly blue solution in approximately 30 seconds. It is suggested that this cloudy/clear transition indicated the formation of a separate polymer phase and may conclude with the end of the coagulation period (*38*). Since a monodisperse latex is eventually obtained, the nucleation period is apparently brief and thus interval I is expected to end shortly after the cloudy/clear transition.

In the higher ionic strength systems reported in this paper the cloudy/clear transition was sometimes difficult to distinguish. This may be due to the large particles formed as a result of the extensive coagulation of higher ionic strength systems. These large particles refract light more efficiently than smaller ones and thus the solution remained cloudy.

Table IV. The effects of changing the addition time of Na_2CO_3

EXPERIMENT #	103	105	106	101
BUFFER				
Deionized Water (g)	10.0	10.0	10.0	10.0
Na_2CO_3 (g)	0.50	0.50	0.50	0.0
Time of addition of Na_2CO_3 (min.)	0	30	45	---
Final pH	4.7	6.0	---	1.9
SIZE AND DISTRIBUTION				
D_n(nm.)	563	338	309	307
Polydispersity	1.037	1.049	1.024	1.034
Standard Deviation (nm)	59.4	40.4	26.3	31.7
Coef. of Variation (%)	10.6	11.9	8.5	10.3

The Effects of Changing the Initiator Concentration on Particle Size. The ionic strength at the beginning of the reaction is important to the particle size and distribution of the latex. The results in Table II were discussed with respect to the changes in ionic strength by the introduction of Na_2CO_3. Table V presents data which show the effect of changing the ammonium persulfate concentration (and thus also the ionic strength) on particle size and size distribution.

The first experiment on Table V (Experiment 101) was performed using a pre-charge of initiator. In the second Experiment (110) was performed as Experiment 101 except the pre-charge of initiator was removed and added to the initiator fed concurrently with the pre-emulsion. Thus, in the second case no initiator was present in the reactor at the start of the reaction but the initiator concentration was slowly increased as the reaction proceeded. In the third Experiment (111), the initiator feed concentration was also reduced; thus the initiator concentration in the reactor was increased more slowly. Significant further reductions in the initiator concentrations did not produce stable latexes since low surface charge densities produce unstable particles, as discussed by Guillaume (*39*).

The particle size was greatly reduced as a result of decreasing the initiator pre-charge. This may be due to a decrease in ionic strength at the beginning of the reaction, which is similar to the reduction of Na_2CO_3 discussed earlier. The decrease in ionic strength allows for greater stability of primary particles. Since the particles are more stable, less coalescence occurs and more particles are produced. This results in a smaller particle size of the final latex.

The decrease in initiator concentration decreases the ionic strength, but also decreases the rate of production of radicals for initiation. Hansen et al. (*40,41*) proposes that the decrease in radical generation rate in a batch system should result in a decrease in surface charge density since fewer oligomers are being generated and thus the size of the particles will increase.

Table V. The effects of changing the initiator concentration on particle size and size distribution

EXPERIMENT #	101	110	111
REACTOR CHARGE			
Deionized Water (g)	225.0	225.0	225.0
Na_2CO_3 (g)	0.0	0.0	0.0
Ammonium persulfate (g)	0.35	0.0	0.0
INITIATOR FEED			
Deionized Water (g)	20.0	20.0	20.0
Ammonium persulfate (g)	1.00	1.35	0.50
SIZE AND DISTRIBUTION			
D_n(nm.)	307	187	159
Polydispersity	1.034	1.032	1.031
Standard Deviation (nm)	31.7	19.0	16.2
Coef. of Variation (%)	10.3	10.2	10.2

This aspect was noted by Dunn and Chong (*20*), who obtained data for a series of vinyl acetate latexes produced by varying the initiator concentration. They kept the ionic strength of the latex constant by the addition of NaCl. The total surface area of the latex increased linearly with increasing initiator concentration. This is expected since the total surface area is a reflection of the concentration of ionic groups on the surface of the polymer particle during the nucleation period. Each ionic end group corresponds to a specific surface area of polymer at a specific ion concentration. When the initiator concentration is increased, the number of ionic groups on the surface increase and therefore there should be a corresponding increase in latex surface area.

In the system studied in this paper it appears that the initiator concentration has a significant effect on the ionic strength of the medium but less effect on the surface charge density of the latexes produced during the nucleation period, since the net effect of added initiator is to increase the particle size.

The Effects of Varying the Ionic Strength on the Particle Size, Particle Number and Surface Area per Particle of the Latex. The data presented in Table VI are the effects of varying the ionic strength by varying the initiator and the Na_2CO_3 concentration on the particle number, particle size, particle surface area, and polydispersity, standard deviation, and coefficent of variation.

The variation of particle size with ionic strength has been demonstrated elsewhere (*4,6,7,14,20,24,33,34,35,36*), but in each case where actual experimental data are presented, only the particle number or the particle diameter is recorded (*4,7,14,21,24,33*), and in some cases log/log graphs are constructed to produce a straight line. It may be that these relationships are in fact not logarithmic.

Table VI. The effects of varying the ion concentration, by varying the initiator and the Na_2CO_3 concentration

EXPERIMENT #	103	102	101	110	111
REACTOR CHARGE					
Na_2CO_3 (g)	0.50	0.10	0	0	0
Ammonium persulfate (g)	0.35	0.35	0.35	0	0
INITIATOR FEED					
Deionized water (g)	20.0	20.0	20.0	20.0	20.0
Ammonium persulfate (g)	1.00	1.00	1.00	1.35	0.50
Ion concentration at 7 minutes (moles/litre)	0.091	0.041	0.028	0.0055	0.0020
Ionic strength at 7 minutes (moles/litre)	0.091	0.041	0.028	0.0055	0.0020
# of particles/liter (x 10^{16})	.0830	.191	.513	2.23	3.66
Diameter (D_n microns)	.563	.426	.307	.187	.159
Surface area/particle (microns^2)	.498	.285	.148	.0555	.0397

It is obvious that the ion concentration in the aqueous phase has a powerful influence on the surface area of the latex particle. It should noted that the ionic strength varies with time in the experiments presented here since initiator is being fed into the reactor. Thus, the ionic strength at the end of the reaction period will be greater than at the beginning. It appears then that the natural plot of the effects of ionic strength of the aqueous phase on the latex particle should be a plot of the surface area per particle of the latex at the end of the nucleation period versus the ionic strength at the end of the nucleation period. This is plausible since during the nucleation period the primary particles coalesce until they achieve a sufficient surface charge density in the form of ionic, initiator-ended polymer chains and adsorbed surfactant to form stable latex particles.

It is very difficult to obtain particle data at the end of the nucleation period due to the difficulty in accurately measuring the radius of the latex particles, by electron microscopy, light scattering, or disc centrifugation. Thus, a plot of surface area per particle at the end of the nucleation period versus ionic strength of the aqueous phase at the end of the nucleation period may not be very accurate. A logical substitute is to plot the surface area per particle of the completed latex versus ionic strength at the end of the nucleation period.

Plots of the particle number, particle diameter, and the surface area per particle versus the ionic strength of the aqueous phase during the nucleation period, (assumed equal to 7 minutes) are illustrated in Figures 2, 3, and 4, respectively. The plot of the final surface area per particle versus ionic strength produces the best linear relation as determined by the correlation factor, R^2 (Figure 3, $R^2 = .95$; Figure 4, $R^2 = .98$). The slope of the graph of surface area

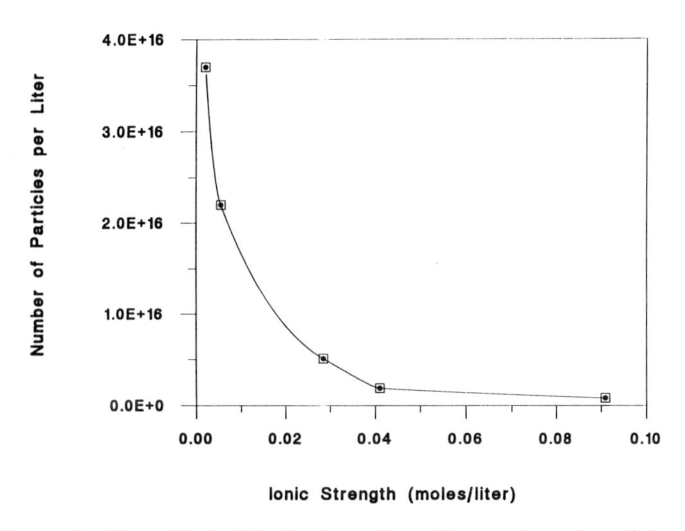

Figure 2. The effect of ionic strength on the number of particles.

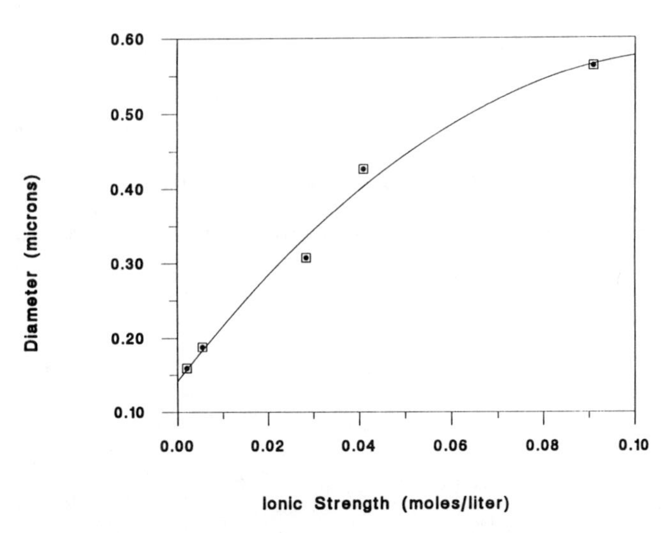

Figure 3. The effect of ionic strength on the diameter of the particles.

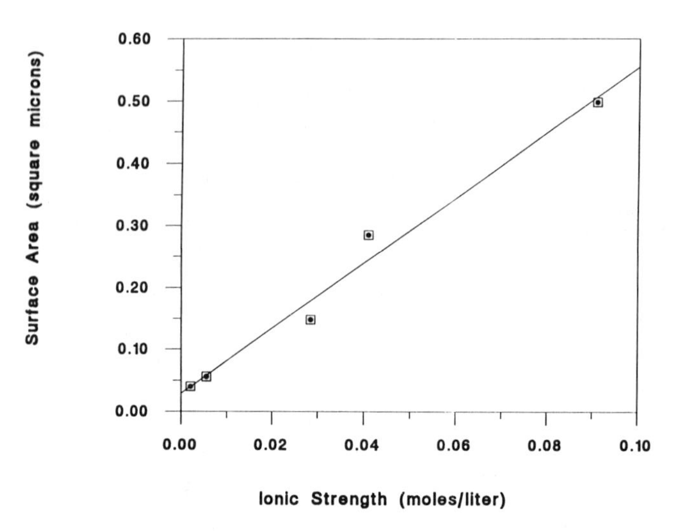

Figure 4. The effect of ionic strength on the surface area per particle.

versus ionic strength of the aqueous phase is expected to change depending on the latex produced, since van der Waals forces and repulsive forces of the polymer and ionic groups, respectively, will change with polymer, surfactant, and initiator type. A positive slope is expected since increases in the ionic strength decreases the repulsive effect of the double layer on the particles. This causes greater coagulation during the nucleation period and therefore fewer but larger particles are produced.

Ingredient Addition Rate Effects.

The Effects of a Batch Addition of a Pre-emulsion on the Particle Size and Distribution. The effects of a batch addition of a pre-emulsion on the particle size and distribution are presented in Table VII. Only the pre-emulsion pre-charge was added to the reactor with the pre-charge of the initiator and water. The rest of the monomer pre-emulsion was not added. As the pre-emulsion pre-charge was increased, the particle size increased.

The theoretical final particle size, had all of the pre-emulsion been slowly added to the reactor after the pre-charge, was calculated using the following equation (*42*);

$$D_p = D_s \left[(W_s + W_1)/ W_s \right]^{1/3}$$

where D_s = seed particle
D_p = projected diameter for the final particle
W_s = seed polymer weight fraction
W_1 = weight fraction polymer to be added on.

Increases in the pre-emulsion pre-charge decreased the theoretical final particle size. The decreasing size with increasing pre-charge is expected since more surfactant has been added at the beginning, stabilizing the initial particles and the initiator has been more diluted lowering the ionic strength. Also, since butyl acrylate reacts faster than vinyl acetate, hydrophobic oligomers were created with shorter critical chain lengths in the aqueous phase. Since the chains were shorter, the particles underwent fewer coagulation events to create stable particles and thus smaller particles were created. The more hydrophobic primary particles also adsorb surfactant more readily (*24*), thus increasing the stability of the primary particles and generating more and smaller particles.

As noted earlier particle size distributions narrow when seeded reactions are performed. Had seed reactions been done with latexes from Experiment 108 & 109 to increase their size to that of the latex from Experiment 107, their particle size distribution would have narrowed. Thus, the particle size distribution was broader for larger batch additions of pre-emulsion than for smaller batch additions when compared at equal particle sizes. This is attributed to a higher surfactant concentration in the reactor during the nucleation period, increasing the length of interval I, and also refluxing of monomer due to larger batch additions of monomer. It appears that excess monomer may have induced coagulation of particles, thus increasing the PSD.

The Effect of Initiator Addition Methods on the Monomer Conversion of the Latex. Significant concentrations of monomer during the reaction causes the polymer composition in the latex particle to drift. Since vinyl acetate monomer is soluble in the polymer, monomer will plasticize the polymer and thus reduce its minimum film forming temperature. Measurements of the latex particle size on the disk centrifuge rely on accurate polymer density values. The density of the polymer will change with changes in monomer content since the polymer is

Table VII. The effects of a batch addition of a pre-emulsion on the particle size and distribution

EXPERIMENT #	107	108	109
REACTOR CHARGE			
Deionized Water (g)	225.0	225.0	225.0
Na_2CO_3 (g)	0.0	0.0	0.0
Ammonium persulfate (g)	0.52	0.52	0.52
PRE-EMULSION PRE-CHARGE			
wt % of total pre-emulsion	11	22	44
SIZE AND DISTRIBUTION			
D_n(nm.)	112	119	133
Theoretical Final Diameter D_n (nm.)	234	197	174
Polydispersity	1.062	1.065	1.065
Standard Deviation (nm)	15.2	16.2	18.5
Coef. of Variation (%)	13.6	13.6	13.9

more dense than the monomer. Therefore, to obtain accurate particle size measurements and homogeneous particles, the reactions must be starve fed and complete.

Table VIII contains data on the effect of initiator addition methods on the monomer conversion of the latex. Initially a latex was produced by the addition of all of the initiator to the reactor at the beginning of the reaction. (Ingredients as Experiment 101 except for initiator.) The final conversion was poor since monomer was pooling on top of the reactor contents. The low conversion may have been due to the half life of ammonium persulfate at 80°C being 2.8 hours, and perhaps much less in the presence of monomer and impurities (*43*).

In the second experiment, most of the initiator was added to the pre-emulsion. The pre-emulsion, which was held at 20°C, was added throughout the reaction. The half life of the initiator was increased and thus the initiator was at a higher concentration at the end of the reaction. This gave a more complete reaction (no visible monomer pooling, 95 percent conversion as analysed by gas chromatography) although the result was still not satisfactory. It appears that the initiator decomposes significantly in the presence of monomer and surfactant even at room temperature. Interestingly no polymer was detected in the pre-emulsion using gravimetric analysis. Significant polymerization was probably inhibited by dissolved oxygen.

In the third experiment most of the initiator was dissolved in deionized water and kept at 20°C. It was added concurrently with the pre-emulsion. This gave an essentially complete reaction. The conversion was 99.75 percent.

Table VIII. The effect of initiator addition methods on the monomer conversion of the latex

EXPERIMENT #	114	115	101
INITIATOR PRE-CHARGE			
Ammonium persulfate	---	0.35	0.35
INITIATOR FEED			
Deionized Water (g)	20.0	20.0	20.0
Ammonium persulfate (g)	1.35	1.00	1.00
Time of addition (min.)	0	continuous in pre-emulsion	continuous separately
Conversion (%)	< 95	95	99.75

Conclusions

Within the constraints of the systems studied here, the influences of reaction variables on the particle size and size distribution of vinyl acetate / butyl acrylate copolymer latexes are many. Increasing the ionic strength of the aqueous phase via additions of Na_2CO_3 decreases the electrical energy of repulsion which in turn increases the particle size and widens the particle size distribution of a latex in the ionic strength region studied. When ammonium persulfate initiator was used to change the ionic strength the same trends were noticed in particle size and distribution. It also appears that there is a linear relationship of surface area to ionic strength of the aqueous phase in the ionic strength range studied. The batch addition of a fraction of the total pre-emulsion at the beginning of the reaction affects the latex by decreasing the size and increasing the size distribution of the particles. Initiator addition methods affected the conversion of the latex. The best method to ensure a complete reaction is to feed the initiator, which is held at room temperature in a water solution, concurrently with the monomer.

Acknowledgments

This research was supported by the Natural Sciences and Engineering Research Council of Canada.

Literature Cited

1 Gardon, J.L. *J. Polym. Sci., Polym. Chem. Ed.* **1968**, *6*, 643.
2 Gardon, J.L. *J. Polym. Sci., Polym. Chem. Ed.* **1968**, *6*, 623.
3 Gardon, J.L. *Br. Polym. J.* **1970**, *2*, 1.
4 Goodwin, J.W.; Hearn, J.; Ho, C.C.; Ottewill, R.H. *Br. Polym. J.* **1973**, *5*, 347.
5 Smith, W.V.; Ewart, R.H. *J. Chem. Phys.* **1948**, *16*, 592.
6 Harkins, W.D. *J. Polym. Sci.* **1950**, *5*, 217.
7 Priest, W.J. *J. Phys. Chem. Soc.*, **1952**, *56*, 1077.

8 Fitch, R.M.; Tsai, C.H. "Polymer Colloids", Plenum Press, New York, 1971
9 Goodall, A.R.; Wilkinson, M.C.; Hearn, J. *J. Polym. Sci., Polym. Chem. Ed.* **1977**, *15*, 2193.
10 Feeney, P.J.; Napper, D.H.; Gilbert, R.G. *Macromolecules* **1984**, *17*, 2520.
11 Song, Z.; Poehlein, G.W. *J. Macromol. Sci. Chem.* **1988**, *4*, 403.
12 Chern, C.; Poehlein, G.W. *J. Polym. Sci., Polym. Chem. Ed.* **1987**, *25*, 617.
13 Hansen, F.J.; Ugelstad, J. *J. Polym. Sci., Polym. Chem. Ed.* **1978**, *16*, 1953.
14 Goodwin, J.W.; Ottewill, R.H.; Pelton, R.; Vianello, G.; Yates, D.E. *Br. Polym. J.* **1978**, *10*, 173.
15 Lichti, G.; Gilbert, R.G.; Napper, D.H. *J. Polym. Sci., Polym. Chem. Ed.* **1983**, *21*, 269.
16 Lichti, G.; Gilbert, R.G.; Napper, D.H. *J. Polym. Sci., Polym. Chem. Ed.* **1977**, *15*, 1957.
17 Feeney, P.J.; Napper, D.H.; Gilbert, R.G. *J. Coll. Interface Sci.* **1985**, *107*, 553.
18 Dunn, A.S.; *Makromol. Chem., Suppl.* **1985**, *10/11*, 1.
19 Goodall, A.R.; Hearn, J. *J. Polym. Sci., Polym. Chem. Ed.*, **1979**, 1019.
20 Dunn, A.S.; Chong, L.C. *Br. Polym. J.* **1970**, *2*, 49.
21 Zollars, R.L. *J. Appl. Polym. Sci.* **1979**, *24*, 1353.
22 Okubo, M.; Mori, T. *Colloid and Polymer Science* **1988**, *266*, 333.
23 Kolthoff, I.M.; Miller, I.K. *J. Am. Chem. Soc.* **1951**, *73*, 3055.
24 El-Aasser, M.S.; Makgawinata, T.; Misra, S. In *Emulsion Polymerization of Vinyl Acetate*, El-Aasser, M.S., Ed.; Vanderhoff, J.W., Ed. Applied Science Publishers, New Jersey, 1981, p. 240.
25 Litt, M.H.; Chang, K.H.S. In *Emulsion Polymerization of Vinyl Acetate*, El-Asser, M.S., Ed.; Vanderhoff, J.W., Ed. Applied Science Publishers, New Jersey, 1981, p. 151.
26 Goodall, A.R.; Hearn, J.; Wilkinson, M.C. *Br. Polym. J.* **1978**, *10*, 141.
27 Nguyen, B.D. Ph.D. Thesis, University of Waterloo, Waterloo, Ontario, 1986.
28 Pichot, C.; Llauro, M.; Pham, Q. *J. Polym. Sci., Polym. Chem. Ed.* **1981**, *19*, 2619.
29 Devon, M.J.; Provder, T.; Rudin, A. In *Particle Size Distribution II*, Provder, T., Ed.; American Chemical Society, Washington, DC, 1991, p. 134.
30 Hansen, F.K. In *Particle Size Distribution II*, Provder, T., Ed.; American Chemical Society, Washington, DC, 1991, p. 134.
31 Sudol, E.D.; El-Aasser, M.S.; Vanderhoff, J.W. *J. Polym. Sci., Polym. Chem. Ed.* **1986**, *24*, 3499.
32 Sudol, E.D.; El-Aasser, M.S.; Vanderhoff, J.W. *J. Polym. Sci., Polym. Chem. Ed.*, **1986** *24*, 3515.
33 Harkins, W.D. *J. Am. Chem. Soc.* **1947**, *69*, 1428.
34 Goodwin, J.W.; Hearn, J.; Ho, C.C.; Ottewill, R.H. *Colloid Polym. Sci.* **1974**, *252*, 464.
35 Chang, H.; Chen, S. *J. Polym. Sci., Polym. Chem. Ed.* **1988**, *26*, 1207.
36 Dunn, A.S.; Al-Shahib, W.A.G.R. *Br. Polym. J.* **1978**, *10*, 137.
37 Suzuki, A.; Ho, N.F.H.; Higuchi, W.I. *J. Coll. Interface Sci.* **1969**, *29*, 553.
38 Brouwer, W.M. *J. Appl. Polym. Sci.* **1989**, *38*, 1335.
39 Guillaume, J.L.; Pichot, C.; Guillot, J. *J. Polym. Sci., Polym. Chem. Ed.* **1988**, *26*, 1937.
40 Hansen, F.K.; Ugelstad, J. *J. Polym. Sci., Polym. Chem. Ed.* **1979**, *17*, 3047.

41 Hansen, F.K.; Ugelstad, J. *Makromol. Chem.* **1979**, *180*, 2423.
42 Devon, M.J. Ph.D. Thesis, University of Waterloo, Waterloo, Ontario, 1987.
43 Chatterjee, S.P.; Banerjee, M. *J. Polym. Sci., Polym. Chem. Ed.* **1978**, *16*, 1517.

RECEIVED December 4, 1991

Chapter 9

Production of Vinyl Acetate–Butyl Acrylate Copolymer Latexes of Narrow Particle Size Distribution

Part 2, Effect of Reaction Variables in Seeded Reactions

Gerald A. Vandezande and Alfred Rudin

Institute for Polymer Research, Department of Chemistry, University of Waterloo, Waterloo, Ontario N2L 3G1, Canada

Seeded reactions using a monomer starved feed method were used to produce a range of narrow particle size distribution (PSD) 85/15 weight percent vinyl acetate / butyl acrylate copolymer latexes at 40 and 55 percent solids. Using ammonium persulfate as the initiator and sodium dodecylbenzene sulfonate (DS-10) as surfactant, a 40 percent solids latex was produced. Seeded reactions produced narrow PSD latexes with diameters between 194 to 1370 nm.

Seeded reactions were also used to produce a range of narrow PSD particle sizes at 55 percent solids with diameters from 250 to 800 nm. These latexes were produced by maintaining the seed latex at 55 percent solids throughout the reaction. Seeded reactions are difficult to carry out due to the need to use nonionic as well as anionic surfactants in the polymerization.

Experiments show that a significant factor in delaying the production of second generation particles is the concentration of the seed latex. Further studies indicate that the ionic strength in the aqueous phase of the seed latex can also be an important factor in delaying the formation of second generation particles.

Latexes of vinyl acetate copolymers are used in a variety of applications such as adhesives and architectural coatings. The particle size and size distribution of the latexes may affect the end use properties of the product. Narrow PSD latexes of vinyl acetate / butyl acrylate copolymers are being studied in this connection in our laboratory.

Various parameters that affect the particle size and size distributions of 85/15 weight percent vinyl acetate / butyl acrylate latexes were discussed in Chapter 8, "Production of Vinyl Acetate–Butyl Acrylate Copolymer Latexes of Narrow Particle Size Distribution: Part 1, Effect of Reaction Variables". Within the constraints of the systems studied, the influences of these parameters on the particle size and size distribution were as follows. Increasing the ionic strength of the aqueous phase via additions of Na_2CO_3 buffer and ammonium persulfate initiator, increases the particle size and widens the particle size

0097–6156/92/0492–0134$06.00/0

distribution of a latex. Buffers are needed to offset the acidity generated by persulfate initiators (*1-8*).

The reactivity ratios of vinyl acetate and butyl acrylate are approximately 0.04 and 5.50, respectively (*9*), and therefore in batch polymerizations heterogeneous copolymers are produced, which is an undesirable effect. As a result of this, monomer starved feed methods are used to produce homogeneous latexes. Batch additions of a fraction of the total pre-emulsion at the beginning of the reaction affects the latex by decreasing the size and increasing the size distribution of the particles.

Larger particles may be produced via monomer starved feed seeded reactions. Seeded reactions require introduction of a preformed latex into the reactor. Monomer is slowly added concurrently with initiator so that the existing particles are increased in size by continual polymerization of monomer in the particle. This results in a narrow PSD latex and an increase in particle size.

In the production of copolymers of vinyl acetate and butyl acrylate a variety of stabilizing agents are employed, the most common ones being sulfonated dodecyl phenols as ionic surfactants, and ethoxylated alkyl phenols as nonionic or steric stabilizers (*8*). The type of initiator affects the stability of the latex particles (*10*). Anionic initiators such as ammonium persulfate produce anionically ended oligomeric radicals. When these oligomers are incorporated into the polymer particle they add stability to the particle. Ionic surfactants play a large part in controlling the latex particle size while the nonionic surfactants allow for the production of vinyl acetate latexes with concentrations above 40 percent solids by providing a steric barrier to prevent particles from coalescing.

A research program in our laboratory comprises a study of the effects of latex particle size and size distribution on the properties of water-based coatings. This article reports practical procedures for production of narrow PSD vinyl acrylic latexes with a range of particle sizes and concentrations similar to those of commercial polydisperse materials that are used in such applications.

Experimental

Latex Preparation and Analysis. The polymerizations were carried out using the apparatus described in the companion paper, "Effects of Reaction Variables in the Production of Narrow Particle Size Distribution Vinyl Acetate/Butyl Acrylate Copolymer Latexes (I)". The pH measurements of the latexes, the monomer concentration determination during the reaction and the NMR characterization of the copolymers were also discussed.

Latex Particle Size and Size Distribution Measurements. The particle size and distribution of the latexes were obtained using an ICI-Joyce Loebl Disk Centrifuge as described in the companion paper.

Particle sizes below 80 nm may not be reliably measured on the disc centrifuge and particles below 40 nm may not even be detected. Smaller particles can be detected however, by the visual observation of the spinning disc soon after the injection of the latex onto the disc. This is especially useful in detecting the onset of a second generation of particles in monodisperse latex systems.

The method used in this work to detect second generation particles was as follows. The highest rotational speed of the disc was selected and the spin fluid, water with a methanol gradient, was injected. The latex, which had previously been diluted to approximately 1 percent using an 80/20 water/methanol mixture, was injected into the spinning disc, onto the spin fluid. After 5 minutes, when the larger particles had migrated to the outer edge of the disc leaving a clear solution, a light blue ring could be seen near the top of the spin fluid indicating the presence of a second of generation particles. This ring also slowly migrated to the outer edge of the disc, but upon migration, the ring was diluted and thus became very faint. The analytical technique was unable to detect these particles. Using this method, the existence of a second generation of particles could be confirmed by the visual observation of the spinning disc, but the size and size distribution could not be recorded.

Surface Tension Measurements. Measurements of the surface tension of selected latexes at 22°C were made using a Du Nouy ring tensiometer.

Narrow PSD Latex and Subsequent Seeded Reactions at 40% Solids

Narrow PSD Seed Latex. In order to produce a small-particle-size narrow PSD seed latex, the following reaction was performed (Table I; Experiment 117). Surfactant was added to the reactor with deionized water. The addition of surfactant slightly increased the particle polydispersity by increasing the length of the nucleation period, but dramatically decreased the particle size. The ammonium persulfate was added to the reactor concurrently with the pre-emulsion, monomer dispersed in water using DS10, and Na_2CO_3 was not added at all. This procedure ensured minimum coagulation during the nucleation period which therefore resulted in the formation of very small particles. Only one sixth of the pre-emulsion was added. The product particle size was not measured, since the particle size was below 80 nm. The disc centrifuge may produce unreliable results below this particle size. The particle size of this seed latex was calculated from the measured particle size of subsequent seeded reactions.

Without the addition of Na_2CO_3 the latex was not buffered, so hydrolysis would take place if the reaction had been taken to completion. In order to circumvent this problem the reaction was only continued until one-sixth of the pre-emulsion was added, which resulted in a 13% solid latex. At this point the reaction was allowed to post react for 1/2 hr. and the pH was measured. The pH had not dropped significantly (pH 6 at the start to pH 5.5 at the end) and thus not much hydrolysis could have taken place. A portion of this product was then used as a seed for a subsequent reaction (Table I; Experiment 118) to increase the particle size and solids concentration. Since the particle size and PSD was determined by the seed produced in Experiment 117 $NaHCO_3$ could be added to the pre-emulsion without affecting the size or PSD of the latex in experiment 118 and thus hydrolysis could be kept to a minimum.

Seeded Reactions. Seeded reactions were performed as listed on Table I. After the initial seeded reaction (Experiment 118), increases in particle size were obtained by adding 200g of the previous latex to the reactor (e.g., Experiment 119). Further reactions, Experiment 120-123 were done as Experiment 119 by using 200 g of the latex of the previous reaction as a seed to produce larger particles. The seeded reactions contained Na_2CO_3 in the pre-emulsion to maintain the pH above 5. A low polydispersity was retained in the products of these seeded reactions (Table III).

The initiator was fed concurrently with the monomer to eliminate the possibility of coagulation of the latex by large additions of ionic initiator. It is very difficult to predict the effects of initiator concentration (*11,12*), and consequently, also the rate of polymerization, in the reactor, especially when the initiator concentration is increased with time. A simple way to ensure a low monomer content in the reactor throughout the reaction time is to vary the

Table I. Final latex and subsequent seeded reactions

EXPERIMENT #	117	118	119
REACTOR CHARGE			
Deionized Water (g)	225.00	168.08	200.00
Sodium dodecyl benzene sulfonate (DS10-Alcolac Chemical) (g)	0.225	---	---
Seed Latex (g)	---	37.0*	200.0**
MONOMER FEED			
Vinyl Acetate (g)	180.0	176.4	180.0
Butyl Acrylate (g)	32.0	31.1	31.9
DS10 (g)	0.225	0.400	0.500
Deionized Water (g)	71.0	74.0	74.0
$NaHCO_3$ (g)	----	0.46	0.46
INITIATOR FEED			
Deionized Water (g)	20.0	12.5	12.5
Ammonium persulfate (g)	0.50	0.41	0.41

* From Experiment 117
** From Experiment 118

emulsion feed rate and keep the initiator feed rate constant so as to keep the monomer concentration in the reactor below 0.5 weight percent. The feed policy for the pre-emulsion which was finally obtained was a step increase of monomer flow rate from 0.6 to 3.6 ml/min over 3 hrs. The initiator was added concurrently with the pre-emulsion at a constant rate over 3 hrs. This feed policy gave monomer concentrations of no more than 0.2 weight percent during the reaction and essentially a monomer starved feed reaction was obtained.

Confirmation that the reaction was monomer starved was obtained from NMR analyses to determine the vinyl acetate centered triad fractions. The reaction was essentially monomer starved for the seeded reactions as can be seen from the NMR results in Table II. The theoretical monomer starved feed and batch values for vinyl acetate centered triad fractions, calculated by Pichot (*10*) for the mole ratio of monomers used in this study, are also given. Deviations may be expected from the feed comonomer composition, since at the beginning of the reaction the polymer formed is richer in butyl acrylate as a result of the differences in reactivity ratios. The monomer is fed into the reactor at 85/15 weight percent, but the actual vinyl acetate concentration in the reactor is very much greater than this, as confirmed by gas chromatography. During the time the monomer composition comes to equilibrium, the monomer composition is expected to deviate from the theoretical composition. Experiment 120 in particular resulted in significant deviations. This may be due to the fact the number of NMR scans to obtain the spectra was only 2/3 of the others. This gave a poorer resolution and thus an inaccurate portrayal of the copolymer composition.

Table II. Microstructure of vinyl acetate / butyl acrylate copolymer latexes produced by monomer starve feed seeded reactions where A = vinyl acetate and B = butyl acrylate in vinyl acetate centered triad fractions from NMR analysis

Triad	A<u>A</u>A	B<u>A</u>A	B<u>A</u>B
theoretical monomer starve fed	0.78	0.20	0.01
theoretical batch	0.78	0.13	0.08
118	0.79	0.18	0.03
119	0.78	0.21	0.01
120	0.86	0.12	0.00
121	0.82	0.16	0.02
122	0.82	0.16	0.02
123	0.81	0.17	0.02

Table III. Particle size and size distributions of seeded reactions with 0.5g DS10 in the pre-emulsion

Latex	**Size (nm.)**	**Poly-dispersity**	**Standard Deviation (nm)**	**Coefficient of Variation (%)**	**Surface Tension (dyne/cm^2)**
117	54	---	---	---	---
118	194	1.034	20.5	10.6	42.5
119	298	1.029	29.2	9.8	38.5
120	452	1.028	42.5	9.4	36.0
121	676	1.02	51.8	7.7	35.2
122	1011	bimodal	---	---	35.2
123	1328	bimodal	---	---	35.0

Successive seeded reactions produced larger narrow PSD particles. The size and size distributions are listed in Table III. In the reaction to produce the 1011 nm latex in Experiment 122, a second generation of particles was obtained. This was probably a result of excess free surfactant and oligomers in the aqueous phase (*13*).

In a subsequent set of seeded reactions the surfactant charge to the pre-emulsion was reduced to 0.07g in the pre-emulsion (experimental procedure as Experiments 118 to 122). This was done in order to delay the onset of the second generation of particles by reducing the amount of surfactant fed to the reactor and thus reduce the coverage of the particles by surfactant (*14*). The results are reported in Table IV. The onset of second generation particle formation was delayed although only enough to produce larger narrow PSD particles to 1050 nm. The initiator plays a significant role in the production of surface active agents and thus decreasing the surfactant can only play a limited

Table IV. Particle size and size distributions of seeded reactions with 0.07g DS10 in the pre-emulsion

Latex	**Size (nm.)**	**Poly-dispersity**	**Standard Deviation (nm)**	**Coefficient of Variation (%)**	**Surface Tension (dyne/cm^2)**
117	54	---	---	---	---
124	206	1.035	21.7	10.5	49.0
125	305	1.033	30.6	10.0	47.0
126	454	1.029	43.5	9.6	45.0
127	690	1.025	61.4	8.9	43.0
128	1050	1.010	61.6	5.9	42.5
129	1489	bimodal	---	---	45.0

role in the delay of second generation particle formation. It is interesting to note that the surface tension of the latexes significantly increased when a second generation of particles was formed in the recipe with low surfactant concentration, but did not change noticeably when a second generation of particles were formed with higher surfactant concentrations.

Finally, seeded reactions were performed without diluting the latex with water after the initial seeded reaction. Thus, by keeping the pre-emulsion at 40 percent by weight monomer, the latex was kept at 40 percent solids throughout the reaction. This resulted in the formation of monodisperse particles from 200 to 1368 nm, presumably due to a higher $NaHCO_3$ and initiator ion concentration as compared to the diluted latex. This concept is more fully explored in the next sections.

Results and Discussion of 55 Percent Solids Latex

Production of 55 Percent Solids Narrow PSD Latex. Since industrial processes produce vinyl acetate / butyl acrylate latexes at 55 percent solids, attempts were made to produce monodisperse latexes at 55 percent solids. Initial attempts to increase the solids of the latex by increasing the sodium dodecylbenzene sulfonate (DS10) and monomer concentration failed. Latexes produced with only DS10 as surfactant did not produce stable latexes at concentrations greater than 44 percent solids. The use of Rexol 25/407 (Hart Chemical; 70 weight percent solids in water; a nonyl phenol ethoxylated surfactant with an average of 40 moles of ethylene oxide), a surfactant that acts as a steric stabilizer, allowed for the production of latexes at 55 percent solids (*15*).

Initially a latex was made by the following method (Table V; Experiment 130). Deionized water and DS10 were charged to the reactor before the addition of any other ingredients. A pre-emulsion of monomer, water, and surfactant (DS10 and Rexol 25/407) was added slowly to the reactor over a period of 4.0 hours. The pre-emulsion feed rate was slowly increased stepwise from 0.6 ml/min to 4.8 ml/min over 4 hours so as to keep the monomer concentration in the reactor very low at the beginning of the reaction. The initiator was added concurrently with the pre-emulsion at a constant rate over 4 hours. The latex produced had a polydispersity of 1.09.

Table V. Seeded reactions of 55 percent latexes

EXPERIMENT #	130	131	132	133	134
REACTOR CHARGE					
Deionized Water (g)	125.0	125.0	115.0	115.0	115.0
DS10 (g)	0.24	0.24	0.24	0.0	0.0
Seed Latex (g) (Exp. 130)	---	100.0	200.0	100.0	100.0
MONOMER FEED					
Vinyl Acetate (g)	316.0	316.0	316.0	316.0	316.0
Butyl Acrylate (g)	56.0	56.0	56.0	56.0	56.0
DS10 (g)	0.58	0.58	0.58	0.38	0.24
Deionized Water (g)	150.8	150.8	150.8	150.8	150.8
Rexol 25/407 (g)	15.0	15.0	15.0	8.8	4.4
$NaHCO_3$ (g)	0.714	0.714	0.714	0.714	0.714
INITIATOR FEED					
Deionized Water (g)	23.6	23.6	23.6	23.6	23.6
Ammonium persulfate (g)	0.714	0.714	0.714	0.714	0.714
SIZE AND DISTRIBUTION					
D_n(nm.)	183	bimodal	bimodal	bimodal	bimodal
Polydispersity	1.09				
Standard Deviation (nm)	30.7				
Coef. of Variation (%)	16.8				

Seeded Reactions. Seeded reactions at 55 percent solids are difficult to carry out since a large amount of nonionic surfactant is needed to stabilize the latex. The nonionic surfactant provides a steric barrier to the migration of oligomers to the surface of the particle compared to the anionic surfactants. This is due to the high viscosity at the surface imparted by the nonionic surfactants (*16*). The initial latex produced at 55 percent gave a unimodal distribution (Table V; Experiment 130). A subsequent reaction, with this latex as the seed latex, produced a bimodal distribution (Experiment 131). A series of experiments was done in order to determine a method to delay the formation of a second generation of particles, but in each case bimodality was discovered very early in the reaction. First, the seed latex charge was increased (Experiment 132). This was done so that the surface area of the seed latex was increased and thus adsorption of oligomers would increase. It was theorized that this would in turn delay the formation of a second generation of particles. A bimodal distribution of particles was obtained. Secondly, the DS10 charge to the reactor was eliminated and the DS10 and Rexol 25/407 concentrations in the pre-emulsion were reduced (Experiment 133). This was done in order to decrease the stability of

Table VI. Seeded reactions with seed at 55 percent and monomer feed at 55 percent

EXPERIMENT #	135	136	137
REACTOR CHARGE			
Deionized Water (g)	116.0	---	---
Seed Latex* (g)	37.0*	100**	100***
MONOMER FEED			
Vinyl Acetate (g)	360.5	360.5	360.5
Butyl Acrylate (g)	63.5	63.5	63.5
DS10 (g)	0.82	0.82	0.82
Deionized Water (g)	173.0	321.0	321.0
Rexol 25/407 (g)	8.50	8.5	8.5
$NaHCO_3$ (g)	0.85	0.85	0.85
INITIATOR FEED			
Deionized Water (g)	25.0	25.0	25.0
Ammonium persulfate (g)	0.81	0.81	0.81
SIZE AND DISTRIBUTION			
D_n(nm.)	253	535	800
Polydispersity	1.023	1.025	1.019
Standard Deviation (nm)	22.3	47.0	61.3
Coef. of Variation (%)	8.8	8.8	7.7

* From Experiment 117; Table I
** From Experiment 135
*** From Experiment 136

the precipitating oligomers, forcing them to coalesce with existing particles. This also produced a bimodal distribution. Thirdly, the DS10 and Rexol 25/407 concentrations were further reduced (Experiment 134). The resulting latex was bimodal and a small amount of coagulum was obtained. Further decreases in surfactant concentration produced a very unstable latex which resulted in massive coagulation. This is expected when the surfactant used is reduced significantly (*14*).

In order to create a more narrow PSD latex and to increase the final particle size, another series of reactions, using less Rexol 25/407 and a latex from Experiment 117 Table I as a seed, were performed. The latex produced in this manner had a polydispersity of 1.023 (Table VI; Experiment 135).

It is interesting to note that for the reactions containing 15 grams Rexol 25/407 a second generation of particles was not detected at the end of the first seeded reaction but they were detected almost immediately after the start of the second seeded reaction. The production of second generation particles in the second seeded reactions may have been due to dilution of the latex at the start of the reaction.

To test the hypothesis that dilution of a latex may be the cause of a second generation of particles, a latex was made without diluting the seed latex, which was at 55% solids, and using a pre-emulsion at 55 percent weight monomer (Table VI Experiment 135 and 137). As a result the latex in the reactor remained at 55 percent throughout the reaction and a narrow PSD product was obtained with a diameter of 535 nm. Another seeded reaction was performed with the 535 nm latex as a seed. Samples were drawn from the reactor at 1/2 hr. intervals and the particle size was analyzed on the disk centrifuge in order to detect the onset of second generation particle formation. Second generation particles were detected when the seed latex reached a diameter of 800 nm. This method of seeding allowed a seeded reaction to produce a monodisperse latex with particle sizes from 250 nm up to 800 nm at 55 percent solids, without changing the surfactant concentration.

The surface area for the seed latexes in experiments 134 (Table V) and 137 (Table VI) were calculated, to determine if the total surface area is responsible for the delay in the formation of second generation of particles when seeded reactions are performed with seed latexes at 55 percent solids (Experiment 137) as opposed to diluted seed latexes (Experiment 134). It is expected that an increase in the surface area would decrease the rate of addition of surfactant per unit surface area and thus facilitate the adsorption of oligomers on the surface of the existing latex particles delaying the production of second generation particles. The seed latex in experiment 134 had a surface area three times that of the seed latex in experiment 137 at the start of the reaction. Since the surfactant concentration in the latex of experiment 134 is only double that in experiment 137, the surface of the latex in experiment 134 is more sparsely covered. Even with half the amount of surfactant in the pre-emulsion feed a second generation of particles was formed immediately in experiment 134, while in experiment 137 no bimodality occurred until a particle size of 800 nm was reached. It appears therefore that the available surface area is not the overriding factor in the delay of production of a second generation of particles in this series of experiments.

It is also probable that the interparticle distance may be a factor in producing a second generation of particles. Poehlein (*17*) gives the following equation for interparticle distance:

$$d_i = [\,.707\,((\,100 \times 2\pi\,)\,/\,(\,3 \times V\,))^{1/3} - 1\,]\,d_p \qquad (1)$$

where d_i is the interparticle distance, d_p is the particle diameter and V is the volume percent of particles in the latex or:

$$d_i = [((\,100 \times 2^{1/2}\,\pi\,p_p\,)\,/\,(\,6 \times W\,))^{1/3} - 1\,]\,d_p \qquad (2)$$

where W is the weight percent polymer in the latex and p_p is the polymer density.

Equation 2 was used to calculate the interparticle distance of the two seed latexes for the experiments 137 and 134. The seed latex particle sizes were 535 and 183 nm, respectively. The 535 nm latex was not diluted and had a interparticle distance of 87.5 nm while the 183 nm latex was diluted with 115 grams of deionized water and after dilution had an interparticle distance of 91.8 nm. It is expected that similar interparticle distances would facilitate similar rates of capture of oligomers.

Clearly it is not just the interparticle distance that is the main influence in the delay of formation of second generation of particles, since the smaller diluted latex has a similar interparticle distance when compared to the larger undiluted latex, yet a second generation of particles were formed in the diluted latex. It was expected that similar distances between particles would maintain similar rates of collapse of oligomers onto existing particles, and thus delay second generation particle formation. The trend does not appears to be the case here.

The only plausible explanation for the delay in formation of a second generation of particles is that the ionic strength in the aqueous phase is a factor in the delay in second generation particle formation. Hansen (*18*) expresses similar ideas when he states that the diffusion rate of oligomers onto polymer particles is reduced by an electrostatic repulsion factor and new particles are formed when repulsive factors are high. The electrostatic repulsive factor is reduced under higher ionic strengths and therefore the formation of second generation particles is delayed. As has been noted earlier in the companion paper "Effects of Reaction Variables in the Production of Narrow Particle Size Distribution Vinyl Acetate/Butyl Acrylate Copolymer Latexes (I)", increasing the ionic strength in the aqueous phase decreases the stability of the primary particles. Dilution of the latex decreases the ionic strength and thus allows for a more stable environment for the ionic ended oligomers and surfactant to form primary particles. This decreases the rate of collapse of oligomers onto existing particles and a second generation of particles can form.

Conclusions

Seeded reactions using a monomer starved feed method may be used to produce a range of narrow PSD vinyl acetate / butyl acrylate copolymer latexes at 40 and 55 percent solids. These reactions are affected by surfactant concentrations. In reactions with ammonium persulfate initiator and sodium dodecylbenzene sulfonate surfactant to produce a 40 percent solids latex, the anionic emulsifier has only a limiting role in the delay of second generation of particle in seeded reactions. The charged polymeric end groups produced by the ionic initiator appears to have a major role. Seeded reactions can produce narrow PSD latexes at 40 percent solids from 194 to at least 1368 nm.

Seeded reactions to produce a range of monodisperse particle sizes at 55 percent solids are difficult to carry out because of the need to use nonionic surfactants as well as anionic surfactants in the polymerization. Experiments show that a significant factor in delaying second generation particles is the concentration of the seed latex. Further studies indicate that the ionic strength in the aqueous phase of the seed latex can be an important factor in delaying second generation particles. Narrow PSD latexes were produced with diameters of 250 to 800 nm without changing the surfactant concentration by maintaining the latex concentration at 55 percent solids throughout the reaction.

Acknowledgments

This research was supported by the Natural Sciences and Engineering Research Council of Canada.

Literature Cited

1 Lichti, G.; Gilbert, R.G.; Napper, D.H. *J. Polym. Sci., Polym. Chem. Ed.* **1977**, *15*, 1957.
2 Dunn, A.S.; Chong, L.C. *Br. Polym. J.* **1970**, *2*, 49.

3 Okubo, M.; Mori, T. *Colloid and Polymer Science* **1988**, *266*, 333.
4 Kolthoff, I.M.; Miller, I.K. *J. Am. Chem. Soc.* **1951**, *73*, 3055
5 El-Aasser, M.S.; Makgawinata, T.; Misra, S. In *Emulsion Polymerization of Vinyl Acetate,* El-Aasser, M.S., Ed; Vanderhoff, J.W., Ed. Applied Science Publishers,
New Jersey, 1981, p. 240
6 Litt, M.H.; Chang, K.H.S. In *Emulsion Polymerization of Vinyl Acetate,* El-Aasser, M.S., Ed.; Vanderhoff, J.W. Applied Science Publishers, New Jersey, 1981, p. 151
7 Goodall, A.R.; Hearn, J.; Wilkinson, M.S. *Br. Polym. J.* **1978**, *10*, 141.
8 Nguyen, B.D. Ph.D. Thesis, University of Waterloo, Waterloo, Ontario, 1986.
9 Pichot, C.; Llauro, M.; Pham, Q. *J. Polym. Sci., Polym. Chem. Ed.* **1981**, *19*, 2619.
10 Fitch, R.M.; Tsai, C. *J. Polym. Sci., Polym. Lett. Ed.* **1970**, *8*, 703.
11 Chatterjee, S.P.; Banerjee, M. *J. Polym. Sci., Polym. Chem. Ed.* 1978, *16*, 1517.
12 Okubo, M.; Mori, T. *Colloid Polym. Sci.* **1988**, *266*, 333.
13 Kiparissides, C.; MacGregeor, J.F.; Hamielec, A.E. *J. Appl. Polym. Sci.* **1979**, *23*, 401.
14 Sudol, E.D.; El-Aasser, M.S.; Vanderhoff, J.W. *J. Polym. Sci., Polym. Chem. Ed.* **1986**, *24*, 3499.
15 Woods, M.E.; Dodge, J.S.; Krieger, I.M.; Pierce, P.E. *J. Paint Tech.* **1968**, *40*, 541.
16 Blackley, D.C.; *Emulsion Polymerization*, Applied Science Publishers, London, 1975.
17 Poehlein, G. *Advances in Emulsion Polymerization and Latex Technology*, 5th Annual Short Course, 1974.
18 Hansen, F.K.; Ugelstad, J. *J. Polym. Sci., Polym. Chem. Ed.* **1979**, *17*, 3033.

RECEIVED December 4, 1991

Chapter 10

Emulsion Copolymer Glass-Transition Behavior and Particle Morphology

J. Guillot

Laboratoire des Matériaux Organiques, Centre National de la Recherche Scientifique, B.P. 24, 69390 Vernaison, France

Taking into account that copolymer T_g are closely connected with the sequence distribution and that individual copolymer chain microstructure is controlled by the composition drift allowed by the emulsion polymerization process, the glass transition behaviour may be accurately predicted. The basic assumption is that any chain keeps its own calorimetric characteristics in the copolymer sample as if it was segrated in an isolated domain. Investigations on the surface chemical composition of latex particles withdrawn at various conversions, in a batch process for instance, show that the chemical composition of the copolymer at the surface is close to the composition of the instantaneous copolymer generated at the conversion. From the kinetics it is also possible, in many cases, to determine the average number of radicals per particle, ñ. A general trend is for ñ to be close to the expected value for a polymerization proceeding in a mixture which is also similar to the instantaneous copolymer composition. All of these independent techniques are in quite fair agreement with particle morphology and a monomer gradient distribution controlled by the comonomer characteristics. So, accurate glass transition investigation of emulsion copolymers is believed to provide more information on emulsion polymerization and latex particle morphology than is usually expected.

Latex particle morphology is interrelated to polymer compatibility. It is well known that many morphologies can be achieved according to the nature, reactivity and polymerization process *(1, 2)*. Electron microscopy, when it is possible to selectively stain polymers, is a very attractive and suitable technique for structural studies. Investigation of the chemical composition of the latex particle surface (adsorption of emulsifier, ESCA, NMR of the latex, etc.) at various conversions was also shown to be useful in this respect. However, there is a lack of quantitative relationships. Multistage emulsion polymerization is the usual process to obtain structured latexes. However, it could be also interesting to know whether some particle structures can

0097–6156/92/0492–0145$06.00/0

occur for other emulsion copolymerization processes such as batch or semi-continuous polymerization. Are the macromolecules in a batch emulsion copolymer perfectly -- i.e. homogeneously -- mixed or, on the contrary, somewhat segregated and how ?

As glass transition behavior (T_g) of copolymers has appeared to be influenced by polymer compatibility *(3,4)*, emulsion copolymer T_g data have been investigated from theoretical and experimental viewpoints and compared with colloidal properties, which also should be influenced by the particle morphology.

It is also well known that experimental thermograms of batch copolymers withdrawn at various conversions can change a lot , as shown for instance, in butadiene/acrylonitrile *(5)*, or even that copolymers with the same overall chemical composition can have very different DSC thermograms according to the polymerization experimental conditions and process.

On the other hand, Bassett et al*(6)* have emulsion polymerized styrene(S)/ethyl acrylate(EA) by their quite elaborate "power feed" process which allowed copolymerization following a huge variety of ways in order to get final latexes with exactly the same overall 50/50 % chemical composition. Dynamic mechanical moduli of films cast from these various latexes were very different from one another. ^{13}C-NMR investigations *(7)* have revealed a huge change in the microstructure (sequence distribution), which corroborated qualitatively quite well with the micro-mechanical measurements. So, it can be inferred that the thermomechanical properties are rather connected with the microstructure than with chemical composition.

This work deals with the glass transition behaviour, which usually presents many similarities with the micro-mechanical moduli (E', E", Tan δ=E"/E') changes vs. temperature. In addition, it is expected that some quantitative relationships can be derived and that a close connection with particle morphology may be established, as was already suggested *(8)*.

When dealing with random copolymers -- as emulsion copolymers are -- a first problem is to know the glass transition temperature of homogeneous copolymer molecules, i. e. without any composition drift. Indeed, owing to the very short life time of a growing radical, a polymer sample, at a given conversion, is a more or less complex mixture of copolymer chains according to the monomer feed composition drift the polymerization process has allowed up to that conversion.

Few systems follow the classical and simple additive rule of the Fox equation *(9)*

$$\frac{1}{T_g} = \frac{W_a}{T_{g,a}} + \frac{W_b}{T_{g,b}}$$

which relates copolymer T_g to the T_g's of "parent" polymers and the overall chemical composition. Except for a few compatible systems it is usually necessary, in addition, to take into account the distribution of the dyads (A-A, B-B, A-B) of monomer units along the chain and the T_g of the ideally alternating copolymer, $T_{g,ab}$. Theoretical considerations of the entropy of configuration also lead to the effect of dyad distribution *(10)*. Several equations have been proposed*(4,11,12)* which all compute T_g as a function of $T_{g,ab}$ and dyads ; the weighting factors differing from one author to another *(13)*. However, all these equations are not equivalent ; if the same alternating copolymer T_g's are used in these equations, obvious deviations are noticed, which means that $T_{g,ab}$ have to be adjusted according to the equation selected (Figure 1). The 50 mol.% copolymer is the most alternating one and $T_{g,ab}$

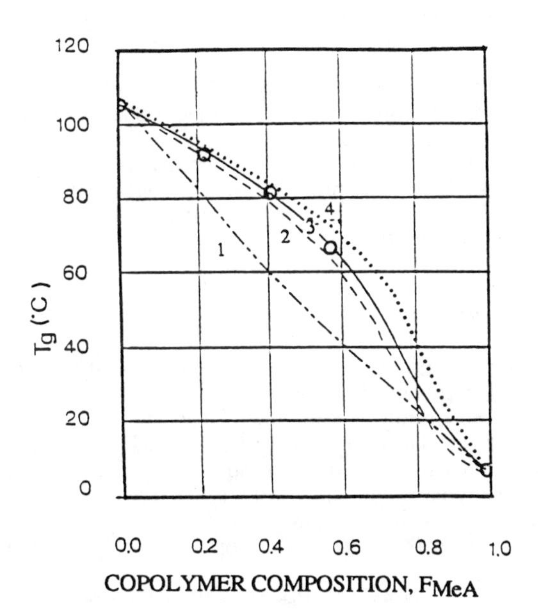

Figure 1 - Glass Transition temperature (Tg) of Styrene(S)-methyl acrylate (MeA) copolymers.
Theoretical curves :
1) Fox ; 2) Barton ; 3) Johnston ;
4) Couchman.
Experimental data (DSC10°C/min) : O

values then have to be estimated from experimental data, following the theory which has been chosen.

For instance, the Johnston equation *(12)* has been selected in this work for it contains a lower number of parameters, among which $T_{g,ab}$ is the only unknown parameter, but one which is easily determinable from experiments. It states that

$$\frac{1}{T_g} = \frac{M_A \quad N_{AA}}{\Sigma n_i M_i \quad T_{g,aa}} + \frac{M_B \quad N_{BB}}{\Sigma n_i M_i \quad T_{g,bb}} + \frac{(M_A + M_B) \quad N_{AB}}{\Sigma n_i M_i \quad T_{g,ab}}$$

where M_i , n_i are molecular weight and mole fraction of monomer "i", respectively ; N_{ij} are the mole fractions of dyad "ij" whose glass transition temperature is T_{gij}, Figure 1.

On the other hand, it is quite easy to know with accuracy the sequence distribution of individual copolymer macromolecules obtained at any conversion (and even in the various polymerization loci, as in an emulsion : water and monomer-swollen particle) from computer simulation programs *(14,15)*. Based on an accurate kinetic mechanism, such software provide instantaneous and average microstructure predictions, regardless of the number of monomers, (including functional monomers, such as carboxylic acids) or the polymerization process (e.g. batch, semi-continuous, composition controlled batch, etc).

Results and Discussion.

In order to get more quantitative data on the actual copolymer sample structural characteristics, in a first step, the simulation of the distribution of the copolymer chains (histogram) according to their microstructure, i.e., their glass transition temperature was derived. The basic assumption is that T_g is determined by the dyad distribution, N_{ij}, following the Johnston equation *(12)*, for instance. From the theory of radical copolymerization the distribution of a dyad "ij" is given by the usual conditional probabilities, which are functions of reactivity ratio and monomer feed composition, only*(16)*,. For instance, with a copolymer of monomers A and B :

$$N_{ab} = (A)*P_{ab} = P_{ba}*P_{ab}/(P_{ba}+P_{ab})$$

where (A) is the mole fraction of monomer A in the instantaneous copolymer and P_{ij} is the conditional probability that a monomer "j" will add a macro-radical ended by an "i" unit during propagation.

For the effect of the composition drift, let us consider a batch emulsion copolymerization of styrene(S)/butyl acrylate(BuA) monomers. The reactivity ratios are r_S=0.71 and r_{BuA}=0.2. The glass transition temperatures, Tg : T_{gS}=370 K, T_{gBuA}=230 K and the glass transition temperature of the alternating copolymer was determined by DSC as $T_{gS\text{-}BuA}$=296 K. When not at the azeotropic composition (i.e., $f_{S,azeot.}$ = [S]/[BuA] =.73) the composition drift is not at all negligible as for a S/BuA=35/65 mol. monomer mixture, which should give a film forming polymer at room temperature, as predicted from the Fox equation. At low conversion, the drift is quite small and a histogram of macromolecular distribution in T_g (micostructure) is narrow as shown in Figure 2a. As conversion increases, the instantaneous copolymer composition deviates more and more from the average copolymer composition, owing to the larger rate of styrene consumption, as shown in Figure 2b. Above 95 % conversion, S is completely depleted and pure BuA molecules are

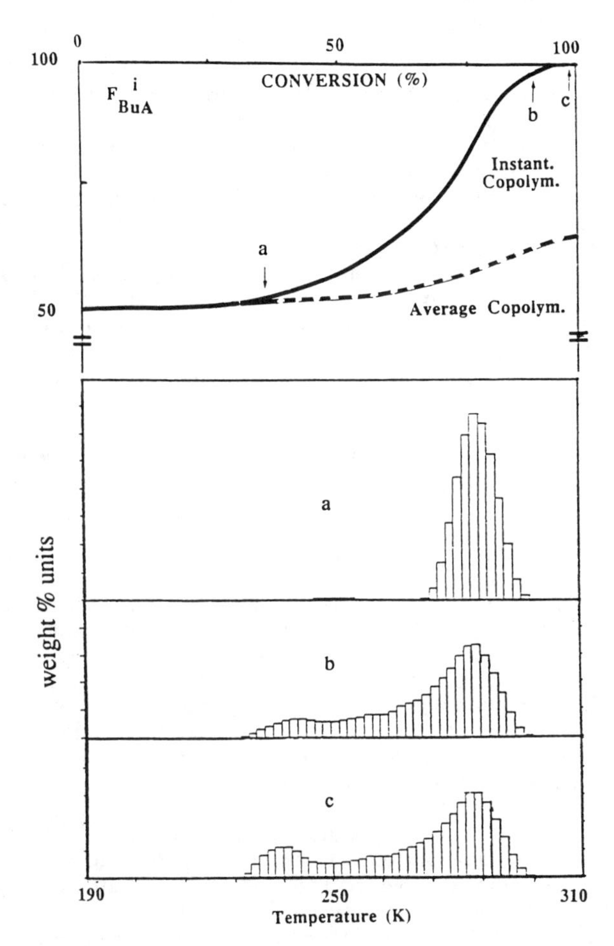

Figure 2 - Styrene-butyl acrylate batch emulsion copolymerization (35/65 mol./mol)
- Instantaneous and average copolymer composition vs. conversion
- Histograms at various conversions :

a - 40 mol. % conversion
b - 91 mol. % conversion
c - 99 mol. % conversion

generated (Figure 2c). As a consequence, theoretical (i.e. computed) histograms are broader and broader. The experimental DSC thermograms (Figure 3) are spread, at low conversion, over a quite narrow temperature range, but over a broader and broader range at higher conversions, and even, above 95 % conversion, a multiple T_g curve is discernible. Without such histograms, the T_g thermograms would be difficult to interpret where glass transition is spread over such a large temperature domain.

If the same monomer feed is now copolymerized by semi-continuous or "composition-controlled" two stage polymerization processes which practically suppress the composition drift, the DSC thermograms recorded at any conversion are spread over a quite limited range (≈15 K) as for pure homopolymers or very homogeneous copolymers as seen in Figure 4.

Such good agreement between the profile of the theoretical distribution histograms and the experimental T_g curves lead us to develop the simulation of DSC/DTA thermograms. The principle is :

- Whatever may be the polymerization process, the computer simulation gives, at any conversion or time, the microstructure of the macromolecules which are instantaneously generated, which alone have a physical meaning, and their corresponding glass transition temperature.
- At a given conversion, the weight fraction of copolymer chains with the same T_g within ± 1 degree Kelvin is computed (histograms).
- Assuming that the overall heat capacity of a copolymer sample, at a given temperature, is an additive rule of the constituting monomer heat capacities corresponding to their state, either glassy (ΔCp,s) or liquid (ΔCp,l), at that temperature, theoretical differential scanning calorimetry (DSC or DTA) curves are computed and plotted, along with the histograms of the weight distribution of the macromolecules according their T_g. The heat capacity changes vs. temperature are included in the simulation*(17)*.
- The theoretical curve is the sum of all the individual DSC data of quasi-instantaneous copolymers, taking into account that for each of them a spreading is observed from experiment, and which is input into the computer simulation. In other words, overlapping of homogeneous copolymer Tg curves is computerized.

The basic assumption now is that any chain maintains its own calorimetric characteristics in the copolymer sample as if it was in isolated domain. For many copolymer systems the simulation programs developed to derive quantitative data based on this theoretical hypothesis lead to theoretical DSC curves and histograms which are in very good agreement with polymer samples withdrawn at any conversion, regardless of the polymerization process (Figures 2-7). This implies that copolymer chain could actually be segregated due to their low compatibilities. Usually the more incompatible the "parent" homopolymers are, the greater is the agreement between theoretical and experimental DSC data.

The same simulation was applied to semi-continuous or composition-controlled batch copolymers. Figure 4 shows the experimental thermograms of S-BuA ; whatever may be the conversion, the transition is spread over a very narrow temperature domain as predicted by the theory. Furthermore, if the semi-continuous polymerization is not well controlled, i.e. when something is wrong in the monomer addition device so that the rate is not perfectly constant, a somewhat broader distribution is expected from the kinetics and the theoretical histogram, resulting in a more complex DSC thermogram ; which is corroborated by experimental data, Figure 5, recorded with BuA -vinyl acetate (VAc). Also, it is well known that the compositional homogeneity of a semi-continuous copolymer depends, to a large extent, on the competition between the rate of monomer addition and the rate of polymerization, i.e. on the rate of monomer accumulation. Simulation, in quite good

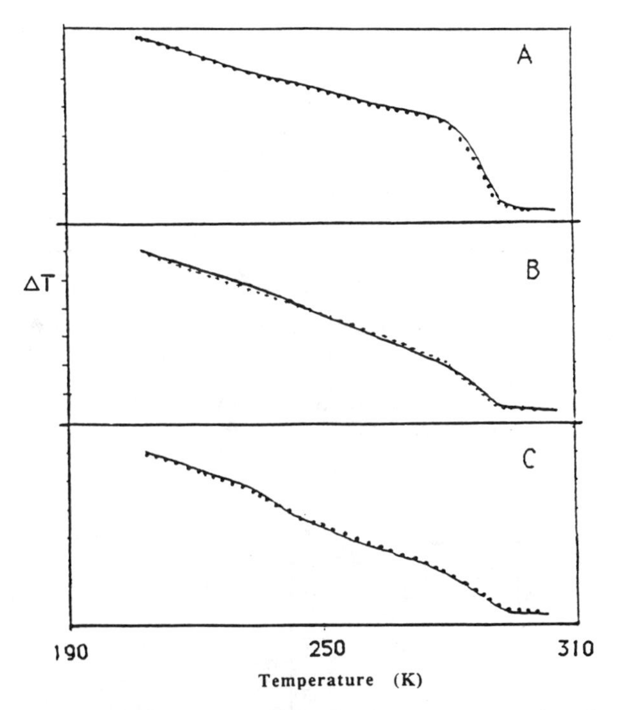

Figure 3 - Experimental DTA thermograms at various conversions for styrene-butyl acrylate (35/65 mol/mol) prepared by batch process
Thermograms (as recorded on SETARAM DSC 101 - Heating rate : 10 K/min)

A - 40 mol. % conversion
B - 1 mol. % conversion
C - 99 mol. % conversion

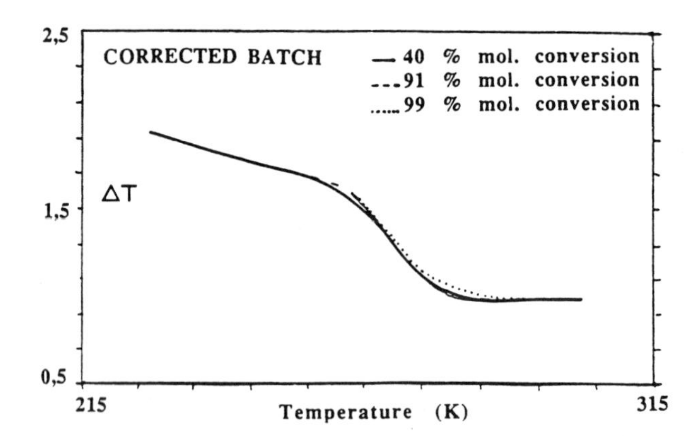

Figure 4 - Experimental DTA thermograms at various conversions for styrene-butyl acrylate (35/65 mol/mol) prepared by "composition controlled" (corrected batch) process.
DTA thermograms (as recorded on SETARAM DSC 101 - Heating rate : 10 K/min) , at various conversions.

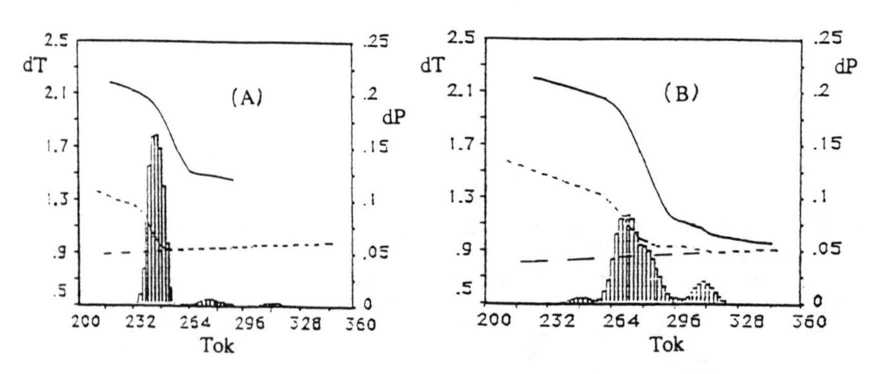

Figure 5.- Vinyl acetate-Butyl acrylate semi-continuous emulsion copolymers.
(A) : Safe control of monomer addition. Monomer feed 18 mol-% Vac.
(B) : Bad control of monomer addition. Monomer feed 67 mol-% Vac.
DSC thermograms : theoretical (dotted line) and experimental (full line) at heating rate = 10 K/min (SETARAM DSC 101). Histogram (see text).

agreement with experiment, shows that the lower the rate of addition, the more homogeneous the copolymer is. This is well illustrated with semi-continuous S-methyl acrylate (MeA) copolymers, Figure 6 ; agreement between theory and experiment being very good for these copolymers certainly owing to the low compatibility of parent homopolymers induced by the huge difference in monomer polarities.

The asumption that copolymer macromolecules of different microstructures are located in isolated domains, which forms the basis for the computer modeling and simulation predictions of the glass transition, leads to a fair agreement with experimental DSC data. As a consequence, a distribution gradient of monomer units should be expected within the latex particles. On the other hand, the shape of this gradient must depend on the composition drift for a given system, given monomer feed and experimental conditions. In VAc-BuA copolymers prepared by batch emulsion polymerization, the huge composition drift induced by large difference in reactivity ratios : $r_{VAc} = 0.07$; $r_{BuA} = 7.0$, results in almost a double T_g system, as for a "core-shell" latex as shown in Figure 7 where the core of the particle is expected to be rich in BuA and the outer layers to be rich in VAc, with a steep gradient.

When r_{ij} values are closer to one another, but an azeotropic composition does exist, as for S-ethyl acrylate (EA) [r_S =.92 ; r_{EA} =.17 *(18)*] or for S-MeA [r =.75 ; r_{MeA} = .17 *(19)*], the drift depends on how far the monomer feed is from the azeotropic composition, as illustrated in Figures 8 and 9 a composition gradient more or less steep should be expected. Agreement with experimental DSC/DTA data is quite fair, including at intermediate conversions, even for systems with hydrophobic monomers as S-ethylhexyl acrylate(EHA) *(20)*.

The model and the simulation of T_g's also assume that the instantaneously generated copolymer molecules are located at the particle periphery during polymerization. As a consequence, the chemical composition of the surface of the average particle changes as polymerization proceeds. This statement was experimentally investigated by determining the specific area occupied by an emulsifier molecule, A_S, from the Maron technique *(12)* ; the larger is A_S, the more polar the surface is. As shown in Figure 10, A_S, for final S-EA batch latexes are larger than for homogeneous semi-continuous S-EA latexes ; moreover, above 80 % EA in the copolymers, A_S is very close to the value measured on pure EA latex particles. This is in quite good agreement with composition drift found by kinetic investigations *(22)*. In this composition range S is consumed much faster than EA and by the end of a batch process EA monomer is accumulated such that the final macromolecules formed consist of pure EA polymer. The same expected trend is observed with S-MeA, Figure 11 *(23)*. A_S measured on cleaned latex samples withdrawn at various conversions are, in addition, quite close to the estimated values for the composition of the instantaneous copolymers generated at the corresponding conversions. Similar investigations have been done on VAc-BuA emulsion copolymers (batch, semi-continuous, controlled composition) and lead to the conclusion that the surface is rich in the more polar VAc monomer at higher conversion *(24)*.

Some ESCA data on BuA-S batch copolymers also agree with accumulation of BuA at the surface*(25)*.

These data, which are derived from a technique independent from DSC seem to be in agreement with a structural model for the latex particles, particularly if the accumulating monomer is the more polar one. Indeed it must be said that the determination of A_S on MMA-ethylhexyl acrylate latex (EHA) does not lead to the safe conclusion that the accumulating monomer (EHA) is at the surface in the final particles, Figure 12 *(20)*. On the contrary, the surface of the average particle

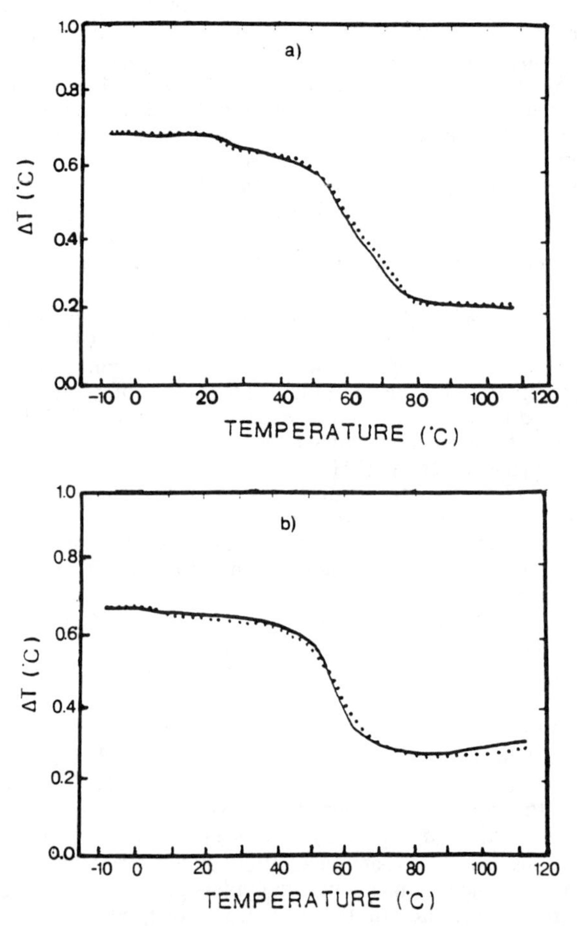

Figure 6 - Emulsion semi-continuous Styrene (S)-Methyl acrylate (MeA) copolymers. . Monomer feed composition MeA/S = 50/50 ; conversion : 52 %. Rate of monomer addition (mole/sec x 10^5) : a) 29 ; b) 2.3 (DSC : 10° C/min - SETARAM DSC 101)

—— experimental thermogram

..... theoretical thermogram

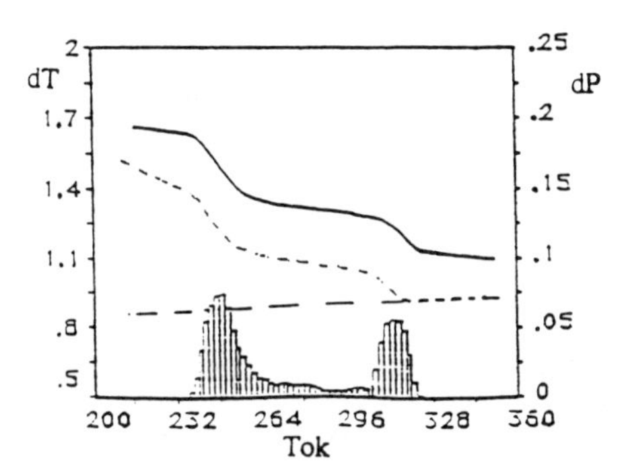

Figure 7 - Vinyl acetate-Butyl acrylate Batch emulsion copolymer ;
Monomer feed : 67 mol-% Vac.; Conversion : 95 %.
DSC thermograms : theoretical (dotted line) and experimental (full line) at heating rate = 10 K/min (SETRAM DSC 101). Histogram (see text)

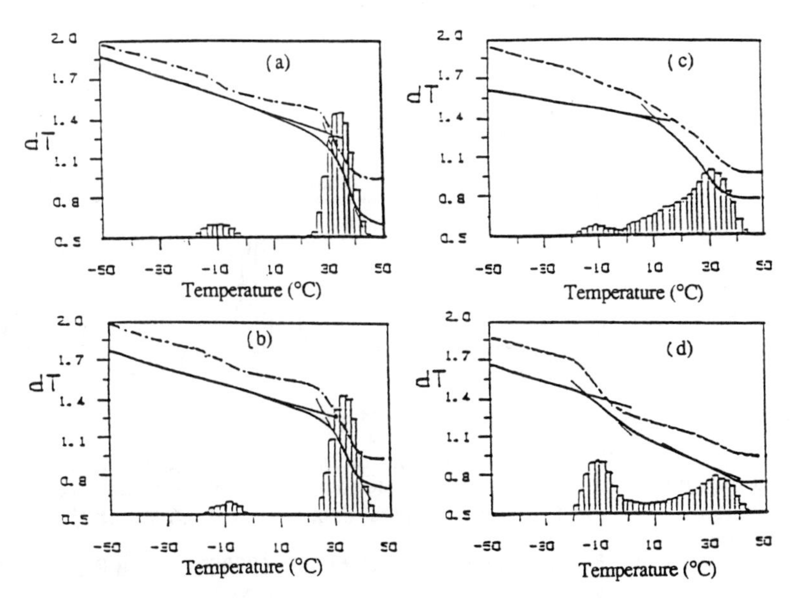

Figure 8 - Styrene-ethyl acrylate batch emulsion copolymers.
DSC curves and histograms of samples withdrawn at various conversions :
(a) : 13 % ; (b) : 22 % ; (c) : 55 % ; (d) : 98 %.
DSC thermograms : theoretical (dotted line) and experimental (full line) at heating rate = 10 K/min (SETRAM DSC 101).
Histogram : weight distribution of chains in which T_g is at temperature T.
Initial monomer feed composition : 80 mol. % ethyl acrylate.

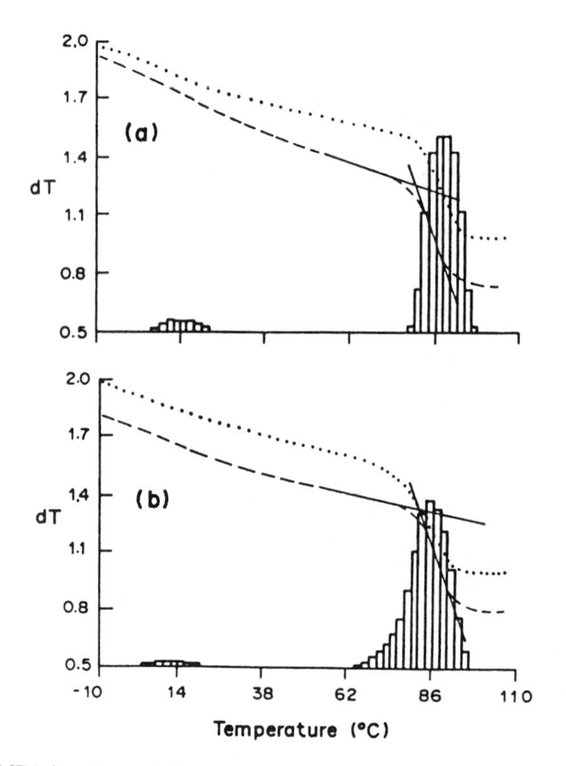

Figure 9 - S/EA batch emulsion copolymerization DSC curves and histograms. Initial monomer feed : 20 mol % EA ; Monomer(g)/water(g) = 0.2. Overall conversion, simulation (···) : Experimental data (---) : (a) 6.6 % ; (b) 97 % conversion.
(Reproduced with permission from ref .18. Copyright 1990 Pergamon Press)

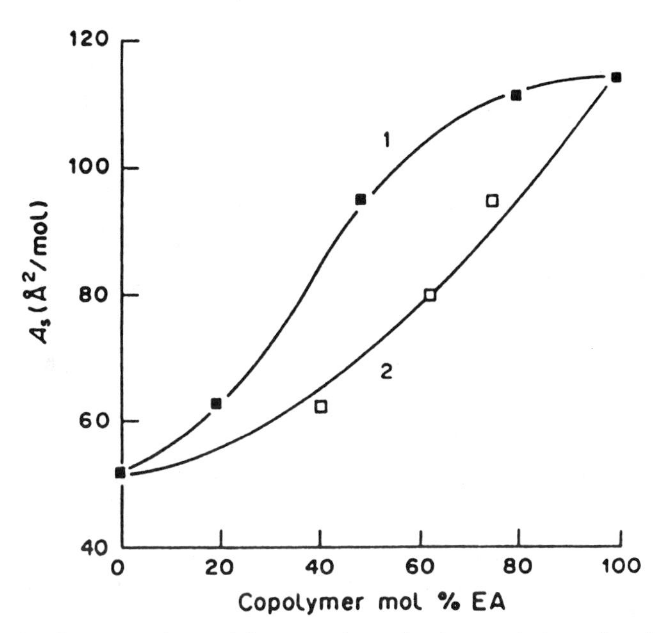

Figure 10 - S/EA batch emulsion copolymerization variation of molecular surface area (A_S) of sodium dodecyl sulfate (SDS) with copolymer composition.
(1) batch ; (2) composition controlled-batch.
(Reproduced with permission from ref .18. Copyright 1990 Pergamon Press)

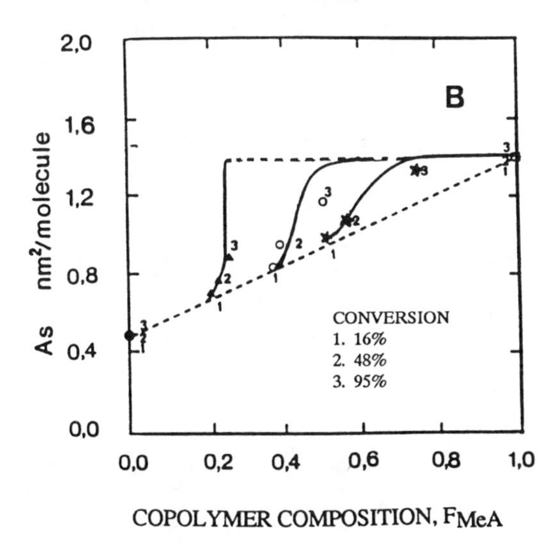

Fig 11 - Styrene (S)-Methyl acrylate (MeA prepared by batch emulsion copolymerization. Change of molecular area of sodium dodecyl sulfate (SDS), A_S, on cleaned latex particles at various conversions and monomer feed composition, F_{MeA} (--).

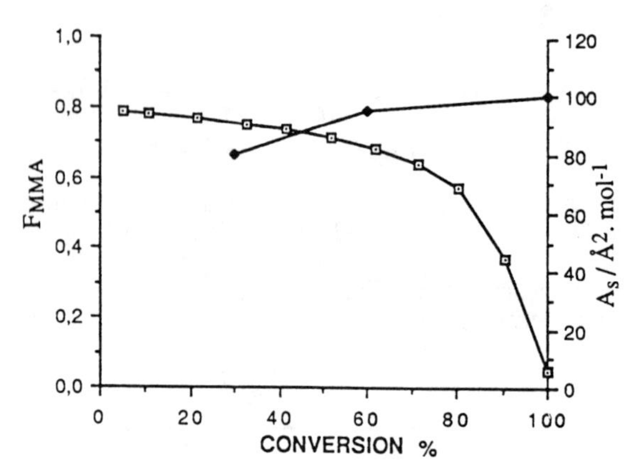

Figure 12 - MMA-Ethylhexyl acrylate batch emulsion copolymers. Instantaneous copolymer composition (-□-) and A_S/Å^2 (-◆-) vs. overall conversion (% mol.).

appeared to be richer in the more polar (hydrophilic) MMA monomer than expected. As DSC data are in relatively good agreement with theoretical histograms and DSC curves, one can infer that EHA rich macromolecules are somewhat buried inside latex particle.

So, when agreement is fair between experimental and theoretical glass transition behavior, the chemical composition of the surface of the average particle is close to the composition of the instantaneous copolymer chains. In addition, as histograms and DSC thermograms given by the simulation are based on the hypothesis that macromolecules of different composition (i.e. microstructure and T_g) are incompatible, i.e., they are not perturbed by the neighbour chains, and segregated in isolated domains, the monomer distribution gradient follows the instantaneous copolymer composition drift. In those cases particle morphology seems to be controlled by copolymerization kinetics.

Another interesting experimental feature of many emulsion copolymerization systems is that changes are observed in some basic kinetic parameters such as, above all, the average number of radicals per particle, ñ. This parameter is derived from the global equation :

$$dM/dt = K_p/[M](\tilde{n}/\mathbf{Na})Np$$

where : K_p is the overall rate constant which is a complex function of reactivity ratios, homopolymerization propagation rate constants, monomer feed composition ; [M] is the global monomer concentration ; N_p is the particle number per unit volume of emulsion and **Na** is Avogadro's number. The computer simulation software can compute at any time/conversion all these values for polymerization taking place within the polymer particle phase ; monomer partition between water and organic phases beeing considered. In many emulsion copolymer systems, ñ was found to vary a lot. As it is illustrated in Figures 13 and 14, in batch S-BuA and VAc-BuA copolymerization, ñ changes in close connection with the instantaneous copolymer composition *(26)*. In the former case, up to 60 % conversion, i.e. when composition drift is low and the S content in the copolymer is high enough, ñ remains close to 0.5, as for homopolymerization of S under the same experimental conditions. ñ then becomes larger and larger as the instantaneous macromolecules are richer and richer in BuA monomer and, finally, ñ reaches the very high values observed when homopolymerizing BuA. In the latter case, ñ decreases from high values corresponding to polymerization of a BuA rich system down to the very low values observed when polymerizing VAc, owing to a radical transfer mechanism. The same kind of behavior has been observed in many other systems.

In such cases, if agreement is good with theoretical glass transition behaviors and surface chemical composition measurements, it is possible to infer that polymerization should proceed mainly in the outer layers of the particles, since the resulting copolymer is likely located at the periphery, at any conversion. As a consequence, radical concentrations should be much higher in the outer layers than inside the particles, as recently proposed *(27)*. On the other hand, the size of the polymerisation domain could be estimated as the order of the radius of gyration of the growing radical up to termination reaction, at least for large enough particles.

Conclusions

Emulsion copolymerization mechanisms, derived from accurate kinetic investigations, lead to the ability to model the microstruture of instantaneous macromolecules and further to develop computer simulation software which aim at the prediction of the glass transition behavior, not only of individual macromolecules, but also of the mixture of all the copolymer chains generated up to a

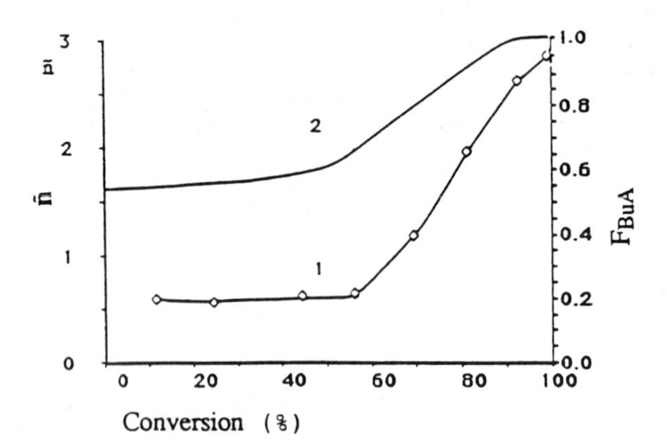

Figure 13 - Latexes prepared by batch emulsion copolymerization Styrene (S)-Butyl acrylate (BuA). Average number of radicals/particle, ñ (curve 1) and instantaneous copolymer composition ; mole fraction of BuA (F_{BuA}, curve 2), as a function of conversion.

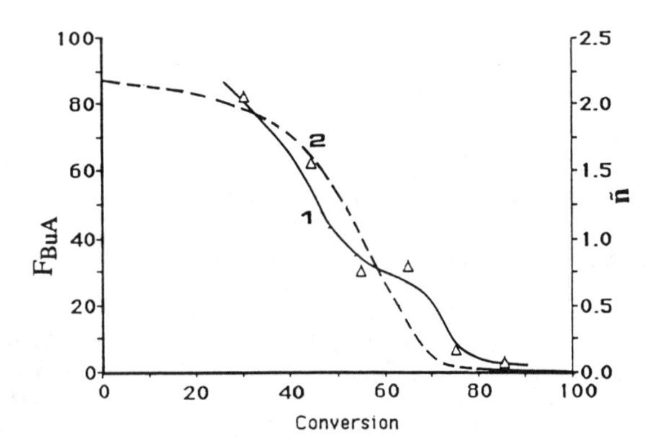

Figure 14 - Latexes prepared by batch emulsion copolymerization vinyl-acetate(VAc)-Butyl acrylate (BuA). Average number of radicals/particle, ñ (curve 1, Δ) and instantaneous copolymer composition ; mole fraction of BuA (F_{BuA}, curve 2), as a function of conversion.

given conversion. Furthermore, theoretical DSC/DTA thermograms, based on the simple assumption that copolymer macromolecules of various microstructure and chemical composition are not compatible and are segregated in isolated domains, can be derived to be matched with DSC/DTA experimental data. The approach, indeed, is not a "best fitting" procedure ; on the contrary, the deviation between theory and experiment is an estimate of how far actual system is from an ideal morphology (i.e. the morphology assumed in the modeling of glass transition). Quite surprisingly, in a large number of cases, theoretical predictions were quite well corroborated by experimental data, so that DSC can be considered, to some extent at least, as a technique to access copolymer macromolecular structure distribution, analogous to GPC for molecular weight determinations.

When the chemical composition of particle surface (A_S) and some important kinetic parameters (ñ) are known and agree with instantaneous copolymer changes, it can be concluded that structure occurs within the latex particle and the change in DSC thermograms can be used to have a good estimate of the shape of monomer distribution gradient. It can also be infered that polymerization mainly proceeds at the periphery or by flocculation of very small newly generated particles ("precursors").

However, deviations are also observed between experimental and theoretical thermograms, as when a semi-continuous or composition-controlled process is poorly run *(18)*. This deviation, then, could be seen as an estimate of copolymer chain compatibilization. In these cases, for instance, if A_S measurements do not agree with DSC/DTA data, that could show that chains rich in a monomer, instead of been located at the surface, are somewhat buried inside (S-EHA).

In any case, there is a close connection between glass transition behaviour and particle morphology. For many copolymer and terpolymer systems, quantitative information has been provided by the simulation programs to predict the glass transition and, to a large extent, to have an estimate of particle morphology. However, it is better to use kinetic data and colloidal parameters, in addition, to get a more reliable view of the actual latex particle morphology and to know how it occurs.

References

1 - Daniel J.C. *Makromol. Chem.***1985** *; Suppl. 10/11*, 359
2 - Okubo.M. *Makromol. Chem. Macromol. Symp.*, **1990** ; *35/36*, 307.
3 - Couchman P.R. *Macromolecules*, **1978** ; *11*, (6), 1156.
4 - Couchman P.R. *Macromolecules*, **1982** ; *15*, 770.
5 - M.R. Ambler. *J. Polym. Sci.*, **1973** ; *11*, 1505.
6 - Bassett D.R., Hoy K.L. In " Emulsion Polymers and Emulsion Polymerization" Edit. D. R. Bassett and H. E. Hamielec. *ACS Symposia Series*, **1981** ; *165*, 371.
7 - Johnston J.E., Bassett D.R., MacCrudy T.B. In " Emulsion Polymers and Emulsion Polymerization" Edit. D. R. Bassett and H. E. Hamielec. *ACS Symposia Series*, **1981** ; *165*, 389
8 - Guillot J. *Makromol. Chem., Macromo. Symp.*, **1990** ; *35/36*, 269.
9 - Fox T. *Bull. Amer. Physic. Soc.*, **1956** ; *1*, 123.
10 - Tonelli A.E. *Macromolecules*, **1975** ; *8* (4),544.
11 - Barton J. *J. Polym. Sci.*, **1970** ; *Part. C 30*, 573 (1970).
12 - Johnston. N.W. *J. Macromol. Sci.*, **1973** ; *A7* (2), 531.
13 - Guillot J.; Emelie B. *Makromol. Chem. Rapid Commun.*, **1991** ; *12*, 117.
14 - Guillot J. *Makromol. Chem. Suppl.*, **1985** ; *10/11*, 236.
15 - Guillot J. In "*Polymer Reaction Engineering*. Ed K.H. Reichert and W. Geiseler, Hütig & Wepf Verlag, **1986** ; p. 147.

16 - Pyun C.W. *J. Polym. Sci*, **1979** ; *23*, 733.
17 - Van Krevelen D.W. In *propertie in Polymers 2nd Ed.*, **1976** ; 81.
18 - Djekhaba S.; Guillot J. *Europ. Polym. J.*, **1990** ; *26 (9),* 1017.
19 - Guillot J., Ramirez W. *IUPAC Meeting, Genova (Italy), Prep.*, **1987** ; 170.
20 - Desroches Th. *Thesis*. Univ. LYON I, **1990** ; to be published.
21 - Maron S.M.; Elder M.E.; Uleritch I.M. *J. Colloid. Sci*, **1954** ; *9*, 89.
22 - Djekhaba S.; Graillat C.; Guillot J. *Europ. Polym. J.,* **1988** ; *24*, 109.
23 - Ramirez W.; Guillot J. *Makromol. Chem.*, **1988** ; *189*, 361.
24 - Kong X.Z.; Pichot C.; Guillot J. *Colloid and Polym. Sci.*, **1987** ; *265*, 791.
25 - Bonardi C.; Guillot J.; Cuirassier F. *AFTPV Meeting, Nice (France). Acta,* **1987** ; p. 283.
26 - Kong X.Z.; Pichot C.; Guillot. *J. Europ. Polym. J.,* **1988** ; *24* (5), 485
27 - De La Cal J.C.; Urzay R.; Zamora A.; Forcada J.; Asua J. *J. Polym. Sci.*, **1990** ; *Part A, 28,* 1011.

RECEIVED December 26, 1991

Chapter 11

Particle Morphology in Vinyl Acetate–Butyl Acrylate Emulsion Copolymers

X. Z. Kong[1], C. Pichot[2], J. Guillot[3], and J. Y. Cavaillé[4]

[1]Department of Chemistry, Shan dong University, Ji Nan 250100, People's Republic of China
[2]UMR 103, Centre National de la Recherche Scientifique, BioMérieux ENSL 46, allé d'Italie, Lyon Cédex 07, France
[3]Laboratoire des Matériaux Organiques, Centre National de la Recherche Scientifique, B.P. 24, 69390 Vernaison, France
[4]CERMAV, Centre National de la Recherche Scientifique, B.P. 53X, 38041 Grenoble, France

Vinyl acetate (VA)-butyl acrylate (BA) emulsion copolymers with various comonomer ratios have been prepared by different emulsion polymerization pathways (conventional batch, semi-continuous addition, with and without emulsifer, composition-controlled batch, core-shell copolymerization) with a view to investigating synthesis-structure-property relationships. Special attention has been paid to the characterization of particle morphology as a function of monomer feed composition and polymerization process. Evidence of particle morphology can be accessed by combining a detailed analysis of emulsion polymerization kinetics of this system with a thorough characterization of the molecular (glass transition temperature, monomer sequence distribution) and colloidal (polarity of particle surface, morphology) properties of the copolymer latexes. In addition, dynamic mechanical measurements have been performed on films cast from these latexes ; their behavior was compared with computer simulated moduli variation vs. temperature and frequency providing useful information on the film morphology in relation with that of particles.

It has long been recognized that emulsion copolymerization of two monomers can lead to the production of polymer particles with a degree of complexity with regards to molecular and colloidal aspects, depending upon the nature of the monomers, the experimental conditions and type of polymerization process (*1*). In the past decade, many academic studies have been devoted on many binary systems, mainly to study the various kinetic and thermodynamic parameters which control their behavior in such an heterogeneous process.

For this purpose, vinyl acetate (VA)-butyl acrylate (BA) emulsion copolymers which are widely used in many technical applications also provide researchers with model latexes for investigating structure-properties relationships (*2,3*). For such a system, large differences are exhibited in the water solubilities of the two monomers (VA = 25 g/l) ; BA = 1-1.5 g/l) (*4*), the propagation rate constants in homopolymerization (k_{pVA} = 4000 l.mole^{-1}s^{-1}; k_{pBA} = 200 l.mole^{-1}s^{-1}

0097–6156/92/0492–0163$07.25/0

(5), reactivity ratios in the copolymerization (r_{VA} = 0.05 ; r_{BA} = 5.5) *(6,7)* and glass transition temperature (32 and -54 °C, respectively). Therefore, a large range of particle morphologies is achievable upon varying the polymerization process and comonomer composition. It was already reported that, for this system, the method of monomer addition played a determining role : a batch process produced core-shell like particles consisting of a BA-rich core surrounded by a VA-rich shell, whereas a semi-continuous one leads to the formation of particles having a more homogeneous distribution of both monomers *(2,3)*. More recently, it was found that these differences in the particle morphology were reflected in the dynamic mechanical behavior of films obtained by coalescence of latexes prepared from various polymerization pathways *(8,9)*.

The objective of this paper is to give more information on the organization of polymer chains within particles, based on a detailed and systematic analysis starting from the synthesis up to the ultimate properties of the latexes. Thc study was schematically developed as follow : first, by combining interpretation of emulsion copolymerization kinetic data as a function of the polymerization pathway and comonomer composition with a complete characterization of molecular properties (Tg, monomer sequence distribution) and colloidal properties of latex particles ; secondly by measuring the micromechanical properties of the correspondent films and comparing them with the simulated ones.

Experimental :

Preparation of polymer latexes. Various series of VA-BA emulsion copolymers were obtained using Aerosol MA 80 (sodium dihexyl sulfosuccinate) and Aerosol OT 75 (sodium dioctyl sulfosuccinate) (both from American Cyanamid) as emulsifiers, ammonium persulfate (98 % purum from Merck) as initiator and sodium bicarbonate (99.5 % from Prolabo) as buffer, according to recipes described elsewhere *(10)*. The following emulsion polymerization processes were performed and briefly described :

i) conventional batch polymerization : in this series, only the initial monomer concentrations were varied .

ii) semi-continuous polymerization : the mixture of comonomers was introduced into the reactor all along the polymerization using 8 wt % of the total charge in a seed initial copolymerization. Emulsifier-free conditions have been selected so as to eliminate of surfactant effects.

iii) composition-controlled batch polymerization : this was achieved by starting the reaction with the less reactive monomer (VA) and the calculated amount of the second monomer (BA) corresponding to the desired copolymer composition. Stepwise addition of the remaining BA is monitored (through on-line analysis by gas chromato-graphy) so as to keep constant the molar ratio of the two monomers .

iv) core-shell (co)polymerization : after preparing seed latex (either PBA or PVA) in the presence of emulsifiers, the second monomer charge (VA or BA) was shell-polymerized in one shot without a preswelling period.

Polymerization kinetics was determined by gravimetry and GC analysis of unreacted monomers.

Characterization of latexes : Particle size was determined by dynamic light scattering (QELS) and transmission electron microscopy (TEM) using cold stage equipment.

The polymer-water interface was characterized using the so-called soap titration method *(11)* with sodium hexadecyl sulfate (SHS) as emulsifier probe according to experimental details provided elsewhere *(12)*.

Latex particle morphology was examined with transmission electron microscopy, and using negative imaging by phosphotungstic acid (PTA).

Copolymer composition was derived from elemental analysis and ^{1}H-NMR spectroscopy after precipitation and subsequent drying of polymer materials.

Dynamic mechanical properties : Polymer film samples were cast from latexes by a standard evaporation technique at room temperature in order to achieve homogeneous solid films of a thickness about 1 mm. Micromechanical measurements were performed on a mechanalyzer (from Metravib Instruments) which is an inverted oscillation pendulum working at fixed frequency (with temperature scanning at a rate of 20 K/hour) or at fixed temperature with frequency scanning (between 10^{-5} and 5 Hz) *(13)*.

Computer Simulation :

Kinetics : the simulations are mainly based on monomer partition coefficients *(14)*. Other assumptions derived from experiments (centrifugation, GC analysis of the phases, swelling, etc) or from the literature data are also introduced in a limited number, whose consequences are verified by experience.

The basic parameters are :

* Reactivity ratios(r_{ij}) and propagation rate constants(k_{ii}):

$$r_{AA} = k_{aa} / k_{ab} \ ; \ r_{BA} = k_{bb} / k_{ba} \quad \text{and} \quad k_{aa} \ , k_{bb}$$

* Partition coefficients :

$$k_i^{w/o} = [M_i]^w / [M_i]^o$$

Only one organic phase (o) is considered : residual monomer + polymer particles ; from experiments *(15,16)* and thermodynamics, the copolymer can change the overall monomer content within the particle (swelling) but has a negligible effect on the monomer feed composition, in general ; in other words, the polymer acts, in a huge number of cases, as a "diluent".

The partition coefficients are experimentally determined. They can be a true constant but also can be a function of monomer feed composition (as already pointed out by Smith *(17)*), of free emulsifier (the more hydrophobic the monomer, the larger the effect on k_i's is), and of temperature.

* Maximum swelling ratio, M/P :

Due to interfacial forces, a limited amount of solvent or monomer can swell a polymer particle, even for a good solvent. For many homopolymer/monomer systems, these monomer/polymer ratios, M/P, are given *(18)*. For copolymers, the M/P is assumed to be a linear combination of the homopolymer M/P's :

$$(M/P)_{copo} = \Sigma \ \phi_i * (M/P)_i \ ;$$

where ϕ_i is the volume or weight fraction of polymer i in the copolymer.

* The main locus of polymerization is the swollen polymer particle, but polymerization in the water phase is also taken into account.

Principle of numerical computations : a small amount of polymer is present at the beginning of the polymerization, say 1/1000 of the total monomers and is assumed to be representative of the latex (composition, particle number, etc).

- At any polymerization increment, a given fraction (10^{-2} to 10^{-3}) of unreacted monomers is polymerized : Sdm_i.
- Copolymerization equations applied in the particle gives dm_i/dm_j.
- dm_i are derived from the previous relations and hence new overall amounts of residual monomer (M_i - dm_i) are calculated.

At any step, the simulation programs compute the many functions of interest : instantaneous values, i.e. related to the individual chains generated at that

conversion, and mean values, for many techniques access only to a global description of the resulting copolymer.
Usually programs give the following parameters conversion versus, time (t/t_{final}) or even absolute time :

. copolymer compositions

. residual monomers within the various phases

. dyad (AA, AB, BB) distribution ; useful for T_g predictions *(19)*

. monomer unit fractions centered in given sequence, for ^1H,^{13}C NMR spectra interpretation

. monomer amounts polymerized in the aqueous and polymer phases and many other parameters, if necessary.

Thermal behavior. The principle of computations applied to predict T_g's is as follows : Whatever may be the process, at any conversion, the computer program calculates the microstructure of instantaneous copolymers (individual molecules) and their corresponding T_g's values from, for instance, the Johnston equation *(20)*

$$1/T_g = \Sigma\,(M_i/Sn_iM_i)*N_{ij}(1/T_{g}ij)$$

where : M_i, n_i are the molar weight and number of moles of monomer "i", respectively, and $T_g$ij, the glass transition temperature of the dyad "ij" whose distribution is N_{ij} *(21)*),
which considers the T_g's to be a function of dyads and not only the overall chemical composition as does the Fox equation*(22)*

$$1/T_g = \Sigma n_iM_i/T_{g}ii.$$

The normalized weight fraction of chains having the same T_g within a given ΔT (say, $\pm$ 1°K), is computed (histograms of T_g's distribution, at the conversion attained). In addition, assuming that the overall heat capacity of the polymer sample in the DSC cell is the sum of the heat capacity of both polymer components at a given temperature (C_i^s and C_i^l *(23)*), a theoretical DSC thermogram can be computed to be compared with the experimental data.
* The experimental observed T_g broadening for pure polymer components is included and a ΔT is calculated between a reference and the sample cells.

Mechanical behavior of films. In this work, mechanical measurements are used not only to obtain parameters such as the shear complex modulus (real part G' and imaginary part G", and the loss angle f, such as tan f = G"/G'), but also as a probe to get information on the material itself. Such accurate measurements were shown to help to understand some features of the molecular mobility, for example, in amorphous polymers *(24,25)* . On the other hand, the mechanical behavior of polyphase systems is extremely dependent on the geometrical arrangement of monophase domains like in semi-crystalline polymers or in polymer blends *(26)*. For instance, and in order to relate the macroscopic behavior of a polymer blend as it is measured by usual mechanical testing machines or by mechanical spectrometers to the behavior of each of the parent polymers, it is necessary to use models such as $M_B = F(M_1, M_2, v_1)$ where M_i is a set of mechanical moduli, for polymer i, M_B, one of the mechanical moduli (either compressive or shear) of the blend and v_1 the volume fraction of polymer 1. F is a more or less complicated function depending upon the manner in which the domains are arranged (matrix with spherical inclusions, co-continuous domains, layers,etc) *(27,28,29)*. In the following, calculations were performed for matrix inclusion systems, on the basis of the Kerner equation modified by Dickie *(28,29)* given by :

$$\frac{G_B}{G_M} = \frac{(1 - v)G_M + (\alpha + v)G_I}{(1 + \alpha v)G_M + \alpha(1 - v)G_I}$$

where $\alpha = 2(4 - 5\ m_M)(7 - m_M)$ and m_M is the Poisson ratio of the matrix, while G_M and G_I refers to the matrix and inclusions, respectively, in which v is empirically replaced by :

$$V_{eff} = v[1 + v \frac{(1 - v_{max})}{(v_{max})^2}]$$

when v approaches v_{max} which is the maximum volume fraction of spherical inclusions (0.69 for spheres with monodisperse size) *(30)*. More recently, a different approach was developed on the basis of a parallel-series model using the percolation concept. It has been shown that calculations lead to similar results except that the latter approach does not require this empirical correction *(31)*.

Results and discussion.

Preparation of latexes.

Kinetic studies of batch copolymerizations. The synthesis of the various series of latexes was precisely followed with regards to polymerization kinetics (monomer conversion, copolymer composition) which were thereafter predicted owing to relatively simple simulation models. Particular attention was paid to the batch copolymerization since during this process, the instantaneous comonomer composition drastically changes as the reaction proceeds, mainly due to the huge difference of the reactivity ratios.

Although both monomers are found to homopolymerize at similar rates, in the case of copolymerization, the overall conversion curves follow a complex behavior especially for a VA molar concentration above 40 mole %. This is well exemplified in Figure 1 which reports the individual monomer conversion versus time (for a recipe with a VA/BA ratio of 40/60 in moles). It is clearly evidenced that BA is the more rapidly consumed, being totally depleted after 40 min. of reaction, whereas VA is polymerized in two stages : i) the first corresponds to effective copolymerization ; ii) the second starts when BA is completely consumed, with a marked increase of the polymerization rate. Consequently the resulting latex particles consist of a complex mixture of macromolecules, which can lead to heterogeneous material by phase segregation.

Simulation of this copolymerization process was investigated. However, parameters such as partition coefficients between water and the organic phase and actual reactivity ratios had to be estimated. Kinetic study taking into account water solubility of VA allowed to determine $r_{BA} = 7.2 \pm 0.2$ and $r_{VA} = 0.08 \pm 0.005$, a set of values slightly higher than recent literature values ($r_{BA} = 5.5$; $r_{VA} = 0.05$) *(6, 7)* for which such a correction was not applied; then,they have to be considered as apparent kinetic parameters. In addition, for modelling absolute monomer conversion vs time, an overall copolymerization rate ($R_p = K_p$ [Mi] [R°] =

$R_{pBA} + R_{pVA}$) was computed taking into account the r_{ij} as above determined and the corrected monomer concentrations with

$$K_p = [\ 1/(r_B/k_{bb} + r_A/k_{aa}\ ([A]/[B])\ )] * \{\ [r_A([A]/[B]) + 1]\ f_A + [r_B + ([A]/[B])]\ f_B\}$$

The change of the overall rate constant K_p as a function of composition of monomer mixture is reported in Figure 2, assuming for the homopolymerization rate constants the following values :

K_{pBA}= 200 l.mole^{-1}.s^{-1}
K_{PVA} = 4.000 l.mole^{-1}.s^{-1} at 60° C *(5)*

It can be seen that the overall rate constant remains rather small as long as the BA content is large enough, then drastically increases as for pure VA homopolymerization.

The radical concentration [R°] was estimated by the usual equation for initiator decomposition $R_i = 2\ fk_d\ [I]°$ and assuming that all polymer chains initiated in the water continue to grow within the particle or generate a primary particle which will grow by capture of other particles. Finally, modelling of the monomer conversions as a function of time was found to provide a quite good account of experimental data (see Figure 1), assuming that polymerization in the aqueous phase is given by a parameter (deduced from experiment) controlling competition between water and particle polymerizations : such a simulation was also applied with success to other batch copolymerizations with different comonomer concentrations.

When considering particle number variation vs conversion, some interesting features were also evidenced according to the composition of the monomer feed (see Figure 3). For a relatively high initial BA monomer concentration, after a more or less large decrease at low conversion, N_p is kept practically constant throughout the polymerization ; on the contrary, with VA rich monomer mixture, N_p is found to increase with the conversion ; a result typical of the emulsion polymerization of many polar monomers *(32)*. Such a behavior might be correlated with the particle growth mechanism, case in which a large compositional drift occurs, causing a change in the rigidity as well as in the interfacial polarity of the particles. Since homogeneous or coagulative nucleation mechanism is prevailing in this system, the newly formed particles do experience various events. In BA-rich mixtures (case II on Fig 3), limited flocculation with BA rich particles (low T_g) is enhanced, whereas in the case of VA - rich mixture, coalescence is less marked (particles are harder) ; which would cause the observed N_p increase in this latter case. Such a result is in favor of the formation of structured particles with a shell containing increasing amounts of VA, as will be shown later on.

From size and kinetic data, the average number of radicals per particle (ñ) was also deduced using the classical equation for the polymerization rate $R_p = K_p$ [M] (ñ/Na) N_p in which K_p is the overall rate constant as mentioned above. Variations of ñ vs conversion or as a function of the comonomer composition exhibit complex behavior reflective of the differences in polymerization of both monomers (Figures 4a,b). In the first case, above f_{BA} = 0.25, ñ is higher than 1 with a sharp increase *(10,11)* for BA-rich and pure P(BA), which could corroborate the assumption of easier flocculation for BA-rich particles . On the contrary, ñ tends to small values for VA-rich composition, as expected, since ñ is very low because of a radical transfer mechanism in the homopolymerization of VA *(33)*.

The variation of ñ during one experiment (Figure 4b) was also found to exhibit a complex behavior, with ñ decreasing from 2.0 to 0.5 as BA is consumed then drastically decreased to very low value (0.1) after BA has be totally depleted. This can be interpreted according to the general trend : at low conversion, the polymerization system would behave like a BA homopolymerization with radical

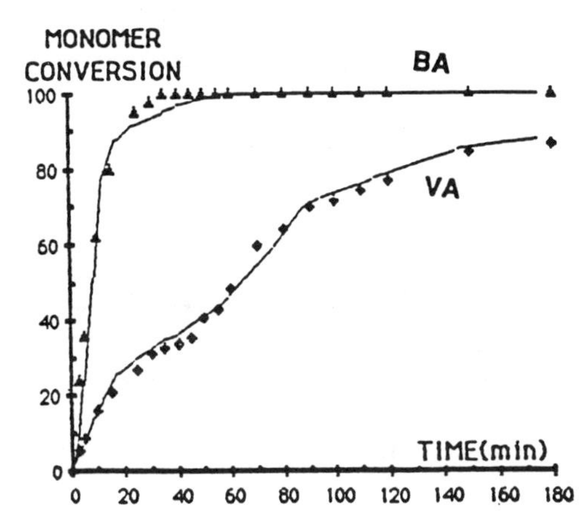

Figure 1 - Kinetics of VA-BA (40/60 in moles) batch emulsion copolymerization. Comparison of experimental and simulated monomer conversion vs time.(Points = experimental ; full lines = simulation) (Reproduced with permission from reference 10. Copyright 1988 Pergamon Press PLC.)

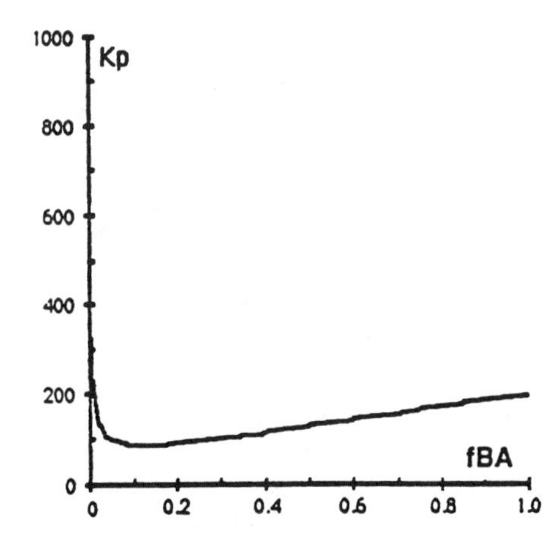

Figure 2 - Evolution of the propagation rate constant (Kp)vs the comonomer feed composition in BA-VA batch copolymerization.
(Reproduced with permission from reference 10. Copyright 1988 Pergamon Press PLC.)

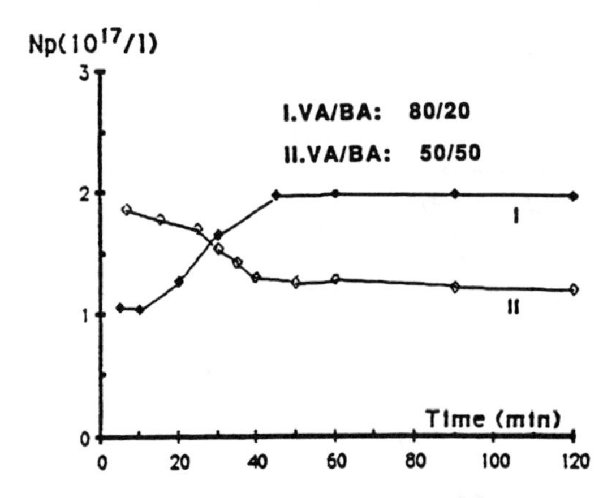

Figure 3 - Particle number variation vs time for VA-BA batch emulsion copolymerization.
(Reproduced with permission from reference 10. Copyright 1988 Pergamon Press PLC.)

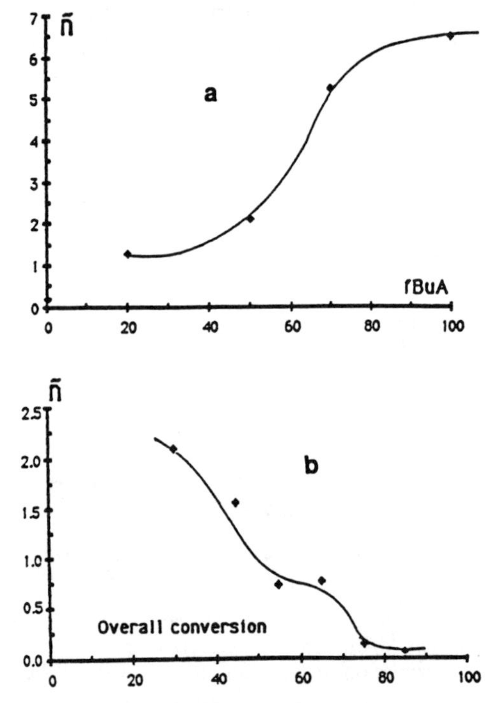

Figure 4a - Average number of radicals per particle (ñ) vs comonomer feed composition (BA mole %).
Figure 4b - Average number of radicals per particle (ñ) vs overall conversion in VA-BA batch emulsion copolymerization (BA = 40 mole %)
(Reproduced with permission from reference 10. Copyright 1988 Pergamon Press PLC.)

accumulation, whereas at higher conversions low ñ values (due to transfer reactions and desorption) would accommodate pure VA polymerization as conversion increases. This can provide an argument for the development of particle morphology in such emulsion copolymerization as was already postulated. Another kinetic argument was recently proposed by De la Cal et al (34) who applied a simulation model of latex particle morphology to the polymerization of VA on a BA-VA copolymer seed (which resembles the situation of a batch copolymerization when all BA had reacted). The prediction of a concentration profile of free radicals in the particle, with a higher concentration of radicals in the outer layer would favor the core-shell morphology.

Studies of other polymerization processes. The same detailed study was achieved when dealing with either composition-controlled one-step polymerization or core-shell copolymerization processes. Several specific features will be emphasized mainly concerning the efficiency of the polymeri-zation method so as to control the compositional homogeneity of the final materials and/or the eventual change in particle size and distribution of latexes.

In the case of the composition-controlled batch process (BC series in Table 1), Figure 5a shows the variation of the average BA copolymer composition vs overall coversion. It is seen that the resulting copolymers are not perfectly homogeneous since the composition (in BA mole %) is fluctuating around the average composition. First, copolymers formed at low conversion are richer in VA than expected from the selected initial feed monomer ratio. As corroborated by simulation, such an effect is likely due to aqueous phase polymerization of VA, what was not taken into account in the determination of the initial VA concentration. Secondly, there is a slight drift as polymerization proceeds leading to a richer BA final copolymer, what assumes rich BA instantaneous copolymer chains. This phenomenon was already evidenced in the case of composition-controlled styrene-acrylonitrile emulsion copolymers *(35)*, which was explained on the basis of thermodynamics : the better apparent diffusion of BA within the growing particles gives rise, indeed, to a higher BA conversion, as was experimentally confirmed for this system.

Concerning the change in the particle number vs conversion (see Figure 6a), results are dependent upon the comonomer composition. With a VA-rich concentration, there is a strong increase in N_p in the low conversion domain, whereas a slight increase is only shown for BA-rich mixtures. Such variations causes the particle size distribution to be broader than for the other latex series (see Table I). It is conceivable that the addition of the more hydrophobic and reactive comonomer might affect the distribution of the emulsifier and, consequently the nucleation mechanism ; but compositional homogeneity is more satisfactorily observed than in the previous latexes, particularly in the SC2 experiment. For the two other runs with a lower composition in BA, a slight BA increase is again observed as polymerization proceeds, which is associated with a higher individual BA conversion. Simulation curves give a good account of these experimental data, and especially they suggest at low conversion a noticeable aqueous phase polymerization of VA.

Particle number variations vs conversion (Figure 6b) provides strong evidence of a limited flocculation mechanism for the particle nucleation which is typical of polymerizations carried out in the absence of emulsifier. A lower particle number is obtained for pure BA polymerization as expected due to higher M_W and the corresponding lower surface SO_4^- charge density (see below). Higher particle numbers are displayed when VA is introduced ; which can be interpreted as a complex nucleation mechanism in which oligoradical formation and subsequent self-precipitation are competing with aqueous phase polymerization.

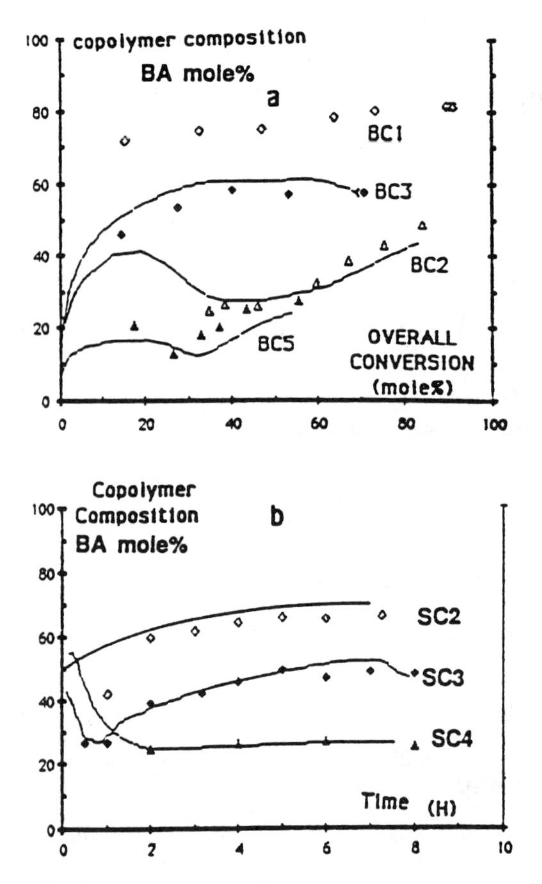

Figure 5 - Variation of the average copolymer composition vs overall conversion or time
a) in composition-controlled batch process
b) emulsifier-free semi-continuous process

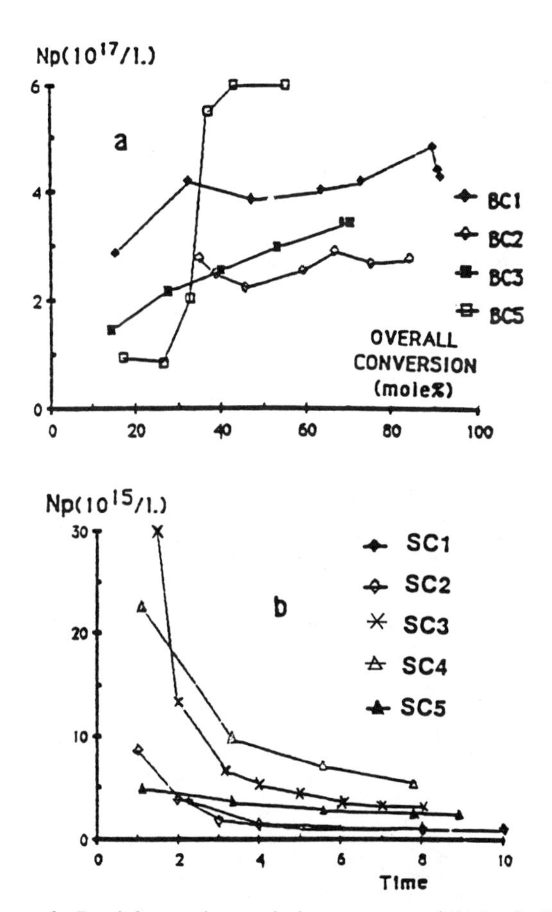

Figure 6 - Particle number variation vs conversion (mole %)
a) composition-controlled batch process
b) emulsifier-free semi-continuous process

TABLE I : CHARACTERISTICS OF VA-BA COPOLYMER LATEXES

Polymerization Process	Latex Code	(BA/VA) mole ratio Initial monomer feed	Final composition from ^{1}H NMR	T_g °C	Particle size and distribution QELS (nm) D_p	TEM (nm) D_n	D_w	PDI = (D_w/D_n)
Batch	BA 1	100/0	-	-49	145	140	148	1.06
	BA 3	61/39	67.7/32.3	-44	160	161	168	1.04
	BA 4	40/60	43/57	-41,36	158	145	156	1.07
	BA 5	31/69	32.7/67.3	-32,36	145	134	145	1.08
	BA 8	14/86	15/85	-20,36	139	-	-	-
	BA 9	0/100	-	+38	120	115	127	1.11
Composition Controlled Batch	BC 1	71.7/28.3	75.2/24.8	-39	93	76.8	89	1.17
	BC 2	61/39	67.7/34.3	-36	90	80.9	92.6	1.15
	BC 3	40.2/59.8	45.3/54.7	-29	93	90.2	96.6	1.07
	BC 4	31/69	36.9/63.1	-17	80	70.4	76.8	1.09
	BC 5	14.4/85.6	25.8/64.2	- 1	62	51.5	61.5	1.19
Core Shell	CS 1	66.8/33.2	80.9/19.1	-49,33	155	150	156	1.04
	CS 2	40.2/59.8	41.0/59.1	-49,33	180	169	180	1.07
	CS 3	18.3/81.7	18.4/81.6	-50,34	200	194	206	1.06
	CS 4(*)	40/60	40/60	-53,34	162	-	-	-
	CS 6(**)	41/59	41/59	-49,32	140	-	-	-
Emulsifier-free semi-continuous	SC 1	100/0	-	-49	540	510	525	1.03
	SC 2	64.8/33.2	64.5/35.5	-36	510	441	488	1.10
	SC 3	41.2/58.9	44/56	-18	410	395	400	1.01
	SC 4	18.3/81.7	21.3/78.7	- 7	290	252	278	1.10
	SC 5	0/100	-	+35	420	385	391	1.02

(*) : inverse core-shell (PVA seed latex) (**) : Three-steps core-shell : PBA seed-VA+BA (40/60 mole ratio)-PVA

When considering the case of two-step polymerizations, Table II clearly shows a change in the evolution of the particle size vs time. It can be seen that there is a fairly good agreement between experimental (QELS) data and calculated particle diameter (using conversion and seed size data). N_p remains constant for all the different runs whether VA is used instead of BA as a seed which is an indication that no significant and observable secondary nucleation does occur during this "shell-polymerization". Final particle size distributions (see Table I) are similar to those of batch latexes.

Molecular properties of copolymer latexes. Differences in the molecular structure of the final copolymers which originated from the various polymerization pathways have been highlighted through the polymer characterization with regards to monomer sequence distribution and thermal behavior.

The microstructure investigation by ^{13}C-NMR was based on a previous work *(6)* which provided the main ^{13}C assignments in both homopolymers and copolymers, particularly those corresponding to the methine and methylene (ranged between 32 and 42 ppm) and allowing access to the monomer sequence distribution related to both units. An example is given in Figure 7 comparing the spectra of copolymers having the same overall composition (55/45 VA/BA in moles) but prepared under various copolymerization processes. Two main features are evident from these spectra, grouping A and B samples (core-shell and batch), and C, D, E, samples (semi-continuous and composition controlled batch), respectively .In the former case, the presence of homogeneous sequences (i.e., AA, AAA, BB, and BBB referring to VA and BA dyads and long sequences is well shown. As expected, this is particularly true in the case of PVA-shell polymerization onto PBA seed (spectrum A). The presence of heterogeneous sequences such as AB or BAB (originated from grafting) is not detected. A similar spectrum is exhibited with the batch sample (BBB at 42 ppm and AA dyads at 40 ppm). In addition, the presence of heterogeneous structures is clearly evidenced (but with a minor contribution than in the homogeneous copolymers) which is unambiguously attributed to the large compositional drift in the comonomer mixture as the reaction proceeds, leading at intermediate conversion to A and B monomer units engaged in alternating BA sequences, BBA sequences and so on at intermediate conversion. In the second case, dealing with homogeneous copolymers, comparison of the spectra (C, D, E) with the previous ones reveals the predominance of heterogeneous structures whereas those related to long BBB sequences are minor but not completely eliminated. This is a good indication that the selected processes would favor the formation of macromolecules having homogeneous compositions ; it is assumed that heterogeneity occurs at the beginning of the reaction, most notably in the semi-continuous process where BA rich particles are first nucleated.

Such microstructure data can be predicted using the above mentioned simulation program, by adding a subroutine as to describe the monomer sequence distribution based on the so-called Markov terminal model *(36)*. An example is reported in Table III showing the variation of the average monomer sequence distribution in a batch copolymer as a function of conversion. Good agreement is observed between experimental and simulated data. The predominant compositional drift is well illustrated particularly in the inverse evolution of the BB and AA dyads respectively, as conversion increases.

The thermal behavior of such copolymers was characterized through DSC measurements. As previously reported *(3)*, the general trend is that copolymers prepared from batch or core-shell processes exhibit two glass transition temperature domains whereas a single transition is observed in the case of composition controlled or semi-continuous processes (see Table I).

TABLE II : Particle number variation vs time in two step VA-BA emulsion polymerization

Latex Code	CS1					CS2				CS3			CS4		
Conversion (wt %)	0	5.5	13	19	63	0	5.7	8.6	97	0	65	98	0	71.5	95
D_P (QELS)	128	129	130	131	139	128	129	150	160	128	180	196	130	155	160
D_P(calculated)	-	130	131	133	139	-	130	150	160	-	178	195	-	153	162
Particle Number (10^{17}/l)	1.78	1.79	1.77	1.75	1.73	1.43	1.44	1.42	1.37	0.65	0.65	0.88	0.88	0.93	0.97

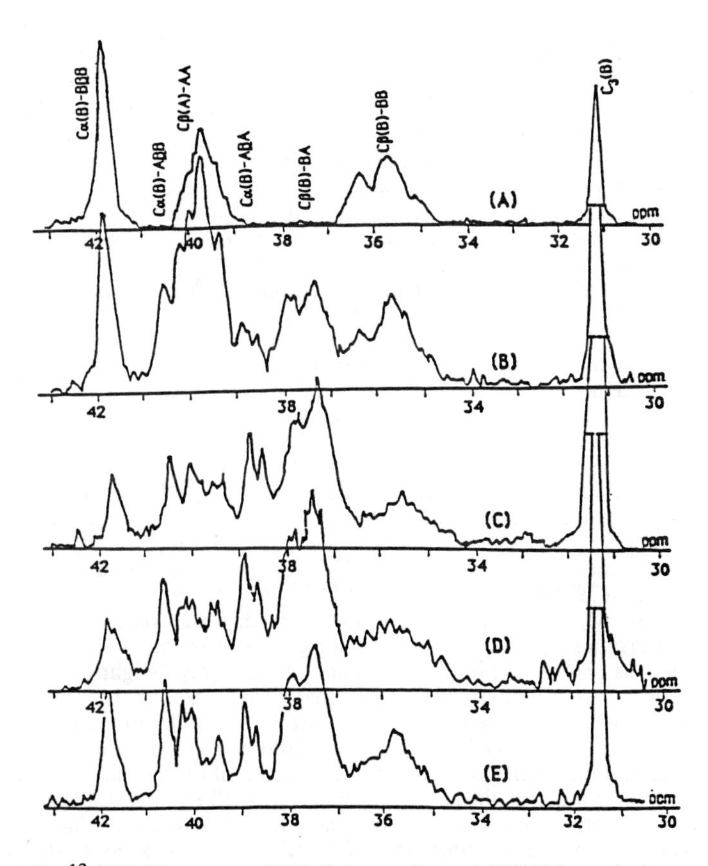

Figure 7 - ^{13}C.NMR spectra of VA-BA copolymers (55/45 in moles) prepared with different processes :
A) core-shell ;
B) batch ;
C) emulsifier-free semi-continuous ;
D) semi-continuous ;
E) composition-controlled batch

More insight was obtained into thermal behavior as a function of polymerization pathway by using the basic kinetic simulation together with the equation to predict T_g's proposed by Johnston *(20)* which takes into account the dyad distribution of copolymer. As detailed elsewhere such a simulation program is able : i) to calculate the T_g of instantaneous macromolecules as the reaction proceeds and then to provide, in histogram form, the weight fraction of chains having the same T_g (within ± 1° K) ; ii) to compute and to describe a theoretical DSC curve. Some representative examples are given in Figure 8.

In the case of semi-continuous or corrected batch copolymers, Figure 8 clearly shows one unique T_g, but differences in the sharpness in the theoretical histogram seems to reveal some heterogeneity in the final copolymers (which was already pointed out in the ^{13}C.NMR spectra). For the batch copolymer, it is obvious that the experimental DSC curve exhibits a multiple and complex transition spectrum, which is quite representative of the copolymer heterogeneity. Such a complexity in the final material is well evidenced in the theoretical histogram which displays two main families of macromolecules, one at low temperature, which is quite broad (corresponding to a wide distribution of copolymers which were formed early in the reaction) and another sharper distribution occuring around 30 °C, corresponding to the PVA-rich copolymers formed at late conversions.

Characterization of the particle morphology. Particle morphology of the various well characterized latexes was examined either directly by transmission electron microscopy using a staining technique or indirectly through the analysis of the water-polymer interface.

In the first case, phosphotungstic acid(PTA) was used in order to improve the TEM contrast of easily deformable particles under the electron beam ,as well as,acting as a staining agent for PVA, probably due to a chemical reaction between the ester group in PVA and the acid. A preliminary study in our group clearly showed the direct staining of PVA by PTA evidenced and recently confirmed by Dimonie et al *(37)* who showed that PTA exclusively ensured darkening of PVA latex particles whereas PBA ones are not affected. Based on these observations, core-shell particles (with a PBA/PVA weight ratio of 50/50) were examined; surprisingly, as shown in Figure 9 the particles are only "slightly" blackened in comparison with pure PVA particles. This may reflect the fact that PVA-rich shell layer (25-30 nm) in this type of core-shell particle would be too thin to be clearly delineated as in PVA homopolymer latex ; in addition, some VA monomer diffusion can occur during the synthesis resulting in a decrease of the available VA-rich shell layer. The reverse is true in the case of inverse core-shell PVA/PBA latexes (with the same overall composition). For these particles, it is clearly shown that, contrary to the polymerization stage order, the BA added in the second stage of the polymerization appears mainly as quasi-spherical inclusions. A few particles in the micrograph exhibit perfect "core-shell" form with PVA in the shell layer, indicating that PBA polymerized onto PVA seed produces phase inversion due to the hydrophobic nature of the second stage monomer with respect to the more hydrophilic PVA. Such a result has been already widely discussed in the literature *(38)*, particularly by Okubo et al *(39)* in the case of PS/PMMA or poly(methyl acrylate) core-shell systems. However, careful observations on the TEM micrographs also reveal the presence of "dumbell" particles. It might be conceivable that these different particle morphologies could be achieved upon aging (the present examination on latex samples being carried out 4 months after the preparation) *(40)*.

More insight about particle morphology was derived with a detailed analysis of the water-polymer interface either using the surface end group titration method or by deducing the polarity of the interface from the soap titration method

TABLE III

Variation of the monomer sequence distribution vs conversion in a batch emulsion copolymerization (BA 4).
(B = BA ; A = VA) ; comparison of experimental and simulated (in brackets) data

Overall conversion (mole %)	38	49	95
Average copolymer	70.9	68.9	45.
composition (BA mole %)			
BAB	0.84 (0.81)	0.89 (0.81)	0.23 (0.25)
BAA	0.16 (0.12)	0.11 (0.13)	0.11 (0.13)
AAA	0.00 (0.04)	0.00 (0.06)	0.66 (0.62)
BB	0.50 (0.51)	0.49 (0.51)	0.25 (0.28)
BA	0.36 (0.37)	0.33 (0.32)	0.28 (0.30)
AA	0.14 (0.12)	0.18 (0.17)	0.47 (0.43)

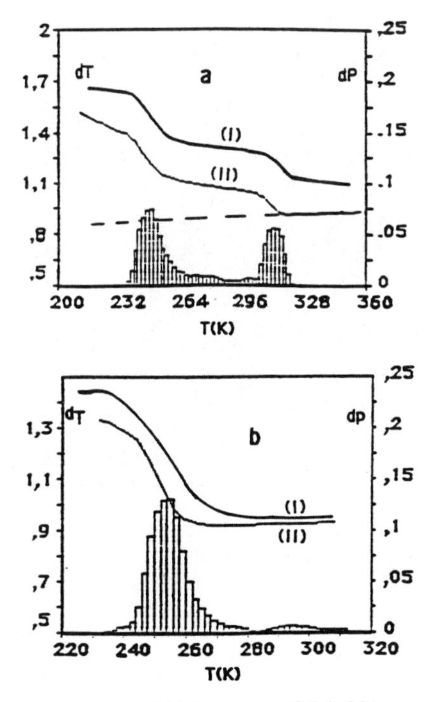

Figure 8 - DSC thermograms and histograms of BA-VA emulsion copolymers.
I (experimental) ; II (simulated).
a) batch process ; b) semi-continuous process

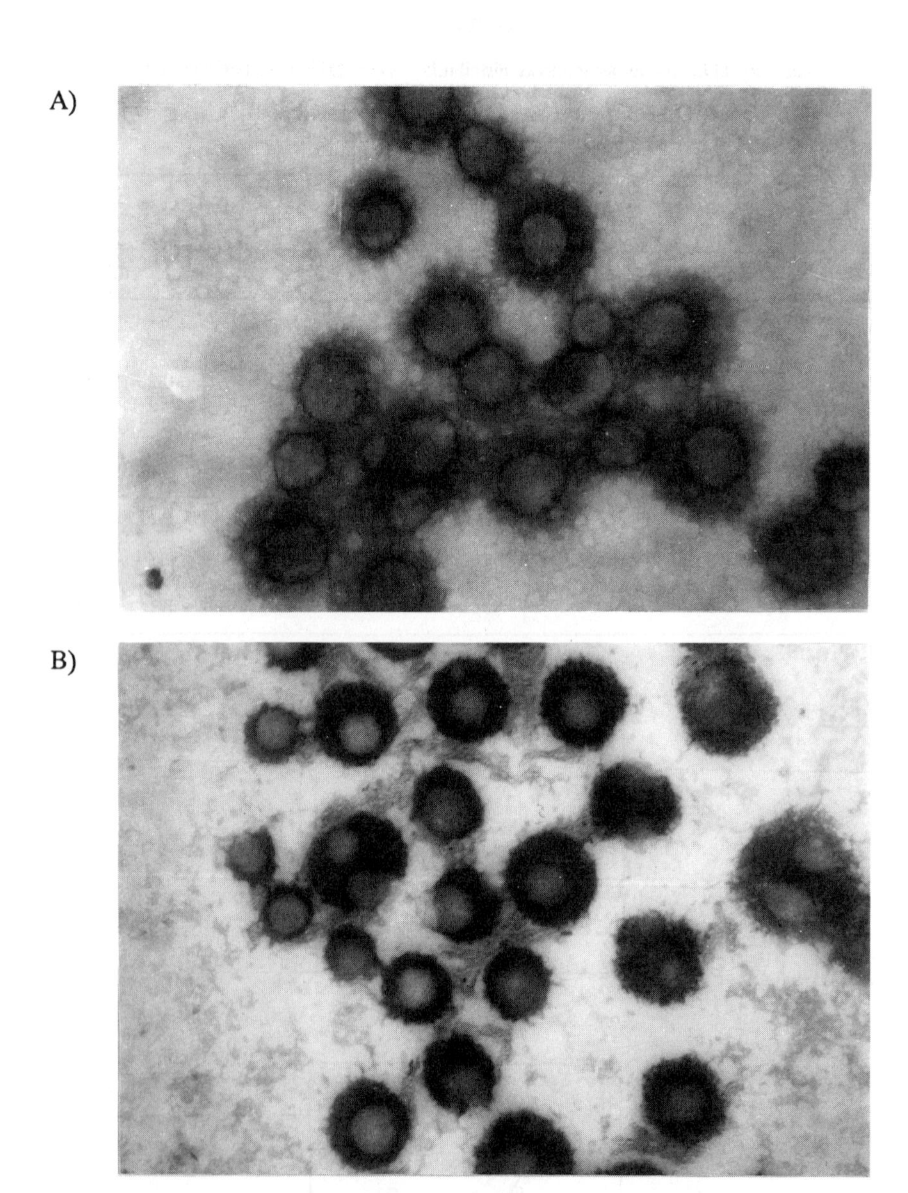

Figure 9 - TEM micrograph of VA-BA copolymer latexes using phosphotungstic acid as the staining agent
A) core-shell sample (CS 2)
B) inverse core-shell sample (CS 4)

(41). In both cases, latex particles prepared under emulsifier-free semi-continuous polymerization conditions were selected as reference materials so as to avoid compositional heterogeneity and surfactant effects.

In these latter latexes, for SO_4^- groups, (see Figure 10) much higher surface charge densities were found than in all other latexes regardless of the composition due to the nature of the particle formation mechanism ; in emulsifier-free polymerization, a homogeneous nucleation mechanism is predominant which subsequently involves limited flocculation of primary particles so that the final particles achieved a sufficient surface charge density to attain the colloidal stability.

On the contrary, when the contribution of micellar nucleation is not negligible, especially in the batch and composition-controlled batch processes (where the emulsifier concentration is largely above the CMC), the contribution of the SO_4^- end charges to the stability of the particles is much less important. For the batch latexes, indeed, the surface charge density is about ten times smaller and the variation upon changing the composition is not significant. In the case of composition controlled batch latexes, the SO_4^- charge density increases which can be attributed to the smaller molecular weights observed with these latexes compared to those of batch latexes and which increases with the BA composition.

The results for the various core-shell latexes (series CS) are quite instructive. For CS2 and CS3, the surface charge density is very similar to that obtained with pure PVA, an indication that during the synthesis of such latexes the mechanism of SO_4^- terminated oligoradical formation is dependent on the amount of VA to be polymerized. When the amount of VA is too small (CS1 samples), the surface charge is closer to that of PBA, what suggests that the complete overcoating of the PBA seed is not ensured.The PVA-PBA (CS4) inverse "core-shell latex" surface charge density was found to be far from the BA latexes, but in between those of both PBA an PVA polymers which confirms the heterogeneous and complex nature of structure in these latexes (with the coexistence of inverted core-shell and dumb-bell particle morphologies).

Analysis of the surface chemical composition was then investigated, which allows one to calculate the molecular surface area (A_S) of the emulsifier probe (SHS) for the various latexes. Figure 11 shows the results from which the following information can be deduced. In composition-constant and semi-continuous polymerized latexes (which lead to a homogeneous distribution of monomer units), there is a continuous change in A_S as the comonomer composition varies, but the obtained A_S seems to reflect a less polar water-polymer interface than expected. On the contrary, in core-shell particles, high A_S values corroborate a PVA-rich shell regardless of the composition and particles produced by batch emulsion polymerization follow the same trend except for BA -rich latexes (BA3).

Of interest was the examination of the emulsifier adsorption on latex samples withdrawn at different conversions during a 60/40 molar ratio VA/BA batch emulsion copolymerization. Figure 12 shows the change in A_S and the instantaneous copolymer composition vs conversion. It is clearly seen that a BA-rich particle is produced at low conversion (low A_S values), thereafter a VA-rich interface progressively takes place, which again confirms the formation of a PVA outer shell layer at high conversion. The efficiency of such a simple method is indeed corroborated when studying the particle morphology in latex systems (containing one polar comonomer) as early pointed out by Okubo *(39)* and also observed in our group on many other emulsion copolymers *(32)*. Analysis of the inverted core-shell latex particles (PVA as the seed) displayed a A_S value (168 $Å^2$/molecule) which would correspond to a copolymer with approximatively 80 mole % VA, which is reflective of a phase rearrangement of the macromolecules so as to expose a much more polar interface in contact with the aqueous phase.

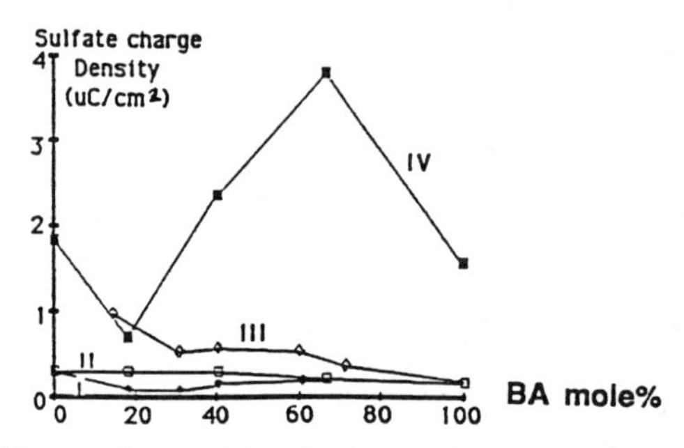

Figure 10 - Sulfate end-group density in VA-BA copolymer latexes vs the overall copolymer composition as a function of the polymerization pathway. (I) Batch ; (II) Core-shell ; (III) Composition-controlled batch ; (IV) Emulsifier-free semi-continuous (Reproduced with permission from reference 12. Copyright 1987 Steinkopff Verlag Darmstadt.)

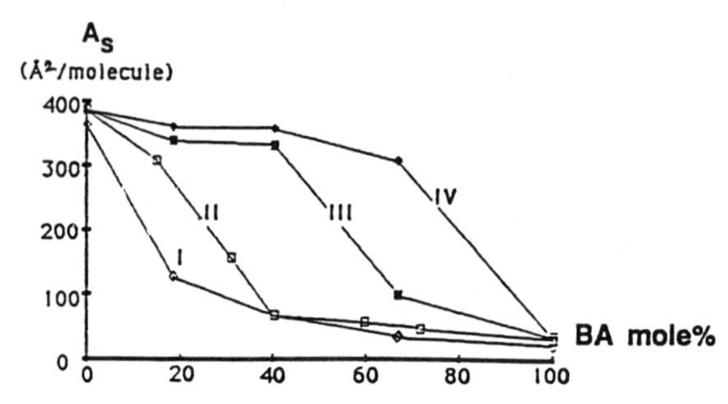

Figure 11 - Variation of the molecular surface area (As) of Sodium hexadecyl sulfate (SHS) with the copolymer composition in VA-BA copolymer latexes as a function of process (same references as in Figure 10). (Reproduced with permission from reference 12. Copyright 1987 Steinkopff Verlag Darmstadt.)

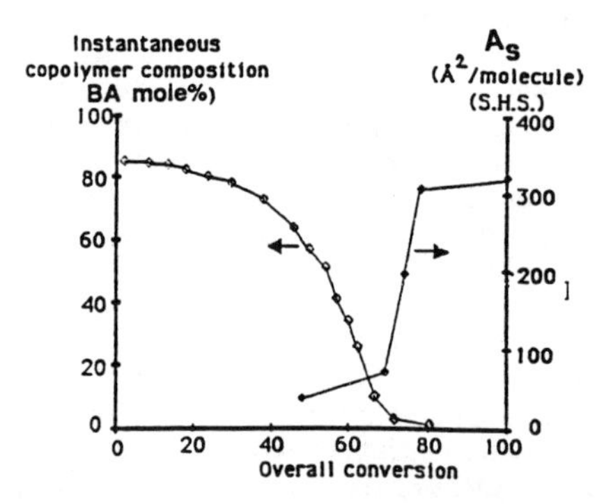

Figure 12 - Variation of the molecular surface area (As) of SHS and instantaneous copolymer composition (BA mole %) with conversion in a 60/40 (mole ratio) VA/BA batch latex. (Reproduced with permission from reference 12. Copyright 1987 Steinkopff Verlag Darmstadt.)

Dynamic mechanical behavior of films As mentioned above, the dynamic mechanical behavior of polymer blends is a way to obtain information on the geometric arrangement of polymer domains. In our case, part of such information should apply to the structure of latex particles as far as this structure remains unmodified during their coalescence. In the following, thermograms of the dynamic mechanical behavior of several systems are shown and discussed. It must be noted that the spectrometer measures in fact the rigidity of the sample and converts it into the shear modulus from the knowledge of its dimensions. In the present study, films were obtained from the different latexes after evaporation of water. Thus, the thickness of achieved films was not exactly constant and therefore the absolute values for G' and G" were not available. As a reasonable assumption, we suggest that G' (T=100 K) is the same for all the samples (~±2 GPa) and is called G_0; thus corresponding thermograms are normalized for plotting. It is worthy to note that $\tan \phi = G''/G'$ is not dependent upon the sample dimensions.

Figure 13 shows the real part (or elastic component) of the modulus G' and tan ϕ, on one hand for pure PVA and PBA (samples BA9 and BA1, respectively), and on the other hand for films obtained from different "core-shell" latexes (CS1 to CS3). Figure 14 refers to films cast from different "batch" latexes (BA3 to BA5 and BA8). Except for the two homopolymers, all the films exhibit two main relaxation peaks (around 230 K, i.e. -40 °C and 310 K, i.e. 40 °C) corresponding to the glass-rubber transitions of PBA and PVA, respectively. This indicates clearly that even in the case of the batch polymerization process, a part of each particle is very rich in BA, while the rest is very rich in VA. Furthermore, the analysis of the decrease in modulus through T_g of PBA remains small (less than one decade) even at high BA content, as for sample BA5. We have shown elsewhere that this behavior corresponds to the case of a continuous rigid matrix (PVA) with soft inclusions (domains rich in BA) *(7)*. This conclusion is consistent with the measured polymerization kinetics which indicates that the cores of these particles are very rich in BA, while their shells are nearly made of pure PVA. The coalescence of such particles at a temperature close to T_g of PVA and in the presence of water (which is known to be a plasticizer of PVA) leads to the formation of a continuous matrix of PVA. However, the comparison of samples "BA" with samples "CS" in Figure 13 indicates a difference : the modulus G' between the two glass transition temperatures decreases more rapidly for the "BA" samples than for the "CS" samples. In accordance to this observation, tan ϕ exhibits a less deep minimum for "BA" than for "CS" samples in the same temperature range.

In order to understand such a behavior, another latex (CS 6) was prepared using three polymerization steps, namely (i) polymerization of pure PBA, (ii) copolymerization of a 50 mole % BA/VA copolymer and (iii) polymerization of pure PVA. Figure 15 shows the corresponding thermograms for films cast from CS2 and CS6, respectively. The same features are observed in the intermediate temperature range. Such a behavior was interpreted by taking into account the presence of the copolymer layer, which in fact introduces a softer gradient in composition from the core to shell polymer than for the classical two step core-shell polymers *(42)*. In a similar manner, the batch polymerization process of BA and VA results, as a first approximation, in particles with three domains: cores with a homogeneous composition copolymer (rich in BA), shells of nearly pure PVA, and an intermediate layer which acts as an interphase with a composition gradient. From the analysis of the dynamic mechanical thermograms, it is possible to roughly determine the profile of this gradient, and in the particular case of "BA", this calculation has led to a curve similar to the one obtained during the polymerization

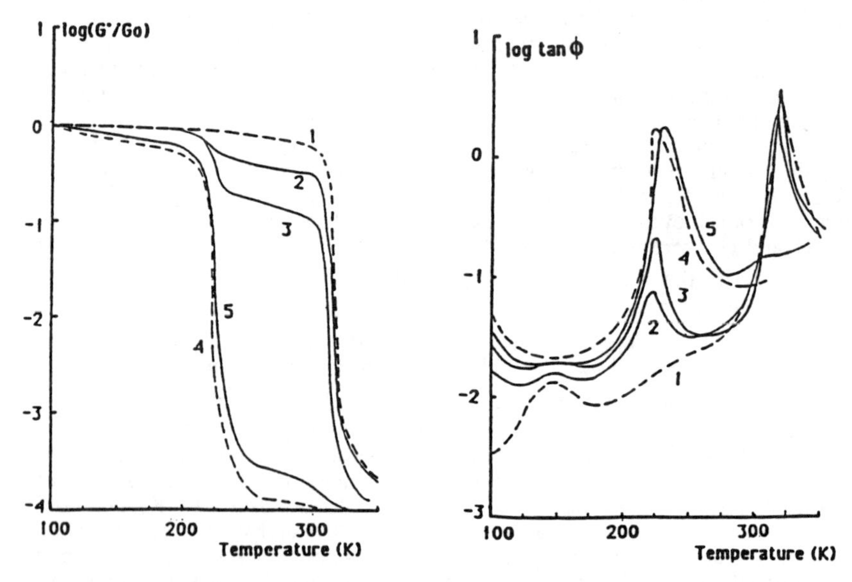

Figure 13 - Micromechanical spectra vs temperature (0.1 Hz) for homopolymers and core-shell VA-BA copolymers.1) BA 9 ; 2) CS 3 ; 3) CS 2 ; 4) BA1 ; 5) CS 1

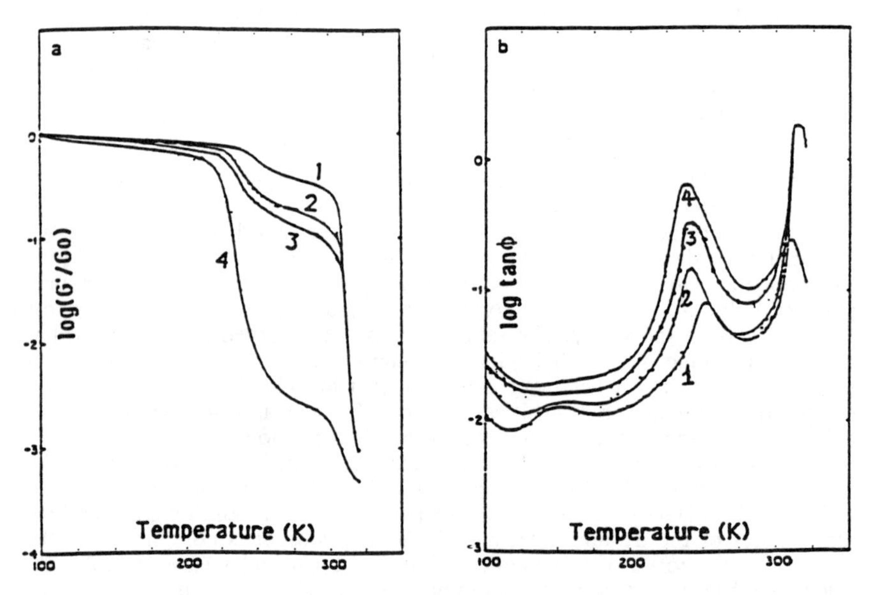

Figure 14 - Micromechanical spectra vs temperature (1 Hz) for VA-BA batch copolymers 1) BA 8 ; 2) BA 5 ; 3) BA 4 ; 4) BA 3

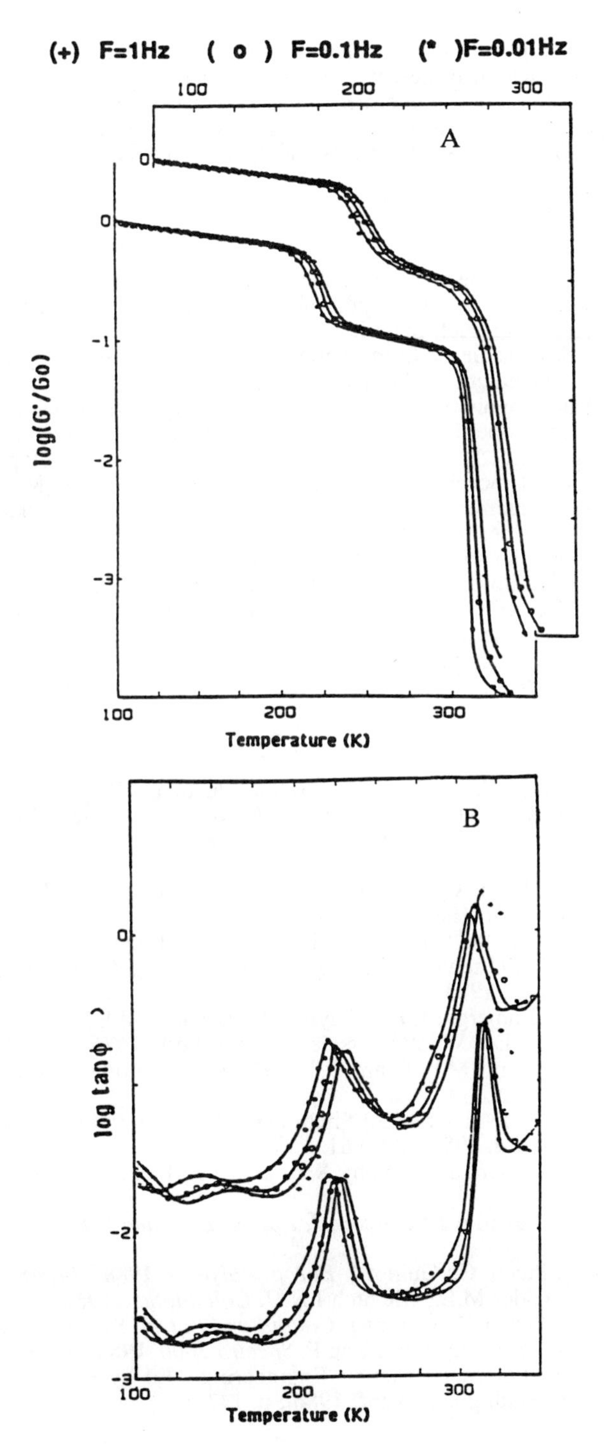

Figure 15 - Micromechanical spectra vs temperature for core-shell copolymers at different frequencies: A) CS 2 ; B) CS 6

(instantaneous composition) *(42)*. It is worthy to note that such information is difficult to obtain once the coalescence is achieved, because of the small size of the domains (the thickness of the intermediate layer was estimated to be less than 10 nm). The consequence of such an observation is that the structure of the latex particles considered in this work remains mainly unchanged during the coalescence step.

Conclusions

This systematic study on the preparation and the characterization of VA-BA emulsion copolymers prepared from various polymerization pathways showed that a quite thorough description of such materials (particularly with regards to the particle morphology) could be achieved from a variety of approachs combining experimental data and theoretical predictions. At first the advantage of an accurate investigation of the kinetics of batch copolymerization process is confirmed. Due to wide differences in physicochemical properties and reactivity of both monomers, a complex kinetic behavior was evidenced, which was thought to favor the development of a composition-gradient morphology in the corresponding particles. Moreover, the knowledge of such parameters was also found to be quite useful to understand the kinetic behavior when dealing with other polymer processes. Secondly, these latexes were examined through various methodologies based on the characterization of either the molecular and colloidal properties of latex particles or the micromechanical behavior of films formed from those latexes. All of these methods were found to well corroborate differences in particle morphology which originated from various types of emulsion polymerization processes.

Literature cited.

(1) Guillot J.; Guyot A.; Pichot C. In "An introduction to Polymer Colloids - F. Candau and R.H. Ottewill Eds, Kluwer Academic Publishers, **1990** ; p 97.
(2) Chujo K.; Harada Y.; Tokuara S.; Tanaka K. *J. Polym. Sci.* C, **1969** ; *27,* 321
(3) Misra S.C.; Pichot C.; El Aasser M.S.; Vanderhoff J.W. *J. Polym. Sci., Polym. Chem. Ed,* **1983** ; *21,* 2383.
(4) Vanderhoff J.W. In "Science and Technology of Polymer Colloids" Vol..1. Poehlein G.W, Ottewill R.H. and Goodwin J.W., Eds, Martinus Nijhoff, Nato Asi Series, **1983** ; 67.
(5) Brandrup J.; Immergut E.H. "Polymer Hanbook" 2nd Ed., J. Brandrup and E.H. Immergut Eds Wiley Interscience, New York, **1975.**
(6) Pichot C.; Llauro M.F.;Pham Q.T.*J. Polym. Sci.,Polym. Chem. Ed,* **1981** *19,* 269.
(7) J. Delgado J.; El Aasser M.S.; Silebi C.A.; Vanderhoff J.W. *J. Polym. Sci., Polym. Chem Ed,* **1986** ; *24,* 861.
(8) Cavaillé J.Y.; Jourdan C.; Kong X.Z.; Perez J.; Pichot C. *Polymer,* **1986** ; *27,* 693.
(9) Jourdan C.; Cavaillé J.Y.; Perez *J. Polym. Eng. and Sci.*, Oct **1988,** ; *28,* 1308.
(10) Kong X.Z.; Pichot C.; Guillot J *Europ. Polym. J,* **1988** ; *24,* 485.
(11) Maron S.; Helder M.E.; Ulevitch I.N. *J. Colloid. Sci.,* **1954** ; *9,* 89.
(12) Kong X.Z.; Pichot C.; Guillot J. *Colloid and Polym. Sci.*, **1987** ; *265*, 791.
(13) Cavaillé J.Y.; Salvia M.; Merzeau P. *Spectra 2000*, **1988** ; *16,* 37.
(14) Guillot J. In Polymer Reaction Engineering". Ed by R.H. Reichert and W.Geiseler., Hüthig and Wepf, **1986** ; p. 147.

(15) Nomura M. *Makromol. Chem.*, **1985** ; Suppl. *10/11*, 25.
(16) Delgado J.; El Aasser M.S.; Silebi C.; Vanderhoff J.W.; Guillot J. *J. Polym. Sci., Part B : Polymer Physics*, **1988** ; *26*, 1495.
(17) J. Smith. J.W. *Am. Chem. Soc.*, **1948** ; *70*, 395.
(18) Yeliseyeva Y.I.; Zubov I.; Malafeyevsbaye V.F. *Vysokomol. Soedin. 7* **1965** ;*7/8*, 1348.
(19) Guillot J.; Emelie B. *Makromol. Chem. Rapid. Comm* **1991** ; *12*, 117.
(20) Johnston N.W..*Polym. Prepr. (Am. Chem. Soc., Div. Polym. Chem.)* **1973** ; *14*, 46.
(21) Coleman B.D.; Fox T.F. *J. Polym. Sci., Part A : Gen. Pap.*, **1963** ; *1*, 3183.
(22) Fox T.G. *Bull. Am. Phys. Soc.*, **1956** ; *1*, 173.
(23) Vankrevelen D.W.and Hofftyzer In "Properties of Polymers" 2nd Edt Elsevier Scientific Publishing Company, Amsterdam, **1976** ; 81.
(24) Perez J. *Rev. Phys. Appl.*, **1986** ; *21*, 93.
(25) Cavaillé J.Y.; Perez J.; Johari G.P. *Phys. Rev. B* **1989** ; *39*, 2411.
(26) Cavaillé J.Y.; Jourdan C.; Perez J. *Makromol. Chemie Symp.*, **1988** ; *16*, 341.
(27) Haskhin Z. J. *Appl. Mechanics*, **1983** ; *50*, 481 (1983).
(28) Kerner E.H. *Proc. Phys. Soc.*, **1956** ; *B69*, 808.
(29) Dickie R.A. *J. Appl. Polym. Sci.*, **1973** ; *17*, 45.
(30) Lewis T.B.; Nielsen L.E. *J. Appl. Polym. Sci.*, **1970** ; *14*, 1449.
(31) Ouali N.; Cavaillé J.Y.; Perez J. 3rd European Symposium on Polymer blends 1990, Cambridge UK, Plastics Rubber and Composites Processing and Applications, **1991.**, *16*, 55.
(32) Ramirez W.; Guillot J. *Makromol. Chem.*, **1989** ; *189*, 361.
(33) Nomura M.; Harada N.; Nakagawara K.; Iguchi W.; Nagata J. *J. Chem. Eng. of Japon*, **1970** ; *4 (2)*, 48.
(34) De la Cal J.C.; Urzay R.; Zamora A.; Forcada J.; Asùa J.M. *J. Polym. Sci., Part A*, **1990** ; *28*, 1011.
(35) Guyot A.; Guillot J.; Pichot C.; Rios Guerrero L. In Emulsion Polymers and Emulsion Polymerization, Eds. A. Hamielec and D. Bassett, ACS Symp. Series, **1981** ; *n° 165*,, 415
(36) Ito K.; Yamashita Y. *J. Polym. Sci.* , **1965** ; *A3*, 2165.
(37) Dimonie V.; Shaffer O.L.; El Aasser M.S.; Vanderhoff J.W. Preprint abstracts of the 33nd IUPAC on Macromolecules, July 8, Montreal **1990**.
(38) Daniel J.C. *Makromol. Chem. Suppl.*, **1985** ; *10/11*, 359.
(39) Okubo M.; Hamada A.; Matsumoto T. *J. Polym. Sci., Polym. Chem. Ed*, **1980** ; *16*, 3219.
(40) Min J.J.; Klein A.; El Aasser M.S.; Vanderhoff J.W. *J. Polym. Sci., Polym. Chem. Ed*, **1983** ; *21*, 2845.
(41) Vijayendran B.R.; Bone T.; Gajria C.J. *Applied Polym. Sci.*, **1981** ; *26*, 1353.
(42) Cavaillé J.Y.; Jourdan C.; Perez J.; Guillot J.*Makromol. Chemie Symp.*, **1989** ; *23*, 411.

RECEIVED December 26, 1991

Chapter 12

Correlation of the Extent of Chain Transfer to Polymer with Reaction Conditions for Emulsion Polymerization of *n*-Butyl Acrylate

Peter A. Lovell[1], Tahir H. Shah[1], and Frank Heatley[2]

[1]Polymer Science and Technology Group, Manchester Materials Science Center, University of Manchester Institute of Science and Technology, Grosvenor Street, Manchester M1 7HS, United Kingdom
[2]Department of Chemistry, University of Manchester, Manchester M13 9PL, United Kingdom

Chain transfer to polymer in emulsion polymerisation of n-butyl acrylate has been studied. Branch points in the structures of the polymers formed have been identified by ^{13}C NMR spectroscopy and reveal that branching results from intermolecular chain transfer reactions in which a propagating chain radical abstracts a hydrogen atom from a tertiary C-H bond in an n-butyl acrylate repeat unit on another polymer chain. Under the usual conditions of high instantaneous monomer conversions, this reaction leads to polymers with high extents of branching (10 - 20 branches per 1000 backbone carbon atoms). The effects of polymerisation conditions upon the extent of branching are discussed, and changes in molar mass distribution during emulsion polymerisation are correlated with observations on the extent of chain transfer to polymer.

Many commercial polymers containing alkyl acrylate repeat units are prepared by emulsion polymerisations performed under monomer-starved conditions, and so throughout most of these reactions the monomer concentration is low and the polymer concentration is high. These conditions favour chain transfer to polymer, which would lead to the formation of branched molecules with consequent effects upon properties (*1,2*). This paper reports results obtained during a research programme into correlations between polymerisation process conditions and properties for water-based adhesives prepared by emulsion polymerisation. Direct evidence for the existence of branches arising from chain transfer to polymer in emulsion polymerisation of n-butyl acrylate is presented and correlated with changes in (i) polymerisation conditions and (ii) the molar mass distributions of the copolymers formed.

Emulsion Polymerisations

Two types of emulsion polymerisation were performed, one (coded BXU) involving the use of ammonium persulphate initiator and a polymerisation temperature of 75 °C, and the other (coded BXUR) employing ammonium persulphate with sodium bisulphite

0097–6156/92/0492–0188$06.00/0

as a redox initiation system at the lower temperature of 55 °C. Both types used comonomer mixtures comprising 87.1 mol% n-butyl acrylate (BA), 6.8 mol% acrylic acid (AA) and 6.1 mol% methyl methacrylate (MMA). Additionally, in the BXU preparations only, t-dodecyl mercaptan was used at a level of 0.035% by weight of the monomers. In each case, 10% of the comonomer mixture was added at the beginning of the reaction for the seed stage of the polymerisation. After appoximately 25 min, the remaining 90% of the comonomer mixture was added uniformly over a period of approximately 150 min. A further 45 min was allowed for completion of the reaction.

Samples removed during the polymerisations were analysed gravimetrically to determine the overall monomer conversion based upon the total quantity of comonomer mixture, and the instantaneous monomer conversion based upon the total quantity of comonomer mixture added at that time.

Characterisation

Solutions for gel permeation chromatography (GPC) analysis were prepared by diluting latex samples with tetrahydrofuran (THF) to give a polymer concentration of approximately 0.2%(w/v). Each solution (70 μL) was injected into the THF flowstream of a GPC system operating at 30 °C, and the polymer separated on a series of three columns of 10 μm PL gel with porosities of 500, 10^4, and 10^6 Å using a flow rate of 1 cm^3 min^{-1}. Polystyrene standards were used for calibration.

Samples for ^{13}C nuclear magnetic resonance (NMR) spectroscopy were prepared by dialysing the latex against distilled, deionised water for a period of at least one week, changing the reservoir of water daily. In this way the majority of the salts and surfactants were removed. The polymer was recovered from the dialysed latex by coagulation induced by freeze-thaw cycling and washed thoroughly with water before drying under vacuum at room temperature. ^{13}C NMR spectra were obtained at ambient temperature using a Varian Associates XL-300 spectrometer operating at 75.5 MHz. Samples were made up as solutions or gels in C_6D_6 containing *c.* 100 mg polymer in 1 cm^3 solvent and chemical shifts were referenced to the solvent peak which was taken to lie at δ_C 128.5. In order to maximise signal-to-noise in a given time, spectra normally were run with continuous proton decoupling using a pulse interval of 0.5 s and a flip angle of 45°. However, under these conditions, the relative intensities do not necessarily reflect the relative abundances of each type of carbon due to differential relaxation times and nuclear Overhauser enhancements (NOE). In particular, the intensities of primary and quaternary carbons are underestimated. Thus in order to obtain accurate relative intensities, some samples also were run with suppression of NOE by gated decoupling and with a pulse interval of 10.5 s to allow complete recovery of all carbons between pulses. The technique of distortionless enhancement by polarisation transfer (DEPT) (*3*) was used to determine the multiplicity of a signal, i.e. whether that carbon is primary, secondary, tertiary, or quaternary. This technique also enabled CH/CH_3 and CH_2 sub-spectra to be derived by respectively adding and subtracting DEPT spectra obtained using final ^1H pulses with flip angles of 45° (i.e. DEPT45) and 135° (i.e. DEPT135). In deriving the sub-spectra, the weighting of each DEPT spectrum was adjusted empirically in order to null resonances known to be entirely of the undesired multiplicity.

Interpretation of the ^{13}C NMR Spectra

The ^{13}C NMR spectra were recorded using C_6D_6 as solvent because it gives a triplet peak which does not conflict with those from the polymers under investigation. However, due to aggregation of carboxylic acid groups from the acrylic acid repeat units, the BA/AA/MMA copolymers do not dissolve in C_6D_6 but merely swell to form relatively stiff gels. Nevertheless, there is enough segmental mobility within the gels to produce spectra which are sufficiently well resolved to facilitate detailed interpretation, as can be seen from the representative ^{13}C NMR spectrum shown in *Figure 1.* In addition to intense peaks from the major BA component, the spectra show a number of small peaks. Some of these signals (marked S in *Figure 1*) are due to residual surfactant molecules that were present because dialysis was not exhaustive. There was no evidence for grafting to the surfactant and, as will be shown in the following discussion, each of the weaker signals can be assigned, self-consistently, to sequences of repeat units known to be present from the monomers employed and to structural features arising from chain transfer to polymer.

Figure 2 shows the sub-spectra corresponding to the normal spectrum presented in *Figure 1.* The CH/CH_3 sub-spectrum (*Figure 2a*) was obtained by adding the DEPT45 and DEPT135 spectra with the relative weightings empirically adjusted to null the CH_2 signal at δ_C 35.6 - 37.0. The CH_2 sub-spectrum (*Figure 2b*) was obtained by subtracting the DEPT135 spectrum from the DEPT45 spectrum, the CH peak at δ_C 42.3 being nulled.

The ^{13}C peak assignments summarised in *Table I* were made using several arguments: (a) comparison with the spectra of homopolymers and of samples prepared with either AA or AA and MMA omitted, (b) determination of the carbon multiplicity using DEPT, and (c) comparison of experimental chemical shifts with those calculated using empirical substituent additivity parameters (*4*). Assignments of the major BA peaks were immediately evident from intensity considerations and from well-established correlations between chemical shift and structure (*4*). Among the smaller peaks, the AA carbonyl signal and the MMA α-CH_3, C_q, and OCH_3 signals were

Table I Assignments of resonances in ^{13}C NMR spectra of BA/AA/MMA copolymers

Peak [a]	Chemical shift, δ_C	Assignment [b]
A	14·3	BA CH_3
B	19·8	BA 3-CH_2
C	21·1	MMA α-CH_3
D	31·4	BA 2-CH_2
E	35·6 - 37·0	BA-BA CH_2
F	38·4	Branch CH_2
G	40·4 - 41·1	Branch CH
H	42·3	BA CH
I	42·5 - 44·0	BA-MMA CH_2
J	46·2	MMA C_q
K	49·1	Branch C_q
L	52·1	MMA OCH_3
M	64·9	BA 1-CH_2
N	172·7 - 173·7	Branch or terminal C=O
O	175·0	BA C=O
P	176·0	MMA C=O + terminal or branch C=O
Q	178·9	AA C=O

[a] The letters refer to the labelling in *Figure 1.*
[b] BA = n-butyl acrylate, AA = acrylic acid, and MMA = methyl methacrylate. The carbons in the n-butyl group are numbered starting at the OCH_2 carbon. The location of branch carbons is described in the text.

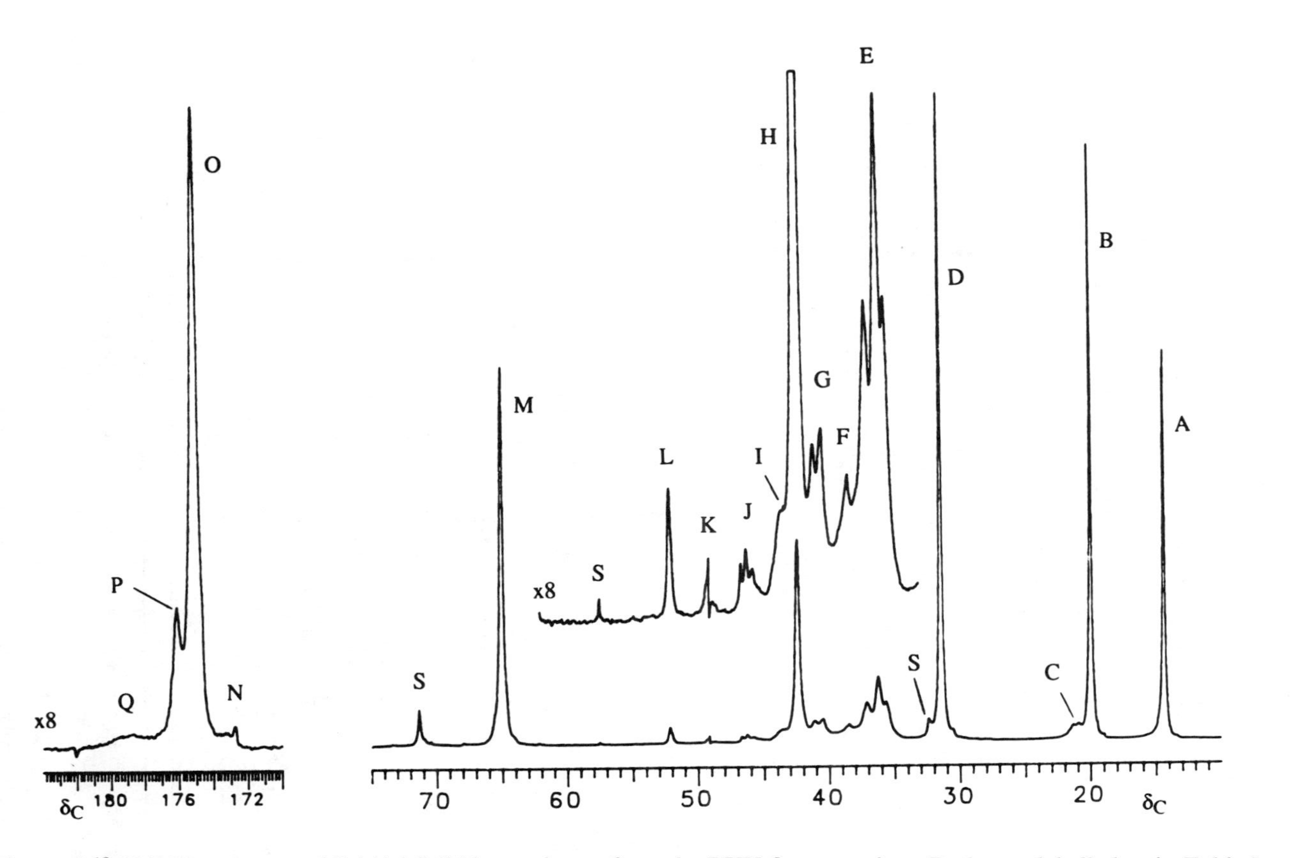

Figure 1 ^{13}C NMR spectrum of BA/AA/MMA copolymer from the BXU-3 preparation. Peaks are labelled as in *Table I*. Peaks marked S are from residual surfactant. Note that there is a small instrumental dispersive spike on peak K.

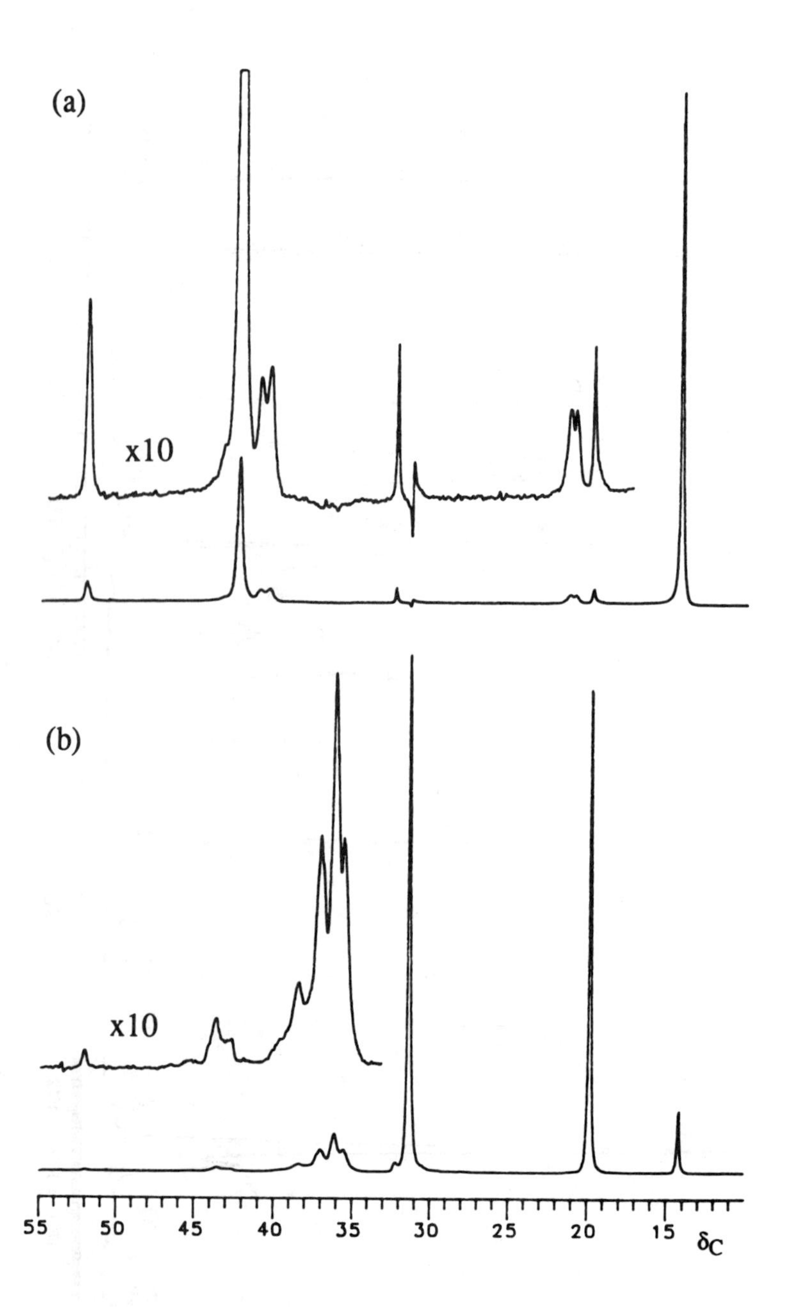

Figure 2 DEPT sub-spectra of the aliphatic carbons of BA/AA/MMA copolymer from the BXU-3 preparation. (a) CH/CH_3 sub-spectrum; (b) CH_2 sub-spectrum.

assigned by virtue of their absence from the spectra of polymers containing no AA or MMA repeat units and (for MMA peaks) by comparison with the spectrum of PMMA (*5*). In the backbone carbon region, the small CH_2 peak at *c.* δ_C 43.5 (partially obscured by the BA CH peak in *Figure 1*, but clearly observed in the DEPT CH_2 sub-spectrum in *Figure 2b*) was assigned to the BA-MMA dyad because of its disappearance in the absence of MMA repeat units and the close correspondence of its chemical shift with the predicted value (δ_C 42, based upon the polyethylene chemical shift of δ_C 30 (*6*) modified by adjacent -COOR and $-CH_3$ groups (*4*)). There was no evidence of CH_2 peaks from MMA-MMA dyads expected at δ_C 55 (*5*), indicating that in spite of the greater reactivity of MMA compared to BA and AA (*7,8*), successive addition of MMA units is suppressed by the low proportion of MMA in the comonomer feed. Further confirmation of the assignment of the BA-MMA dyad CH_2 peak, therefore, is provided by the observation that its relative intensity in the CH_2 sub-spectrum, expressed as a fraction of all backbone CH_2 resonances, is approximately twice the MMA mole fraction.

In the carbonyl region, the small peak at δ_C 176.0 decreased considerably in the absence of MMA repeat units but did not disappear completely. Hence this peak could not be exclusively assigned to the MMA carbonyl carbon. Furthermore, the remaining small peaks, each of which is present in spectra of BA homopolymers, could not be assigned to sequences of normal backbone repeat units. These 'abnormal' signals comprise CH_2 at δ_C 38.4, CH at δ_C 40.4 - 41.1, a quaternary carbon (C_q) at δ_C 49.1 and carbonyl signals at δ_C 172.7 - 173.7 and 176.0 with relative intensities in NOE-suppressed spectra of approximately 3:3:1:1:1. The multiplicities and chemical shifts of these signals are consistent with their origin in a branch point of structure **I**.

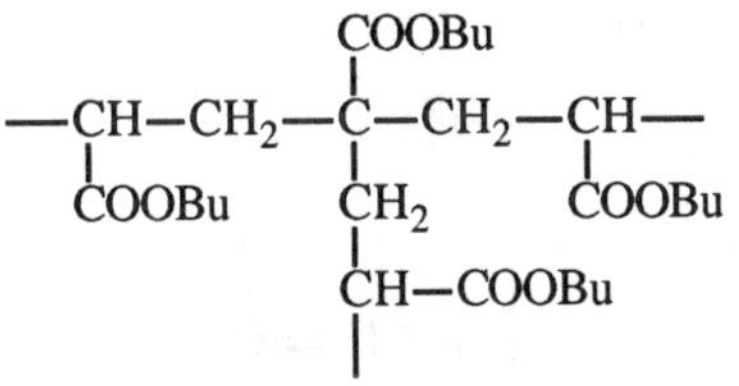

```
                COOBu
                  |
 —CH—CH2—C—CH2—CH—                I
   |              |         |
 COOBu          CH2      COOBu
                  |
                CH—COOBu
                  |
```

The quaternary carbon is the branch point, one of the carbonyl signals is from the ester attached to the branch point, the CH_2 signal is from the carbons adjacent to the branch point and the CH signal is from the next nearest neighbouring carbons. The chemical shifts of the aliphatic carbon signals are also consistent with those calculated on the basis of structure **I** using the chemical shifts of carbons associated with long branches in polyethylene (*6*) and substituent parameters for a -COOR group (*4*), namely: $\delta_C(CH_2) = 38.4$, $\delta_C(CH) = 40.3$; $\delta_C(C_q) = 48.0$.

Branching almost certainly arises via abstraction of a hydrogen atom from a tertiary C-H bond in a BA or AA repeat unit, followed by further propagation from the resulting polymeric radical, as shown in *Scheme 1*. Implicit in *Scheme 1* is the formation of several types of branch point. Formation of branch structure **I**, however, is most probable because of the high proportion of BA repeat units in the copolymers, and the high concentration of BA in the comonomer feed. Other branch points, of similar structure to **I**, may be formed by chain transfer at a tertiary C-H bond in an AA repeat unit and/or by further propagation from the chain transfer product radical by reaction with either AA or MMA. These branch points will only be present at relatively low concentrations because, as a consequence of the small values of the molar ratios AA:BA and MMA:BA, their probability of formation is very much lower than that of branch structure **I**. Each chain transfer to polymer event gives rise to one terminal saturated end-group (indicated by R-H in *Scheme 1*) which in most cases will be

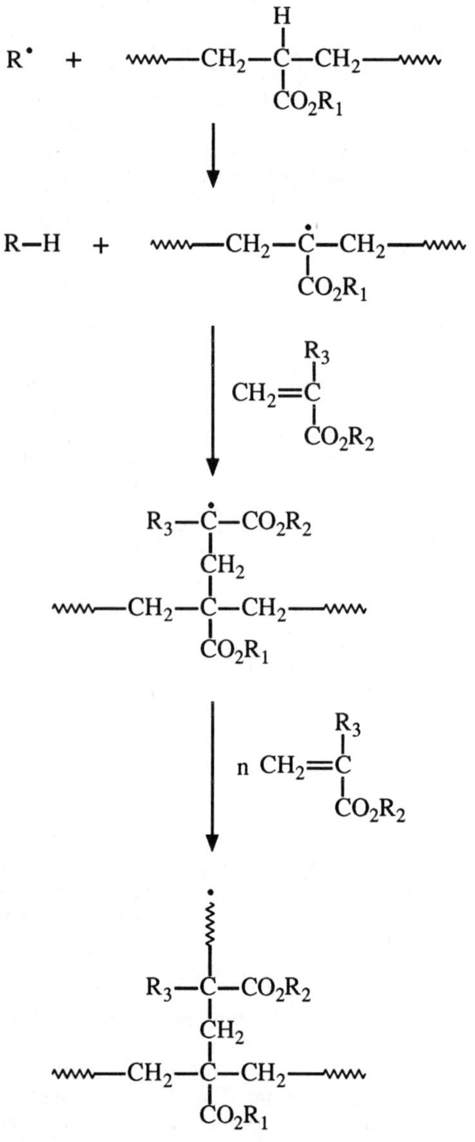

$$R^{\bullet} + \sim\sim CH_2-\underset{CO_2R_1}{\overset{H}{C}}-CH_2\sim\sim$$

$$\downarrow$$

$$R-H + \sim\sim CH_2-\underset{CO_2R_1}{\overset{\bullet}{C}}-CH_2\sim\sim$$

$$\downarrow \; CH_2=\underset{CO_2R_2}{\overset{R_3}{C}}$$

$$R_3-\overset{\bullet}{C}-CO_2R_2 \; ; \; CH_2 \; ; \; \sim\sim CH_2-\underset{CO_2R_1}{C}-CH_2\sim\sim$$

$$\downarrow \; n\ CH_2=\underset{CO_2R_2}{\overset{R_3}{C}}$$

$$\sim\sim^{\bullet} \; ; \; R_3-C-CO_2R_2 \; ; \; CH_2 \; ; \; \sim\sim CH_2-\underset{CO_2R_1}{C}-CH_2\sim\sim$$

Scheme 1 Formation of branches via chain transfer to polymer at sites of tertiary C-H bonds. R_1 = H or $(CH_2)_3CH_3$; R_2 = H, or $(CH_2)_3CH_3$ when R_3 = H; and R_2 = CH_3 when R_3 = CH_3.

$-CH_2CH_2(CO_2Bu)$, again due to the high proportion of BA leading to a high probability for existence of propagating radicals (i.e. $R^•$) with BA terminal active units. If branching is extensive, the number of branches and the number of chain ends will be essentially equal. The chemical shift substituent parameters indicate that the ^{13}C resonances of the ultimate and penultimate main-chain carbons in such a terminal unit should occur at about δ_C 30, i.e. coinciding with (and hence obscured by) the intense n-butyl 2-CH_2 resonance. The presence of saturated BA terminal units also provides an explanation for the existence of the two small carbonyl carbon resonances of approximately equal intensity at δ_C 172.7 - 173.7 and 176.0. One of these signals is due to the $-CO_2Bu$ end-group, and the other to the $-CO_2Bu$ group of the branch point.

Correlation of the Extent of Branching with Emulsion Polymerisation Conditions

A series of emulsion polymerisations were carried out in order to investigate the effects of reaction conditions upon the extent of branching. On the basis of the ^{13}C chemical shift assignments, calculations of the mole percentages of branch points and of the principal repeat units have been made for each of the polymers prepared. These are reported in *Table II* together with information about the methods of calculation, and the reaction conditions and monomer feed compositions employed.

Under the normal, rapid-pulsing conditions the mole percentage of branch points obtained from the branch C_q resonance is significantly lower than that from the carbonyl signals. Also, the mole percentage of MMA calculated from the OCH_3 resonance is lower than that in the feed. These discrepancies arise because the C_q and OCH_3 carbons have relatively long relaxation times and are partially saturated by rapid pulsing. In the NOE-suppressed, fully-relaxed spectra the branch and MMA contents obtained from the aliphatic and carbonyl carbon resonances are identical (within experimental error) and, in the case of MMA contents, in agreement with the monomer feed composition. Since the calculations based upon the carbonyl resonances are reliable under each of the conditions used to record the NMR spectra, they form the basis of the following discussion of the extents of branching.

For each of the polymers prepared the mole percentage of branch points is approximately 2 - 4 mol%, corresponding to 10 - 20 branches per 1000 backbone carbon atoms. This may be compared to that for low density polyethylene (typically 15 - 30 branches per 1000 backbone carbon atoms) which is considered to be highly branched (*1,2,9*). Thus the BA-based polymers reported here may also be considered to have high levels of branching.

The probability of chain transfer to polymer relative to that for propagation is given by the ratio $k_{tr,P}[C\text{-}H]/k_p[M]$ where $k_{tr,P}$ and k_p are the rate constants for chain transfer to polymer and for propagation respectively, and [C-H] and [M] are the total concentrations (within the latex particles) of polymer repeat units with tertiary C-H bonds and of monomer respectively. Under fixed conditions of polymerisation, the ratio $k_{tr,P}/k_p$ can be assumed constant so that in the absence of other competitive chain transfer reactions the ratio [C-H]/[M] controls the extent of chain transfer to polymer and hence the extent of branching. Chain transfer to a particular polymer molecule will depend upon the number of tertiary C-H bonds present in the molecule, and so if all other factors are equal, high molar mass species have a higher probability of undergoing chain transfer to polymer than low molar mass species. In the presence of a chain transfer agent, such as the mercaptan in the BXU preparations, the probability of chain transfer to polymer is reduced and, relative to the probabilities for propagation and chain transfer to mercaptan, is given by $k_{tr,P}[C\text{-}H]/(k_p[M] + k_{tr,SH}[SH])$, where $k_{tr,SH}$ is the rate constant for chain transfer to the mercaptan and [SH] is the concentration of mercaptan in the latex particles.

Figure 3 shows how the instantaneous and overall monomer conversions, and molar mass distribution (MMD) vary with time t from the end of the seed stage ($t = 0$)

Table II Summary of the polymers prepared and the results from ^{13}C NMR spectroscopy

Polymer [a]	Polymerisation temperature, °C	Comonomer feed composition (mol%)			Mole percentages from ^{13}C NMR sprectra [b]			
		BA	AA	MMA	BA+MMA [c]	Branch [c]	MMA [d]	Branch [e]
BXU-3	75	87.1	6.8	6.1	90.3	3.4	5.3 (6.4)	2.4 (4.2)
BXU-17 (at $t = 60$ min)	75	87.1	6.8	6.1	92.5	2.2	5.9	2.3
BXU-17 (at $t = 120$ min)	75	87.1	6.8	6.1	92.5	2.8	6.0	2.4
BXU-17	75	87.1	6.8	6.1	91.5	3.3	5.8	2.6
BXU-15	55-80 [f]	87.1	6.8	6.1	93.0	(1.7)	5.0	2.0
BXU-5	75	93.4	0.0	6.6	97.0 (96.9)	3.0 (3.1)	5.5 (6.8)	2.2 (3.4)
BXU-18	75	100.0	0.0	0.0	96.0	4.0	absent	(4.2)
BXUR-1	55	87.1	6.8	6.1	90.5	1.9	5.0 (5.9)	1.6 (2.2)
BXUR-4	75	87.1	6.8	6.1	90.7	3.0	5.3	2.4
BXUR-5	55	100.0	0.0	0.0	96.9	3.1	absent	(3.2)

[a] At completion of the reaction unless otherwise indicated.
[b] Values calculated from ^{13}C NMR spectra recorded under the normal, rapid-pulsing conditions with the exception of those given in parentheses which were calculated from spectra recorded under conditions of NOE-suppression with a long pulse interval. The estimated accuracies expressed as percentages of the mole percentages are: BA, ±3%; MMA, ±10%; branch, ±15%.
[c] Calculated from the carbonyl carbon integrals $I_1(\delta_C{:}172 - 174)$, $I_2(\delta_C{:}174 - 177)$ and $I_3(\delta_C{:}177 - 182)$ using the equations: mol% BA+MMA = $100I_2/(I_1 + I_2 + I_3)$; mol% branch = $100I_1/(I_1 + I_2 + I_3)$; [with mol% AA = $100I_3/(I_1 + I_2 + I_3)$].
[d] Calculated from the MMA OCH_3 integral relative to half the total backbone carbon integral.
[e] Calculated from the branch C_q integral relative to half the total backbone carbon integral.
[f] See text.

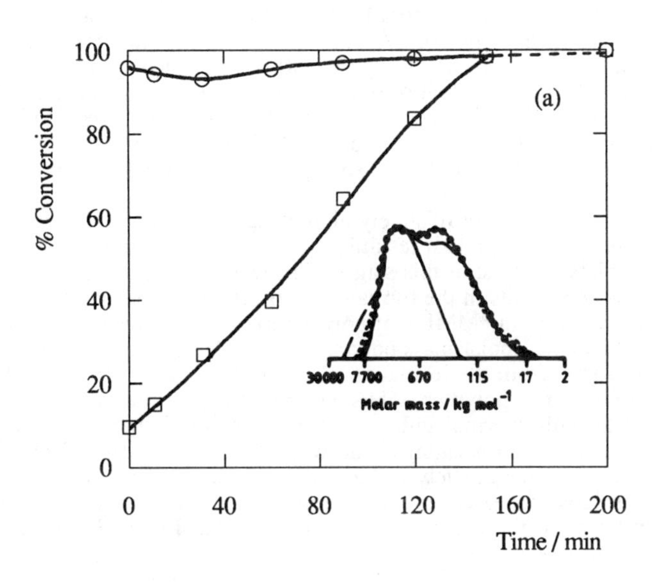

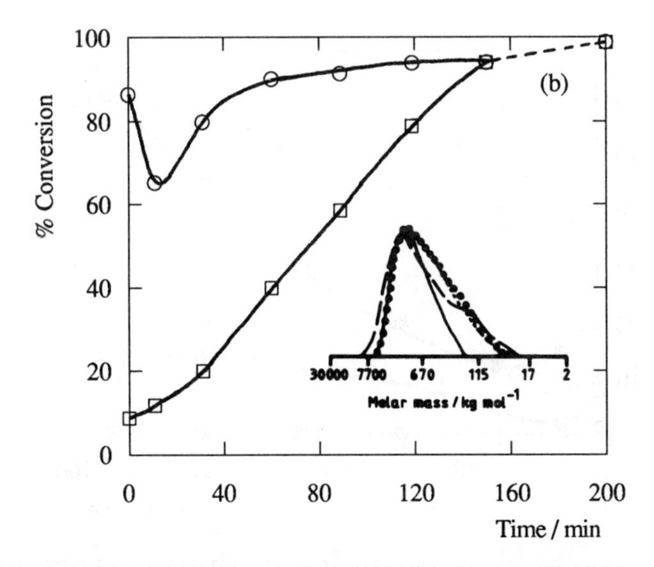

Figure 3 Curves showing the variation of instantaneous (O) and overall (□) monomer conversion with time t following the seed stage in (a) BXU-3 and (b) BXUR-1 preparations of the BA/AA/MMA copolymer. The insets are the molar mass distribution curves of the polymers at $t = 0$ (——), t = 60 min (. . .), t = 120 min (- - -) and the final polymer (— —).

to completion of polymerisation for BXU and BXUR preparations of the BA/AA/MMA copolymer. If it is assumed that the monomers are present predominantly in the latex particles, then with only small errors [C-H]/[M] is given by $f_A.\%IC/(100 - \%IC)$ where f_A is the total mole fraction of BA and AA in the comonomer feed and $\%IC$ is the percentage instantaneous conversion (by weight). Thus, since f_A is high, and $\%IC$ is also high from $t = 0$ to completion of the reaction (*Figure 3*), the ratio [C-H]/[M] is high throughout this period which corresponds to formation of *c.* 90% of the polymer. For both preparations, therefore, most of the polymer is formed under conditions which favour chain transfer to polymer, so explaining the high levels of branching in these polymers.

For BXU-3, polymer of relatively high molar mass is formed in the seed stage due to the comparatively high [M] resulting from the presence of excess monomer in the form of droplets for most of this stage. However, throughout the monomer feed stage [M] is low and results in the formation of much lower molar mass polymer, as can clearly be seen from the MMD's of polymer samples removed during this period (*Figure 3a*). The higher molar mass polymer formed during the monomer feed stage must, therefore, arise from the formation of branches on polymer molecules which undergo chain transfer to polymer. Thus the broadening of the MMD with reaction time is consistent with intermolecular chain transfer to polymer. There also is the possibility of coupling of propagating chains which have resulted from chain transfer to polymer, a process which ultimately will lead to crosslinking. However, since the final polymers were soluble, the extent of such coupling reactions was not high. This can be understood in terms of the expected low mobilities and low concentrations of the propagating chain radicals. The development, late in the polymerisation, of a high molar mass shoulder to the MMD most probably is due to preferential chain transfer

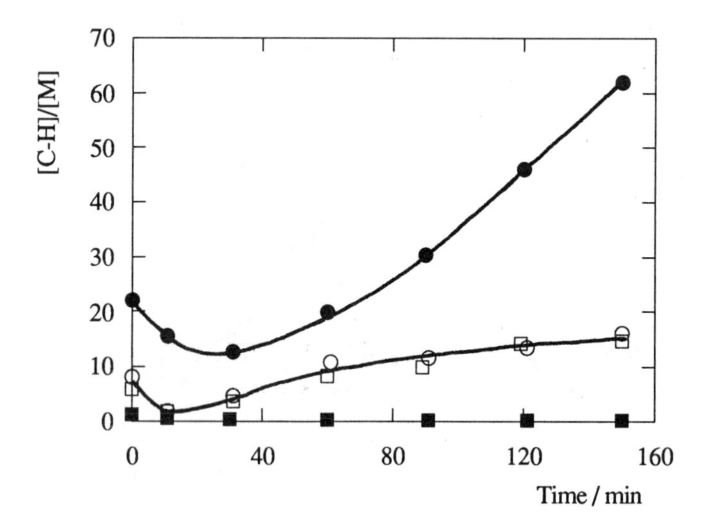

Figure 4 Curves showing the variation of [C-H]/[M] with time t following the seed stage in the BXU-3 (●), BXU-15 (■), BXUR-1 (□) and BXUR-4 (O) preparations of the BA/AA/MMA copolymer. The method of calculation of [C-H]/[M] is described in the text.

to the polymer molecules of high molar mass. Analysis of samples from a repeat preparation (BXU-17) showed that the extent of branching increases with time from $t = 60$ min to completion of the polymerisation (*Table II*), consistent with a steady increase in [C-H]/[M] (*Figure 4*). The close agreement between the extents of branching found in the final BXU-3 and BXU-17 polymers shows that the reproducibility of the reaction is good.

Similar observations to those for BXU-3 can be made for the preparation of BXUR-1 polymer, except that during the monomer feed stage the instantaneous conversions are lower, especially for the period $0 < t < 60$ min (*Figure 3b*). The higher [M] during this period, together with the absence of mercaptan, give rise to the formation of higher molar mass polymer than during the monomer feed stage for BXU-3. Significantly, [C-H]/[M] also is much lower than at equivalent times during the BXU-3 preparation (*Figure 4*). When considered together with the lower polymerisation temperature, which can be expected to reduce the value of $k_{tr,P}/k_p$, these observations explain qualitatively the lower extent of branching detected in the BXUR-1 polymer.

Preparation BXUR-4, which was a repeat of BXUR-1 but performed at the higher polymerisation temperature, gave conversion curves similar to BXUR-1, and an almost identical variation of [C-H]/[M] with time during the monomer feed stage (*Figures 4* and *5*). Thus, the substantially higher extent of branching in the final BXUR-4 polymer must result principally from the increase in $k_{tr,P}/k_p$ corresponding to the increase in polymerisation temperature from 55 to 75 °C. The extents of branching in the final BXU-3 and BXUR-4 polymers (both prepared at 75 °C) are identical within experimental error (*Table II*), despite the much higher [C-H]/[M] existing during the

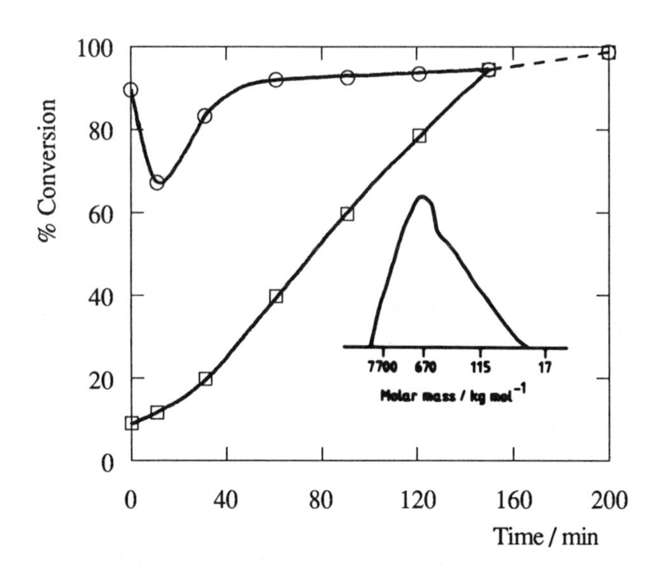

Figure 5 Curve showing the variation of instantaneous (O) and overall (□) monomer conversion with time *t* following the seed stage in the BXUR-4 preparation of the BA/AA/MMA copolymer performed at 75 °C. The inset shows the molar mass distribution curve of the final polymer.

BXU-3 preparation (*Figure 4*). This may be explained, at least qualitatively, on the basis that the extent of branching in the polymers from BXU preparations is reduced by the contribution from the competitive chain transfer reaction with the mercaptan.

In an attempt to provide a comparison with the normal BXUR preparations, a further repeat of the BXU-3 preparation (BXU-15) was carried out using the lower reaction temperature of 55 °C. However, at this temperature, the polymerisation proceeded at a markedly lower rate. The instantaneous conversion was relatively low at the end of the seed stage and decreased further during the addition of the remaining monomer (*Figure 6*) giving rise to relatively low values of [C-H]/[M] (*Figure 4*). On

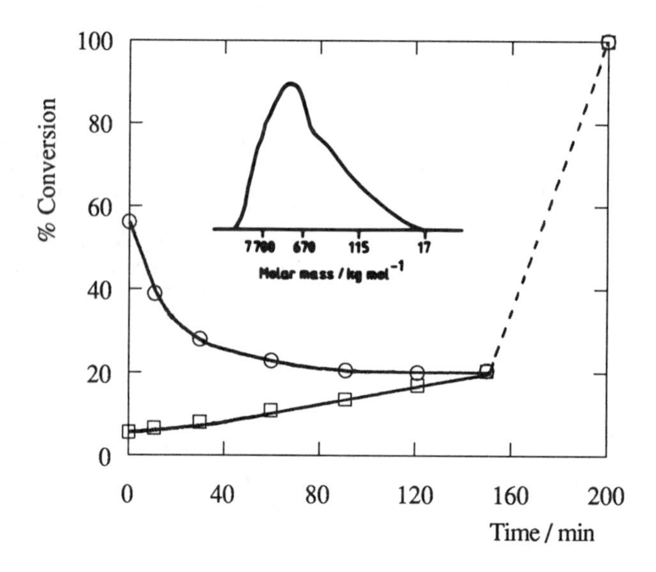

Figure 6 Curve showing the variation of instantaneous (O) and overall (□) monomer conversion with time *t* following the seed stage in the BXU-15 preparation of the BA/AA/MMA copolymer performed at 55 °C. The polymerisation temperature was raised to 80 °C for $t > 150$ min. The inset shows the molar mass distribution curve of the final polymer.

completion of the addition, the temperature was raised to 80 °C in order to convert the unreacted monomer which comprised *c*. 80% of the total monomer charge. Even though the majority of BXU-15 polymer actually was formed at the slightly higher temperature of 80 °C, its formation took place mainly under monomer-flooded conditions with a considerably higher [M] and lower [C-H]/[M] than in a normal BXU preparation, hence yielding a final polymer with a lower proportion of low molar mass polymer and a lower extent of branching than the final polymer from BXU-3.

Experiments aimed at identifying possible differences in the contributions of the three monomers to branching were performed. There was no significant change in the extent of branching when AA was omitted from an otherwise normal BXU preparation (*cf.* data for BXU-5, BXU-3 and BXU-17 in *Table II*). This suggests that there are no major differences in either the H-abstraction capabilities of BA-terminated and AA-terminated propagating chain radicals, or the ease of abstraction of tertiary hydrogen atoms from BA and AA repeat units. However, homopolymerisation of BA (BXU-18 and BXUR-5) leads to increases in the extent of branching which are far greater than the increase expected from the higher proportion of tertiary C-H bonds present in the polymer. Propagating chain radicals terminated in BA units are secondary and so can be expected to be of higher reactivity than the tertiary free-radicals of MMA-terminated propagating chains. Estimates based upon monomer reactivity ratios (for copolymerisation with a reference monomer) and homopropagation rate constants (*7*) suggest that at 60 °C BA-terminated chain radicals are approximately twice as reactive as MMA-terminated chain radicals. Additionally, since chain transfer to polymer produces a tertiary polymeric radical (*Scheme 1*), preferential H-abstraction by the less stable BA-terminated chain radicals is to be expected. These effects should be accentuated by a reduction in temperature, an observation that is in accord with the experimental results which show that the increase in the extent of branching is proportionately greater under BXUR conditions than under BXU conditions (*cf.* BXUR-5 and BXUR-1, and BXU-18, BXU-3 and BXU-17). Thus, the trends in the experimental results can be interpreted in terms of differences in chain-radical reactivity.

Conclusions

The results presented in this paper show that for emulsion polymerisations involving n-butyl acrylate as the principal monomer, the high instantaneous conversions characteristic of semi-continuous batch processes lead to promotion of intermolecular chain transfer to polymer and the formation of highly-branched polymers. In the emulsion polymerisations reported here, the chain transfer reaction proceeds via abstraction of a hydrogen atom from a tertiary C-H bond in either a BA repeat unit or an AA repeat unit in the polymer backbone. Although the presence of branching and the chemistry of formation of the branches was anticipated at the beginning of the research, the existence of high levels of branching in BA-based polymers prepared by emulsion polymerisation has not previously been elucidated. The significance of these results is that they reveal that highly-branched polymers will result from most emulsion polymerisations operated under the usual conditions of high instantaneous conversions if the polymer produced has labile C-H bonds from which hydrogen atoms can be readily abstracted. There are many reports of high levels of branching in polymers prepared by emulsion polymerisation of vinyl acetate (*1,2,10,11*) and it now can be confidently concluded that highly-branched polymers also will be produced from monomer-starved emulsion polymerisations of, for example, acrylic monomers and/or styrene (which gives polymers with labile, benzylic tertiary C-H bonds).

Acknowledgments

The authors thank Rhone-Poulenc Chemicals Ltd for funding this research, and SERC for a grant towards the purchase of the NMR spectrometer.

Literature Cited

1. Sholte, Th.G. in *Developments in Polymer Characterisation*; Dawkins, J.V., Ed.; Applied Science: London,1983 , Vol. 4; p. 1
2. Small, P.A. *Adv. Polym. Sci.* 1977, *18*, 1
3. Doddrell, D.M.; Pegg, D.T.; Bendall, M.R. *J. Chem Phys.* 1982, *77*, 2745
4. Wehrli, F.W.; Wirthlin, T. *Interpretation of Carbon-13 Spectra*; Heyden and Son: London, 1976; p. 37
5. Randall, J.C. *Polymer Sequence Determination: Carbon-13 NMR Method*; Academic Press: New York, 1977; p. 112
6. Randall, J.C. *J. Polym. Sci., Polym. Phys. Edn.* 1973, *11*, 275
7. *Polymer Handbook*; Brandrup, J.; Immergut, E.H., Eds.; 2nd Edn.; Wiley-Interscience: New York, 1975

8. Mori, Y.; Ueda, A.; Tanzawa, H.; Matsuzaki, K.; Kobayashi, H. *Makromol. Chem.* 1975, *176*, 699
9. *Encyclopedia of Polymer Science and Engineering*; Mark, H.F.; Bikales, N.M.; Overberger, C.G.; Menges, G., Eds.; Wiley-Interscience: New York, 1986, Vol. 6; p. 386
10. Hamielec, A.E. in *Emulsion Polymerisation of Vinyl Acetate*; El-Aasser, M.S.; Vanderhoff, J.W., Eds.; Applied Science: London, 1981; p. 49
11. *Encyclopedia of Polymer Science and Engineering*; Mark, H.F.; Bikales, N.M.; Overberger, C.G.; Menges, G., Eds.; Wiley-Interscience: New York, 1989, Vol. 17; p. 393

RECEIVED December 4, 1991

Chapter 13

Alkali Stability of Vinyl Acetate Copolymer and Terpolymer Latexes

Peter S. Martin, Oliver W. Smith, and David R. Bassett

Union Carbide Chemicals and Plastics Company, Inc., P.O. Box 8361, South Charleston, WV 25303

A series of copolymer and terpolymer latexes containing vinyl acetate were synthesized and subjected to alkaline conditions. The various comonomers chosen were as follows: butyl acrylate, 2-ethylhexyl acrylate, vinyl 2-ethylhexanoate, vinyl neo-pentanoate, vinyl neo-nonanoate and vinyl neo-decanoate. The extent of hydrolysis was determined via conductometric titration with standardized HCl following exposure of the latexes to aqueous KOH for periods of 24 and 72 hours. In all cases, enhanced hydrolytic stability correlated with high branched vinyl ester/vinyl acetate and high branched vinyl ester/acrylate molar ratios for vinyl acetate/branched vinyl ester copolymer and vinyl acetate/acrylate/branched vinyl ester terpolymer latexes, respectively. The results were interpreted in terms of a scheme whereby the branched vinyl esters act to guard the vinyl acetate moieties from alkali attack by effectively spacing the vinyl acetates and protecting them via a neighboring group shielding effect.

Owing to its low cost, vinyl acetate (VAc) is an attractive alternative to methyl methacrylate for use in architectural coating latexes. Unfortunately, VAc suffers from poor hydrolytic stability, especially under alkaline conditions and, as such, sees only limited application in exterior coatings.

Alkali resistance is extremely important when a coating is applied over alkaline construction materials, such as cement. It becomes more critical for situations involving exposure to conditions of high humidity.

In Europe, the introduction of the vinyl esters of highly branched carboxylic acids containing nine and ten carbon atoms, vinyl neo-nonanoate (VN-9) and vinyl neo-decanoate (VN-10), respectively, has enabled the formation of all-vinyl emulsion based coatings, which are more suitable for exterior use and possess better alkali resistance than conventional acrylics. Weathering studies in the Netherlands have clearly demonstrated (*1*) that for asbestos cement panels coated with 40% pvc white pigmented 70/30 (w/w) VAc/VN-10 emulsion paints, there was no significant alkaline erosion after ten years of south vertical exposure. By comparison, paints based on an all-acrylic latex experienced approximately 20% erosion over this same time period, while those formulated from vinyl acrylics had completely eroded away after only seven years of exterior weathering.

0097–6156/92/0492–0203$06.00/0

Apart from the cost involved, exterior weathering requires excessive time for evaluation. An alternative, albeit approximate, approach more applicable to a research and development program entails the use of an accelerated stability test.

One type of accelerated alkali stability test (*2*) involves immersion of a pre-weighed latex film in a 2% (w/w) sodium hydroxide solution for a period of fourteen days. Upon removal, the film is rinsed with water and dried to a constant weight. The weight loss is then determined and reported as a measure of alkali stability. While this method has been used to show that all-acrylic latexes are more alkali resistant than VAc/VN-10 latexes, it was not able to distinguish between VAc/VN-10 latexes containing 50 and 70% VAc. One advantage to this method is that it is applicable to pigmented films.

Another approach (*3*) involves the addition of a large excess of one normal potassium hydroxide (1 N KOH) to a latex and the subsequent determination of the unreacted alkali with time by titration with one normal hydrochloric acid (1 N HCl). The degree of hydrolysis is calculated as milligrams of KOH per gram of polymer and expressed as a percentage of the theoretical hydrolysis possible.

Although one may determine the endpoint potentiometrically, both endpoints, i.e., those corresponding to the neutralization of the unreacted KOH and the salt of the weak acid (the saponification product), may be examined successively using conductometric titration. Being independent of pH, the conductometric method enables one to investigate mixtures of acids and bases in a single titration and, unlike the potentiometric technique, yields unambiguous (sharp) endpoints under a wide range of conditions.

In this paper the ability of various branched vinyl esters (BVEs) to provide protection for VAc against alkaline saponification in various VAc-containing vinyl and vinyl acrylic latexes is explored via conductometric titrations. More specifically, this study focuses on the branched vinyl esters of neo-decanoic, neo-nonanoic, neo-pentanoic and 2-ethylhexanoic acid, designated VN-10, VN-9, VN-5 and V2EH, respectively, which are produced from VAc via Union Carbide's transvinylation technology (*4*). Moreover, the effects of 2-ethylhexyl acrylate (2EHA) and butyl acrylate (BA) in combination with both VAc and V2EH in vinyl acrylic terpolymer latexes is also addressed.

Experimental

Emulsion Polymerization. All monomers were used as supplied by Union Carbide. Adequate purity (>98%) was ascertained by gas chromatography and polymerization inhibitors, usually 5 - 55 ppm hydroquinone or monomethyl ether of hydroquinone, were not removed prior to use.

All latexes were prepared using a single emulsion polymerization procedure. The stabilization system comprised 0.4 phm hydroxyethyl cellulose (low molecular weight grade), 3.8 phm nonionic surfactant (25 - 40 mole ethoxylate of alkyl phenol) and 0.1 phm anionic surfactant (alkyl aryl sulphonate). tert-butyl hydroperoxide (0.2 phm) in combination with sodium formaldehyde sulfoxylate (0.2 phm) was employed as the redox activated initiation system and sodium acetate (0.2 phm) as the buffer. Water was included to produce a total solids content of approximately 53%.

Emulsion polymerization was performed in a two-liter kettle equipped with a mechanical stirrer, reflux condenser, thermometer, dropping funnels (3) and heated or cooled via a water bath. Save for 90% of the monomer mixture and 70% each of the redox catalysts as 0.7 wt% aqueous solutions, all ingredients were charged to the reaction vessel. The system was then heated to 60 C while purging with nitrogen. Approximately ten minutes later, the monomer and catalyst feeds (2) were started. A temperature of 68 - 70 C was maintained for the duration of the monomer and catalyst feeds, 3.5 and 4.0 hours, respectively, and for an additional 30 minutes following the completion of the latter feeds. The resulting latex was then cooled and filtered. Monomer conversion was usually 98 - 99 % as determined from the solids content (to nearest 0.01%) of the latex.

Alkaline Hydrolysis. The alkali stability test employed here is similar to one described previously (*3*). The procedure is given as follows: Five grams of latex are added to 50 gms of 1 N KOH, where both are weighed to the nearest 0.01 gm. A 5 gm portion of this mixture, also weighed to the closest 0.01 gm, is diluted to a volume of approximately 80 ml with a 50/50 (w/w) triple distilled water/2-propanol solution and then immediately titrated with standardized 0.5 N HCl. Subsequent 5 gm portions of the latex/KOH mixture are similarly titrated at intervals of 24 and 72 hours. To ensure proper mixing during these time periods, the latex/KOH mixture is agitated on a roller.

Conductometric Titrations. The titrations were performed at room temperature in a conductance cell designed and used to study micelles (*5*). Its essential feature is that the conductance is measured in a region of constant volume, so that the cell constant does not change during the titration. Conductance measurements were performed using a Leeds & Northrup # 4959 Electrolytic Conductivity Bridge operating at 1000 Hz, whose galvanometer null detector provides a sensitivity of 0.1% of the conductance reading.

A typical titration is shown in Figure 1. In this example, 5 gms of a 75/25 (w/w) VAc/V2EH latex were exposed to 50 gms of 1 N KOH for 24 hours. In Figure 1, the conductance (micromhos) of a 5 gm aliquot of the latex/KOH mixture, diluted with a 50/50 (w/w) water/2-propanol solution, is plotted against volume (mls) of titrant (0.5 N HCl). Using a calibrated buret (50 ml - Brinkmann Digital Buret) the mixture was then titrated in increments while being stirred. The conductance was measured after each addition and the titration endpoints were obtained graphically. The initial descending branch in Figure 1 arises from the disappearance of the excess strong base (KOH), the middle portion gives the weak acid content, i.e., the saponification product, and the final increase in conductance results from the excess titrant (HCl). Where the branches intersect mark the endpoints of the titration. The x-component of the middle branch indicates the amount of neutralized weak acid present, be it free acid in the case of VAc or BVE hydrolysis or pendant carboxylic acid produced via acrylate hydrolysis, which thus provides a measure of the extent of alkaline hydrolysis.

Results

The results of alkali stability tests on the vinyl and vinyl acrylic latexes, thirty four in total, are given in Table I. In the table are presented the monomer compositions of the latexes, their relative amounts, given as both weight and mole percents, and the extent of hydrolysis for both 24 and 72 hours of exposure to aqueous KOH. The total amount of saponification is the weak acid increase for these periods expressed as a percent of the theoretical limit. Note that this also includes the contribution arising from acrylate hydrolysis, which gives rise to a pendant weak acid group, as opposed to the free acid resulting from the hydrolysis of the vinyl moieties. Results are also presented in graphical form according to the various families of latexes considered in this study.

In Figure 2 the extent of hydrolysis for VAc/V2EH copolymer latexes exposed to 1 N KOH for 24 and 72 hours is plotted against V2EH content, for V2EH contents of between 10 and 30 wt%. From the diagram it is obvious that

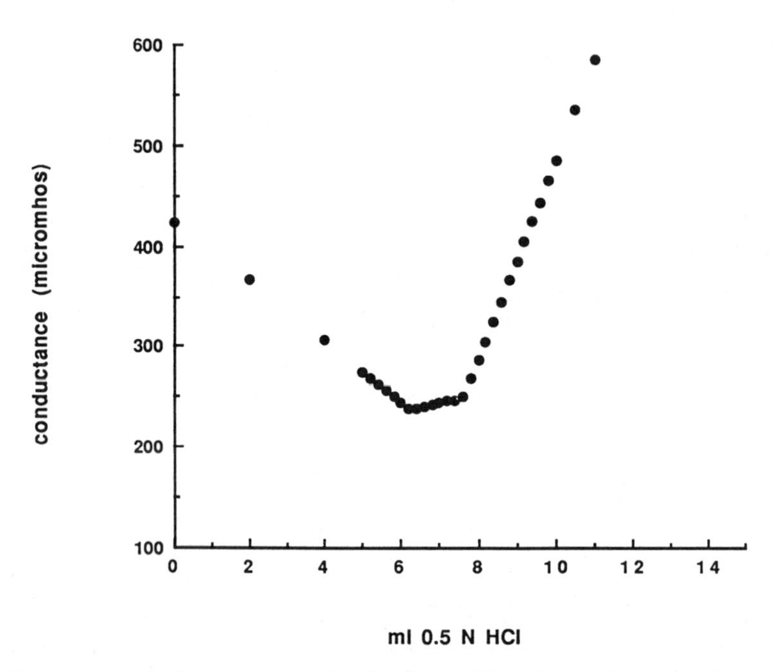

Fig. 1. Conductometric titration of 75/25 (w/w) VAc/V2EH latex following 24 hour exposure to 1 N KOH.

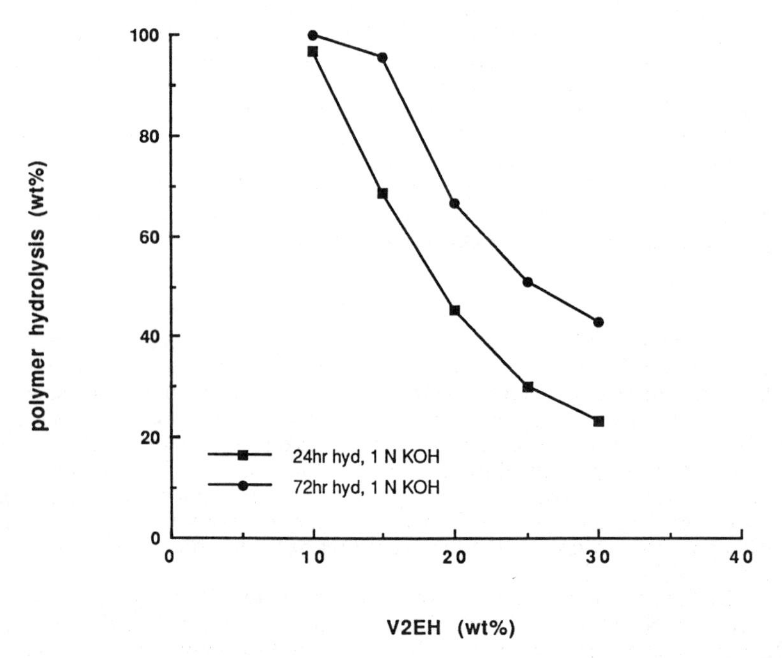

Fig. 2. Hydrolysis of VAc/V2EH latexes following exposure to KOH (aq).

Table I. Hydrolysis of VAc copolymer and terpolymer latexes following exposure to KOH (aq)

Reference	Composition (Monomers)	Composition (wt%)	Composition (mol%)	Hydrolysis (wt%, 24 hr)	Hydrolysis (wt%, 72 hr)
6PSM38	VAc/V2EH	90.0/10.0	94.7/5.3	96.7	100.0
6PSM36	VAc/V2EH	85.0/15.0	91.8/8.2	68.7	95.7
6PSM34	VAc/V2EH	80.0/20.0	88.8/11.2	45.3	66.8
6PSM32	VAc/V2EH	75.0/25.0	85.6/14.4	30.1	51.1
4PSM60	VAc/V2EH	70.0/30.0	82.2/17.8	23.3	43.1
4PSM68	VAc/2EHA	70.0/30.0	83.3/16.7	93.8	100.0
4PSM70	VAc/2EHA/V2EH	70.0/24.0/6.0	83.1/13.3/3.6	85.0	100.0
4PSM72	VAc/2EHA/V2EH	70.0/18.0/12.0	82.9/10.0/7.2	59.4	95.0
4PSM74	VAc/2EHA/V2EH	70.0/12.0/18.0	82.6/6.6/10.7	42.6	71.0
4PSM76	VAc/2EHA/V2EH	70.0/6.0/24.0	82.4/3.3/14.3	34.1	52.9
4PSM50	VAc/BA	70.0/30.0	77.6/22.4	80.2	96.6
4PSM52	VAc/BA/V2EH	70.0/24.0/6.0	78.5/18.1/3.4	71.2	92.9
4PSM54	VAc/BA/V2EH	70.0/18.0/12.0	79.4/13.7/6.9	65.0	91.0
4PSM56	VAc/BA/V2EH	70.0/12.0/18.0	80.3/9.2/10.4	46.7	77.8
4PSM58	VAc/BA/V2EH	70.0/6.0/24.0	81.2/4.7/14.1	31.8	54.6
6PSM74	VAc/BA	57.4/42.6	66.7/33.3	47.8	92.4
6PSM62	VAc/VN-5	57.4/42.6	66.7/33.3	26.2	44.7
6PSM72	VAc/V2EH	50.3/49.7	66.7/33.3	13.4	21.5
6PSM60	VAc/VN-9	48.3/51.7	66.7/33.3	14.4	14.4
6PSM58	VAc/VN-10	46.5/53.5	66.7/33.3	15.[illegible]	18.4
6PSM22	VAc/BA/VN-5	50.2/12.5/37.4	60.0/10.0/30.0	24.6	40.2
6PSM24	VAc/BA/V2EH	44.7/11.1/44.2	60.0/10.0/30.0	13.5	22.9
6PSM18	VAc/BA/VN-9	43.1/10.7/46.2	60.0/10.0/30.0	12.8	14.2
6PSM20	VAc/BA/VN-10	41.7/10.3/48.0	60.0/10.0/30.0	12.2	18.3
6PSM30	VAc/V2EH/VN-5	48.2/15.9/35.9	60.0/10.0/30.0	18.0	26.4
6PSM26	VAc/V2EH/VN-9	41.7/13.7/44.6	60.0/10.0/30.0	9.0	14.8
6PSM28	VAc/V2EH/VN-10	40.3/13.3/46.4	60.0/10.0/30.0	10.9	15.6
6PSM48	VAc/BA/VN-5	40.2/29.9/29.9	50.0/25.0/25.0	23.3	40.8
6PSM50	VAc/BA/V2EH	36.6/27.2/36.2	50.0/25.0/25.0	19.1	24.5
6PSM44	VAc/BA/VN-9	35.5/26.4/38.0	50.0/25.0/25.0	17.0	18.4
6PSM42	VAc/BA/VN-10	34.5/25.7/39.8	50.0/25.0/25.0	14.8	14.8
6PSM56	VAc/V2EH/VN-5	36.6/36.2/27.2	50.0/25.0/25.0	12.1	16.1
6PSM54	VAc/V2EH/VN-9	32.7/32.3/35.0	50.0/25.0/25.0	13.0	17.9
6PSM52	VAc/V2EH/VN-10	31.8/31.5/36.7	50.0/25.0/25.0	8.1	16.2

enhanced hydrolytic stability in VAc/V2EH latexes correlates with high V2EH content. It is interesting to note that in proceeding from a 90/10 to a 70/30 (w/w) VAc/V2EH copolymer, the hydrolytic stability increases by a factor of between two and four.

The results of the alkali stability tests on the vinyl acrylic terpolymer latexes containing 70 wt% VAc are presented in Figures 3 and 4 for the VAc/BA/V2EH and VAc/2EHA/V2EH systems, respectively. In both cases enhanced hydrolytic stability correlates with high V2EH/acrylate ratios. With the exception of the data points for the 24 hour exposure of the 70/30/0 and 70/24/6 (w/w) VAc/BA/V2EH latexes, the extent of saponification appears to be the same within experimental error (~3 wt%) for both systems. One possible explanation for the discrepancy rests with the molecular weight differences between BA and 2EHA, i.e., these two latexes exhibited better hydrolytic stability than their VAc/2EHA/V2EH counterparts simply because, on a molar basis, there is significantly less VAc and more acrylate, which is harder to saponify, in these cases.

The alkali resistance of the VAc/BVE, VAc/BA(10 wt%)/BVE and VAc/BA(25 wt%)/BVE systems with 2/1 VAc/BVE molar ratios are summarized in Figures 5, 6 and 7, respectively. Within experimental error, all three systems showed the same resistance to alkali saponification. More specifically, it is impossible to discern differences among latexes containing V2EH, VN-9 and VN-10, both within a given system and among the three systems considered. Moreover, latexes containing VN-5 displayed poorer stability than those containing V2EH, VN-9 and VN-10 for both 24 and 72 hour exposure to aqueous KOH.

Lastly, in Figures 8 and 9 are displayed the results for the VAc/V2EH(10 wt%)/ BVE and VAc/V2EH(25 wt%)/BVE systems with 2/1 VAc/BVE molar ratios. With the exception of the VAc/V2EH(10 wt%)/VN-5 latex, indistinguishable behavior was observed within and between both systems. More specifically, alkali resistance appears to be independent of both V2EH content in the 0 - 25 wt% range and BVE identity (exception noted) for VAc/V2EH/BVE latexes with 2/1 VAc/BVE molar ratios.

Discussion

There are several factors that contribute to the stability of VAc copolymers. Given that adjacent hydroxyl groups have been shown (*6*) to promote hydrolysis of acetate moieties in partially hydrolyzed p(VAc), one would expect that any comonomer capable of separating the pendent acetate groups on the polymer backbone should effectively decrease the rate of saponification. Since vinyl esters possess very similar reactivity ratios (Table II), copolymerization of VAc with a more sterically hindered ester, i.e., one more resistant to hydrolysis, would be expected to decrease the rate of saponification. In other words, the comonomer would form a random copolymer with VAc, thereby spacing the VAc moieties, which in turn reduces the number and lengths of the VAc sequences, and thus the extent of hydrolysis. This explanation is consistent with the behavior displayed in Figure 2, where alkali resistance correlates with the amount of comonomer, which in this instance is the more robust V2EH.

On the basis of this argument, one would expect ethylene to be the comonomer of choice because its low molecular weight allows for the maximum introduction of nonhydrolyzable segments on a per weight basis. It has in fact been demonstrated that ethylene is an effective stabilizing comonomer for VAc (*3*). Lending to ethylene's effectiveness is the fact that hydrocarbon segments tend to reduce water solubility, thus imparting greater hydrolytic stability. More specifically, polymers with a high oxygen content can more easily facilitate attack by water than can more hydrophobic polymers comprising little or no oxygen (*7*).

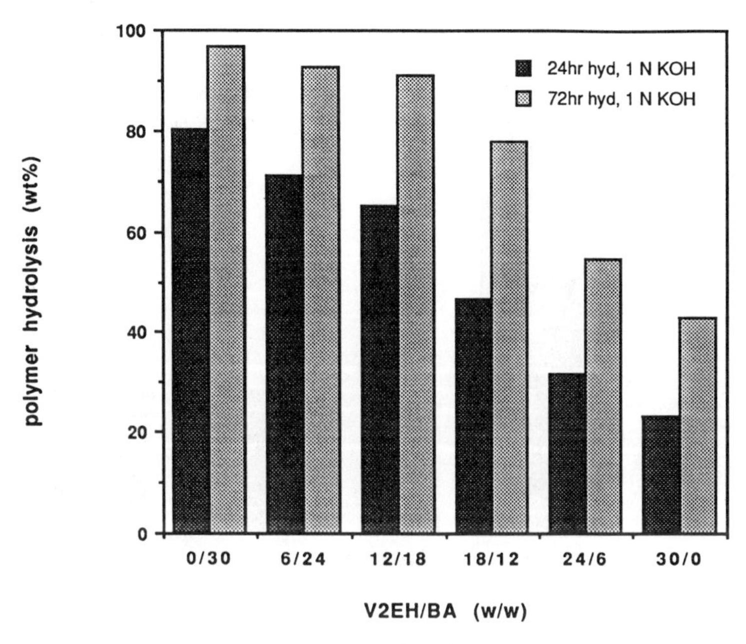

Fig. 3. Hydrolysis of VAc/BA/V2EH latexes containing 70 wt% VAc following exposure to KOH (aq).

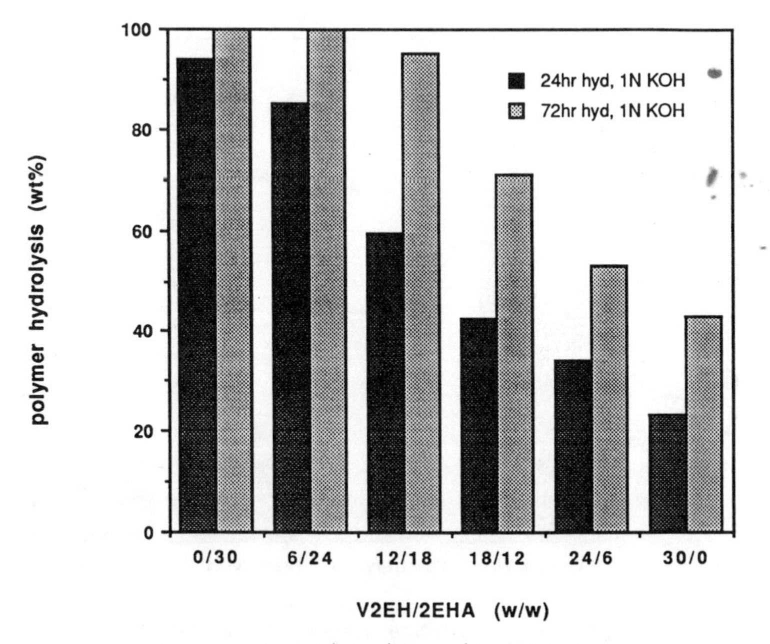

Fig. 4. Hydrolysis of VAc/2EHA/V2EH latexes containing 70 wt% VAc following exposure to KOH (aq).

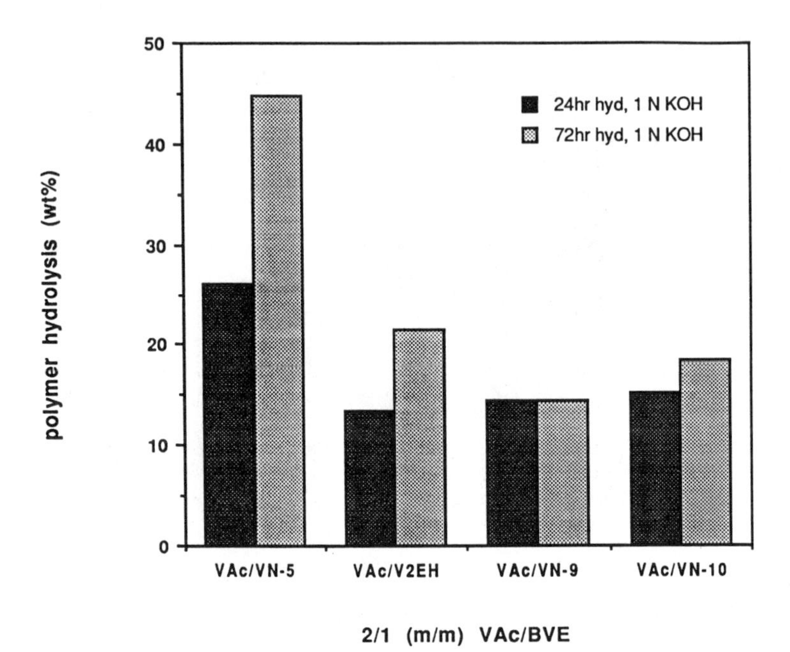

Fig. 5. Hydrolysis of 2/1 (m/m) VAc/BVE latexes following exposure to KOH (aq).

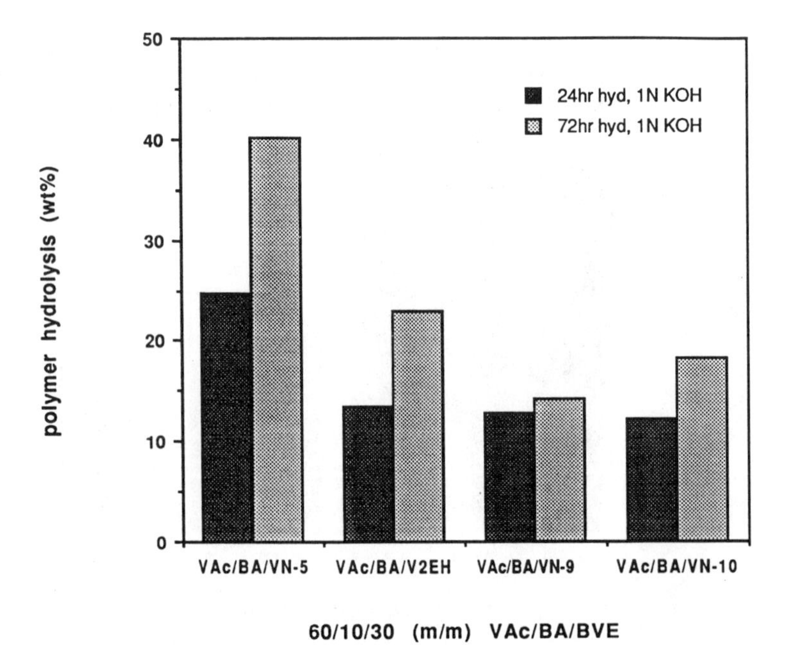

Fig. 6. Hydrolysis of 60/10/30 (m/m) VAc/BA/BVE latexes following exposure to KOH (aq).

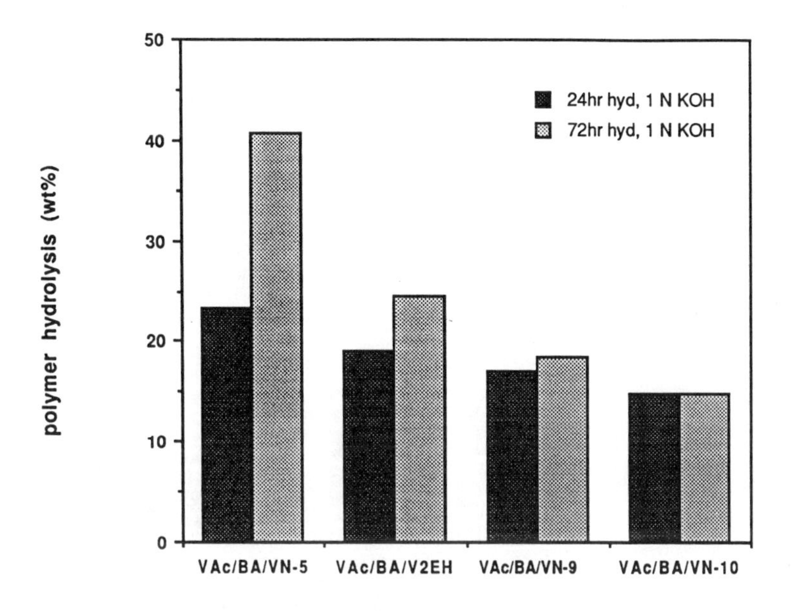

Fig. 7. Hydrolysis of 50/25/25 (m/m) VAc/BA/BVE latexes following exposure to KOH (aq).

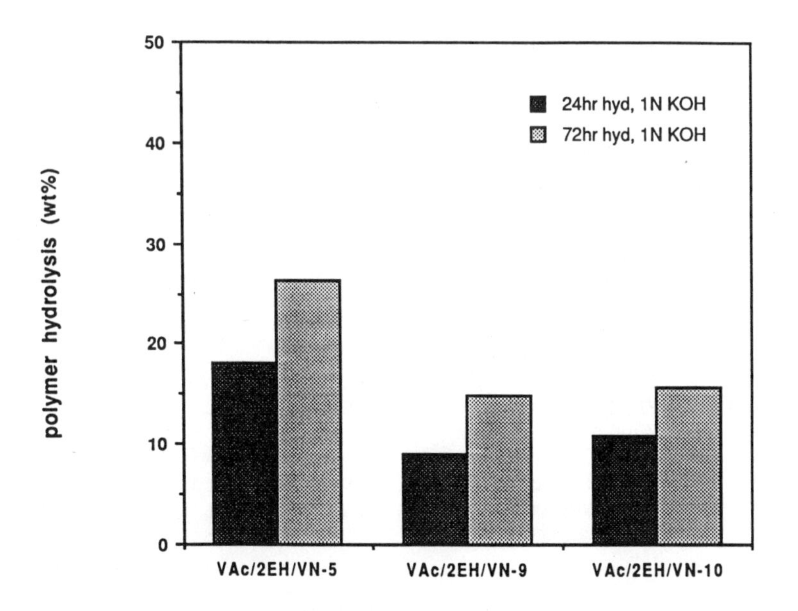

Fig. 8. Hydrolysis of 60/10/30 (m/m) VAc/V2EH/BVE latexes following exposure to KOH (aq).

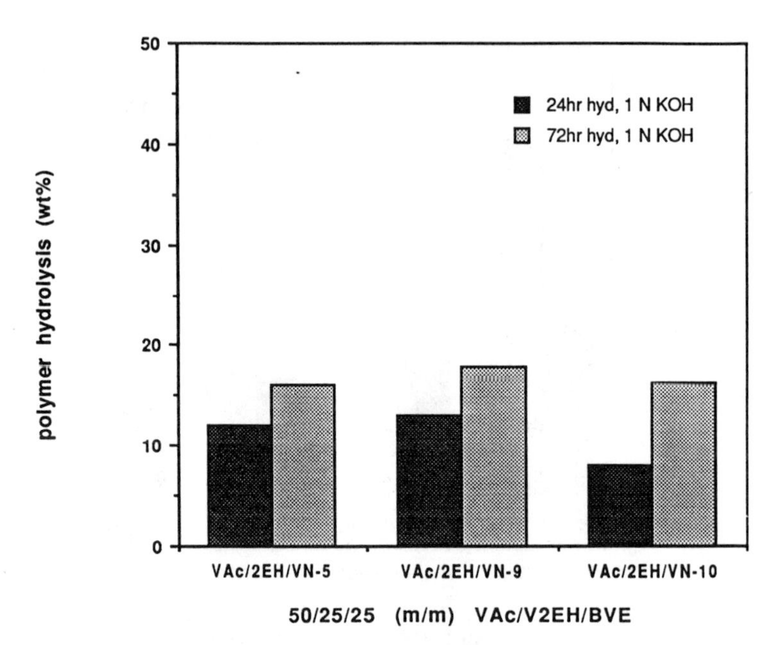

Fig. 9. Hydrolysis of 50/25/25 (m/m) VAc/V2EH/BVE latexes following exposure to KOH (aq).

Table II. Calculated Monomer Reactivity Ratios (r_1, r_2) with Vinyl Acetate (m_1)*

Monomer (m_2)	r_1	r_2
Acrylic Acid	0.02	20.64
Butyl Acrylate	0.05	5.89
Crotonic Acid	0.62	0.27
Ethyl Acrylate	0.02	7.20
Ethylene	0.72	0.59
Methyl Methacrylate	0.03	22.21
Styrene	0.03	24.18
Vinyl Chloride	0.60	1.40
Vinyl Neo-decanoate	0.99	0.92
Vinyl Neo-nonanoate	0.93	0.90
Vinyl Neo-pentanoate	0.79	0.96
Vinyl Propionate	1.06	0.76
Vinyl Laurate	1.40	0.70

* r_1 and r_2 calculated from information in *Polymer Handbook;* 3rd ed., J. Brandrup and E. H. Immergut, Section II, pages 268 - 269, Wiley Interscience, 1989 and, *VeoVa 9* and *VeoVa 10*, VeoVa Technical Manuals, VM 1.1 and 1.2; Shell Chemicals, UK, 1991.

To ascertain whether increased hydrocarbon content is a significant factor, one only needs to compare Figures 3 and 4. Owing to a small molecular weight difference between V2EH and 2EHA, the hydrophobicity of the VAc/2EHA/V2EH series of latexes, as determined by the oxygen content (wt%), remains virtually constant with increasing V2EH/2EHA ratios. On the other hand, for the VAc/BA/V2EH series, oxygen content drops by nearly 2% over the range of BA/V2EH ratios considered. That the VAc/2EHA/V2EH latexes of constant hydrophobicity are no more hydrolytically resistant than the VAc/BA/V2EH latexes, all of which are of lower hydrophobicity, indicates that hydrophobicity is not a dominant factor when V2EH is a comonomer.

At this point some mention should be made of the acrylic comonomer's ability to resist alkaline hydrolysis. From a study of the alkali resistance of various VAc/acrylic copolymer latexes (*3*), it was demonstrated that acrylates do provide protection against hydrolysis. However, it should be pointed out that they do so by their intrinsic resistance to alkaline attack, i.e., because they form block copolymers with VAc, they are not able to effectively space the VAc moieties. Owing to the fact that the carbonyl group, the focal point of attack by hydroxide ion, is directly attached to the polymer backbone in the case of an acrylate monomer, it acquires the steric protection of the backbone and is thus less prone to hydrolysis than VAc, whose carbonyl group is once removed from the main chain. According to the results of that investigation, on a molar basis, ethyl acrylate < n-BA < iso-BA < 2EHA in terms of hydrolytic stability, which indicates the importance of both branching and pendent chain length in providing protection from hydroxide ion attack.

Judging from Figures 5, 6 and 7, it appears that branching and pendent length are also important for vinyl monomers. According to the figures, and entry 6PSM74 in Table I, on a molar basis n-BA < VN-5 < V2EH ~ VN-9 ~ VN-10 insofar as resistance of VAc/BVE copolymer and VAc/BA/BVE terpolymer latexes, with 2/1 VAc/BVE molar ratios, to alkali saponification is concerned. That VN-5 provides less protection than either VN-9 and VN-10 indicates the importance of pendent chain length, i.e., all three possess dimethyl substituted (tertiary) alpha carbons. What is more intriguing are the results involving V2EH, which suggest that pendent chain length takes precedence over a trisubstituted alpha carbon in providing protection against alkali promoted hydrolysis. In the previously cited study (*3*), it was observed that n-BA << vinyl caprate << vinyl ester of Acid 810 (predominantly secondary structures) < VeoVa 911 (mixture of tertiary vinyl esters), which mirrors the findings of this study and also indicates the importance of having a comonomer that possesses at least a monosubstituted alpha carbon atom in order to guard VAc copolymer latexes against alkaline hydrolysis.

From the results it can also be determined that VAc/BVE copolymer and VAc/BA/BVE terpolymer latexes, with 2/1 VAc/BVE molar ratios, are not hydrolyzed to a greater extent than can be attributed to their VAc contents. Thus, VN-5, V2EH, VN-9 and VN-10 may be regarded as virtually nonhydrolyzable monomers, much like ethylene and methyl methacrylate when polymerized. In the aforementioned study (*3*), it was shown that VeoVa 911, on a molar basis, is approximately five times as efficient as ethylene in providing protection from alkaline hydrolysis, which suggests that not only do the long pendent groups protect the carbonyl moieties to which they are bonded, but that they also provide an umbrella type of shield to those of their immediate neighbors. The importance of neighboring group protection of VAc is borne out in the present study by the observation that VN-5 seems to be less effective than V2EH, VN-9 and VN-10. This is consistent with the fact that although all four monomers are capable of shielding their own carbonyl groups from hydroxide ion attack, by virtue of their substituted alpha carbons, only the latter three, owing to their long pendent chains, can provide an umbrella-like protection to their nearest neighbors.

A comparison of Figures 5, 6 and 7 reveals that the degree of protection in VAc/BVE copolymer and VAc/BA/BVE terpolymer latexes, with 2/1 VAc/BVE molar ratios, is insensitive to BA contents between 0 - 25 wt %. This strongly suggests that BA is as resistant to hydrolysis as a VAc group protected by its BVE neighbor. Although one might argue that a vinyl ester moiety with a tertiary (trisubstituted) alpha carbon would be more resistant to attack at its carbonyl carbon than an acrylate moiety, simply because one can consider the polymer backbone to be a disubstituted alpha carbon, it is not obvious as to why the latter ought not to be superior, in terms of protecting ability, than a VAc group utilizing its neighboring BVE's long pendent hydrocarbon chain. Perhaps the backbone configuration is compact enough that adjacent groups are in such close proximity that the steric protection afforded by neighboring group participation becomes as effective as that provided by a disubstituted alpha carbon, e.g., the chain backbone in the case of an acrylate moiety. Since the pendent hydrocarbon group sweeps out an equal region (time averaged) on both sides of the plane defined by the carbonyl carbon, oxygen and vinyl carbon atoms, each BVE comonomer, with the exception of VN-5, is expected to shield two adjacent VAc groups from chemical attack. Therefore, assuming a compact chain structure, one would not predict that a significant improvement could be achieved by reducing the VAc/BVE molar ratio below 2/1. Summarizing the structural/mechanistic aspects of alkali resistant VAc/acrylate/BVE latexes, one can argue that the results are consistent with a scheme whereby the BVEs break up VAc sequences and, when in possession of a pendent hydrocarbon group, they also provide an umbrella-like shield to adjacent VAc moieties within vinyl blocks.

Substituting V2EH for BA in VAc/BA/BVE latexes, with 2/1 VAc/BVE molar ratios, causes the differences between VN-5 and V2EH, VN-9 and VN-10 to disappear for V2EH contents of 25 mol% insofar as alkali promoted hydrolysis is concerned (Figure 9). However, for V2EH contents of 10 mol%, one can still distinguish between VN-5 and the three BVEs that contain a long pendent hydrocarbon group (Figure 8), although the differences have diminished relative to those observed in the case of the VAc/BA/BVE latexes (Figures 5, 6 and 7). That these observations are consistent with the aforementioned scheme, consider the following argument. Although VN-5 lacks a long hydrocarbon chain, its tertiary branched alpha carbon makes it very resistant to hydrolysis. Hence, it can only act as a spacer, which is why it is significantly worse than V2EH, VN-9 and VN-10 in providing protection from hydroxide ion attack in VAc/BA/BVE terpolymer latexes with 2/1 VAc/BVE molar ratios. However, replacing BA with V2EH allows for this protection to be provided, i.e., the 50/25/25 (m/m) VAc/V2EH/VN-5 latex possesses a 2/1 VAc/V2EH molar ratio, which makes it behave similarly to the VAc/BA/VN-9 and VAc/BA/VN-10 latexes with 2/1 VAc/BVE molar ratios. When the VAc/V2EH molar ratio increases to 6/1, as for the VAc/V2EH/VN-5 latex containing 10 mol% V2EH, the degree of protection is necessarily reduced, however; it is still better than its BA counterpart in that VN-5, unlike BA, can at least effectively space the VAc groups.

Finally, the observation that VAc/BA/BVE and VAc/V2EH/BVE latexes containing BVEs with long pendant chains, with 2/1 VAc/BVE molar ratios, cannot be distinguished in terms of hydrolytic stability indicates that reducing the VAc/BVE molar ratio below 2/1 is virtually inconsequential, i.e., if one includes the contribution from V2EH, the actual VAc/BVE molar ratio is 3/2 and 1/1 for the VAc/V2EH/BVE latexes containing 10 and 25 mol% V2EH, respectively.

Conclusions

The ability of comonomers to effectively reduce alkali promoted hydrolysis in VAc copolymer latexes arises from various factors. First, the comonomer must itself be more resistant to alkali attack. This is readily accomplished via increased steric effects, e.g., by the inclusion of a branched alpha carbon in the case of vinyl esters. Second, if the comonomer is able to interrupt the sequence of hydrolyzable VAc units in the copolymer, the rate of hydrolysis will be retarded. This is easily realized with BVEs and/or ethylene owing to their similar reactivity ratios relative to VAc. Third, if the comonomer possesses a long pendent hydrocarbon, it can provide an umbrella type of protection to its two adjoining neighbors, thereby diminishing the ease of hydroxide ion attack. However, it must be said that these steric and interruptive effects, in addition to others not considered in this investigation, all appear to be operating in conjunction to enhance the alkali resistance of VAc copolymer and terpolymer latexes.

Acknowledgements

Grateful acknowledgement is made to J. H. LaPere for synthesizing the latexes and to J. E. Patrick, R. G. Ramsey and J. F. Winter for performing the alkaline hydrolysis analyses.

Literature Cited

1. *VeoVa latices for general purpose emulsion paints* ; VeoVa Technical Manual, VM 3.1; Shell Chemicals: UK, 1988.
2. Scholten, H.; Vermeulen, J.; van Westrenen, W. J. *Polymer Paint and Colour Journal* **1988**, *178*, 852.
3. Davies, R. F. B.; Reynolds, G. E. J. *Journal of Applied Polymer Science* **1968**, *12*, 47.
4. US Patent Pending; Union Carbide Chemicals and Plastics: USA, 1990.
5. Mysels, K. J. *Journal of Physical Chemistry* **1961**, *65*, 1081.
6. Bruice, T. C.; Fife, T. H. *Journal of the American Chemical Society* **1962**, *84*, 1973.
7. Hoy, K. L. Internal Report; Union Carbide Chemicals and Plastics: USA, 1961.

RECEIVED December 4, 1991

Chapter 14

Emulsion Copolymerization of Small-Particle-Size, High-Molecular-Weight Poly(alkylaminoalkyl methacrylate-*co*-alkyl methacrylate) Latexes

J. W. Vanderhoff[1,2], S. H. Hong[3], M. R. Hu[1], J. M. Park[1,5], I. Segall[1], S. Wang[1], and H. J. Yue[4]

[1]Emulsion Polymers Institute and [2]Department of Chemistry, Lehigh University, Bethlehem, PA 18015
[3]Research Directorate, U.S. Army Chemical Research, Development and Engineering Center, Aberdeen Proving Ground, MD 21010–5423
[4]Air Products & Chemicals Company, Inc., 7201 Hamilton Boulevard, Allentown, PA 18195–1501

High-molecular-weight compositionally-homogeneous viscoelastic poly(alkylaminoalkyl methacrylate-co-alkyl methacrylate)'s (e.g., 25:75 (molar ratio) poly(t-butylaminoethyl methacrylate-co-isobutyl methacrylate)) were prepared by semi-continuous emulsion copolymerization using persulfate ion initiator. The results were evaluated in terms of the expected proportionalities of the rate of polymerization to the number of polymer particles and the molecular weight to the number of particles relative to the rate of radical generation. The rates of polymerization were determined gravimetrically or by gas chromatography. The average particle sizes were inordinately small (18-86 nm), and the molecular weights were typically 10^6 daltons or greater. The small particle sizes were first attributed to the high emulsifier concentration and excellent stability of the latexes, and later to the failure of the latex particles containing only 8-11 polymer molecules to grow further. The high molecular weights were attributed to the large number of particles. The copolymer ratio, molecular weight, solubility, and viscoelastic properties in solution were determined, and the effect of the type of initiator (water-soluble or oil-soluble), polymerization temperature, emulsifier type and concentration, and monomer feed (emulsion or neat monomer) were investigated. The properties of laboratory copolymers were compared with those

[5]Current address: Korea Chemical Company, Ltd., C. R. I., San 1–9, Mabook-ri, Kuseong-myun, Yongin-kun, Kyunggi-do, South Korea

0097–6156/92/0492–0216$06.00/0

prepared in larger reactors. Generally, the viscoelastic properties were satisfactory. Analysis of the homogeneity of the copolymers by ^{13}C nuclear magnetic resonance spectrometry, Fourier-Transform infrared spectroscopy, differential scanning calorimetry, and transmission electron microscopy gave conflicting results.

The major cause of fatalities in aircraft crashes is the fire which inevitably results when the jet fuel tanks rupture, and the fuel is thrown into the air and ignites (1). Any organic liquid can be ignited easily, provided it is dispersed in air to a sufficiently fine droplet size; however, viscoelastic polymers dissolved in jet fuel cause the breakup upon atomization to large rather than small droplets, which are much less likely to ignite. The polymers to be used for this purpose must dissolve readily, yet have a high enough molecular weight to produce viscoelastic solutions at low concentrations. Many different compositions can be used, but linear homogeneous copolymers of a protic electron-donating alkylaminoalkyl methacrylate with an aprotic electron-withdrawing dipolar alkyl methacrylate have been shown to be particularly effective (2).

Therefore, the objective of this work was to prepare linear compositionally-homogeneous viscoelastic poly(alkylaminoalkyl methacrylate-co-alkyl methacrylate) with molecular weights of 4×10^6 or greater and good solubility in organic solvents.

Emulsion polymerization is well-suited to the preparation of high-molecular-weight polymers at rapid polymerization rates. As a first approximation, the rate of polymerization R_p and the number-average degree of polymerization $\bar{X}_n$ are given by (3):

$$R_p = k_p[M]\bar{n}N \tag{1}$$

$$\bar{X}_n = 2k_p[M]\bar{n}N/R_i \tag{2}$$

where k_p is the rate constant for propagation, [M] the concentration of monomer in the particles, $\bar{n}$ the average number of free radicals per particle, N the total number of particles, and R_i the rate of radical generation. Thus $\bar{X}_n$ can be increased by increasing N, increasing [M], or decreasing R_i. High emulsifier concentrations give high values of N, but the value of [M] is limited by the equilibrium swelling volume of the latex particles; low initiator concentrations and low polymerization temperatures give low values of R_i. Therefore, the combination of a small latex particle size with a low initiator concentration and polymerization temperature would favor the preparation of high-molecular-weight polymer.

Batch emulsion copolymerization gives a distribution of copolymer compositions, according to the disparity of the copolymerization reactivity ratios. Therefore, semicontinuous emulsion copolymerization, in which the rate of polymerization is controlled by the rate of monomer addition, was used to give random compositionally homogeneous copolymers. The rate of monomer addition was decreased until it controlled the rate of copolymerization. This method gives low values of [M], however, which favors the preparation of low-molecular-weight polymer. This tendency must be compensated for by increasing the value of N further and thus decreasing the value of R_i.

Experimental Details

Except where noted, the inhibitors were removed from the monomers before use; the methyl, ethyl, and isobutyl methacrylates (Rohm & Haas), ethylene glycol dimethacrylate (Rohm & Haas), and t-butylaminoethyl and diethylaminoethyl methacrylates (Alcolac), were passed through an inhibitor removal column (Aldrich) before use. The Triton X-100, X-405, and X-705 (octylphenol-polyoxyethylene adducts of increasing chain length; Rohm & Haas), sodium persulfate initiator (Fisher Scientific), and Wako V-044 (2,2′-azobis-(N,N′-dimethyleneisobutyramidine) dihydrochloride; Wako V-50 (2,2′-azobis(2-amidinepropane) dihydrochloride), and Wako V-65 (2,2′-azobis(2,4-dimethylvaleronitrile) initiators (Wako Pure Chemical Industries) were used as received. Distilled-deionized water was used for all polymerizations.

The CM-120 polymerization recipe had been used earlier (4) for the semi-continuous polymerization of the 25:75 (molar ratio) poly(t-butylaminoethyl methacrylate-co-isobutyl methacrylate); it comprised a 50:50 monomer/water ratio, 9.0% Triton X-405 emulsifier (based on water), and 0.15% sodium persulfate initiator; 7% of the monomer was emulsified at room temperature in the water containing the dissolved emulsifier and potassium persulfate initiator in a four-necked 500-ml round-bottom flask equipped with a stirrer, condenser, nitrogen inlet, and monomer syringe pump, heated to 60°C, and polymerized in batch to form the seed latex. The remaining monomer was then added neat at a constant rate over a four-hour period. After the monomer addition was completed, the latex was heated at 60°C to complete the polymerization. Polymerizations were carried out at 50 or 70°C, and batch polymerizations were also carried out in the same equipment.

Other semi-continuous polymerizations were carried out adding an emulsion of the monomer rather than neat monomer. In this case, the seed monomer was polymerized in batch using a proportionate amount of the water, emulsifier, and initiator. Then, an emulsion of the monomer, prepared using proportionate amounts of the water, emulsifier, and initiator, was added continuously.

The polymer molecular weights were determined from the intrinsic viscosities (Limiting Viscosity Numbers; LVN) in N-methyl pyrrolidone solution; the latex polymers were coagulated with isopropanol, filtered, washed extensively to remove the emulsifier and other ingredients, and dried under vacuum. The latex particle sizes were measured by photon correlation spectroscopy (Nicomp Submicron Particle Sizer), the copolymer ratio by ^{13}C nuclear magnetic resonance spectrometry (Bruker AM500) or Fourier-Transform infrared spectroscopy (Mattson Polaris), the particle morphology by transmission electron microscopy (Philips 400), and the T_g by differential scanning calorimetry (Mettler). The rheological properties were measured using cone-plate geometry (Rheometrics Fluid Rheometer RFR-7800). The attempts to measure the molecular weight by gel permeation chromatography were unsuccessful; the solutions filtered slowly, and the filtrates contained only low-molecular-weight polymer. The final pH of the latexes was 9.40-9.50 except where noted.

Experimental Results and Discussion

Small Particle Size. Table I gives the molar compositions of alkylaminoalkyl methacrylate-alkyl methacrylate mixtures polymerized by semi-continuous polymerization (neat monomer addition) and the average particle sizes of the latexes produced. The number-average particle sizes were inordinately small (18-86 nm). Typically, such semi-continuous polymerizations give number-average particle sizes of 100-150 nm.

In the preparation of high-molecular-weight copolymers, a volume-average particle size of 50 nm and a polymer density of 1.15 g/cm^3 corresponds to 1.3x10^{16} particles/g polymer, a value about as high as is practical to achieve. Therefore, any increase in molecular weight must be made by decreasing the value of R_i.

Table II gives the variation with polymerization time of the number-average and volume-average particle size of a 25:75 poly(t-butylaminoethyl methacrylate-co-isobutyl methacrylate) latex prepared by batch copolymerization at 60°C. The particle size distributions were broad; the polydispersity indexes (PDI's) were 1.36-2.00 (average 1.59). The average particle size increased to 13 nm in 15 minutes and to 32 nm in 20 minutes, and remained constant thereafter (average 25 nm over the range 20-420 minutes).

Table I. Average Particle Sizes of Copolymer Latexes

Sample	Composition	D_n (nm)
116-27	10/90 tBAEMA/EMA	28
116-17	25/75 tBAEMA/EMA	18
116-15	25/75 tBAEMA/EMA	24
116-26	50/50 tBAEMA/EMA	30
116-14	25/75 tBAEMA/iBMA	86
116-21	25/75 tBAEMA/iBMA	27
116-23	50/50 tBAEMA/iBMA	36
116-24	100/0 tBAEMA	76
116-19	25/75 DEAEMA/MMA	24
116-28	10/90 DEAEMA/EMA	44
116-20	25/75 DEAEMA/EMA	44
116-22	25/75 DEAEMA/iBMA	44
116-25	100/0 DEAEMA	38

MMA = methyl methacrylate
EMA = ethyl methacrylate
iBMA = isobutyl methacrylate
tBAEMA = t-butylaminoethyl methacrylate
DEAEMA = diethylaminoethyl methacrylate

Table II. Particle Size in Batch Polymerization

Sample	Polym. Time (min)	D_n (nm)	D_v (nm)	PDI	N/g x10^{-16}
1	15	13	26	2.00	3.1
2	20	32	45	1.41	0.88
3	30	33	45	1.36	1.5
4	40	29	43	1.48	2.0
5	60	29	41	1.41	2.2
6	75	17	34	2.00	3.9
7	90	26	40	1.54	2.5
8	150	26	41	1.58	2.4
final	420	28	44	1.57	1.9

Similar results were observed for semi-continuous polymerizations carried out at 60° and 70°C. Table III shows that the number-average particle size increased to constant values of 32 nm at 60°C and 26 nm at 70°C after 20-30 minutes, which remained unchanged throughout the remainder of the polymerization.

Table IV gives the particle sizes corresponding to one or more poly(t-butylaminoethyl methacrylate-co-isobutyl methacrylate) molecules calculated assuming a molecular weight of 10^6 and a copolymer density of 1.15 g/cm^3. A single copolymer molecule is equivalent to a sphere with a diameter of 14 nm, two molecules, a sphere with a diameter of 18 nm, etc. Thus, the 15-minute batch polymerization sample (Table II) had an average particle size corresponding to a single copolymer molecule, and the succeeding samples had an average particle size of 32 nm corresponding to thirteen molecules. Similarly, the 70°C semi-continuous polymerization had a seed particle size corresponding to two copolymer molecules, and the succeeding samples had an average particle size of 26 nm corresponding to seven molecules.

Thus, the first particles initiated comprised single copolymer molecules. These particles grew until they comprised about 10 copolymer molecules and then, surprising-

Table III. Particle Size in Semi-Continuous Polymerization

Sample	Polym. Time (min)	D_n (nm)	D_v (nm)	PDI	N/g, $x10^{-16}$
	60°C Polymerization Temperature				
1	seed	19	27	1.42	0.58
2	20	28	39	1.39	0.20
3	40	26	36	1.38	0.31
4	60	28	41	1.46	0.29
5	90	29	41	1.41	0.78
6	120	31	44	1.42	1.4
7	140	41	52	1.27	1.1
8	175	37	50	1.35	1.2
final	420	38	51	1.34	1.2
	70°C Polymerization Temperature				
1	seed	18	27	1.50	0.58
2	30	28	37	1.32	1.0
3	60	21	36	1.71	2.9
4	90	24	40	1.67	2.1
5	105	25	39	1.56	2.5
6	130	19	35	1.84	3.7
7	160	34	45	1.32	1.8
final	420	34	46	1.35	1.7

Table IV. Spherical Particle Sizes of Copolymer Molecules

Number of Copolymer Molecules	Spherical Size (nm)
1	14
2	18
3	20
5	24
10	30
20	38

ly, stopped growing. This cessation of particle growth at an early conversion is most unusual. Usually, the particles continue to grow until the supply of monomer or free radicals is exhausted. If particle nucleation ceased early in the polymerization, the final latex would be of uniform size; if the particle nucleation continued for a longer time, the final latex would have a broad particle size distribution. The particle size distributions of these latex samples were relatively broad; the average polydispersity indexes were 1.54 for the batch polymerization (Table II) and 1.22 for the 70°C semi-continuous polymerization (Table III).

Hypothesis for Cessation of Particle Growth. The reason for the cessation of particle growth at ten molecules is not known. However, one possible explanation comprises a reversal of the mode of adsorption of the oligomeric sulfated radicals on the latex particle surface.

The commonly accepted mechanism of initiation using persulfate ion initiator (5,6) comprises the formation of sulfate ion-radicals in the aqueous phase, their slow propagation in the aqueous phase until they become surface-active and adsorb at the particle-water interface with the radical end oriented toward the monomer-polymer phase. The rate of propagation of these radicals is slow in the aqueous phase because of the low concentration of monomer and rapid in the monomer-polymer phase because of the high concentration of monomer.

In the present system, the initial growth of the particles would give an increasing number of positive amino surface groups from the t-butylaminoethyl methacrylate units, so that an acid-base interaction between the surface amino groups and the sulfate groups of the oligomeric sulfated radicals would give a reversed mode of adsorption, with the sulfate end oriented toward the particle surface and the radical end oriented toward the aqueous phase. The radicals oriented toward the particle phase would grow into the particle rapidly because of the high concentration of monomer inside the particles. In contrast, the radicals oriented toward the aqueous phase would grow only slowly because of the low concentration of monomer in the aqueous phase; moreover, they would be vulnerable to termination by oligomeric radicals or primary radicals from the aqueous phase. Termination of the adsorbed radicals by radicals from the aqueous phase would inhibit further polymerization inside the particle. Meanwhile, the nucleation of new particles would proceed concurrently, so that particles would be nucleated, grow to a given size and then stop growing. This cessation of particle growth is a new phenomenon not reported earlier to the authors' knowledge; therefore, this phenomenon is termed limited growth emulsion polymerization.

This mechanism would give a surface layer of different composition from the particle interior, which would be consistent with the indications of inhomogeneity in the particles observed by transmission electron microscopy, which will be described later.

Other Works. Other workers have reported the preparation of small-particle-size latexes, which grew slowly or not at all. Medvedev et al. (7) prepared polysytrene latexes with diameters as small as 60 nm by batch emulsion polymerization of styrene using nonionic fatty alcohol-polyoxyethylene adducts as emulsifiers and oil-soluble benzoyl peroxide or 2,2′-azobisisobutyronitrile initiators. These small particle size latexes were also reported to remain constant in diameter with increasing conversion. The formation of these small particles was attributed to the formation of tiny emulsion droplets and their polymerization "as is" without transfer of monomer from one to another by diffusion through the aqueous phase. There was no indication that the particles stopped growing at any particular stage. Such a mechanism is probably not operative in the present case because it does not account for the cessation of particle growth.

Also, Snuparek and Kleckova (8) found that the semicontinuous emulsion polymerization of acrylate ester/ethylene glycol dimethacrylate mixtures using monomer emulsion feed gave average particle sizes that oscillated between 200 and 600 nm with increasing conversion. This oscillation was attributed to competing processes of particle growth and particle flocculation. Similar oscillations have been observed in continuous emulsion polymerization, where they were attributed to the effect of emulsifier concentration on the number of particles nucleated relative to the competing particle flocculation process (9, 10). This oscillation does not explain the present results, however, where the particles once nucleated grow to a given size and then stop growing, and where there is little or no tendency for the particles to flocculate.

Copolymer Molecular Weight. The molecular weights of the copolymers as measured by the intrinsic viscosity (Limiting Viscosity Number; LVN) were all high, presumably because of the large numbers of particles and the relatively low rate of radical generation. Generally, the molecular weights varied as expected with the number of particles and the rate of radical generation, increasing with increasing number of particles and decreasing rate of radical generation (low initiator concentration; low polymerization temperature). The molecular weight was slightly higher when the monomer was added in the form of an emulsion as compared to when it was added neat; this was attributed to the addition of the initiator dissolved in the emulsion for the emulsion addition as compared to addition in the beginning of the reaction for the neat addition, the former giving a lower concentration of initiator at a given time and hence a lower rate of radical generation. Generally, however, the variation of molecular weight with these parameters, which had been well-established for a different system investigated earlier, was not as great as expected. Therefore, further experiments were carried out.

Table V shows that there were no strong trends in the variation of particle size and molecular weight with sodium persulfate initiator and Triton X-405 emulsifier concentrations for the batch emulsion copolymerization of 25:75 poly(t-butylaminoethyl methacrylate-co-isobutyl methacrylate); however, the averages of the parameters showed that the particle size increased with temperature from 28.9 to 33.7 to 41.6 nm for 50, 60, and 70°C, the width of the distribution decreased (the polydispersity index decreased from 1.65 to 1.48 to 1.29), and the Limi-

Table V. Effect of Emulsifier and Initiator Concentration

Sample	[SPS] (%)	[405] (%)	D_n (nm)	LVN (dl/g)
	50°C Polymerization Temperature			
50#1	0.09	9.0	25	1.35
50#2	0.15	9.0	28	1.27
50#3	0.27	9.0	27	1.50
50#4	0.15	4.5	36	1.29
50#5	0.15	5.8	31	1.37
50#2	0.15	9.0	28	1.27
50#6	0.15	13.5	27	1.32
	60°C Polymerization Temperature			
60#1	0.09	9.0	28	1.04
60#2	0.15	9.0	35	1.39
60#3	0.27	9.0	35	1.32
60#4	0.15	4.5	36	1.25
60#5	0.15	5.8	37	1.20
60#2	0.15	9.0	35	1.39
60#6	0.15	13.5	30	1.30
	70°C Polymerization Temperature			
70#1	0.09	9.0	43	1.44
70#2	0.15	9.0	39	1.64
70#3	0.27	9.0	33	1.61
70#4	0.15	4.5	63	1.58
70#5	0.15	5.8	44	1.40
70#2	0.15	9.0	39	1.64
70#6	0.15	13.5	30	1.43

[SPS] = sodium persulfate concentration
[405] = Triton X-405 concentration

ting Viscosity Number remained about the same (1.34 to 1.27 to 1.53). Thus, the average particle size and molecular weight were not sensitive to temperature in the 50-70°C range. The slight variation of particle size was in the opposite direction to that expected, and the molecular weight did not decrease with increasing temperature as expected.

The molecular weights of the copolymers were generally high, although the correlation between the Limiting Viscosity Number and the number-average or weight-average molecular weight has not yet been established. The best estimate is that a LVN of ca. 2 dl/g is equivalent to number average and weight-average molecular weights of $1x10^6$ and $4x10^6$ daltons, respectively. The copolymer solutions in N-methyl pyrrollidone prepared for the LVN measurements (as well as those in tetrahydrofuran prepared for gel permeation chromatography) were transparent, and without visible gels; however, the response of the solutions to shaking suggested that they were not homogeneous, but comprised swollen gels. Light scattering (11,12) showed that the copolymers associated in solution to give an average aggregate or cluster size of four molecules in isopropylamine and clusters of up to thirteen molecules in higher solvents. This clustering suggested the possibility that the particles were not dissolved completely in any of the solvents. Thus, further work is required to demonstrate the homogeneity of the solutions.

The Presence of Storage Inhibitor. Except where noted, all experiments of this work used monomer from which the storage inhibitor had been removed; however, since the usual industrial practice is to use monomer containing inhibitor for safety reasons, laboratory experiments using inhibited monomer were carried out preparatory to the larger scale polymerizations. Earlier experience with inhibited monomer gave lower molecular weights and, in some cases, slower polymerization rates, according to the residual inhibitor concentration.

Table VI gives the results of experiments based on the CM-120 recipe polymerized at 60°C using monomer containing inhibitor. Generally, the molecular weight decreased with decreasing initiator concentration, decreased with slower monomer feed rates, and decreased with decreasing isobutyl methacrylate concentration in the monomer mixture.

Table VI. Effect of Polymerization Parameters

Sample	[SPS] (%)	Feed Rate (cc/min)	[tBAEMA] (%)	D_n (nm)	LVN (dl/g)
412	0.18	2.0	25	44	1.78
425A	0.18	1.1	21	41	1.90
425B	0.13	2.0	21	43	1.90
418	0.18	1.1	25	40	1.60
426A	0.13	1.1	21	46	1.73
430A	0.13	1.1	18	40	2.08
430B	0.13	0.5	21	45	1.35
426B	0.09	1.1	21	31	1.38
502B	0.09	0.5	18	40	1.32
418-S*	0.18	1.1	25	36	2.04
306-B4**	0.18	1.1	25	55	1.60

* = inhibitor-free monomer
** = batch polymerization; inhibitor-free monomer
[tBAEMA] = t-butylaminoethyl methacrylate concentration

Larger-Scale Polymerizations. Table VII gives the results of the larger-scale polymerizations. In the laboratory, the standard CM-120 recipe gave a particle size of 36 nm and a Limiting Viscosity Number of 2.04 dl/g. The corresponding batch polymerization gave a larger particle size (54 nm) and a lower LVN (1.60 dl/g). Eight five-liter polymerizations of the standard CM-120 recipe gave excellent reproducibility but a larger particle size (average 52; range 50-55 nm) and a lower LVN (average 1.68; range 1.47-1.90 dl/g), as expected. The use of monomer containing inhibitor gave slightly lower Limiting Viscosity Numbers (1.90, 1.85 dl/g). The Morton laboratory polymerizations preparatory to the 100-gallon polymerization using inhibited monomer gave lower Limiting Viscosity Numbers (0.70-1.53 dl/g), but the Morton 100-gallon polymerization gave a higher value (1.78 dl/g). Spray-drying the latex to recover the polymer lowered the LVN; the decrease was greater for the higher temperature (300°F) of spray-drying. The t-butylaminoethyl methacrylate concentrations in the copolymers measured by ^{13}C nuclear magnetic resonance spectrometry were smaller (average 20.1; range 17.5-21.5) than the 25% charged.

Table VII. Larger-Scale Polymerizations

Sample	D_n (nm)	LVN (dl/g)	[tBAEMA] (%)
EPI-315 (CM-120)	36	2.04	----
EPI-306 batch	54	1.60	----
EPI-423A* 8% seed	36	1.90	----
EPI-424A* 12% seed	39	1.90	----
S5-01 5-liters	50	1.47	----
S5-02 5-liters	50	1.58	----
S5-03 5-liters	53	1.96	20.9
S5-04 5-liters	51	1.75	21.5
S5-05 5-liters	52	1.78	19.9
S5-06 5-liters	55	1.50	21.3
S5-07 5-liters	53	1.72	17.8
S5-08 5-liters	53	1.73	20.6
Morton 278*	--	1.00	----
Morton 287*	--	1.53	----
Morton 288*	--	0.70	----
Morton 100-gallons*	39	1.78	21.0
spray-dried 250	--	1.49	20.7
spray-dried 300	--	1.19	17.5
room temp	--	1.45	----

* = inhibited monomer

Miscellaneous Polymerizations. Table VIII shows that the water-soluble cationic Wako V-044 and Vazo V-50 initiators, and the oil-soluble Vazo V-65 initiator, gave lower Limiting Viscosity Numbers than the sodium persulfate initiator. The oil-soluble initiator gave a slightly larger average particle size. The substitution of Triton X-100, Triton X-705, and Triton X-405/sodium lauryl sulfate mixture gave lower Limiting Viscosity Numbers than the Triton X-405 alone. Ethylene glycol dimethacrylate, which was added in 0.05% concentration to introduce branches into the copolymer and thus increase the Limiting Viscosity Number, actually gave a lower value.

Homogeneity of the Copolymers. Homogeneous copolymers of random comonomer sequence distribution were desired for this application. The most widely used method to determine comonomer sequence distribution is ^{13}C nuclear magnetic resonance (NMR) spectrometry, which gives resonance signals corresponding to the diads and triads of the copolymer. The homogeneity of the copolymer can be calculated from the intensity of these resonance signals. The ^{13}C NMR spectra of the 25:75 poly(t-butylaminoethyl methacrylate-co-isobutyl methacrylate) showed no significant changes in the resonance signals compared with the spectra of the two homopolymers, which indicated that no information on the comonomer sequence distribution was obtained. This lack of signals was not unexpected because the backbone structures of the poly(t-butylaminoethyl methacrylate) and poly(isobutyl methacrylate) were the same; the side groups were different, but were located far enough from the carbon backbone skeleton that their electronic energy was insufficient to induce splitting for the corresponding diad or triad resonances. Thus the ^{13}C NMR spectrum for the pro-

Table VIII. Miscellaneous Polymerizations

Sample	D_n (nm)	LVN (dl/g)
Wako V-044 (pH 9.55)	46	0.79
Wako V-50 (pH 9.30)	42	1.01
Wako V-65 (pH 9.00)	87	1.42
Triton X-100	142	0.99
Triton X-405	39	1.24
Triton X-705	51	0.90
Triton X-405/sodium lauryl sulfate	50	0.91
EGDMA (0.050%)	--	1.12

duct was essentially the same as the superposed spectra of the two homopolymers, and the product was a mixture of the two homopolymers or, more likely, a blocky copolymer.

Figure 1 shows a subtle difference, however, between the ^{13}C NMR spectrum of the 25:75 poly(t-butylaminoethyl methacrylate-co-isobutyl methacrylate) and those of the two homopolymers and a physical mixture of the two. The peak at 43.6 ppm in the poly(t-butylaminoethyl methacrylate) spectrum corresponding to the $-CH_2-NH-$ structure was also observed in the spectrum of the copolymer, but was of lower height and shifted. This peak was attributed to the association of secondary amines with carbonyls; it was not observed at all in the physical mixture of the two homopolymers. This suggests that the morphology of the copolymer was different from that of the homopolymer mixture, as expected.

That the polymer comprised copolymers was supported by Fourier-Transform infrared spectroscopy. The secondary amines show a stretching mode at 3400-3500 wavenumbers and a bending mode at 1500-1600 wavenumbers. The association of the amines with the carbonyls shifts the peak for the amine bending mode upfield and that for the amine stretching mode downfield. Figure 2 compares the spectrum of the 25:75 poly(t-butylaminoethyl methacrylate-co-isobutyl methacrylate) with that for a mixture of the two homopolymers. The expected shifts for the amines are seen in the spectrum for the physical mixture but not in that for the copolymer; this result was attributed to the association of the amines with the carbonyls in the copolymer but not in the physical mixture.

The transmission electron microscopy of these latexes gave marginal results because of the small size and poor contrast of these particles in the electron beam. Some micrographs of stained particles, however, showed images of heterogeneous particles, consistent with the presence of two homopolymers; others showed homogeneous particles. Figure 3 shows such an electron micrograph of heterogeneous particles. These findings support the first conclusion of the ^{13}C NMR spectrometry that the particles comprised a blocky copolymer or a mixture of two homopolymers; however, the fact that each particle comprised only a few copolymer molecules suggests that these molecules can arrange themselves so that some of the polar amine functional groups are together, e.g., to form a polar shell around a nonpolar core, and that this morphology gave the heterogeneous appearance observed in a few samples.

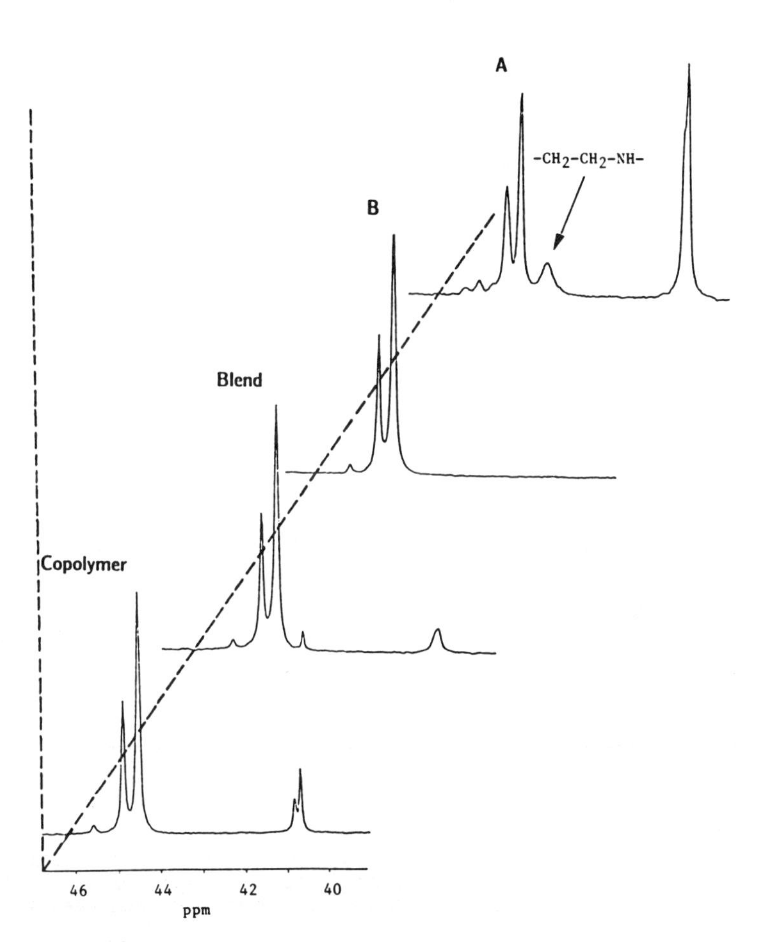

Figure 1. ^{13}C nuclear magnetic resonance spectra of the 25:75 poly(t-butylaminoethyl methacrylate-co-isobutyl methacrylate), the two homopolymers (A and B), and a physical blend of the two.

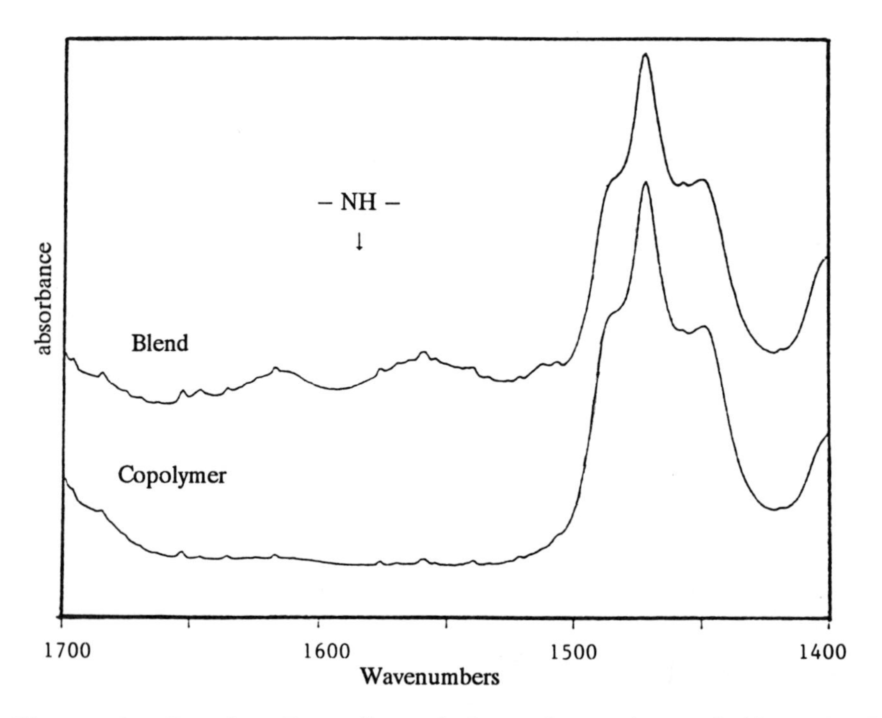

Figure 2. Fourier-Transform infrared spectra of the 25:75 poly(t-butylaminoethyl methacrylate-co-isobutyl methacrylate) and the physical blend of the two homopolymers.

In contrast, differential scanning calorimetry of the purported copolymers suggested that they were copolymers. Table IX gives the values of the apparent second-order transition temperature T_g of these samples, and Figure 4 shows a typical heat flow-temperature curve. All copolymers gave single incisive T_g values of 60-63°C compared to the values of 38° and 70°C for poly(t-butylaminoethyl methacrylate) and poly(isobutyl methacrylate), respectively, consistent with homogeneous copolymers.

Table IX. Apparent Second-Order Transition Temperature

Sample	Composition	T_g (°C)
116-44	poly(isobutyl methacrylate)	70
116-55	25:75 copolymer	63
140-61	25:75 copolymer	63
140-62	25:75 copolymer	63
140-64	25:75 copolymer	62
140-66	25:75 copolymer	62
140-71	25:75 copolymer	60
140-73	25:75 copolymer	60
140-72	poly(t-butylaminoethyl methacrylate)	38
	physical blend of 116-44 and 140-72	38,70

Finally, semi-continuous polymerization should give homogeneous copolymers of t-butylaminoethyl methacrylate and isobutyl methacrylate because the copolymerization reactivity ratios $r_1 = 1.2$ and $r_2 = 0.8$ (calculated from the Q and e values) were not too disparate. In contrast, it was shown earlier (13,14) that, for the poly(vinyl acetate-co-butyl acrylate) system, which has far more disparate copolymerization reactivity ratios of $r_1 = 0.05$ and $r_2 = 5$, batch copolymerization gave a mixture of a butyl acrylate-rich copolymer with polyvinyl acetate, independent of the monomer ratio, whereas semi-continuous polymerization gave a single-peaked distribution of copolymer compositions, the T_g of which varied according to the monomer ratio. Thus, semi-continuous polymerization gave a homogenous copolymer of a monomer pair with far more disparate copolymerization reactivity ratios than the 25:75 poly(t-butylaminoethyl methacrylate-co-isobutyl methacrylate). Therefore, it is expected that homogeneous copolymers would be formed in the present system.

In summary, the single T_g values of the polymers, the spectroscopic evidence for the association of the secondary amine groups with the carbonyl groups of the ^{13}C nuclear magnetic resonance and Fourier-Transform infrared spectra, and the use of semi-continuous polymerization suggested that the polymers were indeed homogeneous copolymers. In contrast, absence of peaks for the diads and triads of the ^{13}C nuclear magnetic resonance spectra and the heterogeneous appearance of the particles in some transmission electron micrographs suggested that the polymers are mixture of homopolymers. Further work is required to resolve this apparent discrepancy; however, on

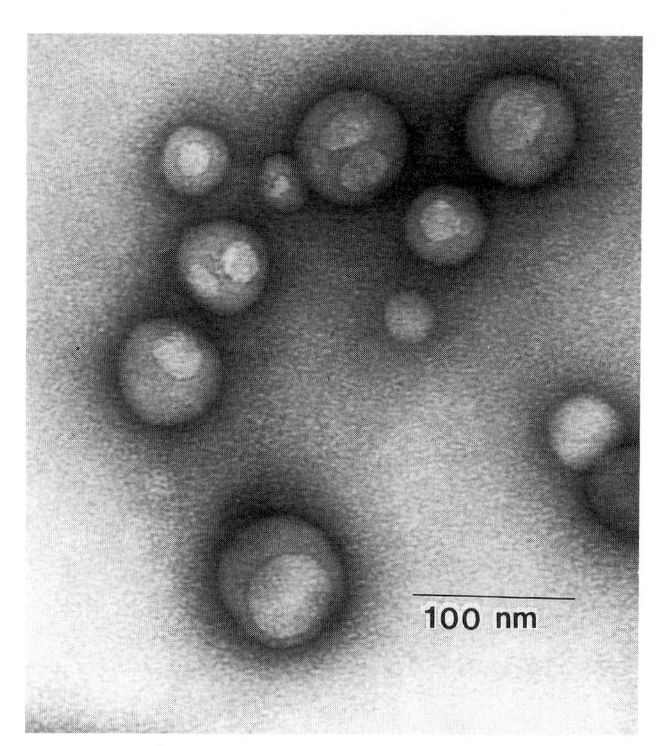

Figure 3. Transmission electron micrograph of Sample 116-55 25:75 poly(t-butylaminoethyl methacrylate-co-isobutyl methacrylate).

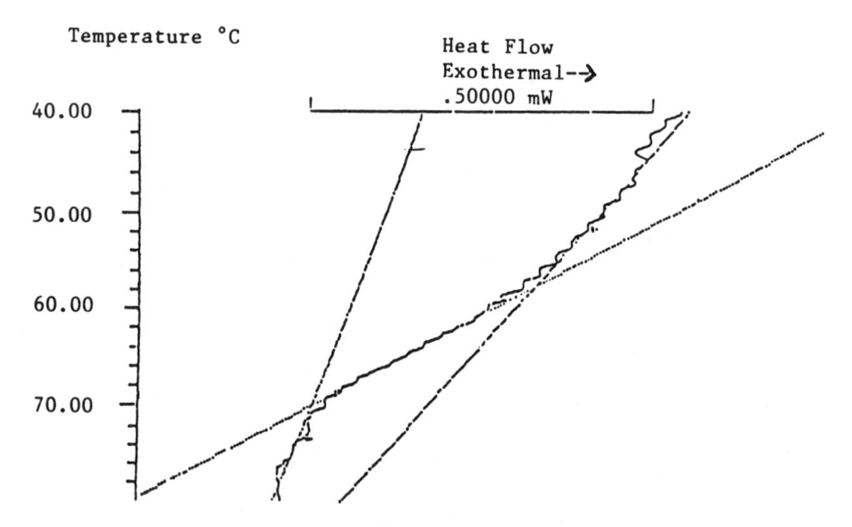

Figure 4. Differential scanning calorimetry heat flow-temperature scan of 25:75 poly(t-butylaminoethyl methacrylate-co-isobutyl methacrylate).

balance, the formation of homogeneous copolymers is considered to be the more likely result.

Rheological Properties of Dilute Solutions. Table X shows, for dilute solutions of the larger-scale samples, the variation of the apparent viscosity and first normal stress difference (FNSD)) with shear rate. The values of the apparent viscosity of the five-liter samples were generally within the acceptable range, with the exception of sample S5-03, which showed an unexpectedly high apparent viscosity, but a FNSD value in the acceptable range; this sample also had the highest LVN value of the five-liter samples. The cause of this aberration is not known. The FNSD values of all five-liter samples indicated that these samples had a high enough viscoelasticity in the given solvents. Both the apparent viscosity and the FNSD value of the Morton 100-gallon sample were almost identical to those of the original CM-120 sample, which indicated that the 100-gallon polymerization was quite successful.

Table X. Rheological Properties of Dilute Solutions

Sample	Solvent	Concn.	FSND (dyne/cm^2)				Apparent Viscosity (poise)			
			Shear Rate/sec				Shear Rate/sec			
		(g/dl)	100	250	500	1000	12.5	25	50	99
CM-120	Mix 1	2.0*	155	530	1215	----	2.1	2.0	1.9	1.7
	Mix 2	3.0	---	---	702	2296	4.9	3.2	2.8	2.4
		3.0	---	---	204	1722	2.8	2.7	2.5	2.3
		3.1	(68)	464	1232	2891	3.9	3.4	2.9	2.5
S5-01	Mix 1	2.0	427	878	1220	1329	2.3	2.0	1.6	1.4
	Mix 2	3.0	---	---	350	497	1.7	1.6	1.5	1.3
S5-02	Mix 1	2.0	(74)	229	476	570	1.9	1.8	1.6	1.4
	Mix 2	3.0	---	252	731	1663	3.4	2.9	2.4	2.2
S5-03	Mix 1	2.1	488	971	1489	2043	9.0	7.1	5.6	4.0
	Mix 2	3.0	(39)	364	1009	2310	5.9	4.8	4.1	3.5
S5-04	Mix 1	2.1	445	1062	1857	2760	4.9	4.3	3.7	3.0
	Mix 2	3.1	181	282	857	2055	4.2	3.6	3.1	2.7
S5-05	Mix 1	2.1	520	853	1391	1767	2.9	2.6	2.3	2.0
	Mix 2	3.0	---	324	893	2058	3.8	3.4	2.8	2.5
S5-06	Mix 1	2.1	125	351	724	988	2.0	1.9	1.7	1.6
	Mix 2	3.1	---	---	193	928	3.5	2.9	2.7	2.4
S5-07	Mix 1	2.0	115	373	766	1148	2.2	2.0	1.8	1.6
	Mix 2	3.0	---	(87)	544	1399	3.6	3.1	2.7	2.3
S5-08	Mix 1	2.0	181	608	1115	1399	2.7	2.4	2.1	1.9
	Mix 2	2.9	(40)	435	1111	2411	4.0	3.5	3.0	2.6
Morton	Mix 1	2.0	150	511	1034	1555	2.5	2.3	2.1	1.8
100-gal	Mix 2	3.0	---	280	824	1971	3.5	2.9	2.5	2.3

* estimated values from measurements of 2 g/dl solutions at 20, 30, and 40°C
FNSD = first normal stress difference
Mix 1, Mix 2 = proprietary organic solvent mixtures

Conclusions

High-molecular-weight 25:75 poly(t-butylaminoethyl methacrylate-co-isobutyl methacrylate) was prepared by semi-continuous emulsion copolymerization. The number-average and weight-average molecular weights measured by Limiting Viscosity Number were ca. $1x10^6$ and $4x10^6$ daltons, respectively, and the viscoelastic properties of the solutions were satisfactory; however, the polymer molecules appeared to associate in solution, forming clusters of up to thirteen molecules in some solvents, and the molecular weights did not vary as sensitively with particle number and rate of radical generation as expected. Moreover, the latex particle sizes were inordinately small (18-86 nm) instead of the expected 100-150 nm; the small sizes were attributed to the rapid nucleation and growth of the particles to a size equivalent to about 10 copolymer molecules, followed by cessation of growth. The compositional homogeneity of the copolymers has not yet been demonstrated unequivocally: transmission electron gave evidence of heterogeneous particles, and ^{13}C nuclear magnetic resonance spectrometry showed no peaks for the expected diads and triads; however, differential scanning calorimetry gave single incisive T_g's of the expected values, both ^{13}C nuclear magnetic resonance spectrometry and Fourier-Transform infrared spectroscopy gave evidence for the association of the secondary amine groups with the carbonyl groups for the copolymers but not for the physical blend of the two homopolymers, and the semi-continuous polymerization method used has given homogeneous copolymers for other systems with more disparate copolymerization reactivity ratios.

Literature Cited

1. Videocassette, Why Planes Burn, PBS, WGSA Educational Foundation, NOVA, 1988.
2. Chu, B.; Wang, J.; Pfeiffer, D. G.; Shuely, W. J. Macromolecules **1991**, 24, 809.
3. Smith, W. V.; Ewart, R. H. J. Chem Phys. **1948**, 16, 592.
4. Hong, S. H. Review of Emulsion Polymerization: Optimization for Production of Ultrahigh Molecular Weight Poly(alkylaminoalkyl Methacrylate-co-Alkyl Methacrylate), U.S. Government Report CRDEC-TR-88146, UNCLASSIFIED Report, **1988.**
5. van der Hoff, B. M. E. Polymerization and Polycondensation Processes, ACS Adv. Chem. Ser. **1962**, 34, 6.
6. Vanderhoff, J. W.; van den Hul, H. J.; Tausk, R. J. M.; J. Th. G. Overbeek In Clean Surfaces: Their Preparation and Characterization for Interfacial Studies, Goldfinger, G., Ed., Marcek Dekker: New York, NY, 1970, p. 15.
7. Medvedev, S. S.; Zuikov, A. V.; Gritskova, I. A.; Dudukin, V. V. Vysokomol. Soyed (English transl.) **1971**, A13, 1397.
8. Snuparek, Jr., J.; Kleckova, Z. J. Appl. Polym. Sci. **1984**, 29, 1.
9. Gershberg, D. B.; Longfield, J. E. Symposium on Polymerization Kinetics and Catalyst Systems, 45th AIChE Meeting, New York, NY, 1961, preprint 10.
10. Greene, R. K.; Gonzales, R. A.; Poehlein, G. W. In Emulsion Polymerization, Piirma, I., Ed., ACS Symp. Ser. **1976**, paper 22.
11. Chu, B.; Wang, J.; Shuely, W. J. Macromolecules **1990**, 23, 2252.

12. Chu, B.; Wang, J.; Shuely, W. J. Polymer **1990**, 31, 805.
13. El-Aasser, M. S.; Misra, S. C.; Pichot, C.; Vanderhoff, J. W. J. Polym. Sci., Polym. Lett. Ed. **1979**, 127, 567.
14. Misra, S. C.; Pichot, C.; El-Aasser, M. S.; Vanderhoff, J. W. J. Polym. Sci., Poly. Chem. Ed. **1983**, 21, 2382.

RECEIVED January 24, 1992

Chapter 15

Control of Core–Shell Latex Morphology

S. Lee and Alfred Rudin

Institute for Polymer Research, Department of Chemistry, University of Waterloo, Waterloo, Ontario N2L 3G1, Canada

Composite latex particles with core-and-shell structures made via staged emulsion polymerization have gained interest in recent years. Control of particle structures has been a challenge, particularly when the desired core polymer is more hydrophillic than the shell polymer. The structure of a PMMA/PS system can be controlled by altering the thermodynamic and kinetic variables. These variables include particle surface polarity, stage ratio, core particle size, the mode of monomer addition, and the degree of crosslinking.

Latexes made via two consecutive emulsion polymerization stages are commonly referred to as "core-shell" latexes, implying a particle structure with the initially-polymerized material at the center and the later-formed polymer as the outer layer. By polymerizing different polymer types in each separate stage, composite polymer particles can be obtained which are neither like random copolymers nor like polymer blends.

Such two-stage latexes have a wide range of potential applications in a variety of technologies, from paints and organic opacifiers, to impact modifiers for plastics, to carriers for biomolecules. For examples, two-stage latexes can be prepared for use as binders in latex paints. In such an application, the paint properties can be modified by varying the balance of properties of the first and second-stage polymers, e.g. the glass transition temperature of each stage (*1*). In addition, microvoid particles have been made by preparing particles with a base-sensitive hydrophillic core polymer surrounded by a hydrophobic shell polymer. These particles become hollow upon treatment under conditions of elevated temperature and pH, so that they contain mainly water at the center. Such hollow particles have been used as organic opacifiers in paints (*2*). Core-shell particles with elastomeric centers and rigid shells have also been made, which can be useful as impact modifiers in high impact plastics.

Early workers largely assumed that two-stage polymerizations led to core-shell structures. However, upon more thorough investigation, and with the

0097–6156/92/0492–0234$06.25/0

advent of electron microscopy techniques, it was discovered that the morphology of two-stage latexes could vary greatly. Okubo et al. have reported a number of morphologies other than the core-shell type exhibited by two-stage latex particles, including structures which they have designated "raspberry-like", "confetti-like", "snowman-like", "octopus ocellatus-like", and "mushroom-like" (*3*). Cho and Lee, in addition, reported two-stage latex particles with "sandwich-like" and "halfmoon-like" arrangements (*4*). Lee and Ishikawa obtained "inverted" particles, with the second-stage polymer at the particle center and the first-stage polymer on the outside (*5*). A few of these configurations are illustrated in Figure 1. It is notable that in the majority of cases where these unexpected morphologies occur, the first-stage polymer is more hydrophillic in nature than the second-stage polymer.

To a large degree, the physical properties of two-stage latexes depend on the morphology of the particles. Since such wide variation in morphology is possible, it is necessary to have an understanding of how particle morphology is controlled in order to control product properties. The control of two-stage latex particle morphology can be understood in terms of two major types of influences in the system, the thermodynamic forces and the kinetics of the morphological development. These two types of influences combine to determine the final particle structure. The effects of specific thermodynamic and kinetic factors will be discussed in this paper to illustrate this point.

An analysis of the thermodynamics of two-stage particle formation has been developed by Sundberg et al. (*6*) in which the system was considered simply in terms of the free energy changes at the interfaces of a three phase system (i.e. polymer 1, polymer 2, and water) based on the following equation:

$$G = \Sigma \gamma_{ij} A_{ij} \quad (1)$$

where: G is the Gibbs' free energy of the system; γ_{ij} is the interfacial tension between phases i and j, and A_{ij} is the interfacial area between phases i and j. According to this analysis, each particular morphological configuration will have a different value for G, and the arrangement with minimal free energy will be the one which is more thermodynamically favoured. For example, the free energy of the two extreme morphologies, the core-shell and the inverted configurations, can be calculated using equation 1 as:

$$G_{core\text{-}shell} = \gamma_{12} 4\pi r_1^2 + \gamma_{2w} 4\pi r_2^2 \quad (2)$$

$$G_{inverted} = \gamma_{12} 4\pi r_2^2 + \gamma_{1w} 4\pi r_1^2 \quad (3)$$

where r_1 and r_2 are the appropriate radii shown in Figure 2. Using these equations, Sundberg et al. demonstrated the importance of the interfacial thermodynamics, particularly the interfacial tensions γ_{ij}, in predicting the morphology of the particles. Y-C. Chen and coworkers have developed a thermodynamically-based mathematical model to describe free energy differences between different possible particle structures (*7*).

Additionally, another process for achieving composite polymer latex particles of controlled morphology prepared by the combination of preformed polymer particles has been reported recently by Waters (*8*). In this process, particles of polymer 1 are combined with particles of polymer 2 dispersed in a continuous medium, phase 3, at temperatures above the functional glass

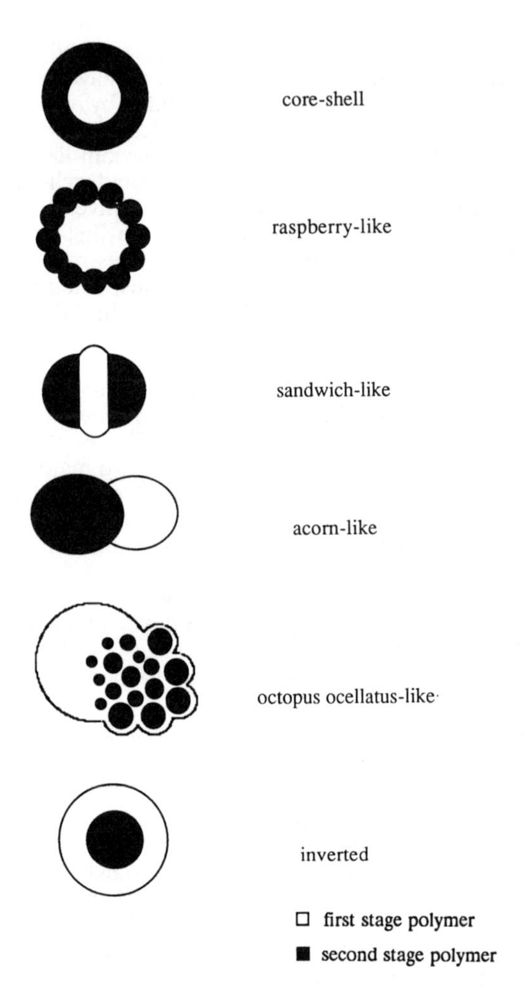

Figure 1. Examples of possible composite latex particle morphologies.

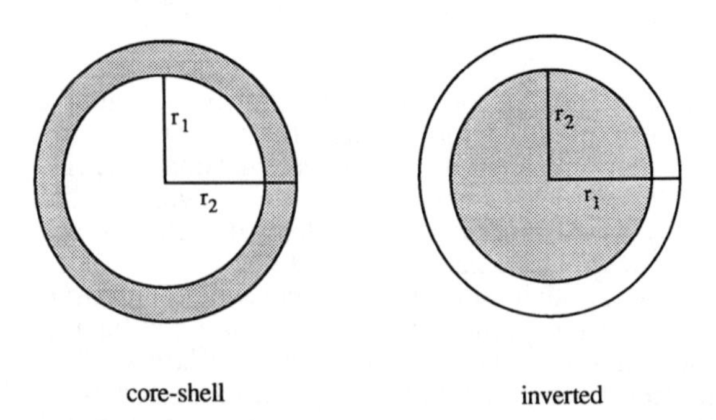

Figure 2. Radii of core-shell and inverted morphologies.

transition temperature of polymer 2. Starting from the same fundamental equations as those of Sundberg et al. as cited above, further equations expressed in terms of phase volume instead of interfacial area were derived. It was reported that the final particle morphology could be controlled by adjusting the relative phase volumes of polymers 1 and 2. For example, core-shell morphology would be expected to be more favorable than inverted morphology if $G_{core\text{-}shell} < G_{inverted}$, which when expressed in terms of interfacial tensions and polymer phase volumes becomes:

$$\frac{\gamma_{1w} - \gamma_{2w}}{\gamma_{12}} > \Phi_1^{2/3} - \Phi_2^{2/3} \qquad (4)$$

where ϕ_1 and ϕ_2 represent the volumes of polymers 1 and 2.

All the models cited are mutually consistent. Waters' theory is particularly convenient for present purposes and is considered further below.

Although the thermodynamics of the two-stage latex system determines which configuration will be ultimately the most stable, the kinetics of the morphological development determine the degree to which such a low energy arrangement is realized. Since the polymeric phases are subject to serious diffusional limitations, a kinetic barrier may exist between the initial state and the thermodynamically favoured state. Therefore, kinetic factors also play a role in determining two-stage particle morphology.

In this paper, a number of poly(methyl methacrylate)/polystyrene two-stage latexes (PMMA/PS) are described which demonstrate the effects of both thermodynamic forces and kinetic considerations on particle morphology. The parameters affecting the thermodynamics of the system include the particle surface polarity, the relative phase volumes, and the core particle size. The parameters affecting the kinetics of the morphological development include the mode of monomer addition and the use of crosslinking agents.

Experimental

Materials. Styrene (Aldrich) was purified by reduced pressure distillation, and was stored in a refrigerator. Water was obtained from Milli-Q purification system. All other materials were used as received without further purification. Reagent grade methyl methacrylate, allyl methacrylate, ammonium persulfate, potassium persulfate, and technical grade divinylbenzene were also obtained from Aldrich; 2,2'-azobisisobutyronitrile (AIBN) from Fluka; ethylene glycol dimethacrylate from Polysciences, and the surfactants Igepal CO-890 (nonylphenol ethoxylate with an average 40 ethylene oxidants) from Domtar and DS-10 (sodiumdodecylbenzene sulfonate) from Alcolac.

Polymerizations. All polymerizations were performed in a 1L glass kettle reactor suspended in a thermostated water bath and equipped with a condenser, mechanical stirrer, and, for semi-batch polymerizations, a monomer pump. The temperature was maintained at 80°C for methyl methacrylate polymerizations with semi-batch addition of monomer to the reactor, and at 60°C for styrene polymerizations with either batch or semi-batch addition of the monomer. Typical recipes are listed in Tables I to III.

Table I. Recipe for PMMA Core Latex using Ionic Initiator

Ingredient	Amt. (g)
water	375
methyl methacrylate	150
ammonium persulfate	1

Table II. Recipe for PMMA Core Latex using Nonionic Initiator

Ingredient	Amt. (g)
water	225
DS-10 surfactant	1.3
methyl methacrylate	150
AIBN	1.3

Table III. Typical Recipe for Second Stage Polymerization

Ingredient	Amt. (g)
PMMA core latex (dry weight)	10
water	90
Pre-emulsified feed:	
water	100
Igepal CO-890	0.2
styrene	10
AIBN	0.2

PMMA seed latexes were used as prepared, without purification.

Characterization. Dried PMMA/PS latex pieces were embedded in a mixture of Poly/Bed 812 and Araldite 502 epoxy resins (Polysciences) and cured at 60°C for 48 h. The samples were ultramicrotomed to 900 Angstrom thin sections using a Reichert ultramicrotome. The polystyrene domains were preferentially stained by exposure to RuO_4 solution vapours (Polysciences). The stained sections were examined and photographed using a Phillips 300 TEM at an accelerating voltage of 60 keV.

Results and Discussion

Alteration of Particle Surface Polarity. It is important to note that since the PMMA/PS system is a hydrophillic/hydrophobic system, one can predict *a priori*

that generally the most thermodynamically favorable arrangement will be an inverted morphology rather than a true core-shell structure. This is because γ_{2w} is greater than γ_{1w}. Therefore, one would expect the following relationships to apply:

$$\gamma_{12}A_{12} + \gamma_{2w}A_{2w} > \gamma_{12}A_{12} + \gamma_{1w}A_{1w} \quad (5)$$

and so:

$$G_{core-shell} > G_{inverted} \quad (6)$$

The formation of inverted morphology eliminates the most unfavorable hydrophobe-water surface and replaces it with a hydrophile-water surface. This situation may be represented by the diagram in Figure 3. This type of system is the one which is most complex in terms of morphological control and in which it is most difficult to achieve core-shell morphology.

Cho and Lee (*4*) reported that the use of an ionic initiator in the second stage of a PMMA/PS latex system resulted in anchoring of the polystyrene phase by the ionic surface groups generated from decomposition of the initiator. The ionic groups at the polystyrene chain ends rendered the polystyrene surface more polar, and thus the polystyrene phase could remain on the outside of the particle.

PMMA/PS latex particles made using an ionic initiator in the first stage and a nonionic initiator in the second stage have a very strong tendency to invert due to the inherent difference in the polarity of the two polymers coupled with the presence of ionic groups from the initiator on the PMMA surface. Particles containing many polystyrene domains within the PMMA core were obtained, as shown in Figure 4. A core-shell arrangement for the PMMA/PS system can be made more favorable by using a nonionic initiator such as AIBN in the first-stage polymerization and an ionic initiator in the second stage. Figure 5a shows that indeed such particles do exhibit a core-shell morphology. In addition to the thermodynamic effects, the use of a nonionic initiator led to increased PMMA molecular weight (M_w = 517,000; M_n = 227,000) compared to the PMMA made with an ionic initiator (M_w = 81,500; M_n = 38,900). Higher molecular weights should result in reduced phase mobility. Thus the formation of core-shell morphology can be promoted by using a nonionic initiator.

Furthermore, PMMA/PS particles were made using AIBN initiator in both stages. Under semi-batch conditions, a true core-shell structure was obtained with no polystyrene domains within the PMMA core, as shown in Figure 5b. Nevertheless, when nonionic initiators were used in both stages, and the shell polymerization carried out under batch conditions, a core-shell structure with many polystyrene inclusions was again obtained. The significance of batch versus semi-batch polymerizations will be discussed later. Apparently, eliminating covalently bound ionic groups from the PMMA surface can greatly reduce the core particle surface polarity and make a PMMA/PS core-shell structure more favorable thermodynamically.

The Effect of Stage Ratio. To evaluate the applicability of Waters' thermodynamic equations to a real two-stage latex system, and to investigate the possibility of using the relative amounts of each polymer, or the stage ratio, to counterbalance the influence of the polymer polarities, which tend to favour inversion, the following latexes were prepared. PMMA/PS two-stage particles were made with polymer 1/polymer 2 weight ratios of 3:1 and 1:3 prepared via

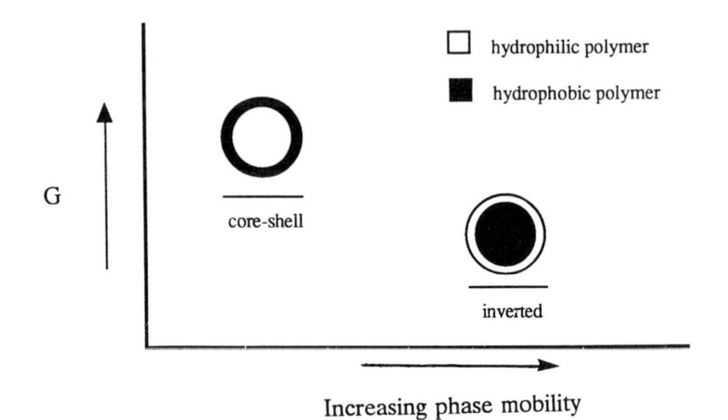

Figure 3. Free energy (G) of core-shell and inverted morphologies.

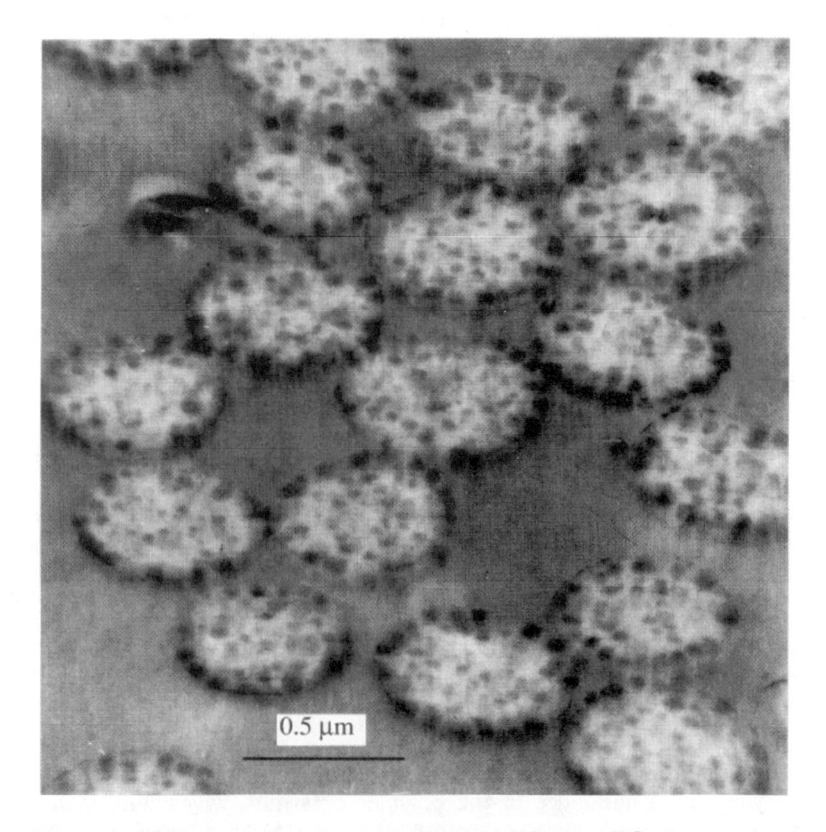

Figure 4. Microtomed section of 1:1 PMMA/PS latex (PS phase stained black with RuO_4). Initiators: first stage ionic; second stage nonionic; semi-batch second-stage polymerization.

(a)

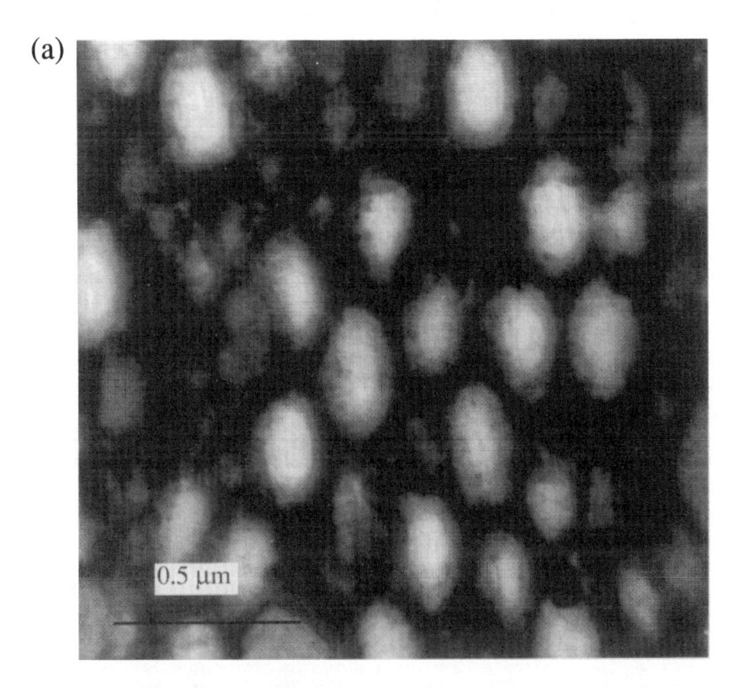

(b)

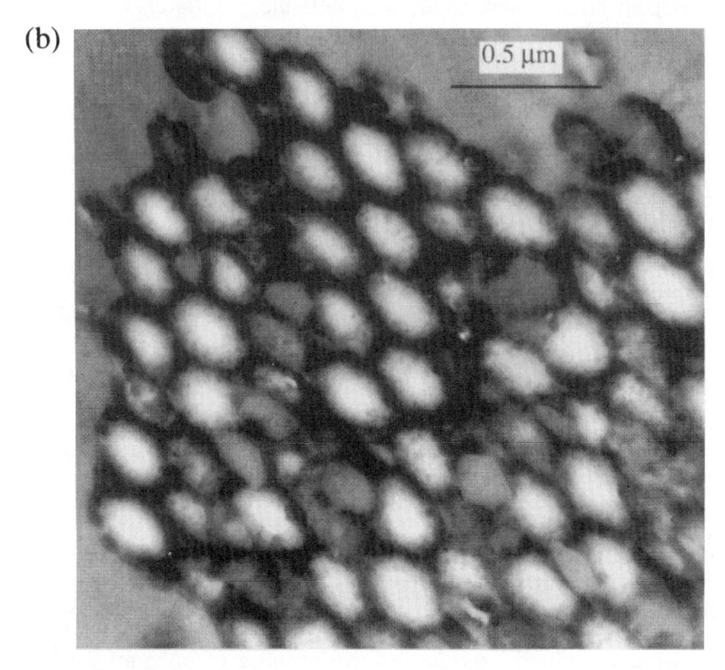

Figure 5. 1:1 PMMS/PS. First-stage initiator: nonionic. Semi-batch second-stage polymerization (a) second-stage initiator: ionic, (b) second-stage initiator: nonionic.

semi-batch polymerizations. These particles were made under the same conditions as those shown in Figure 4, in which a 1:1 stage ratio was used, where the first-stage initiator was ionic and the second-stage initiator was nonionic. It can be seen from Figures 4, 6a and 6b that as the relative amount of polymer 2 increases, core-shell morphology becomes more favorable. This corresponds to the results predicted by thermodynamics, from equation 4 above. That is, as the volume of polymer 2 increases, the value of the right hand side of equation 4 decreases, implying that a core-shell should become more stable.

The Influence of First-Stage Particle Size on Two-Stage Particle Morphology. In the preceding section, Waters' equation regarding the influence of relative phase volumes was discussed and experimental support for the applicability of that equation to real two-stage latexes was presented. This section will demonstrate the implications of the same equation to the role of core particle size on two-stage particle morphology.

In Waters' patent, the relative phase volumes were controlled by one of two methods: (i) By varying the number of polymer 1 and polymer 2 particles with both particle types having approximately equal diameters (Figure 7), or (ii) by varying particle size for a given ratio (1:1) of polymer 1 and polymer 2 particles (Figure 8).

Furthermore, as was shown in the preceding section, equation 4 can apply to two-stage latex formation as well. The two-stage particle morphology changes with stage ratio (Figure 9). Interestingly, the thermodynamic equations also appear to hold true for the situation where ϕ_1/ϕ_2 is held constant, but the particle size is varied, as demonstrated by the following experiments. A series of two-stage PMMA/PS latexes were made with persulfate-initiated PMMA first-stage particles and AIBN initiation of the second stage, under semi-batch conditions. The monodisperse first stage latex diameter was different in each polymerization, varying from 100 to 900 nm. A 1:1 stage ratio was used in all the polymerizations. Figure 10 shows a schematic diagram of the morphologies obtained. Using first-stage particles of small diameter, the two-stage particles obtained had an inverted morphology (Figure 11a). The same morphology was obtained using 200 nm first-stage particles. Again with a 1:1 PMMA/PS stage ratio, using first-stage particles of intermediate diameter (400 or 600 nm), the final particles had "raspberry-like" structures with polystyrene inclusions, as in Figure 4. Finally, with larger PMMA first-stage particles of 900 nm diameter, a core-shell morphology was formed with no inclusions of PS inside the PMMA cores, as shown in Figure 11b. Thus, under the polymerization conditions used, and for core particles in the range 100 to 900 nm, the larger the core particle diameters, the more favorable is formation of core-shell morphology.

Such behaviour may be explained using the equation derived by Waters. From equation 4, a core-shell morphology becomes more thermodynamically favoured as ϕ_2 increases, or conversely as ϕ_1 decreases. This equation, however, was derived to describe the phase behaviour of the combination of preformed particles. Thus, the phase volume was a convenient parameter to consider, since it is a parameter directly related to the surface area of each phase. The fundamental equation from which equation 4 was derived, however, was expressed in terms of interfacial areas.

Let us now consider Waters' experiments in terms of phase surface areas rather than phase volumes. Firstly, particles of polymers 1 and 2, of similar diameters were combined. As the relative number of polymer 2 particles was increased, core-shell morphology became more thermodynamically favorable. Therefore, as A_2, the total surface area of polymer 2 was increased, relative to

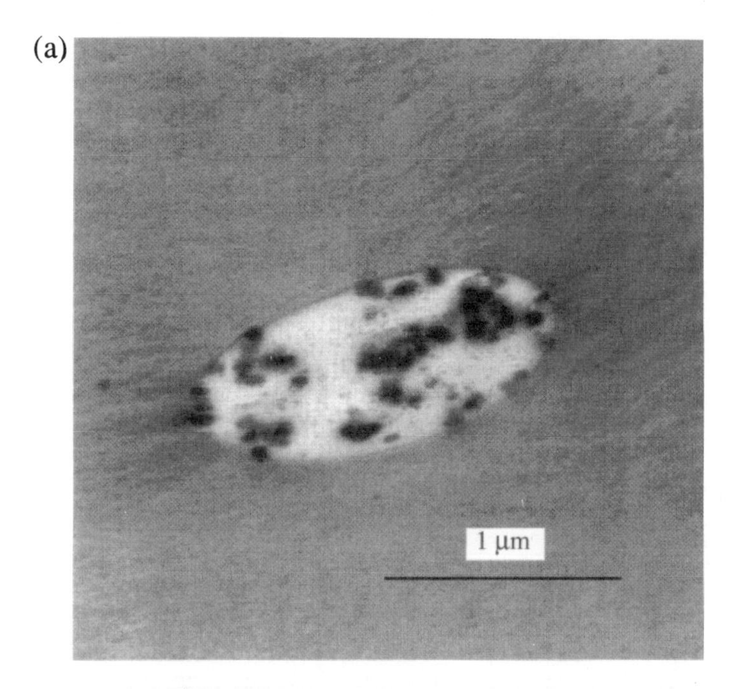

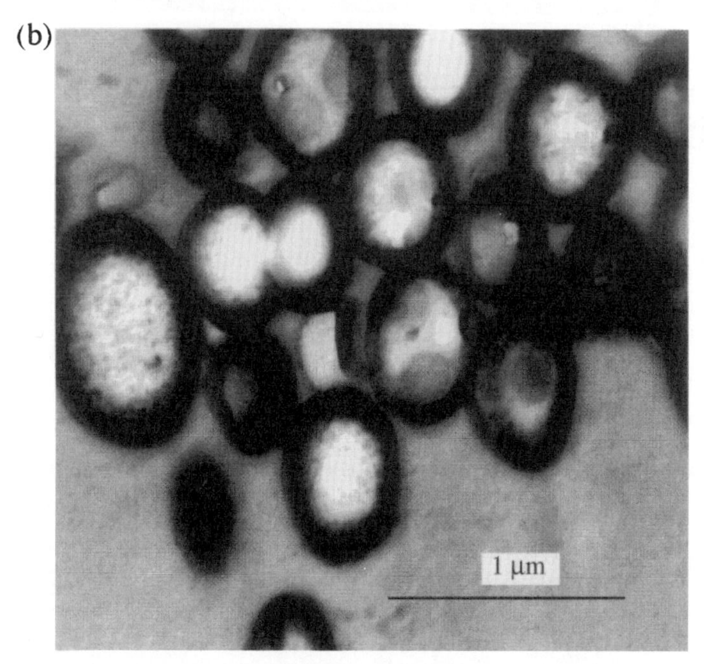

Figure 6. Semi-batch second-stage polymerizations with ionic first-stage initiator and nonionic second-stage initiator. (a) 3:1 PMMA/PS (b) 1:3 PMMA/PS.

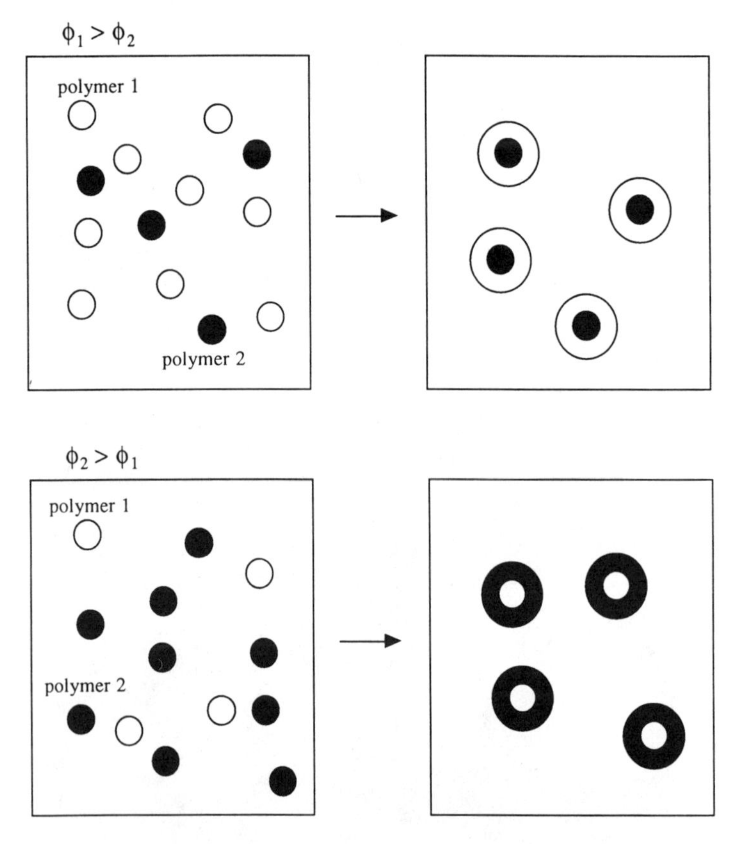

Figure 7. Varying phase volume (ϕ) by altering particle numbers.

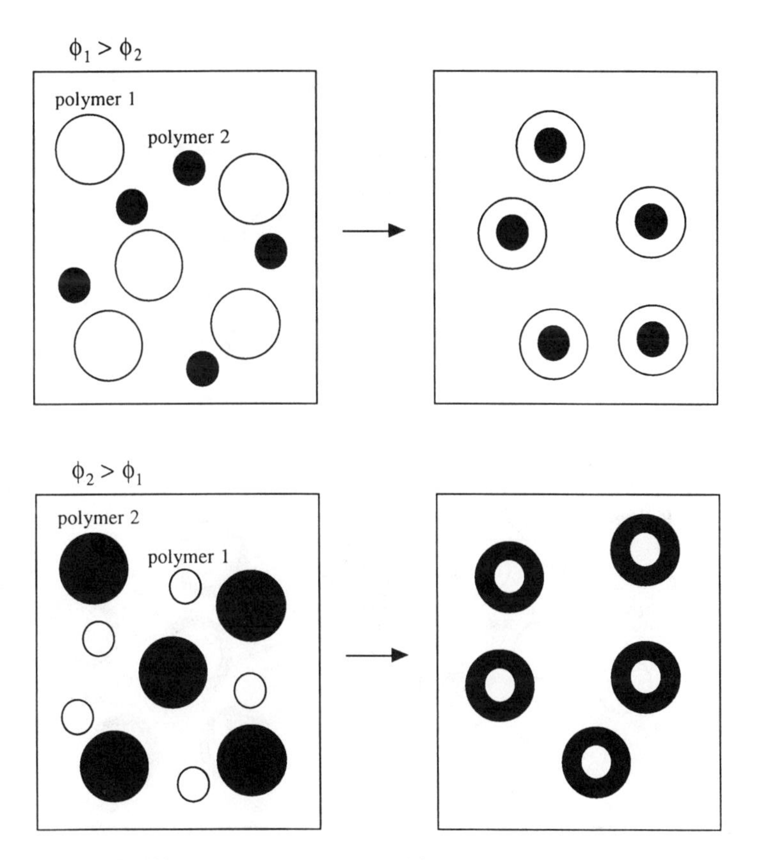

Figure 8. Varying phase volume (ϕ) by altering particle sizes.

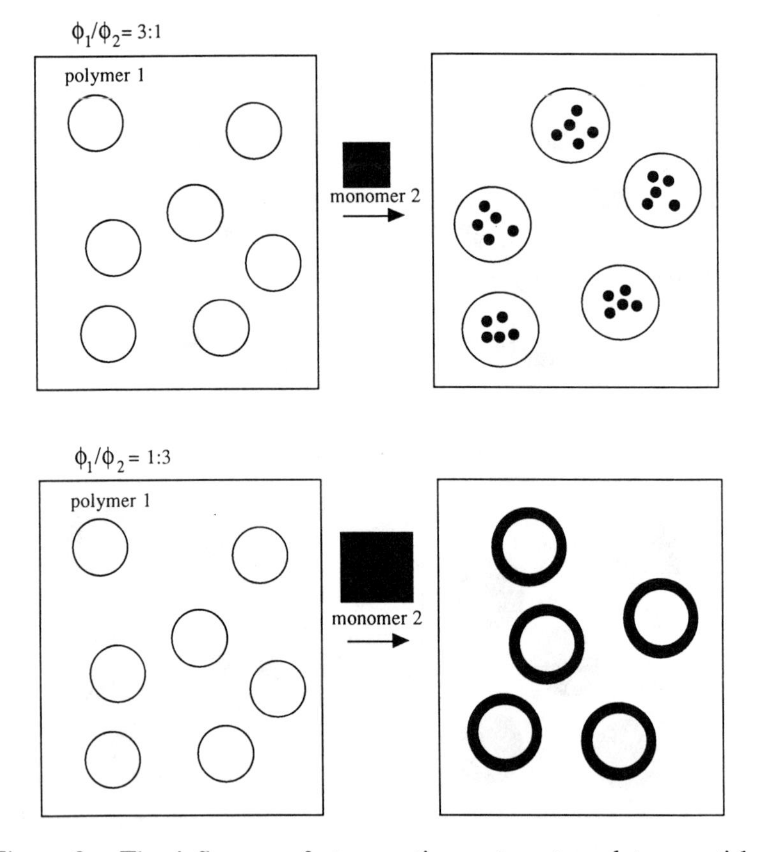

Figure 9. The influence of stage ratio on two-stage latex particle morphology.

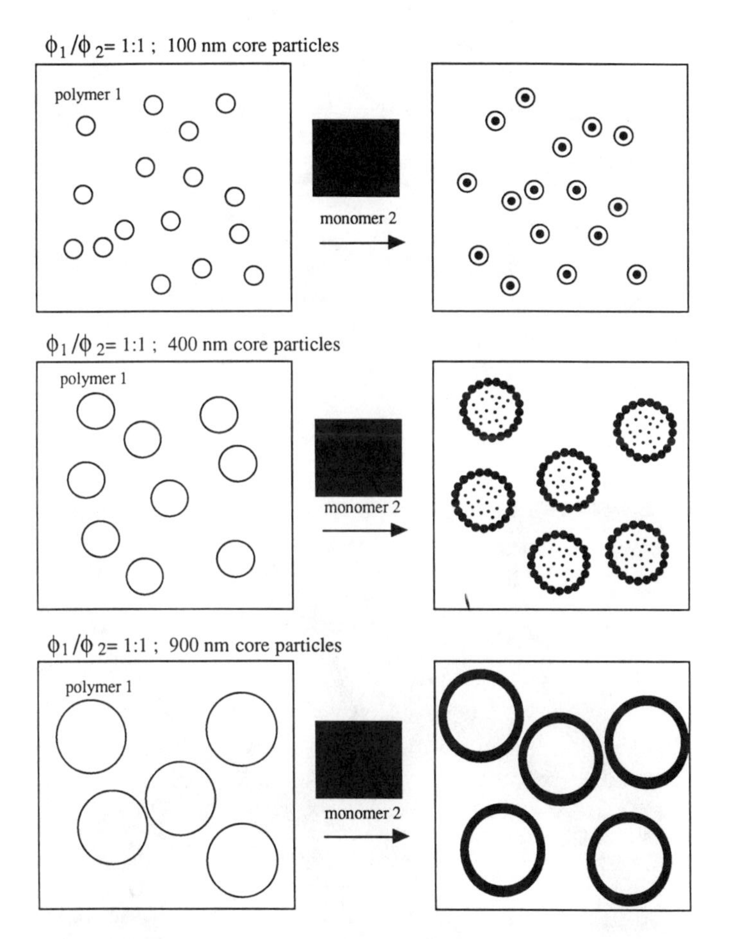

Figure 10. The influence of core latex particle size on two-stage particle morphology.

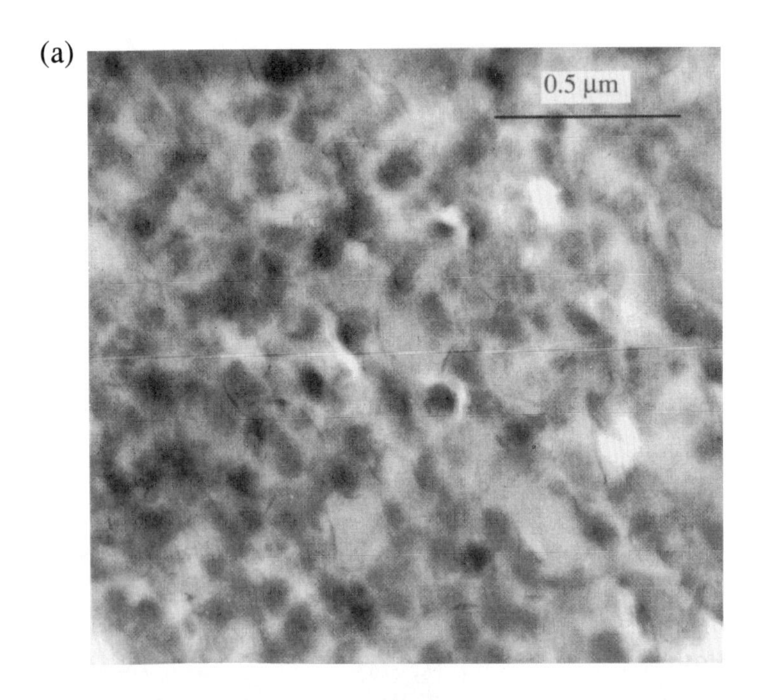

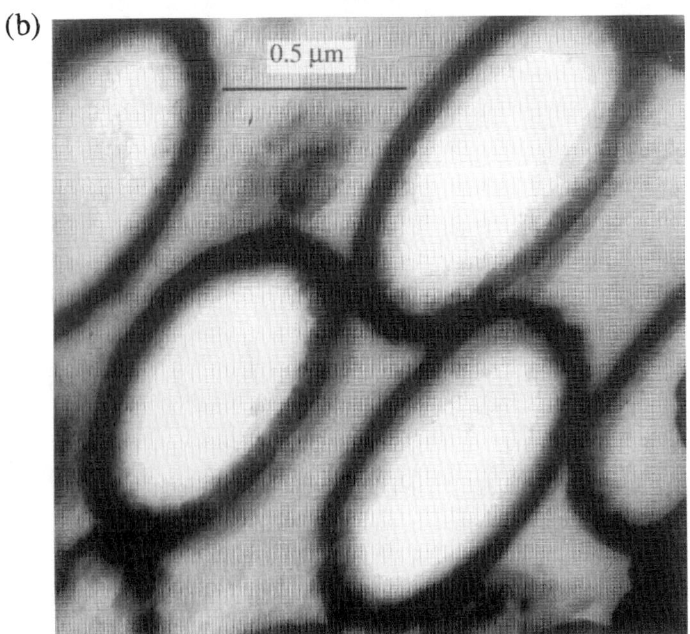

Figure 11. Microtomed morphology of 1:1 PMMA/PS particles: (a) 100 nm core (b) 900 nm core.

the total surface area of polymer 1, core-shell morphology became more favorable.

Secondly, particles of polymers 1 and 2 of dissimilar diameter were combined in a 1:1 ratio (i.e. equal numbers of each particle type). As the diameter of polymer 2 particles increased, core-shell morphology became more favorable. Here again, as the total surface area of polymer 2 was increased relative to the total surface area of polymer 1, core-shell morphology became more favorable. From equation 1, it is evident that the thermodynamics of morphological development are actually dependent on the minimization of the interfacial free energy, as determined by the interfacial tensions and interfacial *areas* in the system. In short, the morphologies obtained using a constant stage ratio, but different first-particle sizes can be explained by replacing ϕ_1 and ϕ_2 with kA_1 and kA_2 in equation 4, where A_1 is the total surface area of the polymer 1 particles and k is a constant equal to $(36\pi)^{-1/3}$. The physical meaning of A_2 is ill-defined for a two-stage polymerization, but this will be discussed later. Equation 4 can be re-expressed as follows:

$$\frac{\gamma_{1w} - \gamma_{2w}}{\gamma_{12}} > k(A_1 - A_2) \text{ when } G_{core\text{-}shell} < G_{inverted} \text{ given that}$$

$$A_i = \Sigma\, 4\pi r_i^2 \text{ and} \qquad (7)$$

$$\phi_i = \Sigma\, \frac{4}{3}\pi r_i^3$$

In experiments with pre-formed particles, core-shell morphology became more favorable as ϕ_1 decreased. As a corollary to those results, for a fixed amount of polymer 1, larger particle size translates to lesser total surface area, A_1. As shown in Figures 11a and 11b, it was found experimentally that core-shell morphology became more favorable as A_1 decreased. Figure 4 is an example of PMMA/PS particles made with 400 nm seeds.

As for the physical meaning of A_2, one might consider A_2 to be the total surface area of particles of polymer 2 polymerized in the absence of polymer 1 under the same conditions as the second-stage polymerizations. Such a hypothetical value for A_2 should be constant for a given set of stage 2 polymerization conditions. Therefore, for the two-stage latexes made with varying core particle size but constant phase volume, the controlling parameter is the value of A_1.

The Kinetics of Two-Stage Particle Development. In addition to the interfacial thermodynamics which determine the overall driving forces governing morphological development, the kinetics of that development is of utmost importance. Core-shell morphology can be made more favorable for a hydrophillic/hydrophobic two-stage polymer system through the methods described above. Such thermodynamic approaches essentially involve lowering the free energy of the core-shell morphology relative to the inverted morphology (Figure 12a). Alternatively, the kinetics of the morphological development can be used to control morphology. In order for the particles to achieve inverted morphology, the phases must overcome the kinetic barrier. The first-stage polymer must diffuse to the outside and the second phase must diffuse to the inside of the particles. Core-shell morphology can be preserved if the kinetic barrier to inversion is raised (Figure 12b).

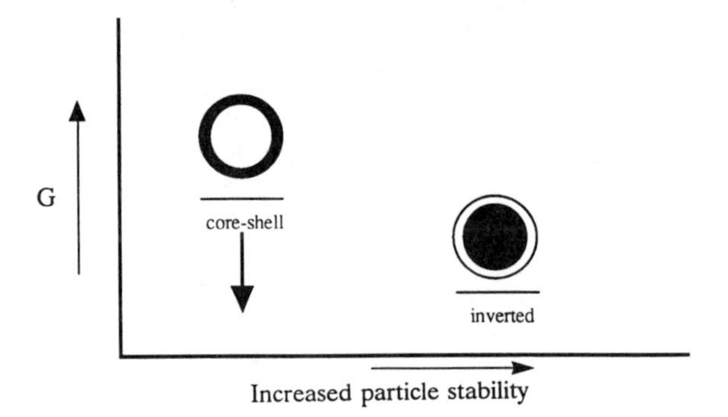

(a) Lowering free energy of core-shell morphology

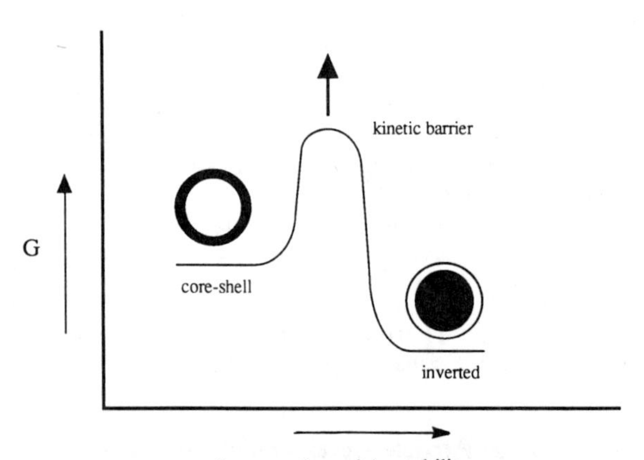

(b) Raising kinetic barrier

Figure 12. Methods to achieve core-shell morphology.

Semi-batch versus Batch Polymerizations. The mode of second stage monomer addition is highly influential in determining the morphology of two-stage composite latex particles. In a batch second-stage polymerization process, where all the monomer is added to the first-stage particles at the outset, relatively high levels of second-stage monomer are present initially. These monomer molecules have much greater freedom to diffuse than macromolecules. Moreover, the presence of monomer molecules in the first-stage particles enhances the mobility of the first-stage polymer molecules, such that migration of these molecules is also more likely.

In contrast, when semi-batch polymerization methods are employed where the second-stage monomer is added gradually to the first-stage particles, the level of monomer present at any time during the polymerization can be kept to a minimum. In so doing, the second phase will be present primarily in the form of polymer molecules, which are relatively hindered in mobility compared with monomer molecules. Therefore, batch polymerization conditions will decrease the kinetic barrier to inversion; whereas, semi-batch polymerizations impose a higher kinetic barrier.

This effect can be observed, for example, in Figures 13a and 13b showing micrographs of PMMA/PS particles prepared under batch and semi-batch conditions respectively, using ionic initiators in both polymerization stages. The particles made under batch conditions in the second stage show PS inclusions throughout the PMMA core, but the particles made under semi-batch conditions have few inclusions located primarily at the periphery of the PMMA core. As a general observation for all the PMMA/PS polymerizations made, core-shell morphology was difficult to achieve when the second-stage was carried out under batch polymerization conditions.

Control of Particle Morphology using Crosslinking Agents. Another method for increasing the kinetic barrier to inversion is by introducing crosslinking agents during the polymerization of two-stage particles. If the level of crosslinking is sufficiently high, the phases can be locked into a core-shell configuration indefinitely. For example, a 1:1 PMMA/PS latex was prepared under semi-batch conditions with an ionic initiator for the PMMA core and a AIBN initiator for the PS stage, a preparation similar to that used for the particles shown in Figure 4. However, in this case, crosslinking agents were incorporated into both first and second stages, 5 percent allyl methacrylate in the PMMA phase and 5 percent divinylbenzene in the PS phase, by weight. The resulting latex, shown in Figure 14a exhibited a true core-shell morphology, indicating that the mobility of the phases was decreased to such an extent that migration of the phases was prevented.

The use of crosslinking agents in both core and shell phases was found to be necessary to maintain a core-shell configuration. Figure 14b shows PMMA/PS particles made in the same way as those of Figures 4 and 14a, but containing 5 weight percent allyl methacrylate in the first stage and no crosslinking agent in the second stage. This time the particles had a "raspberry-like" structure. The core and shell phases did not mix, but the shell phase was separated into polystyrene subdomains. Possibly, this was due to the movement of PS chains to reduce the amount of PMMA-PS contact.

Another PMMA/PS latex was prepared with 5 weight percent divinylbenzene in the second stage but no crosslinking agents in the core. Figure 14c shows the resulting morphology, a PMMA core which contains many PS subdomains, with the whole particle surrounded by a PS shell. Although the PMMA core was essentially immobilized, a significant amount of polystyrene was

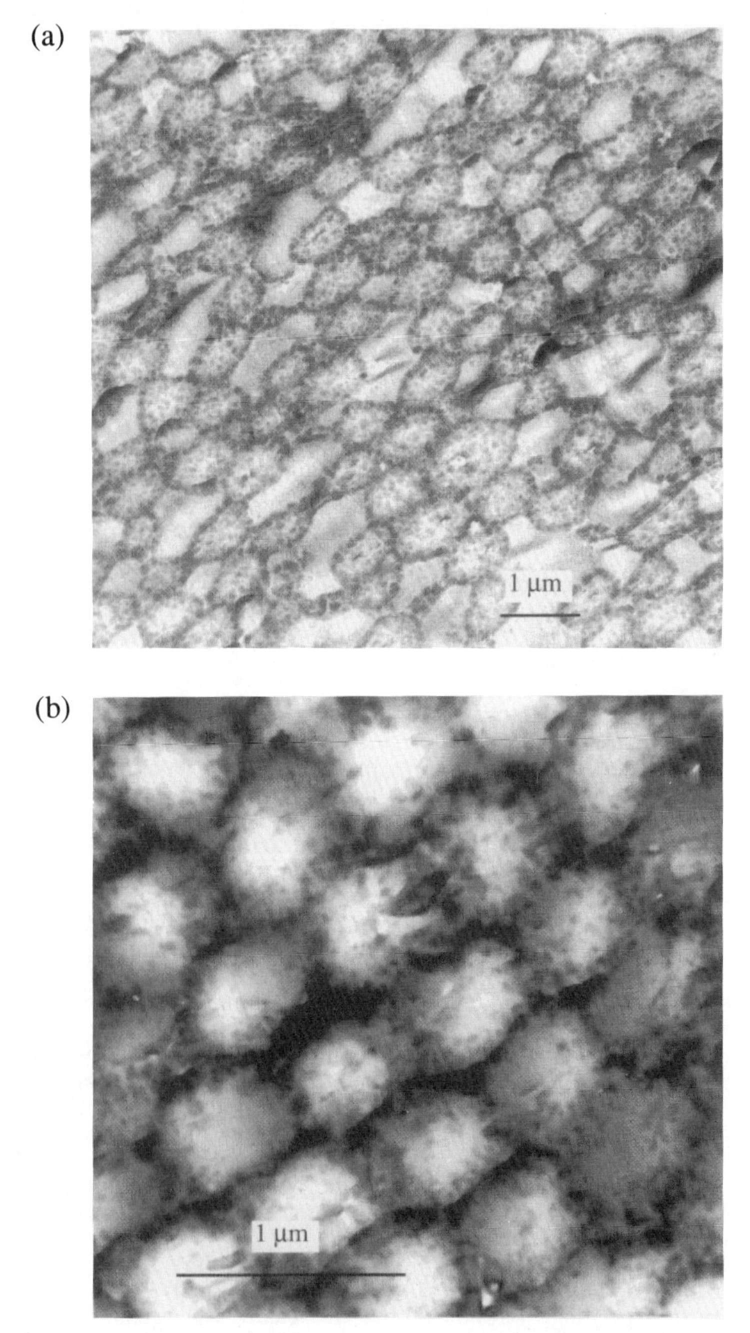

Figure 13. Morphology of microtomed PMMA/PS latexes prepared by: (a) batch second-stage polymerization (b) semi-batch second-stage polymerization.

(a)

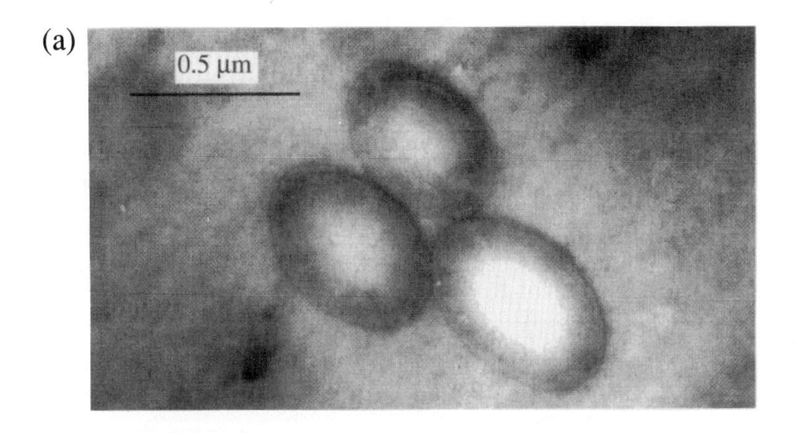

(b)

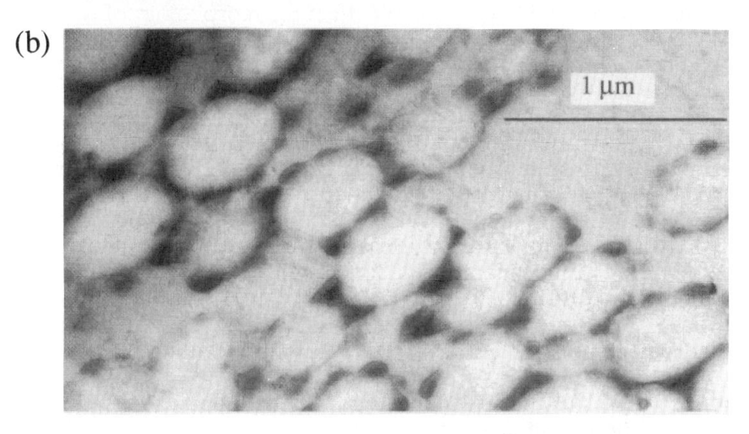

(c)

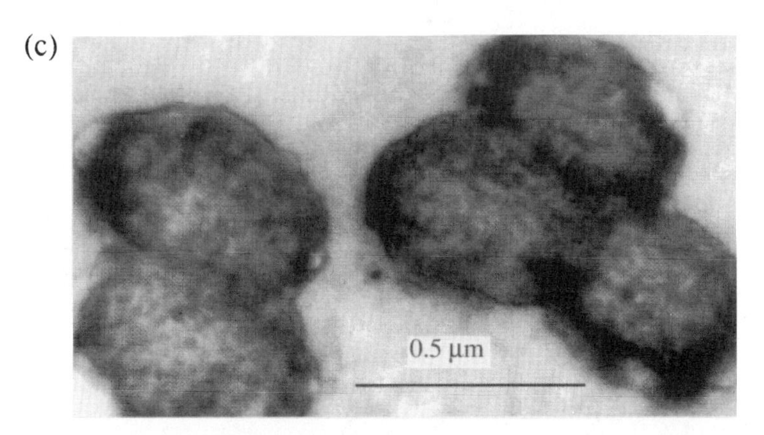

Figure 14. Morphology of microtomed sample of 1:1 PMMA/PS composite latex using a process: semi-batch second-stage polymerization. (a) 5 wt% allyl methacrylate in core; 5 wt% divinylbenzene in shell (b) 5 wt% allyl methacrylate in core only (c) 5 wt% divinylbenzene in shell only.

mobile enough to diffuse throughout the PMMA phase. Thus, crosslinking agents in both stages were found to be necessary here to eliminate phase mixing.

Lower levels of crosslinking agents also yielded core-shell morphologies. No phase mixing was observed in particles made with 0.2 weight percent allyl methacrylate in the core and 5 weight percent divinylbenzene in the shell, nor in particles made with 0.2 weight percent allyl methacrylate in the core and 0.5 weight percent ethylene glycol dimethacrylate in the shell. Further lowering of the crosslinking levels resulted in phase mixing. Moreover, phase mixing was observed when 5 weight percent allyl methacrylate was incorporated in the first stage and 5 weight percent divinylbenzene was incorporated in the second stage, but batch polymerization conditions were used. Thus, as discussed above, semi-batch conditions were necessary for kinetic considerations to dominate the morphological development.

Conclusions

Both thermodynamic and kinetic factors affect the morphology of two-stage latex particles. By altering the surface polarities and the surface areas of the latex particles, core-shell morphology can be made more thermodynamically favorable. By raising the kinetic barrier, the rearrangement of phases to an inverted morphology can be prevented. For the most part, in a hydrophillic/hydrophobic two-stage system, neither thermodynamics nor kinetics along will determine the final morphology. Under batch polymerization conditions, thermodynamics predominates initially, due to the high degree of mobility, but kinetics become important when the monomer is depleted. When the phases are highly crosslinked, phase mixing can still occur if batch polymerization conditions are employed. Thus, thermodynamics and kinetics together determine the two-stage particle morphology.

Acknowledgement

The authors would like to thank the Natural Sciences and Engineering Research council of Canada for financial support of this research.

Literature Cited

1. Devon, M.; Gardon, J.; Roberts, G; Rudin, A. *J. Appl. Polym. Sci.* **1990**, *39*, 2119.
2. Kowalski, A.; Vogel, M.; Blankenship, R.M. US patent 4 468 498, **1984**.
3. Okubo, M. *Makromol. Chem. Macromol. Symp.* **1990**, *35/36*, 307.
4. Cho, I; Lee, K.W. *J. Appl. Polym. Sci.* **1985**, *30*, 1903.
5. Lee, D.I.; Ishikawa, T. *J. Polym. Sci. Polym. Chem. Ed.* **1983**, *21*, 147.
6.(a) Berg, J.; Sundberg, D.; Kronberg, B. *Polym. Mat. Sci. Eng.* **1986**, *54*, 367.
(b) Berg, J.; Sundberg, D.; Kronberg, B. *J. Microencapsulation* **1989**, *6*, 327.
(c) Sundberg, D.; Casassa, A.P.; Pantazopoulos, J.; Muscato, M.R.; Kronberg, B.; Berg, J. *J. Appl. Polym. Sci.* **1990**, *45*, 1425.
7. Chen, Y-C.; Dimonie, V.; El-Aasser, M.S. *J. Appl. Polym. Sci.* **1991**, *42*, 1049.
8. Waters, J.A., Eur. Pat. Appl. EP 327 199, **1989.**

RECEIVED December 4, 1991

Chapter 16

Two-Stage Emulsion Polymerization of Acrylonitrile and Butadiene

Samuel Laferty[1] and Irja Piirma

Institute of Polymer Science, University of Akron, Akron, OH 44325–3909

The morphology of particles produced in a two stage aqueous phase polymerization of butadiene-acrylonitrile (BD-AN) and acrylonitrile-methyl acrylate copolymer (AN-MA) was studied by varying polymerization parameters. Three distinctly different latex particle morphologies were observed. These were in appearance one phase particles, two phase hemisphere morphology particles, and two phase coreshell morphology particles. The polymer produced by coagulating the two phase particles had a heterogeneous two phase matrix with the (BD-AN) copolymer phase being the dispersed phase in a (AN-MA) dispersion medium. The crosslink density of the BD-AN seed particles was found to control the formation of two stage particles. When the gel content of the seed exceeded 50% two phase particles resulted in the second stage particles.

Numerous publications (*1-7*) have shown that particle morphology in two stage (seeded) polymerizations depend on many variables in the polymerization procedure, including the monomer types and interactions between them. Probably the most important property is the interfacial energy associated with the various interfaces between phases (*2*). Other important factors are particle viscosity, molecular weight, crosslink density, degree of grafting between phases, and the ability to transport monomers, radicals, and oligomers between phases (*1*).

Thermodynamically, the more hydrophilic monomers (and subsequently polymer) will initially orient preferentially at the aqueous interface. One factor that can alter this orientation is the presence of a surfactant or other species at the interface which can change the interfacial tension and compatibility (*1*).

[1]Current address: Transohio Building, Suite 808, 156 South Main Street, Akron, OH 44308–1304

0097–6156/92/0492–0255$06.00/0

Thus, when studying the thermodynamic stability of an interface, we need to look at the polymer substrate and any surface active material that may modify the interfacial energies of the surface. According to Berg and Sundberg (*2*), the morphology of two phase dispersed particle systems depends upon minimizing the interfacial tension. The authors studied the morphology of particles formed from methyl methacrylate polymerization in the presence of simple oils (like decane and hexadecane) with a variety of surfactants and initiators (oil or water soluble). To develop predictive models, it is necessary to sum up the interfacial energies associated with the interfaces between different components of the system. Several different morphologies typically found are shown in Figure 1. Such calculations were carried out by Jonsson et al. (*1*) on a two phase system of styrene and methyl methacrylate. They expected the morphology with minimal energy to be thermodynamically most stable, and suggested that a change to this thermodynamic state occurs only if the path to this morphology does not lead through a less stable thermodynamic state.

Another factor affecting particle morphology of phase separated systems is the kinetics of phase separation. With small molecules and low viscosities, phase separation occurs more quickly than with high molecular weight polymers. Researchers have either plasticized seed latices with solvents or swollen the latex during post-polymerization procedures (*1*). The presence of a crosslinked component in particles has also been shown to change the morphology. Vanderhoff et al. (*3*) have recently shown that in a two stage polymerization an increase in the crosslink density of the polystyrene seed latex causes the formation of nonspherical latex particles during the second stage polymerization. With large particles (5.2 µm in diameter) morphology changes occurred during swelling in the absence of polymerization. Crosslinked polymers would be expected to swell less than linear polymers, and the viscosity within a swollen gel would be higher than in systems without crosslinks.

Lee and Rudin (*4*) compared batch and semi-batch polymerizations of styrene in the presence of poly(methyl methacrylate) seed particles. They found a number of different morphologies depending on the type of polymerization and the surfactant used. Min et al. (*5*) studied a system where poly(butyl acrylate) seed particles were used for the polymerization of styrene. They found that grafting occurred between the two polymers and that this affected the particle morphology and the changes in morphology with aging. Dimonie et al. (*6*) used polystyrene seed latex in the polymerization of styrene and acrylonitrile. They found that initiation, which occurred predominantly in the aqueous phase, sometimes caused the nucleation of a second crop of particles. Poly(methyl methacrylate) seed latex was used by Cho and Lee (*7*) in the polymerization of styrene. A significant effect on morphology was observed from the ionic terminal groups of the initiator.

The study reported in this paper relates the morphology of particles formed in a two stage aqueous phase polymerization to different controllable parameters in the polymerization. It differs from other studies in that the seed particles are composed of butadiene copolymer, which can achieve various crosslink levels by reactions of the residual double bonds in the polymer backbone.

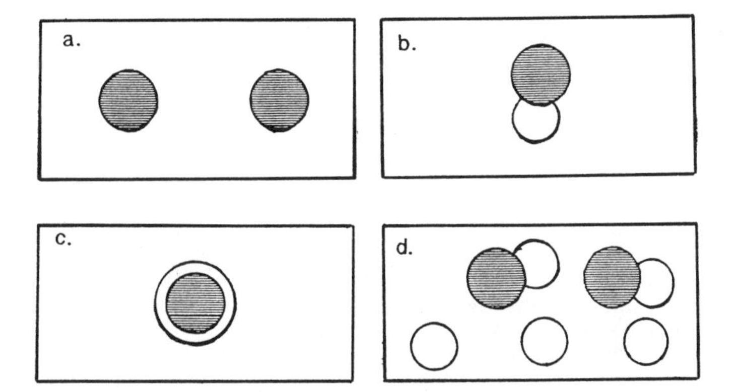

Figure 1. Particle morphologies achieved during second stage polymerizations. a. Apparent one phase blend, b. hemisphere two phase morphology, c. core-shell morphology, and d. two phase particles along with some one phase particles.

Since acrylonitrile is used in both the seed and as second stage monomer, it provides compatibility through strong hydrogen bonding between the different acrylonitrile molecules. The butadiene-acrylonitrile combination provides some important advantages in the particle morphology study. Among the most important is the fact that in electron microscopy studies, it is possible to stain only one copolymer, poly(butadiene-co-acrylonitrile), thus providing contrast on the electron micrographs.

Experimental

Materials. Butadiene (BD) was purified by passing it in gaseous form through 20% NaOH solution to trap polymerization inhibitor, then dried over $CaCl_2$, and finally condensed in a trap cooled by dry ice-isopropanol. Acrylonitrile (AN) and methyl acrylate (MA) were purified by vacuum distillation. The water used was deionized to a resistance of 17 MΩ·cm. The other chemicals were used as received from their respective manufacturers. These include surfactants, sodium dodecyl sulfate (SDS) (Fisher Scientific) and Emulphogene BC-840 (GAF).

Polymerizations. All polymerizations were carried out in 100 cm^3 bottles tumbled end over end at 38 RPM in a constant temperature water bath. The bottle charging procedure for the first stage polymerization started with the introduction into the bottle of the water, initiator, and surfactant. The air was then purged from the bottles and replaced with argon gas. The content of the bottles were then frozen and then acrylonitrile and butadiene added. An excess of butadiene was used to flush the headspace with butadiene gas replacing the oxygen. The bottles were then sealed and the dodecyl mercaptan was added with a syringe. The polymerization recipe was based on 100 g of monomers, with a BD to AN weight ratio of 65:35, with 200 g of water, 1.67 g sodium dodecyl sulfate, 0.27 g AIBN, and 0.77 g dodecyl mercaptan.

In the second stage of the polymerization, the amount of seed latex used was based on 10 g seed polymer solids for every 100 g of second stage monomers. The weight ratio of acrylonitrile to methyl acrylate was 75:25 in all experiments. The other ingredients used were 200 g water, 0.42 g sodium dodecyl sulfate, 0.133 g potassium persulfate, 0.77 g dodecyl mercaptan, and 2×10^{-4} g sodium hydroxide. All polymerizations were ran at 50 °C unless otherwise stated.

Percent Gel and Swell Index Measurements. These were performed on the first stage seed latex samples to determine the extent of crosslinking. Samples were prepared for this test by adding an polymeric p-nonyl phenol antioxidant to the latex, coagulating the latex with methanol or alum solution, washing the crumb (coagulated latex) with distilled water, and drying one hour in a vacuum oven. Then 0.5 g of the dry crumb was weighed into 100 cm^3 of methyl ethyl ketone (MEK) and allowed to dissolve and swell over 72 hours. The polymer and MEK was then filtered through a preweighed 100 mesh screen. The screen and swollen gel were weighed to determine the weight of the swollen gel. The screen and gel were dried to determine the weight of the dry gel. The weight of the dry gel divided by the sample weight gives the percent gel. The weight of the swollen gel divided by the weight of the dry gel gives the Swell Index.

Dynamic Light Scattering (DLS). This method of particle sizing was used for the examination of the swelling of the particles with second stage monomers still within the particles. An experiment was set up with different concentrations of surfactant, both sodium dodecyl sulfate and mixtures of sodium dodecyl sulfate and BC-840, using seed latex with different crosslink densities. The amount of surfactant was varied to assure that the DLS equipment, Malvern Autosizer 2c, was measuring only monomer swollen particles and not other species formed from the surfactant and monomers.

Thin Sections for TEM Examination. Compression molding was carried out at 160 °C with approximately 500 psi pressure. The latex had been coagulated, washed, and dried in a vacuum oven at 60-70 °C before molding. Thin sections of the molded samples were cut with a microtome at -20 °C sample temperature and -40 °C knife temperature from the compression molded samples. Following the process of thin sectioning, the samples (now deposited on a copper grid) were stained in osmium tetroxide fumes created by a 1 wt.% osmium tetroxide solution in water. The sections were believed to be about 90 nanometers thick based upon previous thin sections and meter readings on the microtome micro feed system.

Dynamic Mechanical Thermal Analysis. Samples of the various morphologies were melt compression molded (under similar conditions to thin section molding) at 160 °C for dynamic mechanical thermal analysis. The instrument used for this test was a Polymer Laboratories Mark II with variable frequency and temperature controlled test chamber. The specimens were mounted in a single cantilevered conformation, 5 mm separation between clamps, 1 Hz frequency, scanning from -80 to 120 °C. It was hoped that the presence or absence of T_g values characteristic of the copolymer phases could affirm or disaffirm phase separation.

Results and Discussions

Formation of Seed Particles. The seed latex butadiene-acrylonitrile (BD-AN) copolymer was made from the azeotropic composition, 35 wt.% acrylonitrile, so there would be no drift in the chemical composition during the polymerization. Several different formulations were used where the initiator level and surfactant level were varied and the effect on particle size and polymerization rate was studied. The recipe was then optimized to give 90-100 nm diameter particles with minimum gel. Percent gel and Swell Index are controlled by crosslinking reactions and were characterized using a method described in the experimental section.

Crosslinks are formed in BD polymer by the addition of growing free radical chain ends to the residual double bonds in the BD polymer backbone (*8*). The reaction is competitive with the addition of monomer to the growing chain ends. The activation energy for crosslinking is usually larger than the activation energy for monomer addition. Thus, lower temperature polymerization favors monomer addition over crosslinking and the polymer formed at lower temperatures is less crosslinked. As the polymer becomes crosslinked its ability to swell with solvents or monomers decreases due to physical restraints on swelling caused by the crosslinks (*8*).

Crosslinked polymer can be separated from less crosslinked polymer by centrifugation or filtration. Since developing crosslinking restricts swelling with solvents or monomer, and polymers tend to be more dense than their monomers, a difference in density between swollen crosslinked polymer and swollen polymers without crosslinks exists. This difference in density can allow separation of the two different components by centrifugation. Also crosslinked polymers tend to form gel which is a continuous polymer network. At high crosslinked levels this gel can be macroscopic in size and can be separated from non-gel by filtration through coarse screens. The gel of molecular weight (MW) sufficient to give dimensions larger than the screen pore size is trapped on the screen while lower MW polymers pass through the pores in the screen. This separates gel above a certain MW but allows smaller lower MW gel to pass through.

Once polymer molecules are incorporated into gel, further crosslinking simply increases the frequency of crosslinks along the polymer backbone. This increasing frequency of crosslinks lowers the molecular weight of the segments between crosslinks. This decrease in molecular weight between crosslinks can be characterized by the swell index, defined as the weight of the polymer swollen with a specified solvent (MEK in this study) divided by the weight of polymer.

In this study dodecyl mercaptan was used to modify the molecular weight of the butadiene-acrylonitrile copolymer. High molecular weight polymer can be crosslinked into a gel network at low levels of crosslinking. One can demonstrate this by using the average degree of polymerization of a polymer chain. If one uses a polymer of degree of polymerization 10,000, then each crosslinking reaction could potentially tie approximately 10,000 monomer units into the gel network. If one reduces the average degree of polymerization to 100 then one crosslink could only add 100 monomer units into the network. Thus 100 times as many crosslinks would be necessary to tie 10,000 monomer units of a polymer of degree of polymerization of 100 into the gel network.

The first stage latex samples made by the typical recipe had a gel content of 2% or less gel (defined as that material retained on 100 mesh screen after swelling in methyl ethyl ketone, MEK) at approximately 80-90% conversion of monomer to polymer. These latex samples were not stable with regards to percent gel. On storage at room temperature the gel content increased until the measured gel content was 50-60% of the sample. A study was initiated to determine variables controlling growth of the gel portion in latex with aging. Low gel content latex was made by emulsion polymerization and stored under various conditions. Figure 2 shows the Percent Gel and Swell Index changes on two samples aged at 25° C. Latex exposed to air and aged at 25°C crosslinked quickly to about 50-60% gel and stabilized there. Latex inhibited for free radical reactions with hydroquinone developed gel very slowly. The swell index also changed more slowly in the presence of hydroquinone. Reducing the temperature also slowed the development of gel through crosslinking reactions. After these effects were noted, all first stage latex samples, where the presence of low gel was required, were stored at refrigerator temperature with a slight amount of hydroquinone added.

Development of Two Phase Morphology. The percent gel in the seed latex was found to affect the morphology of particles in the second stage polymerizations. When low gel content seed latex (usually less than ten percent gel) was used in the second stage polymerization, each latex particle developed into a larger sized particle that appeared homogeneous by TEM examination. A micrograph of these second stage particles is shown in Figure 3. The seed latex with a higher gel content (usually 50-60% gel) in the second stage always developed particles that had two phases, as shown in Figure 4. Both latex samples were approximately ten percent BD-AN copolymer and ninety percent by weight AN-MA copolymer. The same latex can be used as a low gel seed latex and then allowed to crosslink further during aging. The aged samples with higher gel and crosslinking will give two phase particles in the second stage while their low crosslinked seed counterpart will always give one phase polymer particles.

The mechanism of formation of these two phase particles was of interest, since a better understanding of the formation mechanism would allow better control of the morphology development. Work by Vanderhoff et al. (*3*) has shown that morphology changes in the crosslinked particles take place during swelling and prior to polymerization. Unfortunately, experiments with TEM can only be carried out with particles in the non-swollen state, because the TEM is operated under high vacuum. Thus by taking samples at various conversions, one only sees the polymer particle morphology after monomer is removed. A study was done looking at the morphology of the particles as a function of conversion but no definitive results were found.

Dynamic light scattering (DLS) studies of the swelling of these seed latices with the acrylonitrile-methyl acrylate monomers showed that lightly crosslinked seed latex particles could swell four times their original volume. Highly crosslinked seed latex particles (like those that give two phased second stage latex particles) could swell to only twice their original volume.

It could also be assumed that following the polymerization of AN-MA inside the polymer particles that the newly formed AN-MA polymer could swell with their monomers. The acrylonitrile-methyl acrylate copolymer used could swell in the second stage monomer to 1.5-1.6 times its original weight. All observations of the second stage polymerization bottles, prior to polymerization, indicated that there was excess monomer present. Monomer droplets were seen even with the low crosslink density seed latex samples. The DLS data on limited swelling of the seed particles and monomer droplets indicate that in all the polymerizations studied, a substantial portion of the second stage monomers exists outside of the BD-AN copolymer (note that 100 g of second stage monomers were used for each 10 grams of BD-AN polymer in the seed latex). These monomers could polymerize in the aqueous phase and later coagulate onto the

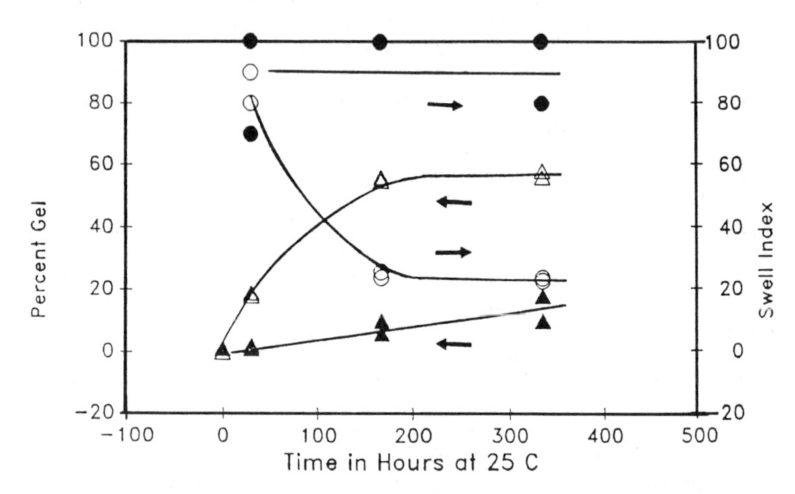

Figure 2. Development of percent gel (△) and swell index (○) in seed latex with aging time at 25 °C. (filled symbols) in latex stored with hydroquinone free radical scavenger, (open symbols) in latex stored without a free radical scavenger.

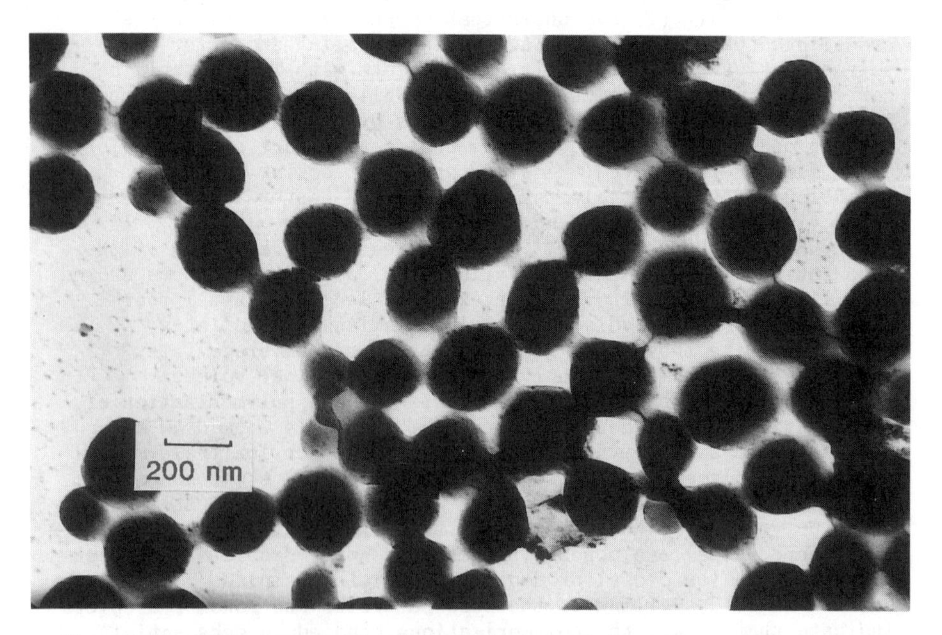

Figure 3. TEM micrograph of second stage latex, from seed with low crosslinking density. Visually appears as a one phase system after staining with osmium tetroxide.

BD-AN particles. The lightly crosslinked seeds could imbibe a more significant portion of the monomers allowing more polymerization to occur within the particles.

Other Two Phase Morphologies. A two phase core-shell morphology was made by changing polymerization conditions slightly from the two phase hemisphere polymerization conditions. Eventually two different methods of causing the morphology to go to core-shell with highly crosslinked seed latex were developed. The first involved doubling the dodecyl mercaptan concentration from 0.77 g/100 g monomers to 1.55 g/100 g monomers in the second stage polymerization of acrylonitrile-methyl acrylate. The second method was the addition of a nonionic Emulphogene BC-840 (1.67 g/100 g monomers, 13 carbon alkyl chain with 15 ethylene oxide units, GAF product) surfactant to the second stage recipe. Figure 5 shows the morphology with added dodecyl mercaptan while Figure 6 shows a similar morphology caused by the addition of BC-840. Again the particles are ten percent BD-AN and ninety percent by weight AN-MA. That two phases still existed after melt compression molding was confirmed by TEM examination of thin sections and dynamic mechanical thermal analysis (DMTA). The thin section of coagulated polymer made from the core shell morphology particles (shown in Figure 5) is shown in Figure 7.

Dispersed Phase in Coagulated Polymers. Although it is interesting to make two phase polymer particles, the enhanced performance of a new polymer blend often depends on converting the latex into a film or bulk polymer. After removal of water, the temperature of the polymer blend was raised to effect fusion and flow. During this melt compression molding the system could potentially convert to a more thermodynamically favored state. With respect to the system in this study so far, the two phase particles formed in the second stage have remained two phases, while the one phase particles formed in the second stage remain one phase during melt compression molding at 160 °C for 5 minutes. This does not mean these states are the most thermodynamically favorable in terms of surface energy, but merely they are the only morphologies achievable under the conditions chosen.

The above conclusion regarding morphological changes in melt compression molded samples is based upon the evaluation of thin sections and dynamic mechanical thermal analysis of several samples. When one phase polymer particles of 90 wt.% AN-MA on 10 wt.% BD-AN seed were processed, there was some gradual variations in contrast (indicating possible variation in butadiene content of polymer) in the thin sections at 5000 magnification, but the sample was very homogeneous at 30,000 magnification. The latex sample exhibiting hemisphere morphology with the same weight composition, whose thin section is shown in Figure 8, when melt compression molded and thin sectioned, showed two phase morphology. A sphere, or in two dimensions a circular shape, minimizes the amount of interfacial area which would seem to be more thermodynamically stable. The irregular shape of the dispersed phase in the thin section might be from the irregular interface between the two copolymers in the hemisphere shape components of the original latex particle.

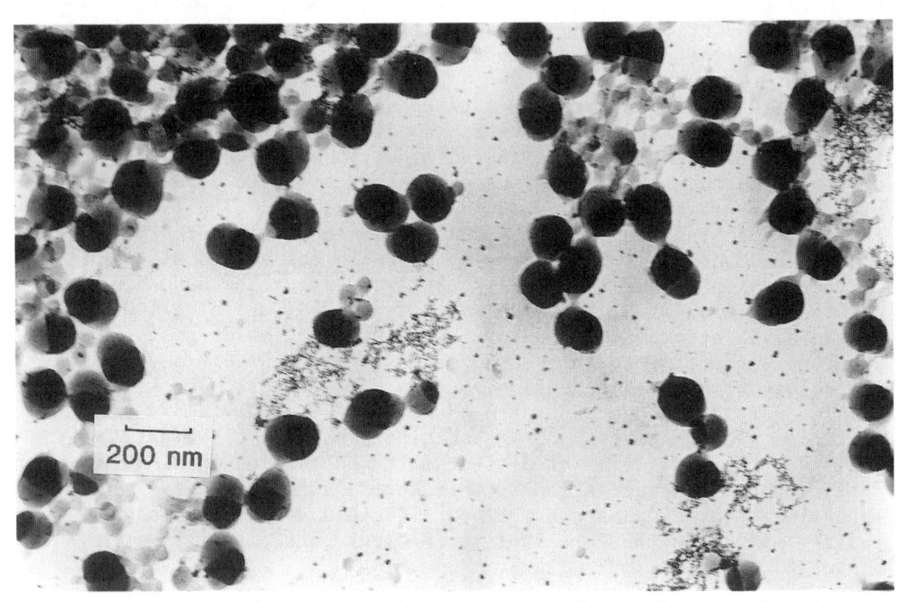

Figure 4. TEM micrograph of second stage latex from seed with high crosslink density showing hemisphere morphology after osmium tetroxide staining.

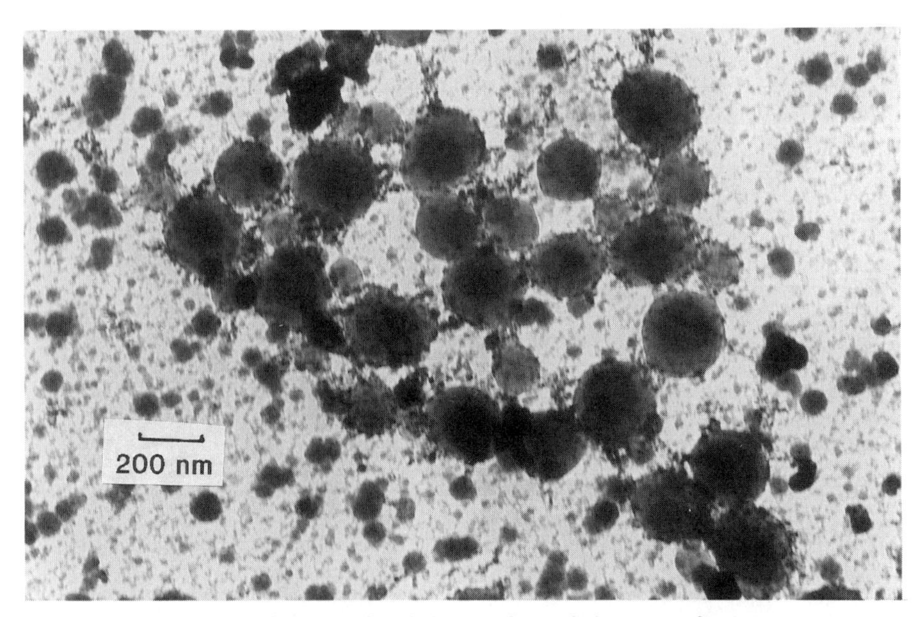

Figure 5. TEM micrograph of latex formed in second stage polymerization when mercaptan concentration increased from 0.77 to 1.55 g/100 g monomers. Morphology is characterized as being core-shell.

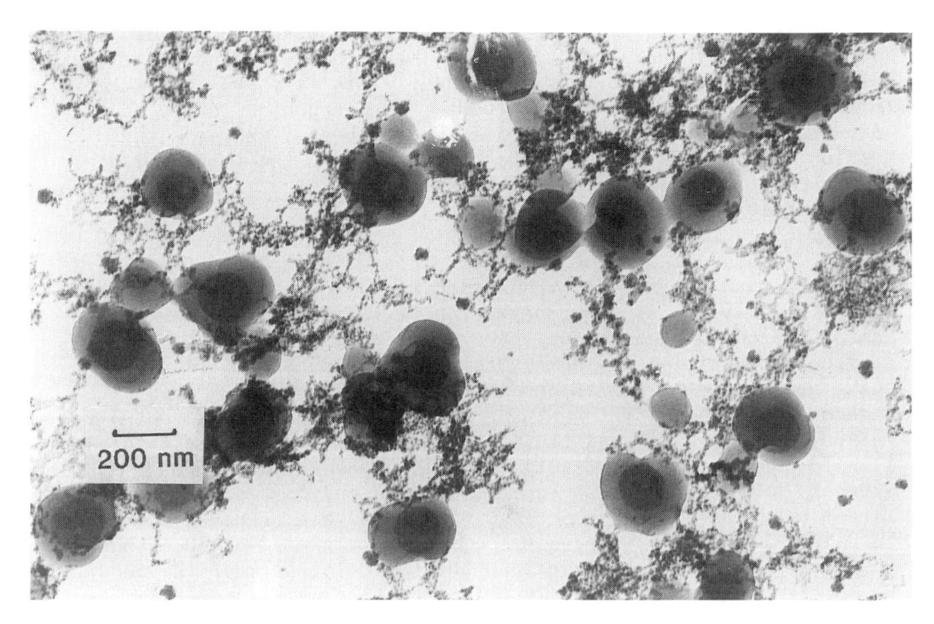

Figure 6. TEM micrograph of latex from second stage polymerization when Emulphogene BD-840 nonionic surfactant added to second stage recipe at 1.67 g/100 g monomers. The morphology is characterized as being core-shell.

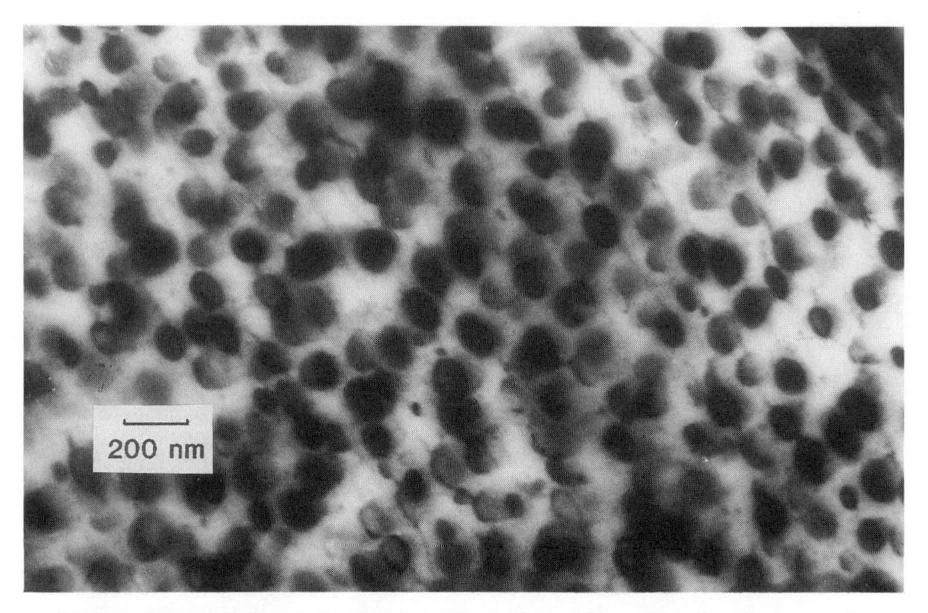

Figure 7. A thin-section of melt compression molded polymer made from a two phase core-shell morphology latex. Material stained with osmium tetroxide so darker areas are richer in BD-AN copolymer.

A core-shell latex made by a using higher mercaptan level keeping the polymeric composition the same in the second stage polymerization was also thin sectioned. This is shown in Figure 7. In this instance the BD-AN rich phase had a spherical shape in the core-shell structure and a circular shape when viewed as a thin section in TEM. The size of the BD-AN phase in the thin section was comparable to its size in the unsectioned core-shell structured latex particles.

Dynamic Mechanical Testing for Phase Separation. A one phase AN-MA copolymer gave a single T_g transition at 100 °C as seen in Figure 9. Similarly, an AN-MA copolymer on BD-AN seed (10 wt% BD-AN) that appeared to produce one phase polymer particles gave only a single T_g transition at 100 °C, as seen in Figure 10. The two phase hemisphere particles produced from AN-MA on BD-AN seed with high crosslink density (90 wt% AN-MA and 10 wt% BD-AN) gave two T_g values, one at -20 °C and the other at 100 °C, as seen in Figure 11. To confirm that the small tan delta peak at -20 °C was actually a T_g of the phase separated BD-AN polymer, a sample with a larger percentage BD-AN phase (20 wt% BD-AN) was prepared and run on DMTA. The tan delta peak associated with the T_g of BD-AN phase on the new sample was larger than the tan delta peak of the 10 wt.% BD-AN blend but at the same location, as seen in Figure 12. Obviously this is indicative that the peak was correctly termed a T_g, and was correctly identified as that of BD-AN rich phase.

Control of Other Polymerization Variables. Other factors besides crosslink density of the seed latex might affect the development of two phase polymer particles and the morphology of these particles. Several factors listed in the literature are interfacial tension between various phases (*2*); particle viscosity; molecular weight of polymers involved; potential grafts between phases; and rate of transport of monomers, radicals and oligomers (*1*). Controllable parameters studied in the second stage polymerization in this paper were: i) the polymerization temperature, ii) ratio of seed and monomer to water, iii) surfactant concentration, and iv) molecular weight of BD-AN copolymer before crosslinking (as controlled by the concentration of chain transfer agent in the first stage polymerization).

i) When polymerizations with a highly crosslink seed latex were conducted at 25 °C using potassium persulfate, sodium bisulfite,and iron initiating system (instead of the 50 °C standard polymerization temperature with potassium persulfate only) little difference in second stage particle morphology was detected.

ii) When the ratio of water to monomer and seed latex was increased by a factor of two (thereby decreasing the concentration of seed particles, but keeping the ratio of monomer to seed constant) a new crop of one phase AN-MA latex was formed along with the two phase particles. The presence of two types of particles (particles with dark BD-AN phases and particles with only AN-MA which are lighter and smaller) can be observed in Figure 13. As noted previously this polymerization recipe, ten parts BD-AN seed per 100 parts AN-MA monomers, seems to have excess monomer beyond that required to swell the seed polymer particles to their equilibrium swelling concentration. This excess monomer can

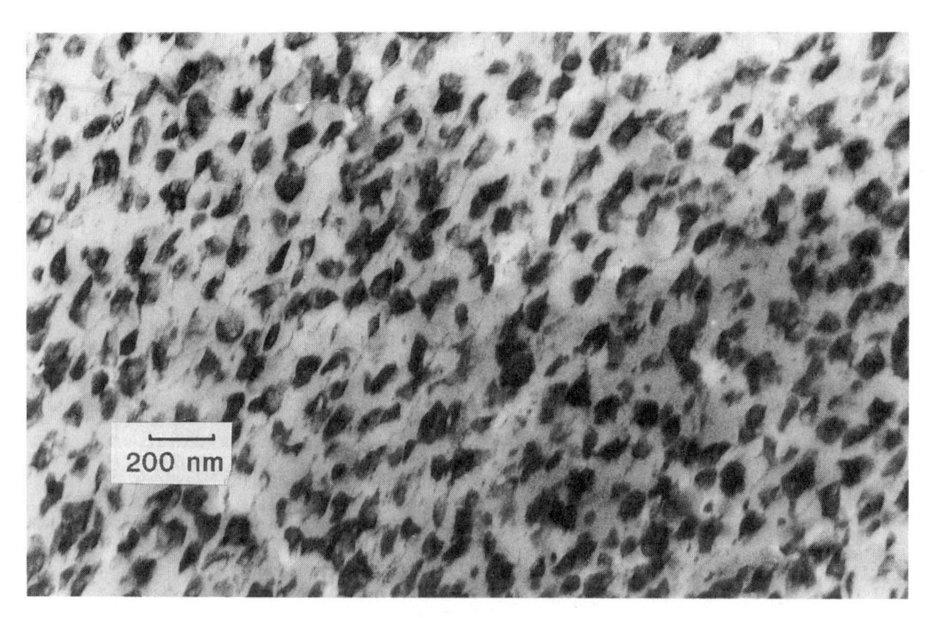

Figure 8. A thin-section of melt compression molded polymer made from two phase hemisphere morphology latex. Material stained with osmium tetroxide so darker areas are richer in BD-AN copolymer.

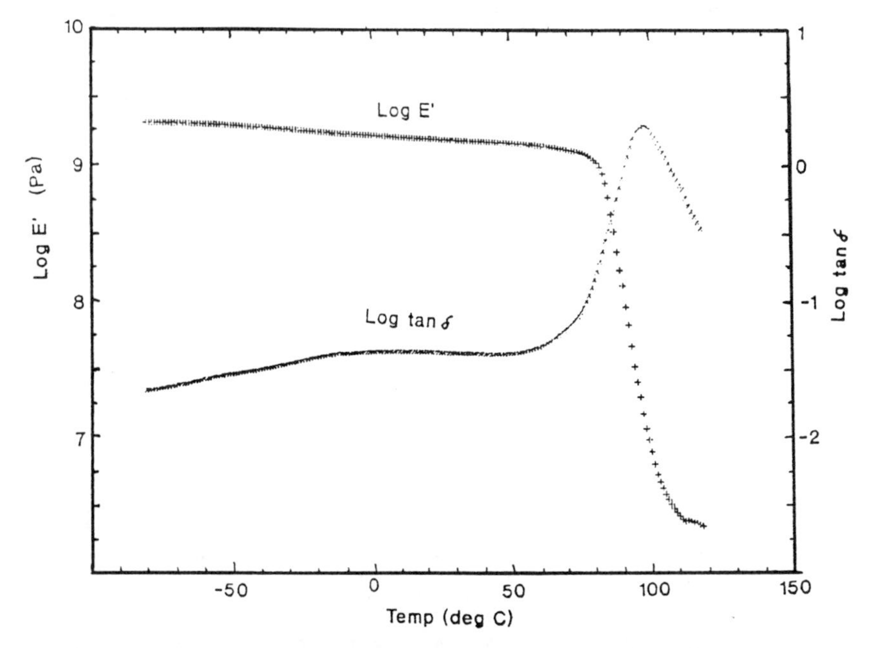

Figure 9. Dynamic Mechanical Thermal Analysis (DMTA) plot of log E' and log tan δ of a copolymer of 75:25 weight ratio AN-MA copolymer.

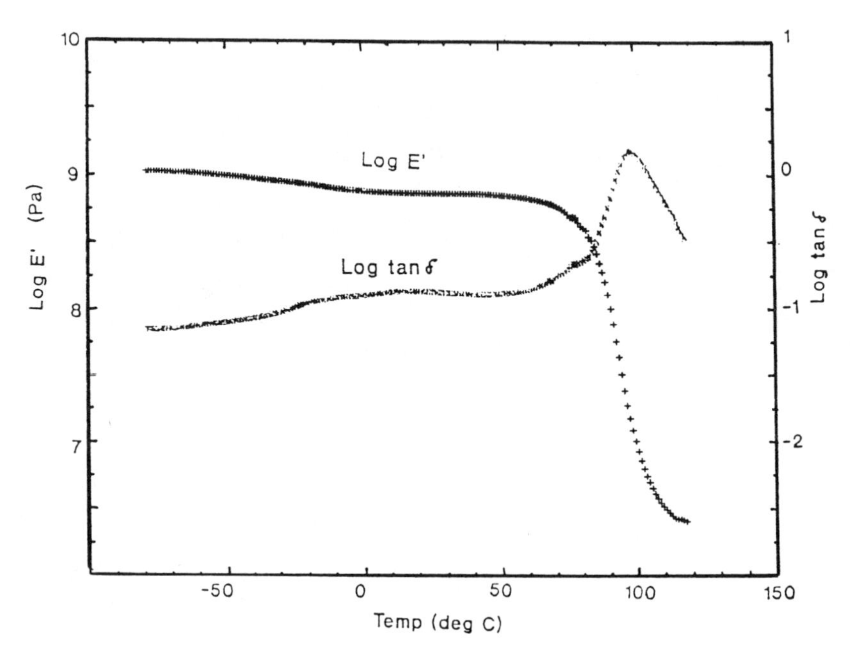

Figure 10. DMTA plot of log E' and log tan δ on melt compression molded sample of one phase second stage latex particles.

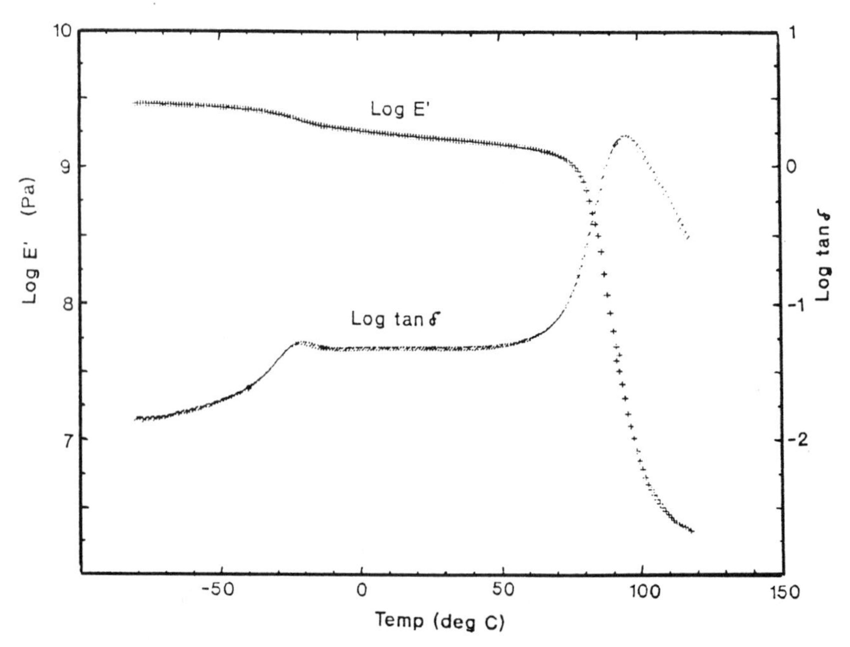

Figure 11. DMTA plot of log E' and log tan δ on melt compression molded sample of two phase latex particles of hemisphere morphology.

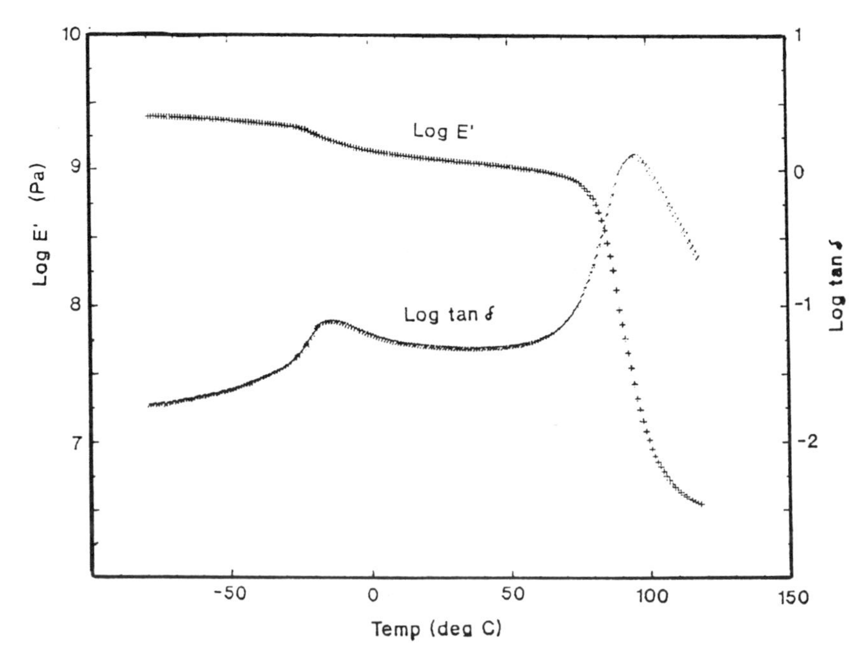

Figure 12. DMTA plot of log E' and log tan δ on melt compression molded sample of two phase latex particles of hemisphere morphology when weight percent BD-AN increased from 10 to 20.

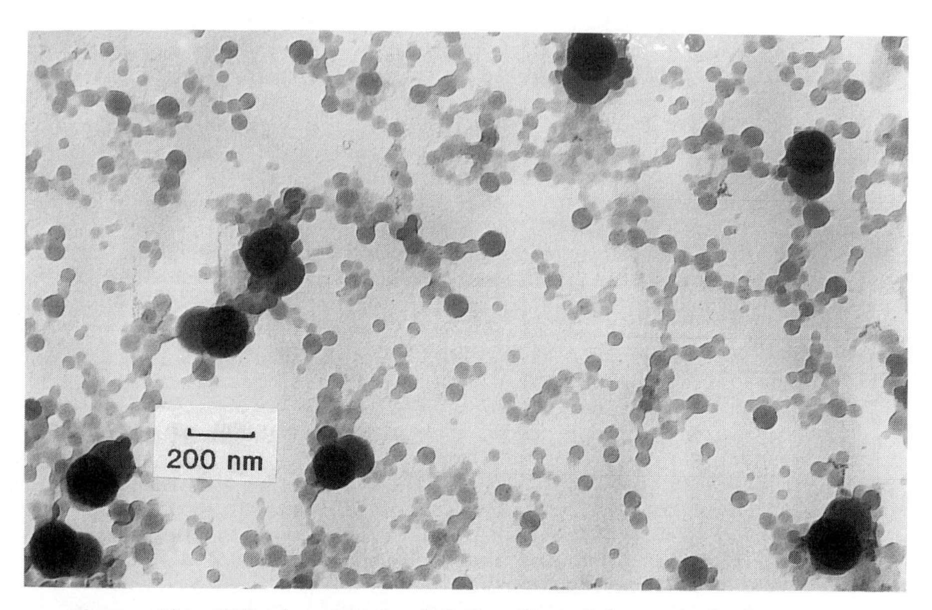

Figure 13. TEM micrograph of latex formed in second stage polymerization. Sample was stained with osmium tetroxide so the darker areas are richer in BD-AN copolymer.

nucleate new particles under favorable conditions, such as low amounts of seed particles or increased surfactant concentration.

iii) When the surfactant concentration was increased beyond that amount estimated to be 100% surface coverage of the seed particles, then a new crop of AN-MA polymer particles was formed along with the two phase particles. This could be explained by the increased level of surfactant stabilizing otherwise unstable particle species in the aqueous phase.

iv) Experiments were run where the dodecyl mercaptan concentration was varied in the first stage of the polymerization. This would produce BD-AN polymer of higher molecular weight than produced with the standard recipe. It was found that the molecular weight of the BD-AN copolymer, before crosslinking in the seed latex, seems to have little observable effect on the second stage morphology when the gel content is 50-60% prior to the second stage polymerization. Equivalent gel contents were produced by allowing oxidative crosslinking of the low molecular weight BD-AN latex prior to the second stage.

Conclusions

Apparently homogeneous one phase and two phase particles were observed in blends of butadiene-acrylonitrile (BD-AN) and acrylonitrile-methyl acrylate (AN-MA) copolymers made by two stage polymerization. When the BD-AN seed particles could swell to four times their original volume, most of the polymerization seems to have occurred within the seed particle leading to an apparently uniform distribution of both copolymers within the particle. When a more highly crosslinked BD-AN seed latex was used, swelling of the original seed particle was limited, and two phase polymer particles were observed at the end of the polymerization. Osmium tetroxide staining indicated one phase rich in butadiene and one phase with less butadiene existed as separate domains in the phase separated particles.

When the latex samples were converted to bulk polymer melts and tested by DMTA the presence of phase separation (if present in the second stage latex particle) was maintained. If the latex was one phase, it remained in one phase during melt compression molding. Thin sections made on a microtome and stained with osmium tetroxide confirmed again that one phase particles melted into one phase bulk samples, while two phase particles formed bulk polymer with two phases. The domain sizes of the rubbery phase in thin sections seemed similar to the domain sizes of the second stage polymer particles. The hemisphere shaped particles produced bulk polymer with non-spherical domains while core-shell particles produced BD-AN domains of spherical shape.

All of the particles produced by a single recipe in the second stage polymerization were of similar morphology using the standard recipes. However, when the number of seed particles was reduced or the surfactant concentration was increased above a fixed value a second crop of particles that were primarily AN-MA copolymer was produced. This was attributed to the aqueous phase polymerization of AN-MA and the competition between nucleation of new particles and coagulation of less stable precursor particles with seed particles.

Acknowledgments. The authors would like to acknowledge the financial support of BP America Extra Mural Research Award.

Literature Cited.

1. Jonsson, J.L.; Hassander, H.; Jansson, L.H.; Tornell, B. *Macromolecules*, **1991**, *24*, 126-131.
2. Berg, J.; Sundberg, D.; Kronberg, B. *Polym. Mat. Sci. & Eng.*, **1986**, *54*, 367-369.
3. Vanderhoff, J. W.; Sheu, H. R.; El-Aasser, M. S. In *Scientific Methods for the Study of Polymer Colloids and Their Applications*; Candau, F.; Ottewill, R.H., Ed.; Nato ASI Series Vol 303; Kluwer Academic Publishers: Dordrecht/Boston/London, 1990; pp. 529-565.
4. Lee, S.; Rudin, A. *Makromol. Chem. Rapid Commun.*, **1989**, *10*, 655-661.
5. Min, T.I.; Klein, A.; El-Aasser, M.S.; Vanderhoff, J.W. *J. Polym. Sci., Polym. Chem. Ed.*, **1983**, *21*, 2845-2861.
6. Dimonie, V.; El-Aasser, M.S.; Klein, A.; Vanderhoff, J. W. *J. Polym. Sci., Polym. Chem. Ed.*, **1984**, *22*, 2197-2215.
7. Cho, I.; Lee, K. *J. Applied Polymer Science*, **1985**, *30*, 1903-1926.
8. Flory, P.J. *Principles of Polymer Chemistry*, Cornell University Press: Ithaca and London, 1953.

RECEIVED December 4, 1991

Chapter 17

Preparation of Particles for Microvoid Coatings by Seeded Emulsion Polymerization

Soft Hydrophilic Polymer Core–Hard Hydrophobic Polymer Shell

J. W. Vanderhoff[1,2], J. M. Park[1,4], and Mohamed S. El-Aasser[1,3]

[1]Emulsion Polymers Institute, [2]Department of Chemistry, and [3]Department of Chemical Engineering, Lehigh University, Bethlehem, PA 18015

Hollow particles for microvoid coatings were prepared by polymerizing a crosslinked hydrophobic polymer shell around a hydrophilic polymer core, then recovering and drying the particles. Typical cores were soft carboxyl-containing copolymers which have been neutralized; typical shells were hard crosslinked polystyrene or poly(methyl methacrylate). The volume of the core must be great enough so as to shrink and leave a void upon drying; the shell must be strong enough to maintain its integrity during drying. These core/shell particles were prepared by sequential addition of the different monomer mixtures; however, the morphology of the particles did not always match the order of monomer addition, i.e., the hydrophilic polymer generally comprised the shell and the hydrophobic polymer, the core, the opposite of the order of addition. The preparation of the requisite particles by seeded emulsion polymerization is described, in particular, the techniques required to prepare the desired particles with soft hydrophilic polymer cores and hard crosslinked hydrophobic polymer shells.

Hollow polymer particles for use in microvoid coatings have been sold under the trade names Ropaque OP-42 and OP-62 (Rohm & Haas), Spindrift (Dulux Australia), and Voncoat (DaiNippon Ink & Chemicals). The Ropaque particles were prepared by sequential emulsion polymerization: the preparation of soft alkali-swellable (carboxyl-containing) latex particles, the encapsulation of these core particles with hard crosslinked polystyrene or poly(methyl methacrylate) shells, the neutralization of the alkali-swellable polymer cores with ammonia or a tertiary amine, and the drying of the particles to evaporate the water, leaving voids within them (1,2). The volume of the core must be

[4]Current address: Korea Chemical Company, Ltd., C. R. I., San 1–9, Mabook-ri, Kuseong-myun, Yongin-kun, Kyunggi-do, South Korea

0097–6156/92/0492–0272$06.00/0

great enough to shrink upon drying and leave a void inside the particle; the shell must be strong enough to maintain its integrity during swelling and drying of the core.

The application requires the polymerization of a hard, rigid hydrophobic polymer shell around a soft, hydrophilic carboxyl-containing polymer core. Thus, a hydrophobic monomer is polymerized in the hydrophilic carboxyl-containing seed latex. Generally, however, in core/shell emulsion polymerization, the more hydrophilic monomer polymerizes preferentially in the shell and the more hydrophobic monomer in the core, independent of the order of monomer addition. Thus, this polymerization is likely to give a hydrophilic polymer shell around a hydrophobic polymer core. This morphology is the opposite of that required for this application. Therefore, the objective of this work was to develop methods for the preparation of the requisite hydrophilic core/hydrophobic shell morphology, which can be dried to form hollow particles with voids large enough to serve as microvoids.

Experimental Details

The styrene (Polysciences) and methyl methacrylate (Rohm & Haas) were washed with 10% aqueous sodium hydroxide to remove the inhibitor, dried with anhydrous calcium sulfate, and distilled under nitrogen and over cuprous chloride at low temperature and pressure. The sodium styrene sulfonate (Dow Chemical) was recrystallized once from aqueous sodium chloride and twice from methanol at low temperature, then dried at room temperature under vacuum. The divinylbenzene (Dow Chemical), COPS-1 (sodium 1-allyloxy-2-hydroxypropyl sulfonate) and sodium dodecylbenzene sulfonate (both Alcolac) were used as received, as was the ammonium persulfate and sodium hydroxide (both Fisher Scientific). Distilled-deionized water was used for all polymerizations.

The carboxyl-containing latexes were prepared by single-stage (batch) or two-stage (batch and semi-continuous) polymerization. The batch polymerizations were carried out under nitrogen in 12-oz capped glass bottles rotated end-over-end in a thermostated bath or 500-cm^3 round-bottom glass flasks, equipped with a reflux condenser, nitrogen inlet, and monomer syringe pump, and stirred at 200 rpm with a two-blade paddle. The semi-continuous polymerizations were carried out in the stirred glass flasks. The second-stage and third-stage polymerizations were carried out by semi-continuous polymerization in stirred glass flasks. The three-stage, and some two-stage, polymerizations were carried out by adding the different monomer mixtures in sequence without stopping. The latexes were neutralized after the second stage by adding ammonium hydroxide at high temperature over a 30-minute period and stirring for another 30 minutes.

The latex particle sizes were determined using transmission electron microscopy (Philips 300) or photon correlation spectroscopy (Nicomp Submicron Particle Sizer).

Experimental Results and Discussion

Many attempts were made to polymerize a hydrophobic polymer shell around a hydrophilic seed polymer core. These experiments were generally unsuccessful; they gave either eccentric core/shell particles with the hydrophilic core exposed to the aqueous phase or aggregated nonspherical composite particles with the shells attached to the seed particles as several small separated particles. The thermodynamic analyses of the combined effects of interfacial free energies and phase separation between the two polymer phases (to be published later) showed that the expected result was the formation of these nonspherical particles rather than the desired spherical concentric core/shell particles. In addition, it was shown experimentally that the particle morphology was determined by the composition of the seed latexes and the viscosity of the monomer-swollen polymer phase.

Figure 1 shows one hypothetical pathway to prepare concentric hydrophilic core/hydrophobic shell particles based on the heterocoagulation of newly formed oligomers onto the seed particles to form high-viscosity "patches," which grow until they merge with one another and form a fused shell, with the viscosity maintained at a high level to prevent the expected phase inversion. This postulated hypothetical pathway is based upon the interfacial tension as the thermodynamic driving force for phase migration and the viscosity as the kinetic resistance to this migration. The important points are as follows.

1. The secondary particles formed in the aqueous phase of an emulsifier-free polymerization have a lower monomer/polymer ratio than those formed in conventional emulsion polymerization because there is no emulsifier in the aqueous phase for nucleation in micelles.
2. The viscosity of the monomer-swollen polymer phase formed by heterocoagulation of the secondary particles onto the seed particle surfaces is relatively high.
3. The partitioning of the second-stage monomer in the seed latexes (the monomer/polymer ratio) is as low as possible to give a more viscous shell layer. This suggests that a methyl methacrylate-based seed latex may be better than a styrene-based seed latex for the seeded polymerization of styrene.
4. The instantaneous conversion during the second stage is maintained as high as possible to increase the viscosity of the shell layer. This is accomplished by decreasing the rate of monomer addition as well as the initiator concentration.
5. The high viscosity of the shell layer works against the contraction of this layer to form eccentric core-shell particles.
6. Higher initiator concentrations enhance the encapsulation of the hydrophilic core with a hydrophobic shell because of the anchoring effect of ionic endgroups of the polymer chains (3), as well as the lower polymer-water interfacial tension resulting from the higher surface charge density.

"Emulsifier-free" latexes have been prepared without conventional emulsifier by using sulfate polymer endgroups introduced by persulfate ion initiator or a functional monomer to stabilize the latex particles. The latexes prepared without conventional emulsifier using methacrylic acid were too unstable for this application. Therefore, the methacrylic acid was augmented by addition of sodium styrene sulfonate or COPS-1 (sodium 1-allyloxy-2-hydroxypropyl sulfonate) to stabilize the particles.

Table I gives the polymerization recipe for the preparation of such a composite latex at 80°C beginning with a seed latex made by emulsifier-free polymerization.

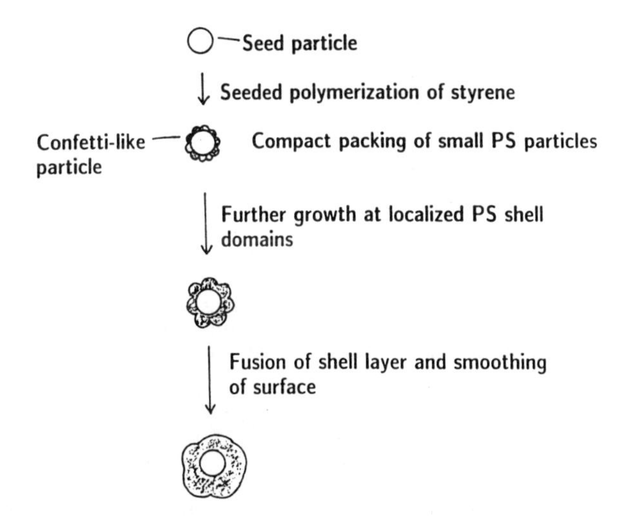

Figure 1. Hypothetical pathway to prepare hydrophilic core/hydrophobic shell particles.

Table I. Emulsifier-Free Polymerization Recipe

Ingredient	Weight (g)
Stage 1: emulsifier-free seed latex	
methyl methacrylate	2.00
methacrylic acid	0.05
ammonium persulfate	0.40
distilled-deionized water	220.00
Stage 2: carboxyl-containing core latex	
carboxyl-containing core latex (Stage 1)	as above
methyl methacrylate	16.00
methacrylic acid	6.80
ethylene glycol dimethacrylate	0.25
ammonium persulfate	0.40
distilled-deionized water	15.00
Stage 3: core/shell latex	
styrene	166.00
ammonium persulfate	0.80
distilled-deionized water	30.00

Figure 2 shows transmission electron micrographs of the evolution of particle morphologies during the seeded emulsion polymerization of styrene (Stage 3 in Table I): A. poly(methyl methacrylate-co-methacrylic acid) seed latex (Stage 1); B. crosslinked poly(methyl methacrylate-co-methacrylic acid) seed latex (Stage 2); Stage 3 at C. 6%; D. 17%; E. 61%; F. 96% conversion. Table II shows the particle size measurements made by photon correlation spectroscopy. The percent conversion in Stage 3 increased near linearly from the beginning to more than 90% in 240 min; the instantaneous conversion increased slowly from about 75% at 20 min to 96% at 300 min. Thus, the monomer/polymer ratio was small and relatively constant.

Table II. Particle Size of Latexes of Table I

Sample	D_n (nm)	D_v (nm)	Distribution
A Stage 1	183	191	unimodal
B Stage 2	324	347	unimodal
C Stage 3 - 6% conv	394	427	unimodal
D Stage 3 - 17% conv	340	412	bimodal
E Stage 3 - 61% conv	139,503	154,554	bimodal
F Stage 3 - 96% conv	214,646	224,675	bimodal

Figure 2 shows that the growth pattern of the composite latex particles closely resembled the hypothetical pathway of Figure 1. During Stage 3, most of the styrene polymerized in the aqueous phase to form secondary particles, which then flocculated with the hydrophilic seed particles (Figure 2C); this heterocoagulation was attributed to the lack of emulsifier and the low partitioning of styrene in the seed particles. This heterocoagulation was followed by further polymerization of styrene in localized domains near the particle surface (Figure 2D). The uneven surfaces of the larger composite particles also showed this unique pattern of particle growth (Figures 2E,2F) and the fusion of the shell layer to give a smooth surface, even though the monomer feed rate and monomer/polymer ratio were not sufficient to give a completely smooth sur-

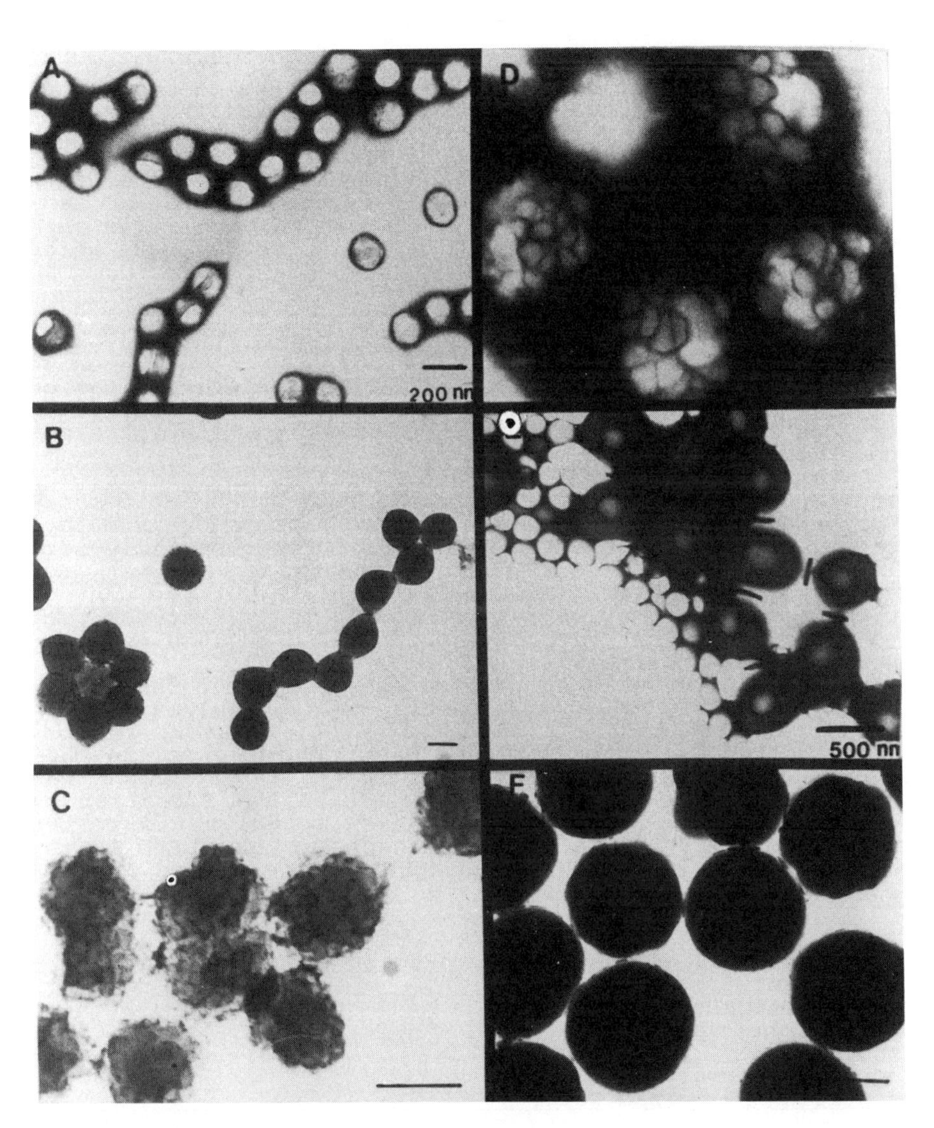

Figure 2. Transmission electron micrographs of the evolution of particle morphologies during the seeded emulsion polymerization of styrene (Stage 3): A. poly(methyl methacrylate-co-methacrylic acid) seed (Stage 1); B. crosslinked poly(methyl methacrylate-co-methacrylic acid) seed latex (Stage 2); Stage 3 at C. 6%; D. 17%; E. 61%; F. 96% conversion.

face. Even though about 90% of the styrene was polymerized in the seed particles, many seconadry particles were also formed in the aqueous phase (Figures 2E,2F).

Table II confirmed that bimodal particle size distributions were observed in samples D-F, i.e., not all secondary particles flocculated with the hydrophilic seed latex particles. The efficiency of this heterocoagulation depended upon the ionic strength and the rate of radical generation of the system. According to Frenkel (4), the driving force for the coalescence of the secondary particles with the seed particles by viscous flow is the interfacial tension between the particles and the medium; the resistance to this coalescence is the viscosity of the polymer, which is high because of the high instantaneous conversion. The Frenkel theory was applied (5) earlier to the film formation of latexes. The relationship between interfacial tension and particle viscosity was reported earlier by Okubo et al. (6,7), who found that the morphology of composite particles formed during the seeded emulsion polymerization of styrene in a more hydrophilic acrylate ester seed latex was controlled by the monomer concentration (i.e., the monomer/polymer ratio).

The foregoing method was successful in the preparation of composite particles with a hard rigid hydrophobic shell surrounding a soft hydrophilic core; however, this method is best-suited to the preparation of large-particle-size latexes, and it was difficult to prepare composite latex particles smaller than 400 nm diameter. Therefore, conventional sodium dodecylbenzene sulfonate emulsifier was used to prepare smaller particles, which could be used as a hydrophilic seed latex.

Table III gives a polymerization recipe for the preparation of the seed latex by batch emulsion polymerization in capped bottles for 12 hours at 80°C. Methacrylic acid was used because it partitioned more favorably in the particle phase than acrylic acid. Methyl methacrylate was chosen over styrene because it hindered the formation of water-soluble polymer, and gave less coagulum and better morphology. Ethylene glycol dimethacrylate was used to inhibit the swelling of the hydrophilic core during the second-stage polymerization.

Table III. Polymerization Recipe for Seed Latex

Ingredient	Weight (g)
methyl methacrylate	2.45
methacrylic acid	0.05
sodium dodecylbenzene sulfonate	0.0050-0.050
ammonium persulfate	0.020
distilled-deionized water	47.50

Table IV shows that the particle size of the carboxyl-containing particles varied with sodium dodecylbenzene sulfonate concentration from 63 to 163 nm. The particles were not monodisperse, as was shown by the coefficient of variation (σ/D_n) determined by photon correlation spectroscopy and the polydispersity index (PDI) determined by transmission electron microscopy.

Table IV. Seed Latex Particle Size

[SDBS]* (%)	Nicomp D_n (nm)	Nicomp σ/D_n (%)	TEM D_n (nm)	TEM PDI
0.20	163	43	145	1.21
0.40	105	17	102	1.18
0.60	87	16	79	1.11
0.80	85	9	73	1.13
1.20	78	13	71	1.10
1.50	72	12	64	1.18
2.00	63	20	52	1.29

* sodium dodecylbenzene sulfonate based on monomer

Seeded batch emulsion polymerization of the styrene shell gave eccentric core/shell particles with the cores exposed to the aqueous phase; semi-continuous seeded emulsion polymerization gave aggregated nonspherical particles with well-defined polystyrene and carboxylated-polystyrene regions. These results demonstrated that the morphology of these particles depended on the thermodynamic and rheological properties of the growing latex particles as well as the sequence of monomer addition. The desired morphology was attained by minimizing the partitioning of the second-stage monomer in the seed particles, attaining a high instantaneous conversion, and using heterocoagulation as the mechanism of particle growth.

Table V gives the multistage polymerization recipe using the carboxyl-containing latex as the seed latex. The semi-continuous polymerizations were carried out at 80°C using a slow monomer feed rate to prepare a more homogeneous carboxyl-containing copolymer, localize the polymerization at the particle surface during the formation of the hydrophobic shell, and minimize the diffusion of growing polystyrene chains into the interior of the hydrophilic core particles. The carboxyl-containing core particles were prepared by a two-stage polymerization to give better control of the latex particle size (and microvoid size), obtain more monodisperse particles, and minimize the probability of the formation of water-soluble polymer during the preparation of the carboxyl-containing particles.

An intermediate shell of poly(butyl acrylate-co-methyl methacrylate-co-methacrylic acid) was used to increase the efficiency of encapsulation of the hydrophilic carboxyl-containing core with the shell and to control the T_g at about 80°C so that the neutralization could be carried out without increasing the temperature. Also, the emulsifier concentration was adjusted after the second-stage polymerization according to the surface tension of the latex; insufficient emulsifier gave larger nonspherical particles from the coagulation of smaller particles.

The presence of closed voids in the resulting particles was shown by transmission and scanning electron microscopy. The collapse of the shell, which was observed in some latexes, was prevented by polymerizing another shell layer of crosslinked polystyrene after neutralization.

Using these methods, hollow latex particles with void diameters of 134-765 nm were prepared. Table VI shows that the size of the voids was determined mainly by the size of the seed particles.

TABLE V. Multistage Polymerization Recipe

Ingredient	Weight (g)
Stage 2: carboxyl-containing latex	
seed latex (Table IV; 10% solids)	40.0
methyl methacrylate	27.9
methacrylic acid	11.9
ethylene glycol dimethacrylate	0.20
ammonium persulfate	0.16
distilled-deionized water	360.0
Stage 3: first shell layer	
carboxyl-containing core latex (stage 2)	48.0
butyl acrylate	8.6
methyl methacrylate	33.8
methacrylic acid	0.88
sodium dodecylbenzene sulfonate	0.020
ammonium persulfate	0.54
distilled-deionized water	18.0
Stage 4: neutralization	
26% aqueous ammonium hydroxide	3.40
Stage 5: second shell layer	
styrene	42.4
divinylbenzene	0.85
sodium dodecylbenzene sulfonate	0.050
ammonium persulfate	none
distilled-deionized water	18.0

Table VI. Void Diameters of Hollow Polymer Particles

Latex	D_{seed} (nm)	D_{void} (nm)	$D_{particle}$ (nm)
C-525	54	134	303
C-530	81	284	465
C-604	112	384	582
C-605	198	765	1070

Most of the hollow latex particles were spherical, with a well-defined shell and a void in the center; however, the shells of the largest particles (1070 nm diameter) had collapsed slightly because they were thin relative to the large void size. The shells of the smaller particles were spherical and well-defined, with no signs of collapse.

Conclusions

Hollow latex particles were prepared by forming soft hydrophilic polymer core/hard hydrophobic polymer shell particles using the following procedure: 1. batch polymerization of a methyl methacrylate-methacrylic acid mixture to give a poly(methyl methacrylate-co-methacrylic acid) seed latex core; 2. sequential seeded polymerization of this latex with a methyl methacrylate-methacrylic acid-ethylene glycol dimethacrylate mixture to complete the cross-linked poly(methyl methacrylate-co-methacrylic acid-co-ethylene glycol dimethacrylate) seed latex; 3. neutralization of the carboxyl groups with ammonium hydroxide; 4. polymerization of a methyl methacrylate-butyl acrylate-methacrylic acid mixture in the seed latex to give the poly(methyl methacrylate-co-butyl acrylate-co-methacrylic acid) first shell layer; 5. polymerization of a styrene-divinylbenzene mixture to give the crosslinked polystrene second shell layer; 6. drying of the latex to remove the water from the internal voids of the particles, to give hollow particles. These reactions can be carried out in sequence, without stopping.

Literature Cited

1. Kowalski, A.; Vogel, M.; Blankenship, J. European Pat. 22,633 (to Rohm & Haas), 1980.
2. Hook, J. W.; Harren, R. E. In Organic Coatings Technology, Parfitt, G. D.; Patsis, A. V., Eds.; Marcel Dekker: New York, NY, 1984, p. 299.
3. Cho, I.; Lee, K. J. Appl. Polym. Sci. **1985**, 30, 1903.
4. Frenkel, A. J. Phys. USSR **1945**, 9, 385.
5. Dillon, R. E.; Matheson, L. A.; Bradford, E. B. J. Colloid Sci. **1951**, 6, 108.
6. Matsumoto, T.; Yamada, A.; Okubo, M. Kobunshi Ronbunshu **1979**, 36(7), 459.
7. Okubo, M; Katsuta, Y.; Matsumoto T. J. Polym. Sci., Polym. Lett. Ed. **1980**, 18, 481.

RECEIVED January 24, 1992

Chapter 18

Production of Multihollow Polymer Particles by Stepwise Alkali–Acid Method

Alkali Treatment Process

M. Okubo, K. Ichikawa, and M. Fujimura

Department of Industrial Chemistry, Faculty of Engineering, Kobe University, Rokko, Nada, Kobe 657, Japan

Polymer particles having many hollows can be produced by treating submicron-size styrene-butyl acrylate-methacrylic acid (74.3/17.0/8.7, molar ratio) terpolymer particles stepwise with alkali and acid. The effects of initial pH value, temperature, and time in the process of alkali treatment as the first step on the multi-hollow structure within the particles are examined in detail.

We have been studying about the production of carboxylated polymer particles produced by emulsion polymerization, in which carboxyl groups are predominantly localized at the surface layer (*1-5*). For the purpose, we have proposed the surface hydrolysis method of polyethyl acrylate particles (*1*) and the localization method (*2-4*) of carboxyl groups from the inside to the surface of carboxylated polymer particles by alkali treatment. Moreover, in order to produce the desired carboxylated polymer particles directly in a polymerization process, the latter technique was developed to the two-stage emulsion polymerization technique (*5*) in which carboxylated polymer seed particles are first prepared at low pH and, subsequently, seeded emulsion polymerization without unsaturated acid monomer is continued at high pH.

Throughout these experiments, it was found that styrene-butyl acrylate-methacrylic acid terpolymer (P(S-BA-MAA)) particles were changed to those containing many hollows (see Fig. 1) by treaing stepwise with alkali and acid (*6*). We named it the "stepwise alkali/acid method".

The purpose of this work is to clarify the effects of alkali treatment conditions as the first step on the formation of multi-hollow polymer particles.

Experimental

Materials Styrene (S), butyl acrylate (BA) and methacrylic acid (MAA) were purified by distillation under reduced pressure in a nitrogen atmosphere and stored in a refrigera-

NOTE: This chapter is Section III of Part CXXXIV of the series "Studies on Suspension and Emulsion".

0097–6156/92/0492–0282$06.00/0

tor. Analytical grade potassium persulfate was purified by recrystallization. Analytical grade hydrochloric acid and potassium hydroxide were used without further purification. Commercial grade nonionic polyoxyethylene sorbitan monooleate emulsifier (Tween 80, Kao Atlas Co.) was used without further purification. Deionized water was distilled.

Preparation of polymer particles. P(S-BA-MAA) terpolymer particles were prepared by emulsifier-free emulsion polymerization at 70°C under conditions listed in Table 1, where 5% of all monomers were preliminarily polymerized for 1.2 h and then the remainder was added to a glass reaction flask from a micro-feeder for 14.5 h. The residual monomers were not detectable by gas chromatography. The molar ratio of P(S-BA-MAA) terpolymer was calculated to be 74.3/17.0/8.7 from the polymerization recipe.

Stepwise Alkali/Acid treatment. The original P(S-BA-MAA) colloid was diluted to 1.5 g/l in a small glass vessel, and treated stepwise with alkali and acid conditions, as follows:

Alkali treatment. First, as a stabilizer to prevent the coagulation of particles, Tween 80 (10 wt% of polymer solid) was added to the diluted colloid. The initial pH of the colloid was adjusted to various values in the range from 7.0 to 13.0 with 0.2 N KOH and then heated at various temperatures for different times. After the treatments, each colloid was rapidly cooled to room temperature by dipping the vessel in running water.

Acid treatment. Tween 80 (10 wt% of polymer solid) was added to the alkali-treated colloids to prevent the coagulation of particles. In this experiment, the acidic treatment as the second step was always carried out under the same conditions: initial pH value of 2.2 (with 0.2 N HCl); 3 h; 75°C. After the treatment, each colloid was rapidly cooled as above.

Electron microscopy. Each colloid was diluted to about 50 ppm, and a drop was placed onto a Formvar film-coated grid and allowed to dry at room temperature in desiccator. Transmission electron micrographs were taken with a JEM 200CX transmission electron microscope (TEM), Nihon Denshi Co.

Hollow size and the number of hollows per particle. The number-average particle and hollow sizes were, respectively, determined by measuring those sizes for 30-50 particles on the TEM negative films (Okubo, M.; Ichikawa, K.; Fujimura, M. *Colloid Polym. Sci.*, in press). In the case that the number of hollows per particle was less than about 10, the number-average value was determined by directly counting it for 30-50 particles, whereas in the case that it exceeded about 10 it was calculated by dividing the expansion volume by the hollow size. The latter two volumes were, respectively, calculated from corresponding sizes.

Results and discussion.

Figure 2 shows the relationship between the initial pH value in the alkali treatment at 75°C for 3 h on the expansion of particle volume. The particle volume was not changed at the initial pH values below 9.0, whereas in the initial pH values from 10.0 to 12.2, it increased remarkably with an increase in the intial pH value, and then decreased above that. The decrease seems to be based on the suppression of alkali swelling due to an increase in the electrolyte concentration. This will be discussed in a further article.

Figure 3 shows the relationships between the initial pH value in the alkali treatment and the hollow size and the number of hollows per particle. The hollow size increased with the increase in the initial pH value above 10.0. On the other hand, the number of hollows per particle decreased with the increase in the initial pH value above 10.0, to only one at the initial pH values of 12.5 and 13.0.

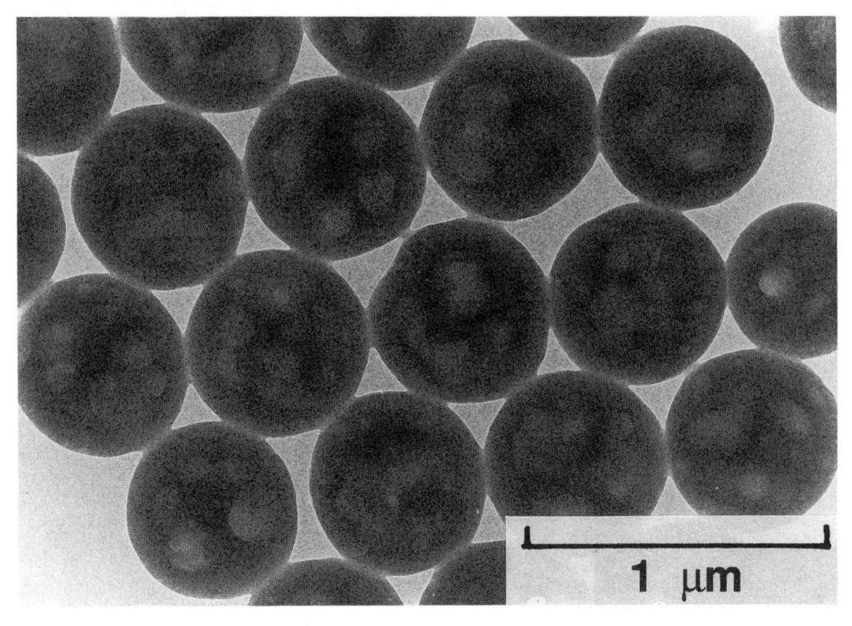

Fig. 1. TEM photograph of multi-hollow P(S-BA-MAA) (74.3/17.0/8.7, molar ratio) particles produced by the stepwise alkali/acid method.

Table I. Stage-type Emulsifier-Free Emulsion Polymerization[a]

Ingredient		First stage[b]	Second stage[c]
S	(g)	2.55	48.5
BA	(g)	0.72	13.7
MAA	(g)	0.24	4.55
KPS[d]	(g)	0.375	
Water	(g)	223	

[a]Conducted at 70°C for 26 h under a nitrogen atmosphere;
[b]For 1.2 h; [c]For 14.5 h; [d]Potassium persulfate

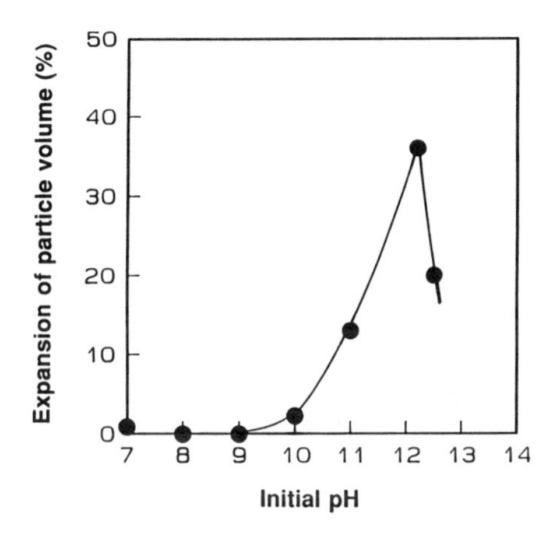

Fig. 2. Relationship between the initial pH value in the alkali treatment at 75°C for 3 h and the expansion of the particle volume after the stepwise alkali/acid treatments (initial pH= 2.2; 75°C; 3 h).

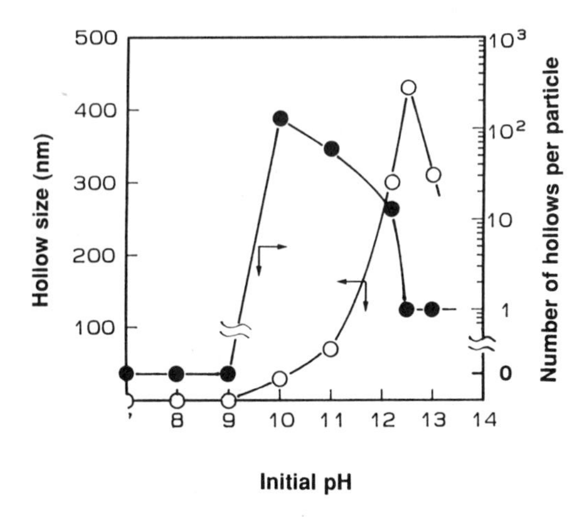

Fig. 3. Relationships between the initial pH value and the hollow size and the number of hollows per particle obtained in the experiment shown in Fig.2.

Figure 4 shows the relationship between the alkali treatment temperature and the particle size, where the alkali treatment was carried out for 3 h at the initial pH 12.2. The particle size increased markedly with an increase in the alkali treatment temperature, especially above 65°C, where multi-hollow structures were observed. The particle expansion was observed even below the minimum temperature at which multi-hollow structure was observed. This suggests that these treated particles may have very small hollows in the inside, although it was not observed using the TEM.

Figure 5 shows the relationships between the alkali treatment temperature and the hollow size and the number of hollows per particle. The hollow size increased linearly with the alkali treatment temperature above 65°C. On the other hand, the number of hollows was about 200 at 65°C and decreased with the increase in the alkali treatment temperature above 60°C. The minimum temperature at which the hollow structure was observed was lower than the glass transition temperature (Tg=78°C) of the base terpolymer measured by differential scanning calorimetry. This suggests that the carboxylated polymer particles can swell under alkali conditions even at a lower temperature than the Tg of the base polymer. Actually, we have confirmed this using a dynamic light scattering method (Okubo, M.; Fujimura, M., The Kobe University, unpublished data).

Figure 6 shows the relationships between the alkali treatment time and the hollow size and the number of hollows per particle, where the alkali treatment was carried out at 75°C at the initial pH of 12.2. As the treatment time increased, the hollow size increased and its number decreased to one.

Figure 7 shows the relationship between the particle size and the hollow size obtained in the experiment shown in Fig. 6. As will be seen, there was a good linear relationship between those sizes. This suggests that the alkali swelling state controls the hollow size.

From the above results, it is concluded that the multi-hollow structures formed by the stepwise alkali/acid method can be controlled by the alkali treatment conditions. We assume that the formation of a hollow structure is similar to that of an asymmetrical polymer membrane consisting of skin and sponge layers formed by wet cast method (*7*).

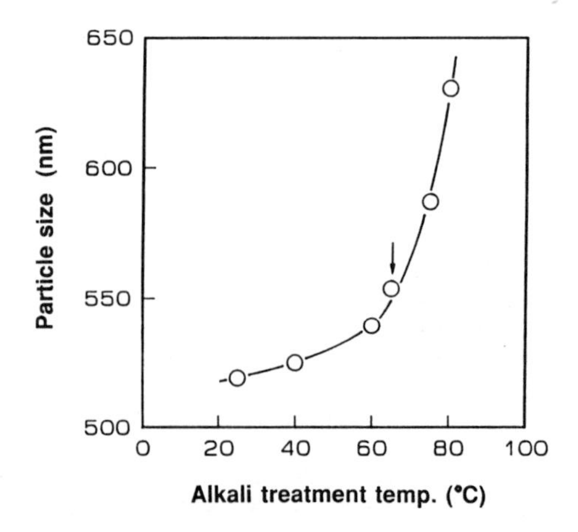

Fig. 4. Relationship between the alkali treatment temperature at pH 12.2 for 3 h and the particle size after the stepwise alkali/acid treatments (initial pH= 2.2; 75°C; 3 h). The arrow shows the minimum temperature at which the multi-hollow structure was observed by TEM.

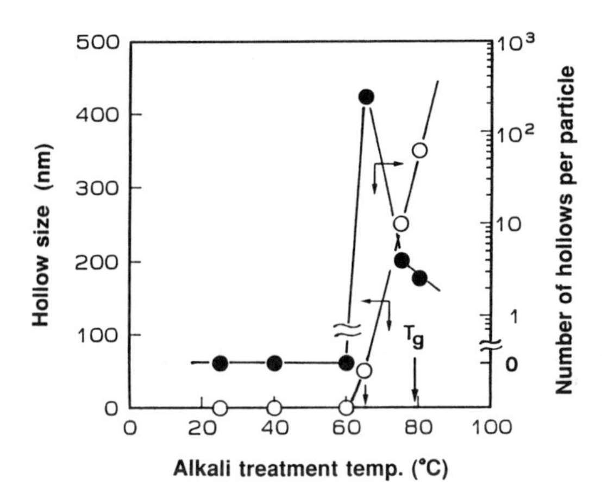

Fig. 5. Relationships between the alkali treatment temperature and the hollow size and the number of hollows per particle. The small arrow shows the minimum temperature at which the multi-hollow structure was observed by TEM.

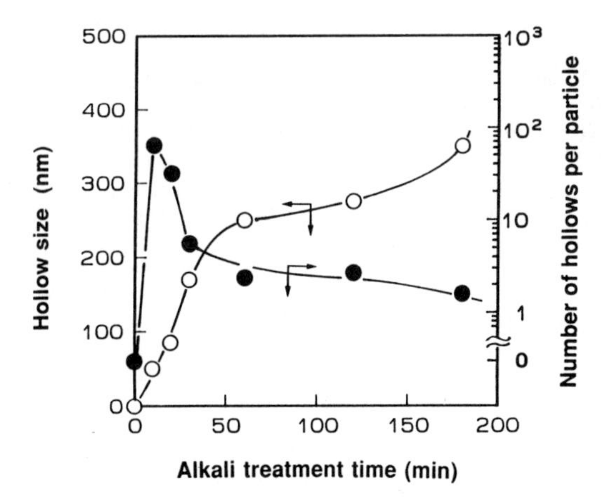

Fig. 6. Relationships between the alkali treatment time at initial pH 12.2, 75°C and the hollow size and the number of hollows per particle after the stepwise alkali/acid treatments (initial pH=2.2; 75°C; 3 h).

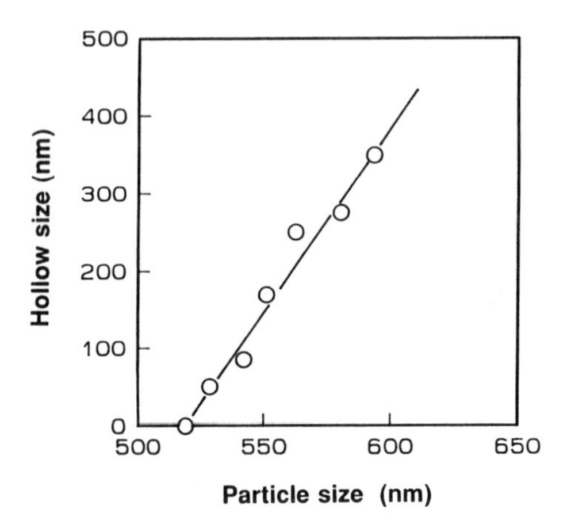

Fig. 7. Relationship between the particle size and the hollow size obtained in the experiment shown in Fig.6.

Acknowledgements

This work was partly supported by a Grant-in-Aid for Developmental Scientific Research from the Ministry of Education, Science and Culture (No. 02555185).

Literature Cited

1. Matsumoto, T.; Okubo, M.; Yasui, M. *Kobunshi Ronbunshu* **1974**, *31*, 112
2. Okubo. M.; Nakamura, Y.; Matsumoto, *T. Kobunshi Ronbunshu* **1976**, *39*, 473
3. Okubo, M.; Miyanaga, M.; Nakamura, Y.; Matsumoto, T. *Kobunshi Ronbunshu* **1983,** *40*, 707
4. Okubo, M.; Xu, D.-H.; Kanaida, K.; Matsumoto, T. *Colloid Polym. Sci. Polym. Let. Ed.* **1987,** *265*, 246
5. Okubo, M.; Kanaida. K.; Matsumoto, T. *J. Appl. Polym. Sci.* **1987,** *33*, 1511
6. Okubo, M.; Kanaida, K.; Fujimura, M. *Chemistry Express* **1990,** *5*, 797
7. Matsumoto, T.; Nakamae, K.; Ochiumi,T.; Horie, S. *J. Membrane Sci.* **1981,** *9*, 109

RECEIVED December 27, 1991

Chapter 19

Measuring Particle Size and Size Distribution from Micrograph Images

Jason J. Ruan

Reichhold Chemicals, Inc., P.O. Drawer K, Dover, DE 19903

A method is developed for measuring agglomerated and dispersed particle size and size distribution from electron micrographs. The chord lengths of horizontal lines scanned through the micrographs are measured. The chord distribution is smoothed and converted to particle size distribution as if all particles were deagglomerated. A mathematical procedure is derived for converting chord distribution to particle size distribution. Non-linear regression curve fitting is applied using a Gaussian or Lorentzian function. The technique has been verified by computer simulation and implemented on a variety of latex samples. All results show that this technique gives excellent representations of the actual particle size distributions. The whole process is completely automated and takes less than half a minute for each frame of an image. This method is specially designed to resolve agglomerated particles and also applicable to micrographs of fracture surface for the core-shell type structure and/or solid foam analysis.

Particle size distribution (PSD) is one of the most important parameters in characterizing a wide variety of process materials and final products such as emulsions, paints, coatings, adhesives, rubbers, and ceramics. Rheology, porosity, elasticity, solubility, viscosity, stability, permeability, and opacity are some of the variables strongly affected by particle size and size distribution. For instance, two emulsions may have the same composition and average particle diameters and yet exhibit quite dissimilar behavior because of differences in their size distribution.

For this reason, many methods have been developed to measure particle size and size distribution*(1)*, including the more recent ones such as Sedimentation Field-Flow Fractionation*(2)* and Capillary Hydrodynamic

0097–6156/92/0492–0289$06.00/0

Fractionation*(3)*. Microscopic techniques probably are still the most reliable and widely used methods. However, microscopic methods have been slow and tedious until the advent of the computer-image analyzer. The major problem of this technique is that it is not able to analyze coagulated or agglomerated particles because they are treated as a single object. Even the very sophisticated systems with the feature of deagglomeration of spheres can only analyze barely touching and very narrow distribution particles.

To separate particles from overlapping ones is an extremely complicated and slow process even using robust mathematical operations and normally involves image enhancement, transformation, segmentation, and reconstruction*(4)*. The separation often occurs at the wrong place and the treated image does not accurately represent the original particles.

An alternative method (patent pending) for particle size analysis was developed based on the correlation between the circle diameter distribution (CDD) and chord length distribution (CLD). Even though the correlation is straightforward, the CLD being an integral transform of the CDD, a mathematical procedure is derived for inverting the CLD to obtain the CDD. The computer automation is performed by a two step process. In the first step, chords are measured by defining two end points of each peak and selected by comparing against the predefined criteria. In the second step, the chord data is assembled in a histogram, smoothed using binomial convolution or non-linear regression, and then converted to the CDD. Since chord lengths are much easier to measure than diameters, and because it lends itself readily to computer automation, the approach presented here as shown in the following examples is an effective technique for the quick determination of particle size and size distribution, especially for resolving agglomerated particles.

Mathematical Analysis

In particulate micrograph images, the circles are the projection of particles, and hence PSD is the same as CDD. Let $F(x)$ represent the normalized PSD, where x is the diameter. The probability, $G(x)dx$, that circles in the diameter range $(x, x+dx)$ intersect a random line is proportional to their diameters, hence

$$G(x)dx = \frac{xF(x)dx}{\int_0^\infty xF(x)dx} = \frac{x}{<x>}F(x)dx \qquad (1)$$

where $<x>$ is the mean diameter. For a line which intersects a circle of diameter x, the probability that its distance from the center lies in the range $(r,r+dr)$ is $2dr/x$, giving a chord length $v=(x^2-4r^2)^{1/2}$. Therefore, the probability that circles in the range $(x,x+dx)$ are intersected to produce chords of length in the range $(v, v+dv)$ is

$$G(x)dx\frac{vdv}{x\sqrt{x^2-v^2}} = \frac{vF(x)dxdv}{<x>\sqrt{x^2-v^2}} \tag{2}$$

Any circles so long as their diameters are greater than or equal to chord length, v, might be intersected by a line and produce a chord, v. Summing all the possibilities, we have the CLD,

$$g(v) = \frac{v}{<x>}\int_v^{\infty}\frac{F(x)}{\sqrt{x^2-v^2}}dx \tag{3}$$

which is an integral transform of CDD, and the inverse transform as shown in Appendix I is

$$F(x) = -\frac{2<x>x}{\pi}\int_x^{\infty}\frac{1}{\sqrt{v^2-x^2}}\left[\frac{g(v)}{v}\right]'dv \tag{4}$$

Since the denominator of the integral function in equation 4 becomes zero as v approaches x, a computer program can not be directly applied. Integrating equation 4 by parts to remove the singularity, and approximating the integration as the summation, provided that $f(v)$ is a smooth function, we have

$$F(x) = 2x\frac{<x>}{\pi}\sum_{v=x}^{N}\frac{\sqrt{v^2-x^2}}{v}\left[f(v+1)-2f(v)+f(v-1)-\frac{f(v+1)-f(v-1)}{2v}\right] \tag{5}$$

where $f(v)=g(v)/v$, x and v assume the discrete values *1, 2, ... N*, and N is the maximum chord length.

Computer Automation

Micrograph images can be transferred to a computer with a frame grabber either directly from an electron microscope or from a video camera. The analog signals are then digitized into digital intensities called gray levels at each pixel according to their brightness: 0 for completely dark and 255 for completely bright.

Figure 1 shows gray level variance along a horizontal line across an image. Each downward peak is corresponding to the intersection of a particle. Hence, chord lengths can be measured by acquiring the distances between the two end points of each peak. The flow-chart as shown in Figure 2 is an outline for auto chord data collection. The program reads pixel data row by row. After locating the left end point, *I0*, at which the gray-level, *P(I)*, is just below the peak threshold, PTH, it looks for the minimum point, and continues to find the right end point where *P(I)* is just above PTH. The derivative analysis can also

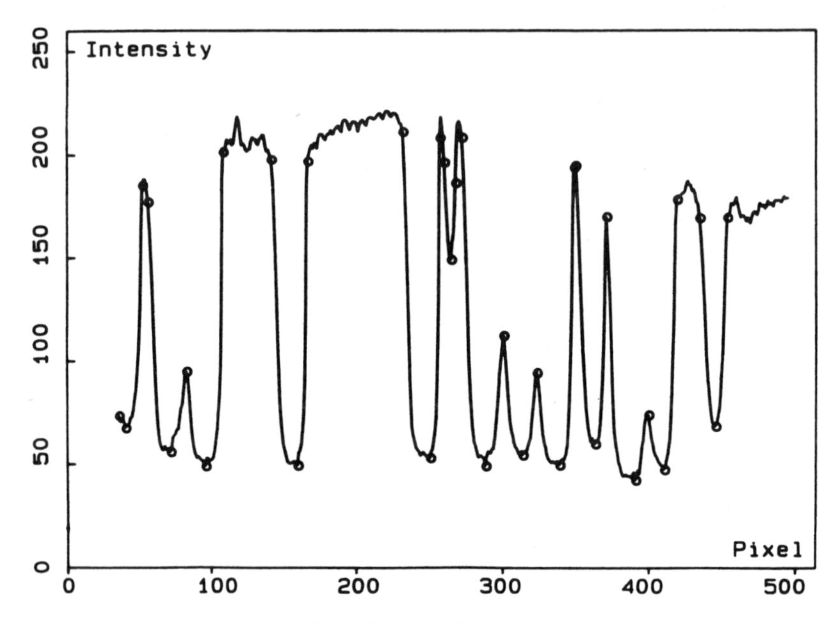

Figure 1. Gray-level variance along a line.

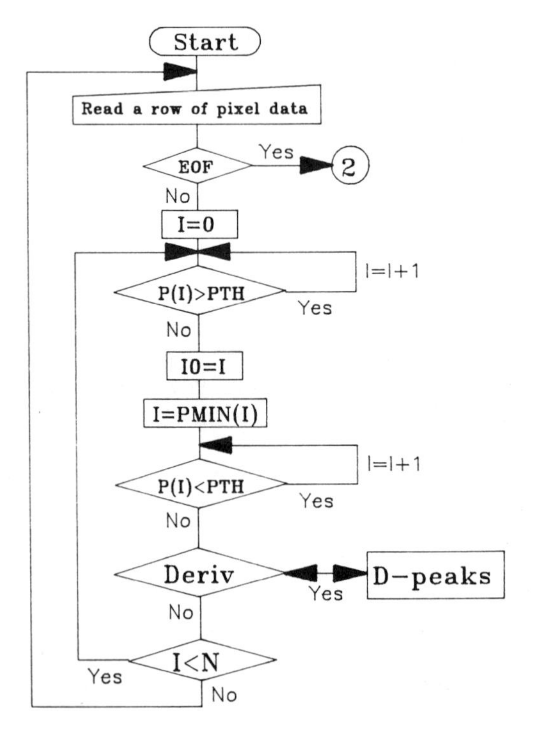

Figure 2. Flow chart of auto chord data collection.

be carried out to further define or separate chords from the possible overlapped particles. This process was repeated until the end of a line where $I=N$. It then reads the next line and continues. After collecting all the chords, it goes to the second program, the flow chart of which is shown in Figure 3. The chord data was assembled into a histogram, smoothed by Fourier analysis*(5)*, binomial convolution*(6)*, Gaussian, or Lorentzian curve fitting, and finally converted to the PSD.

Experimental

A microscope or video camera and a personal computer with a frame grabber and Reichhold Advanced Particle Size Analysis(RAPSA) menu-driven software are all required in order to implement this technique. RAPSA is available from Reichhold Chemicals, Inc. Depending on the sample form and applications, this technique can be applied using different methodologies as shown in Figure 4.

In the case of section micrographs of bulk or solid foam, PSD, $P(u)$, is no longer the same as CDD, but their correlation is the same as that of CDD and CLD. Therefore, PSD can be computed via double inversions with the same transform or single inversion with a different transform*(7)*.

$$g(v)=\frac{v}{\langle x\rangle}\int_v^{\infty}\frac{1}{\sqrt{x^2-v^2}}\frac{1}{\langle u\rangle}\int_x^{\infty}\frac{P(u)}{\sqrt{u^2-x^2}}dudx \qquad (6)$$

$$=\frac{2v}{\langle u^2\rangle}\int_v^{\infty}P(u)du \qquad (7)$$

This paper, however, concentrates on latex suspension samples. The latexes were diluted to roughly 1:2000 with distilled water and cured by bromine, osmium tetroxide, or phosphotungstic acid(PTA) if it was necessary. The sample was spotted on a formvar film supported by a copper grid, and dried at ambient temperature. Micrograph images were transferred directly from a JEOL 100 TEM with installation of a Gatan 673 wide angle TV camera to a computer with a MetraByte MV1 frame grabber in it. RAPSA software was used in analyzing particles. Over 50,000 chords from about 10 frames of images for each sample were accumulated before inversion. It takes about 20 seconds to transfer and analyze each frame. Normally, no image enhancement (though available) is required, hence, no image distortion or artifact occurs. The direct measurement of circle diameters was performed using a mouse pointer with RAPSA software. The data was then assembled into a PSD histogram.

Results

Figure 5 displays a micrograph of polystyrene latex standard along with the corresponding CLD and PSD. The chord data was collected using the algorithm outlined in Figure 2, assembled into CLD, and smoothed by averaging

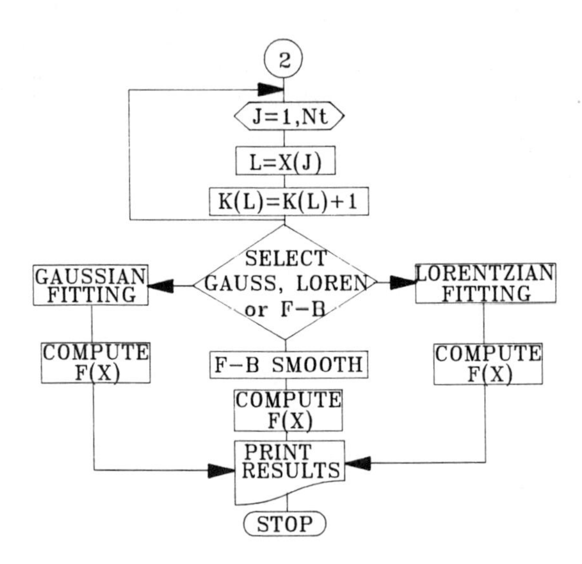

Figure 3. Outline for assembling, smoothing, and converting CLD to PSD.

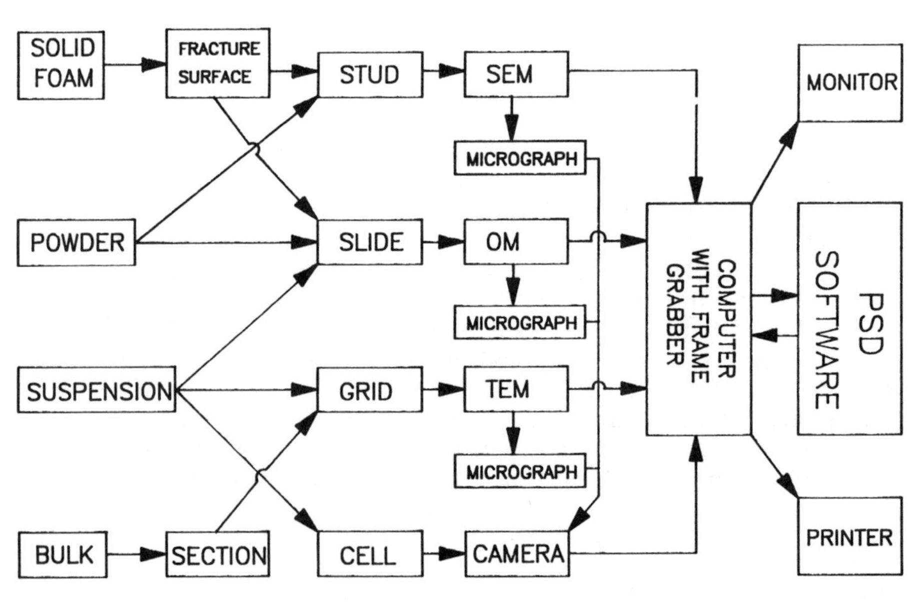

Figure 4. Schematic methodology flow chart.

over adjacent values with a binomial distribution spanning 9 channels. It was then inverted to PSD using equation 5, which agrees very well with the PSD by manual measurement(Figure 6).

Figure 7a gives a broad base of latexes with severe overlapping, which is extremely difficult, if not impossible, to separate using conventional deagglomeration. With this technique, the PSD is readily determined in reproducible quantities as shown in Figure 7b, which also agrees well with the manual measurement(not shown here).

This technique has been implemented on a variety of latexes including multimodal, heavily overlapped, broad base, and negatively stained particles. They all give excellent representations of the actual PSDs. However, since the error from manual measurement could be quite significant, it is difficult to evaluate the true accuracy in reference to the manual method. One way to test the accuracy is to use computer simulation as described below.

Computer Simulations

To test the accuracy and sensitivity of this technique, circles with a given distribution were simulated at random locations within a video screen. The screen image (treated as a micrograph image) was analyzed with exactly the same procedure used in micrographs. After collecting all the chord data and inversion, the computed PSD was compared against the originals. Broad Gaussian and delta (uniform) distribution were employed in the simulation.

Figure 8 shows the given and computed PSD along with the simulated circles. They agree very well. The difference in number or weight average is less than 0.2%.

Figure 9 shows computed PSD from simulated uniform circles. Again, it is very close to the originals, the diameter of which was set to 2070 Å.

Smoothing by Regression

Gaussian and Lorentzian distribution are probably the two most important functions in describing the distribution of random observations for most experiments. PSD, in general, can be represented as the weighted sum of N modified Gaussian functions,

$$F(x)=\sum_{n=0}^{N} a_n F_n(\frac{x}{x_n}) \tag{8}$$

with

$$F_n(\frac{x}{x_n})=\frac{2(n+1/2)^n+1}{n!x_n}\left(\frac{x}{x_n}\right)^{2n+1}\exp\left[-\frac{(n+1/2)x^2}{x_n^2}\right] \tag{9}$$

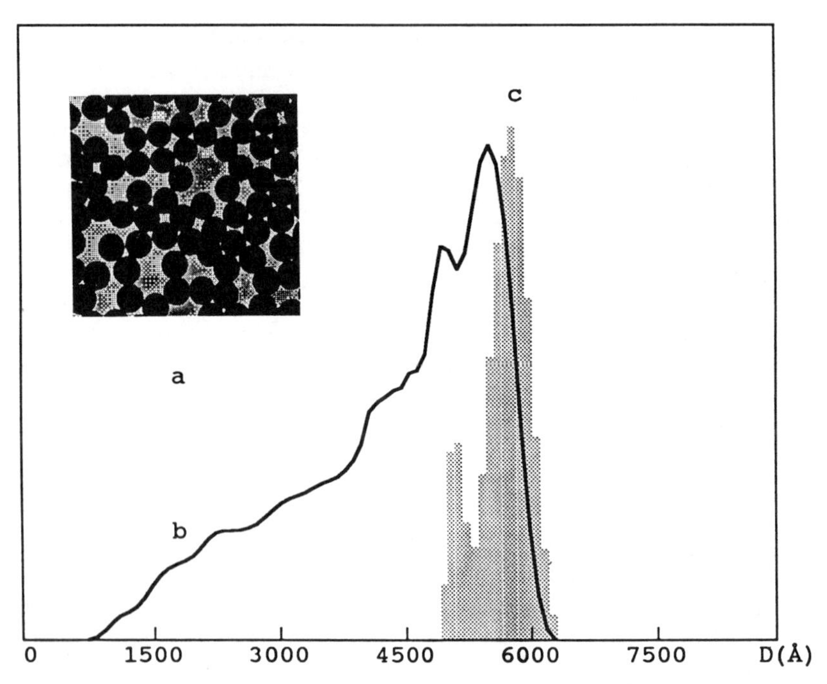

Figure 5. a) Micrograph of polystyrene latexes; b) CLD; c) PSD.

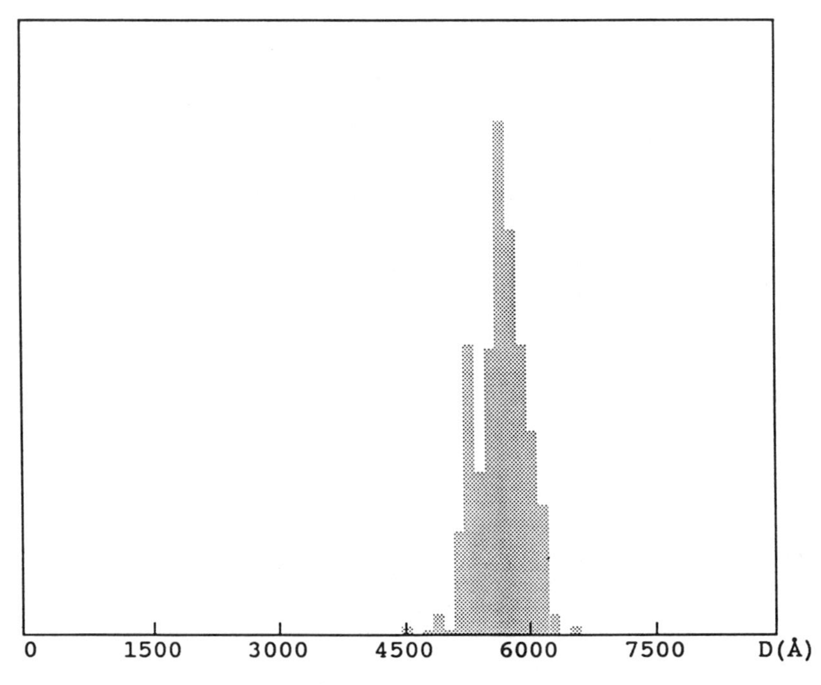

Figure 6. PSD by manual measurement.

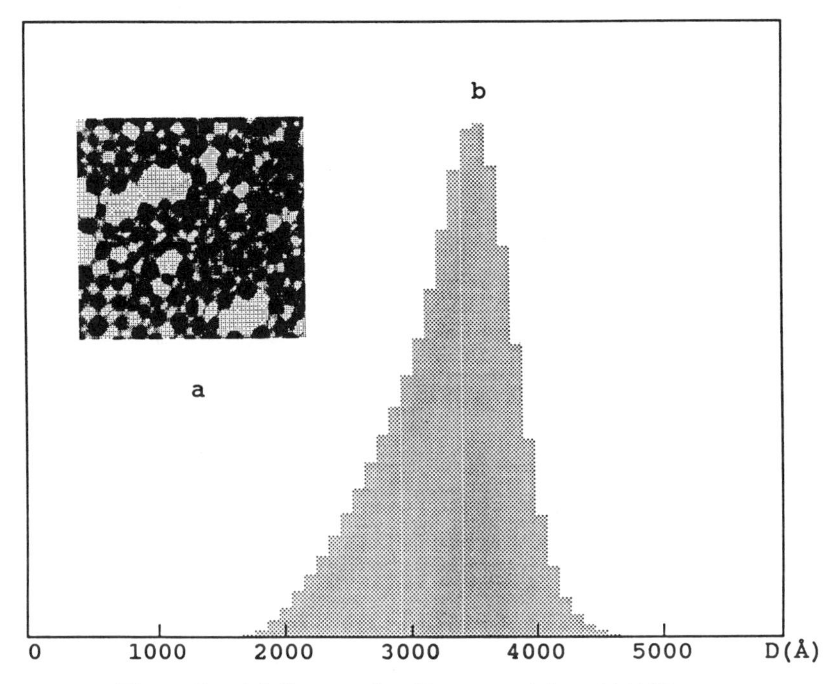

Figure 7. a) Micrograph of latex particles; b) PSD.

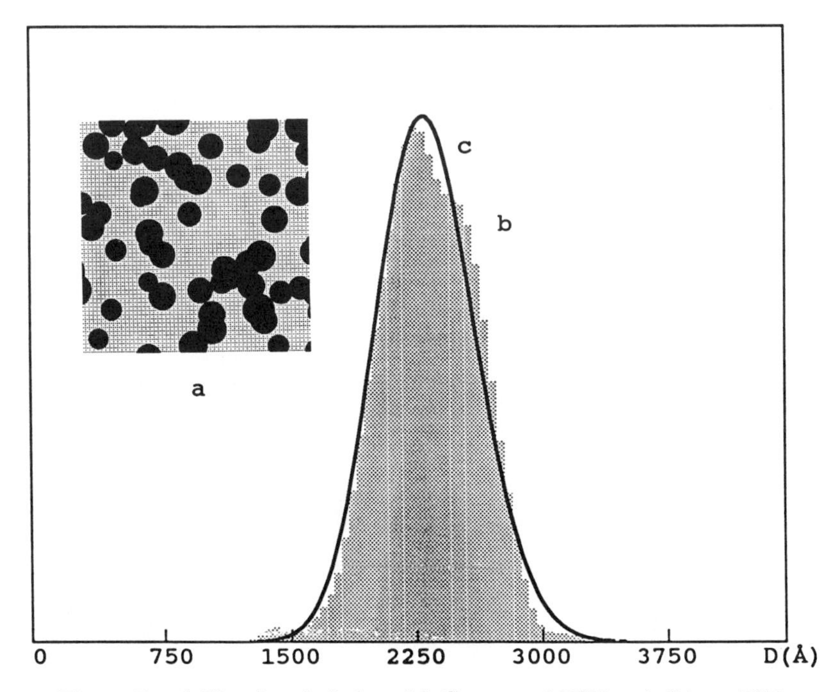

Figure 8. a) Simulated circles; b) Computed PSD; c) Given PSD.

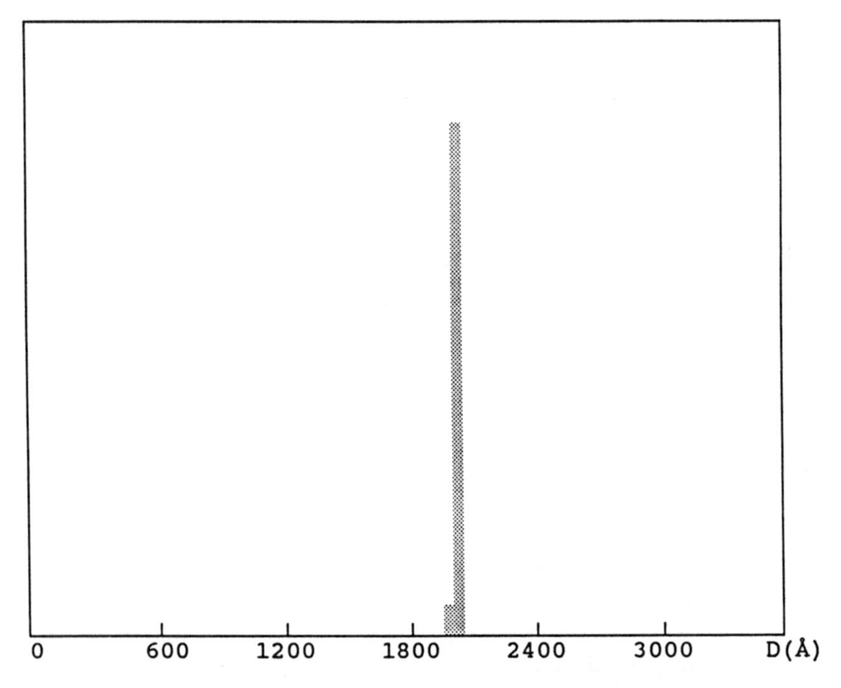

Figure 9. Computed PSD from simulated uniform circles.

maximizing at $x=x_n$. $F_n(x/x_n)$ is so chosen that peak half width decreases with n as $c(2n+1)^{-1/2}$ (see Appendix II) and CLD can be expressed as sum of N explicit functions (see Appendix III),

$$g(v)=\sum_{n=0}^{N} a_n g_n(v/x_n) \tag{10}$$

with

$$g_n(\mu)=\frac{\mu\sqrt{\pi}}{<x>}\exp(-(n+1/2)\mu^2)\times\sum_{k=0}^{n}\frac{(2n-2k-1)!!}{k!(2n-2k)!!}[(n+1/2)\mu^2]^k \tag{11}$$

Similarly, PSD can be represented as the sum of N Lorentzian functions,

$$F(x)=\sum_{n=0}^{N} a_n L_n(\frac{x}{x_n}) \tag{12}$$

where

$$L_n(x,x_n,w_n)=\frac{1}{\pi}\frac{w_n}{(x-x_n)^2+w_n^2} \tag{13}$$

is again centered at x_n with the peak half width $2w_n$. The corresponding CLD as shown in Appendix IV,

$$g(v)=\sum_{n=0}^{N} a_n g_n(x,x_n,w_n) \tag{14}$$

with

$$g_n(x,x_n,w_n)=\frac{1}{\pi<x>}\frac{1}{abc}\left[\frac{a+c}{4}\log\frac{a+b+c}{a-b+c}+\frac{b(a-c)}{\sqrt{4ac-b^2}}\arctan\frac{\sqrt{4ac-b^2}}{c-a}\right] \tag{15}$$

where

$$a=\sqrt{(v+x_n)^2+w_n^2},\qquad b=\sqrt{2(ac+x_n^2-v^2+w_n^2)},\qquad c=\sqrt{(v^2-x_n^2)^2+w_n^2}.$$

By adjusting the center positions, x_n, peak width factor, $1/(2n+1)^{1/2}$ or w_n, and weighting factors, a_n, the chord data can be fitted to equation 10 or 14, and the parameters can be used to compute the PSD using equation 8 or 12 depending on if it is Gaussian or Lorentzian fitting. Figure 10 exhibits an overlapped bimodal Gaussian distribution of negative stained polyacrylate latex particles.

While it is difficult to quantitatively evaluate the influence of the degree of aggregation on the particle size analysis, the influence should be minimized by the following facts. First, since the correlation between CDD and CLD is based on statistics, ignoring any overlapped chords does not affect CDD so long as the overlapping is random. Second, during the inversion, CDD may become negative at some diameter range due to experimental error of chord data. Resetting them to zero is equivalent to auto chord error reduction so that CDD is meaningful. Therefore, initial error of chords could be significantly reduced after inversion.

Conclusions

An automatic method was developed for determining particle size and size distribution. Chord data were selected and smoothed with binomial convolution or regression using a sum of modified Gaussian or Lorentzian functions. The results from a variety of latexes and computer simulations indicate that this method gives excellent representations of the actual PSDs. The mathematical inversion, computer automation, and the smoothing techniques developed in this report make a quick and accurate procedure for obtaining particle size and size distributions from micrographs, especially for resolving agglomerated particles.

Appendix I

The following is the derivation of converting CLD to CDD. Integrating equation 3 by parts

$$\frac{g(v)}{v} = \frac{-1}{<x>}\int_v^{\infty}\sqrt{x^2-v^2}\left[\frac{F(x)}{x}\right]' dx \tag{I1}$$

$$\left[\frac{g(v)}{v}\right]' = \frac{1}{<x>}\int_v^{\infty}\frac{v}{\sqrt{x^2-v^2}}\left[\frac{F(x)}{x}\right]' dx \tag{I2}$$

$$\int_r^{\infty}\left[\frac{g(v)}{v}\right]'\frac{dv}{\sqrt{v^2-r^2}} = \frac{1}{<x>}\int_r^{\infty}\int_v^{\infty}\left[\frac{F(x)}{x}\right]'\frac{v\,dx\,dv}{\sqrt{(x^2-v^2)(v^2-r^2)}}$$

$$= \frac{1}{<x>}\int_r^{\infty}\left[\frac{F(x)}{x}\right]'\int_r^{x}\frac{v\,dv\,dx}{\sqrt{(v^2-r^2)(x^2-v^2)}}$$

$$= \frac{1}{2<x>}\int_r^{\infty}\left[\frac{F(x)}{x}\right]'\int_{-\alpha}^{\alpha}\frac{du}{\sqrt{\alpha^2-u^2}}dx = \frac{-\pi}{2<x>}\frac{F(r)}{r} \tag{I3}$$

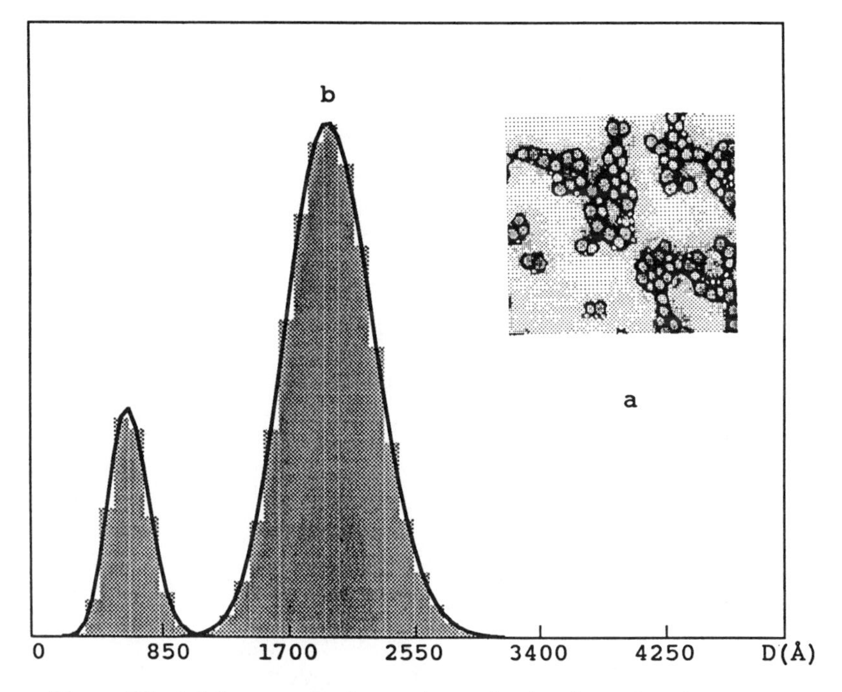

Figure 10. a) Micrograph of negative stained polyacrylate latexes; b) PSD by Gaussian regression.

Therefore,

$$F(x) = -\frac{2x\langle x\rangle}{\pi}\int_x^{\infty}\left[\frac{g(v)}{v}\right]'\frac{dv}{\sqrt{v^2-x^2}} \tag{I4}$$

Appendix II

This shows that the peak half width decreases as $c(2n+1)^{-1/2}$.

$$F_n(x) = \frac{2(n+1/2)^{n+1}}{n!x_n}\left(\frac{x}{x_n}\right)^{2n+1}\exp\left[-(n+1/2)\frac{x^2}{x_n^2}\right] \tag{II1}$$

Let $F_n'(x) = 0$, we have $x=x_n$. At the half maximum,

$$\frac{F_n(x)}{F_n(x_n)} = \left(\frac{x}{x_n}\right)^{2n+1}\exp\left[-(n+1/2)\frac{x^2-x_n^2}{x_n^2}\right] = \frac{1}{2} \tag{II2}$$

At large n, $x \approx x_n \pm \alpha$ with $\alpha << x_n$,

$$(2n+1)\log(1\pm\frac{\alpha}{x_n})-(n+\frac{1}{2})(\pm\frac{2\alpha}{x_n}+\frac{\alpha^2}{x_n^2}) = -\log(2) \tag{II3}$$

or

$$\alpha = \pm\frac{\sqrt{2\log(2)/3}}{\sqrt{2n+1}} \tag{II4}$$

Thus, the peak half width

$$2|\alpha| = 2\frac{\sqrt{2\log(2)/3}}{\sqrt{2n+1}} \tag{II5}$$

Appendix III

The derivation of chord distribution from the modified Gaussian distribution(equation 7) is shown below.

$$g_n(v) = \frac{1}{\langle x\rangle}\int_v^{\infty}\frac{F_n(x,x_n)}{\sqrt{x^2-v^2}}dx \tag{III1}$$

$$=\frac{1}{<x>}\int_{v}^{\infty}\frac{1}{n!}\left[(n+1/2)\frac{x^2}{x_n^2}\right]^n \exp\left[-\frac{(n+1/2)x^2}{x_n^2}\right](n+1/2)\frac{2xdx}{x_n^2} \qquad \text{(III2)}$$

Let

$$u=(n+1/2)\frac{x^2-v^2}{x_n^2}, \quad \text{and} \quad \alpha=(n+1/2)\frac{v^2}{x_n^2},$$

then,

$$g_n(v)=\frac{2v}{<x^2>}\int_0^{\infty}\frac{1}{n!}(u+\alpha)^n\exp(-\alpha-u)du$$

$$=\frac{2v}{<x^2>}\exp\left[-(n+1/2)\frac{v^2}{x_n^2}\right]\sum_{k=0}^{n}\frac{1}{k!}\left[(n+1/2)\frac{v^2}{x_n^2}\right]^k \qquad \text{(III3)}$$

Appendix IV

The following is the derivation of the chord distribution from the Lorentzian distribution(equation 11).

$$g_n(v,x_n,w_n)=\frac{1}{\pi<x>}\int_v^{\infty}\frac{dx}{[(x-x_n)^2+w_n^2]\sqrt{x^2-v^2}} \qquad \text{(IV4)}$$

$$\text{Let} \quad t^2=\frac{x-v}{x+v}, \quad \text{then} \quad x=\frac{2v}{1-t^2}-v, \quad dx=\frac{4vtdt}{(1-t^2)^2}$$

$$x-x_n=\frac{(v-x_n)+(v+x_n)t^2}{1-t^2}, \qquad \sqrt{x^2-v^2}=\frac{2vt}{1-t^2}$$

$$g_n(v,x_n,w_n)=\frac{2}{\pi<x>}\int_0^1\frac{(1-t^2)dt}{((v-x_n)+(v+x_n)t^2)^2+w_n^2(1-t^2)^2}$$

$$=\frac{2}{\pi<x>}\int_0^1\frac{(1-t^2)dt}{(at^2+bt+c)(at^2-bt+c)} \qquad \text{(IV2)}$$

where

$$a=\sqrt{(v+x_n)^2+w_n^2},\quad b=\sqrt{2(ac+x_n^2-v^2+w_n^2)},\quad c=\sqrt{(v-x_n)^2+w_n^2}.$$

Thus,

$$g_n(v,x_n,w_n)=\frac{1}{abc\pi\langle x\rangle}\left[\frac{a+c}{4}\log\frac{a+b+c}{a-b+c}+\frac{b(a-c)}{\sqrt{4ac-b^2}}\arctan\frac{\sqrt{4ac-b^2}}{c-a}\right] \tag{IV3}$$

Acknowledgements

The author would like to express his gratitude to Dennis Parker for preparing the electron micrographs and counting particles. The author also thanks Jan Grossman and Dr. John Vincent for their help and support throughout this work. Special thanks go to Dr. Eugene Elzy, Dr. Charles Boone, and Dr. Dan Biggerstaff for their encouragement and instructive suggestions.

References

1. Collins, E. A.; Davidson, J. A.; Daniels, C. A. *J. Paint Technol.* **1975**, 47, 35.
2. Lee, S.; Myers, M. N.; Beckett, R.; Gidding, J. C. *Anal. Chem.* **1988**, 60, 1129.
3. DosRamos, J. G.; Silebi, C. A. *J. Colloid Interface Sci.* **1990**, 135, 165.
4. Hewitt, F. C.; Laferty, D. R.; Mahran, H. E. *American Laboratory* **1990**, April 42.
5. Goertzel, G. *Mathematical Methods for Digital Computers*; Wiley, NY, 1965.
6. Bevington, P. R. *Data Reduction and Error Analysis for the Physical Science*; McGraw-Hill Book Company, NY, 1969.
7. Ruan, J. Z.; Litt, M. H.; Krieger, I. M. *J. Colloid and Interface Science* **1988**, 126, 93.

RECEIVED December 4, 1991

Chapter 20

Rheological Study of Ordered Polymer Colloids

Y. P. Lee[1] and F. J. Micale

Zettlemoyer Center for Surface Studies, Sinclair Lab No. 7, Lehigh University, Bethlehem, PA 18015

Rheological techniques by means of continuous forced oscillation and stress relaxation, which have been rarely used on ordered latex systems, were successfully employed and found to yield useful and fruitful information. The viscoelasticity of the structured dispersions was studied as a function of particle concentration, temperature, salt valency and salt concentration. It was found that the storage modulus G' decreases and the phase shift $\tan^{-1}(G''/G')$ increases with decreasing particle concentration, or increasing temperature or salt concentration. This is ascribed to the corresponding reduction in the strength and range of the interparticle Coulombic repulsion force. Thermal motions of particles thus overcome the suppressed electrostatic repulsion and disintegrate the ordered dispersion structure. Efforts were also made to compare the experimental shear modulus with the theoretical high frequency limiting modulus ($G(\infty)$) calculated from a crystal lattice model and pairwise additive theory proposed by Goodwin and coworkers. Good agreement was obtained when the "corrected" Debye length and a realistic Stern potential derived from the relaxation modulus were used in the calculation of $G(\infty)$.

As formed by emulsion polymerization, a typical synthetic latex has a broad particle size distribution and displays a milky white color. However, cleaned monodisperse latexes show iridescence due to Bragg diffraction of visible light from the ordered arrays of particles (*1*). It is believed that the strong electrostatic repulsion force between particles overcomes the Brownian motion and freezes the particles within the crystal lattice.

The rheological behaviors of ordered dispersions are very interesting because of the dynamic viscoelasticity caused by the screened electrostatic repulsion forces between particles. The viscoelasticity of structured dispersions has been studied by many authors. Crandall and Williams(*2*) obtained the Young's modulus by measuring the vertical compression of the ordered structure due to gravity. Goodwin and

[1]Current address: Miami Valley Laboratories, The Procter & Gamble Company, P.O. Box 398707, Cincinnati, OH 45239

0097–6156/92/0492–0305$06.75/0

coworkers (*3*) measured the high frequency limit of the storage modulus and the low-shear limit of viscosity by shear wave propagation and by creep compliance measurements, respectively. The rigidity and viscosity of monodisperse latexes were measured by the torsional quartz crystal method by Mitaku et al. (*4, 5*) at high excitation frequencies (40 kHz and 70 kHz), and an abrupt change was found at the order-disorder transition. Pieranski et al. (*6*) proposed a simple method to determine the elastic modulus of the ordered dispersion from the spectrum of their mechanical vibrations. Okubo (*7*) claimed that the stress-strain relationship for the crystal holds in the range of small shear rate values where the crystal-like structures are distorted but do not flow by the shear. Therefore, he evaluated the elastic modulus of ordered dispersions using a rotational viscometer to be 330-1900 Pa from plots of shear stress against shear rate times time. Russel et al. (*8*) used a home-made membrane osmometer and an oscillatory viscometer and found the osmotic pressures and shear moduli increase by orders of magnitude with increasing particle concentration and decreasing particle radius or ionic strength.

It is very important to be able to relate the macroscopic rheological properties of a dispersion to interparticle forces which are controlled by parameters, such as particle concentration, temperature, salt valency, and salt concentration. In this investigation, the dynamic modulus and stress relaxation of the ordered latex dispersions are studied as a function of the above variables. The applicability of rarely used rheological techniques on the structured dispersions are explored. The rheological data are rationalized within the framework of DLVO theory. Efforts are also made to compare the experimental shear modulus with the theoretical high frequency limiting modulus calculated from a crystal lattice model and pairwise additive theory proposed by Goodwin and coworkers (*9*).

Experimental

Material. A Dow polystyrene latex was used in this study, and designated latex A. Previous studies (*10*) by transmission electron microscopy, conductometric titration and Fourier Transform infrared spectroscopy showed that the latex particles have a mean diameter 233 nm with a standard deviation of 14 nm and carry sulfate and carboxyl functional groups on their surfaces. In view of the small coefficient of variation on the mean particle diameter, the particle size distribution was considered to be monodisperse. The serum replacement technique was employed to remove electrolyte, oligomers, and emulsifiers from the latex. The water used for the latex cleaning and for dispersion and salt solution preparation was distilled and deionized by the Milli-Q water system. A rotary evaporator under reduced pressure was applied to concentrate the latexes. Particle volume fractions were measured by drying weighed samples at 100°C to constant weight. The salt concentrations in the latexes reported here are molarities (mole/l) based on the total volume of the dispersions.

Electrokinetics. Electrophoretic mobilities of latex A were measured using the Pen Kem 3000 (Pen Kem, Inc., New York, U.S.A.) at infinite dilution over a range of temperature and salt concentrations. The corresponding zeta potentials for the measured electrophoretic mobilities of latex A as a function of salt concentration are given in Figure 1. Zeta potentials for the latex particles at different salt concentrations were obtained from Ottewill's numerical tables (*11*) which take into account the electrophoretic retardation due to backward flow of the solvent and the relaxation effect due to distortion of the ionic atmosphere surrounding the particle. The zeta potential decreases monotonically with increasing ionic strength or decreasing double layer thickness $1/\kappa_0$, as predicted by electrical double layer theory, for both cases (NaCl and $MgCl_2$). Lower zeta potentials were obtained with higher counterion valency. The zeta potential vs. temperature is shown in Figure 2. The zeta potential was calculated

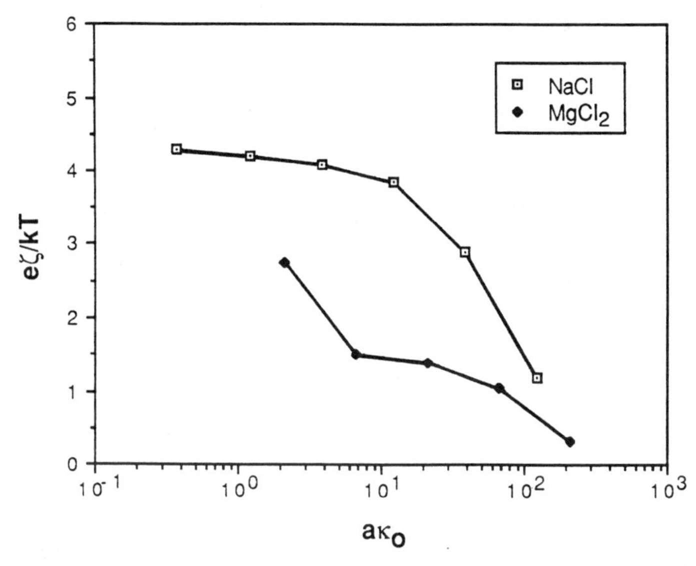

Figure 1. Dimensionless zeta potential vs. $a\kappa_0$ for latex A in various aqueous salt solutions at 25 °C (e : electronic charge, ζ : zeta potential, k : Boltzmann constant, T : absolute temperature, a : particle radius, $1/\kappa_0$: Debye length).

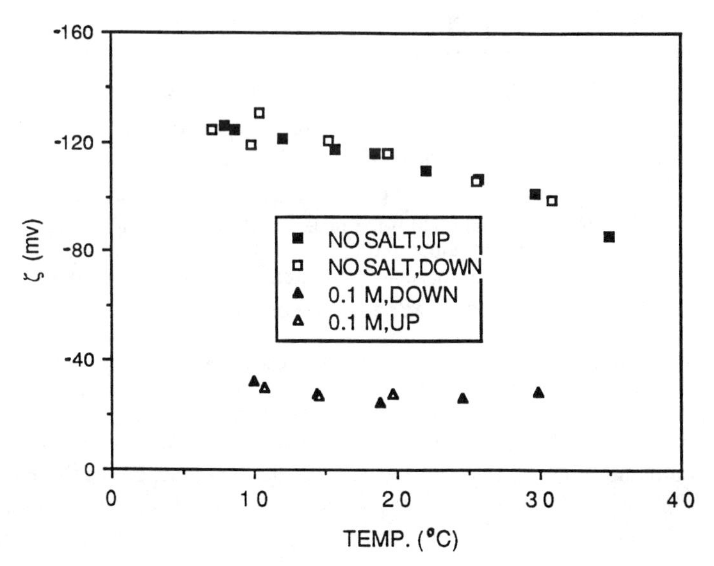

Figure 2. Zeta potential vs. temperature for latex A at various aqueous NaCl concentrations (mole/l).

from the electrophoretic mobility by using the Smoluchowski and Huckel equation (*12*) and taking into account the temperature dependence on the viscosity and dielectric constant of water. The results show that the zeta potential decreases as the temperature is increased for the latex without added salt. Whereas for the latex with high NaCl concentration (0.1 M), the zeta potential is essentially independent of temperature. The pronounced dependency of the zeta potential on temperature can be understood by recognizing the corresponding thicker electrical double layer around the latex particles.

Rheometry. A Bohlin VOR rheometer (Bohlin Reologi, Lund, Sweden) was used for rheological measurements including dynamic oscillation, strain sweep, flow oscillation, and stress relaxation. A couette measuring geometry (C25) with a moving cup of diameter 27.5 mm and a fixed detector bob of diameter 25 mm was used for dynamic and transient measurements. Calibration of the instrument was carried out using a Newtonian mineral oil (Canon Instrument Co., PA, U.S.A.)

Dynamic Oscillation, Strain Sweep and Flow Oscillation. In the oscillatory measurements, one applies a sinusoidal strain with a certain amplitude and angular frequency ω ($\omega=2\pi f$, f in Hz) to the cup of a concentric cylinder and then simultaneously compares the measured stress with the strain. If the system is purely elastic, within this amplitude and frequency range, the stress and the strain will be exactly in phase, i.e. the phase angle shift $\delta=0^o$. But, if the system is purely viscous, then $\delta=90^o$. Viscoelastic systems fall between the above two extremes and have $0^o<\delta<90^o$. The phase shift is automatically computed from the time displacement between the sine waves of stress and strain, i.e. $\delta=\omega\Delta t$. From phase shift (δ) and amplitude (τ_o) and strain (ε_o) one can obtain various rheological parameters (*14*), i.e. complex modulus ($G^*=\tau_o/\varepsilon_o$), storage modulus ($G'=G^*\cos\delta$, a measure of energy stored per cycle of deformation representing the elastic contribution), loss modulus ($G''=G^*\sin\delta$, a measure of energy dissipated per cycle of deformation representing the viscous contribution), phase shift ($\delta=\tan^{-1}(G''/G')$), and dynamic viscosity ($\eta'=G''/\omega$). However, for the above relationships to be applicable, one should be in the linear viscoelastic region, i.e. where G*, G' and G" are independent of the applied strain (ε_o). Strain sweep is thus carried out at a fixed frequency. In this work, linear dynamic moduli were always obtained if the peak strain was less than 7 %.

The flow oscillation technique used to follow the structure recovery of the latex dispersion involves an oscillation measurement after steady shearing for a finite period of time. G', G", and δ are displayed as a function of time.

Stress Relaxation. In the stress relaxation experiment, the material is subjected to a rapidly applied small strain which is held constant for the remainder of the experiment. The relaxation behavior is then studied by continuously monitoring the delayed stress vs. time, t. The stress divided by the constant strain gives the relaxation modulus G(t), and the relaxation spectrum $H(\lambda)$ can be readily calculated from $[-dG(t)/d\ln t]_{t=\lambda}$ by applying Alfrey's rule (*19*). The relaxation was conducted with the strain rise time of 0.02 sec in this work.

Results and Discussion

Dynamic Oscillation. The viscosity measurements probe the strength of the dispersion microstructure by subjecting samples to relatively large stresses while dynamic and transient techniques are more useful in dealing with the state of the dispersion due to the unperturbed structure at small deformations. Concentrated dispersions usually display a viscoelastic response upon subjecting them to an oscillatory strain (or stress) within a certain frequency range. Typically, the storage modulus G' dominates at high frequency and reaches a limiting plateau modulus $G(\infty)$ with increasing frequency, whereas the loss modulus G" prevails at low frequency. If the Deborah number (De) is defined as the ratio of the relaxation time of the material to the experimental time, then the above results indicate that when De>>1 the material will behave more like a solid whereas it will perform like a liquid when De<<1.

Oscillation results obtained for the cleaned polystyrene latex A with a volume fraction of 26.97% at 35 °C are given in Figure 3. G' dominates over G" throughout the frequency range employed. The small amplitude oscillation induces weaker hydrodynamic interactions and maintains the ordered structure. Electrostatic energy is thus stored and generates the observed elasticity in G'. The data also show that logG' and logG" increase linearly with the log frequency according to the relationships $G'=13.59(2\pi f)^{0.09}$ and $G''=2.43(2\pi f)^{0.09}$ while the phase shift δ maintains a constant low angle of 10° which indicates that this structured dispersion is highly elastic and G' and G" are equally sensitive to the frequency. It is interesting to see that there also exists a power law shear-thinning relationship between dynamic viscosity and frequency analogous to that found between the steady state viscosity and shear rate $(\eta=m(\dot{r})^{n-1})$.

Strain Sweep. Strain sweep is also an oscillation measurement which is carried out at a fixed frequency but at various strain amplitudes. This test is useful for establishing the range of linear viscoelastic behavior where dynamic moduli are independent of the strain amplitude. Figure 4 shows the particle concentration dependence of the strain sweep obtained for the cleaned polystyrene latex A. It can be clearly seen that G' remains constant with applied strain ε, until a critical value ε_m is reached above which G' decreases with a further increase in ε. In small amplitude oscillations, the particles never move out of their potential wells so the electrostatic energy is stored and appears as the elasticity of the dispersion, G'. Above ε_m, the ordered structure is perturbed and the dispersion shows a nonlinear response. G' increases with increasing particle concentration at the same strain amplitude because the electrostatic interactions between particles become stronger and numerous. Surprisingly though, higher ε_m is obtained with lower particle concentration. The longer linear region at low particle concentration is possibly ascribed to the fact that the particles are far from point masses. When the particles are closer together, the potential well becomes anharmonic at smaller displacements.

The effect of particle concentration on the storage modulus G', loss modulus G", and phase shift $\tan^{-1}(G''/G')$ of the cleaned latex A is illustrated in Figure 5. While G' rises sharply, G" has an unexpected mild increase with increasing particle concentration. An increased possibility of having more crystal defects and disordered regions at higher particle concentrations might explain the higher dissipated viscous energy which appears as G". For the case of water, the phase shift is 90°. Hence, a dramatic decrease in phase shift from 90° to 3° was observed for latex particle concentrations varying from 0-31%. This indicates that the latexes with higher particle concentrations behave more like a perfect Hookean elastic body and store most of the electrostatic energy upon deformation.

Figure 6 illustrates the temperature dependence of the strain sweep for the cleaned polystyrene latex A. The latex dispersion with the lower particle concentration

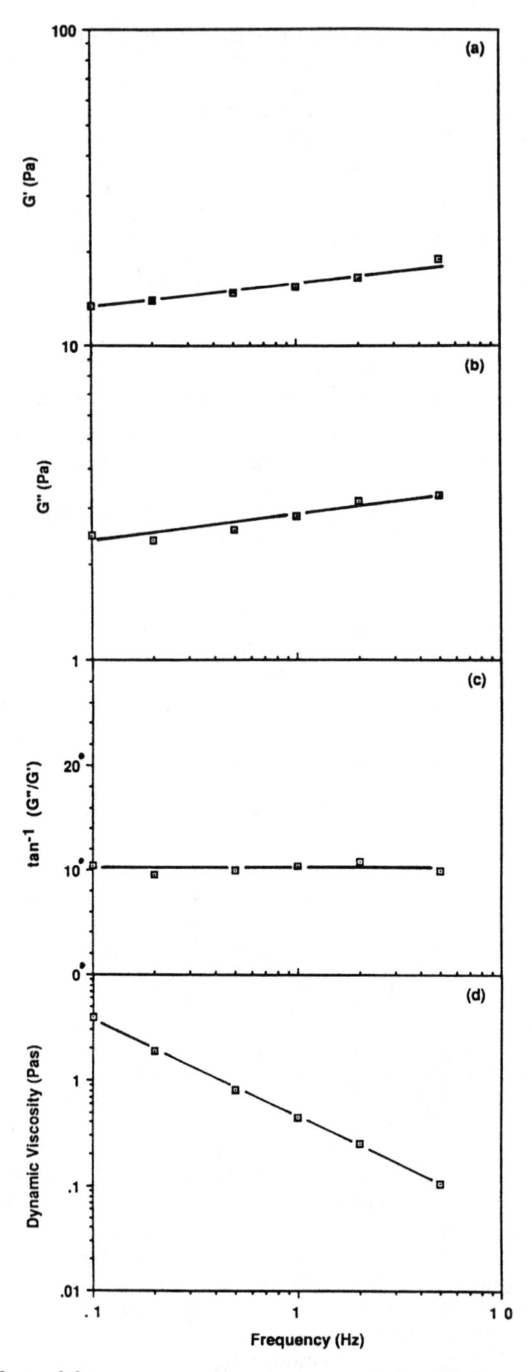

Figure 3. Effect of frequency on the storage modulus G', loss modulus G", phase shift $\tan^{-1}$(G"/G'), and dynamic viscosity η' of cleaned latex A with ϕ = 26.97% at 35 °C.

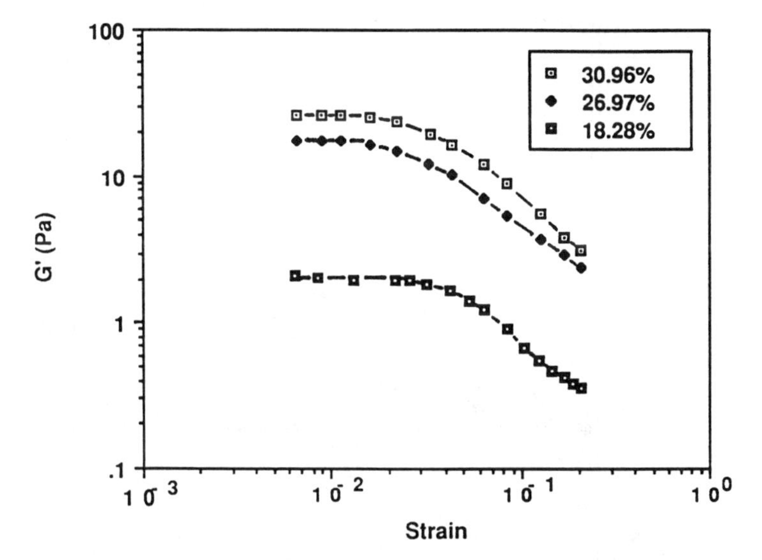

Figure 4. Effect of strain and particle concentration (vol.%) on the storage modulus of the cleaned latex A at 25 °C and frequency 2 Hz.

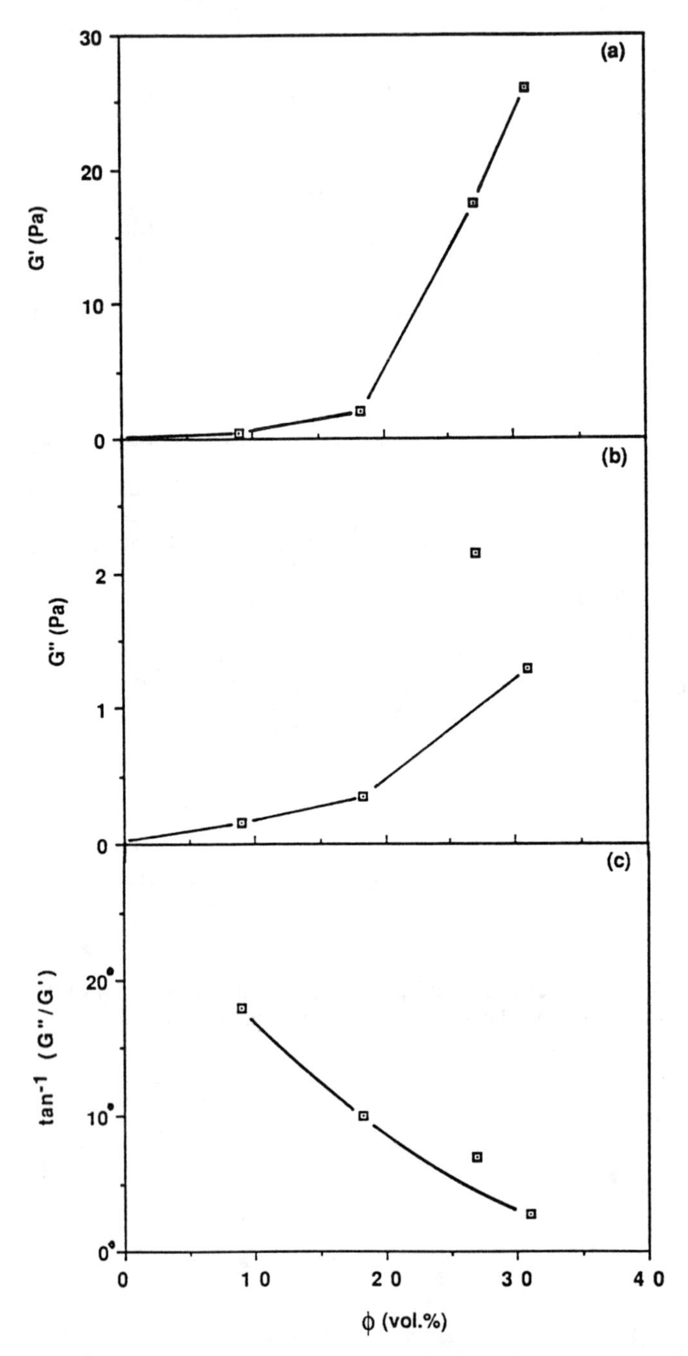

Figure 5. Effect of particle concentration on the storage modulus G', loss modulus G", and phase shift $\tan^{-1}$(G"/G') of the cleaned latex A at 25 °C and frequency 2 Hz.

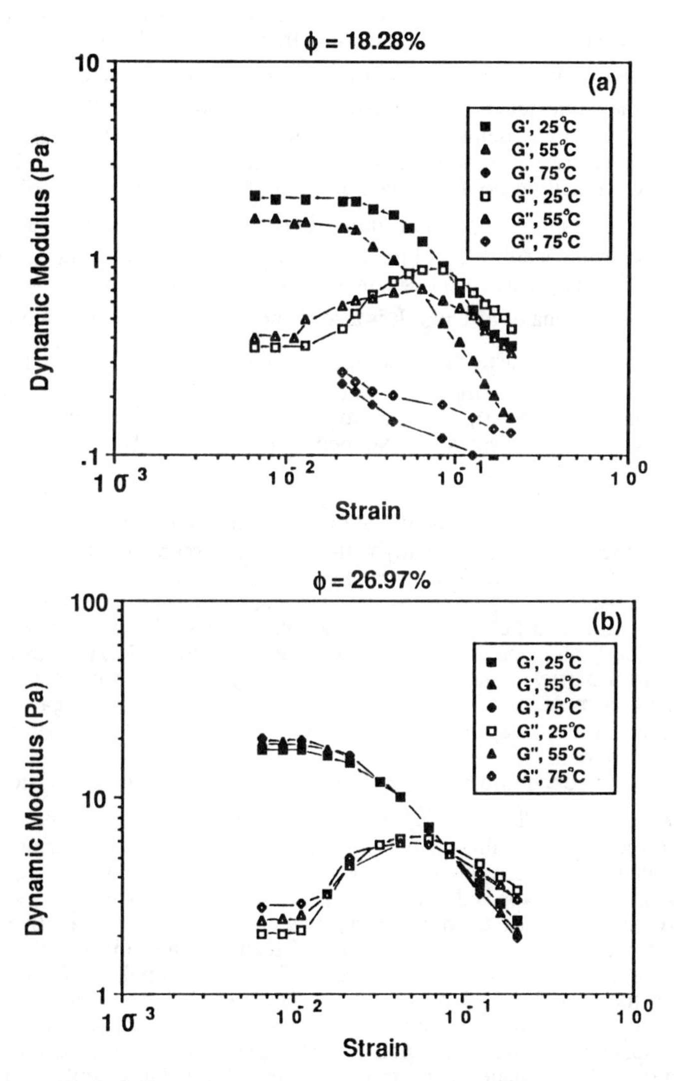

Figure 6. Effect of strain and temperature on the storage and loss modulus of the cleaned latex A at frequency 2 Hz.

is first examined in Figure 6(a). The elasticity (G') of the structured dispersion decreases with increasing temperature (T) due to the corresponding reduction in electrostatic repulsion. It is also seen that a linear viscoelastic plateau can be observed at low temperature as long as the dispersions possess an ordered structure. The elasticity drops dramatically at high temperature since the ordered structure is perturbed by the thermal motion of the particles. This represents a temperature induced order-disorder phase transition, i.e. a melting transition of colloidal crystals. On the other hand, upon increasing the strain (ε) G" shows an initial plateau, climbs to a maximum, then intersects with G' and drops thereafter. G', G" vs. ε behavior resembles G',G" vs.T behavior of polymeric materials. The glass transition temperature interpreted as the onset of long-range, coordinated molecular motion, represents the transition point between glassy state and rubbery state of polymeric materials. Whereas the intersection strain (ε_c) between G' and G" of structured latex dispersion indicates a solid-liquid transition point as a result of long-range particle diffusion. This interpretation is based on the observation that G' is dominant over G" as $\varepsilon<\varepsilon_c$ and G" dominates as $\varepsilon>\varepsilon_c$. It is also found that the strain at which the system gives a nonlinear viscoelastic response (ε_m) for G' is higher than that (ε_n) for G". ε_m decreases with increasing temperature. For the latex dispersion with a higher particle concentration (Figure 6(b)), temperature has less impact on the elasticity and the ordered structure of the dispersion. No melting transition is observed throughout the temperature range employed.

Figure 7 exhibits the effect of temperature on the storage modulus G', loss modulus G" and phase shift $\tan^{-1}$(G"/G') of the cleaned latex A. While G" and $\tan^{-1}$(G"/G') rise, G' decreases slightly with increasing temperature for the latex with a lower particle concentration. The increased temperature compressed the Debye length, decreased the magnitude of the electrostatic repulsion between particles, and fueled the Brownian motion of particles. Therefore, the weaker dispersion microstructure results in lower elasticity and more electrostatic energy is dissipated. With regards to the latex with the higher particle concentration, an unexpected minimum in G' was observed due to the experimental artifact of water evaporation at the higher temperatures. The particle concentration dependences of G', G", and $\tan^{-1}$(G"/G') agree with those observed in Figure 5.

The salt concentration dependence of the strain sweep of latex A with ϕ=20.70% at 25 °C is shown in Figure 8. A linear viscoelastic response was always obtained if the applied strain was below 0.02. As may be seen, the various curves at different salt concentrations appear to reach a common horizontal asymptote as strain increases. Zero G' is expected to be obtained eventually upon applying a sufficiently large deformation because the ordered structure of the dispersion is totally destroyed. G' decreases with increasing salt concentration because of the corresponding reduction in the strength and range of the interparticle Coulombic repulsion force. Thermal motions of the particles then overcome the suppressed electrostatic repulsion and disintegrate the crystal-like structure of the dispersion. Closer examination reveals that within the nonlinear viscoelastic region G' for the latex with a lower salt concentration decreases at a faster pace with strain than that for a latex with a higher salt concentration. This can be understood because the structured dispersions become less sensitive to the strain when the dispersions become more disordered at higher electrolyte concentration.

Figure 9 illustrates the effect of salt valency and concentration on the storage modulus G', loss modulus G" and phase shift $\tan^{-1}$(G"/G') of latex A. At low added salt concentrations, plateau storage and loss moduli were observed because the total ionic strength dominated by the counterion density was constant. With a further

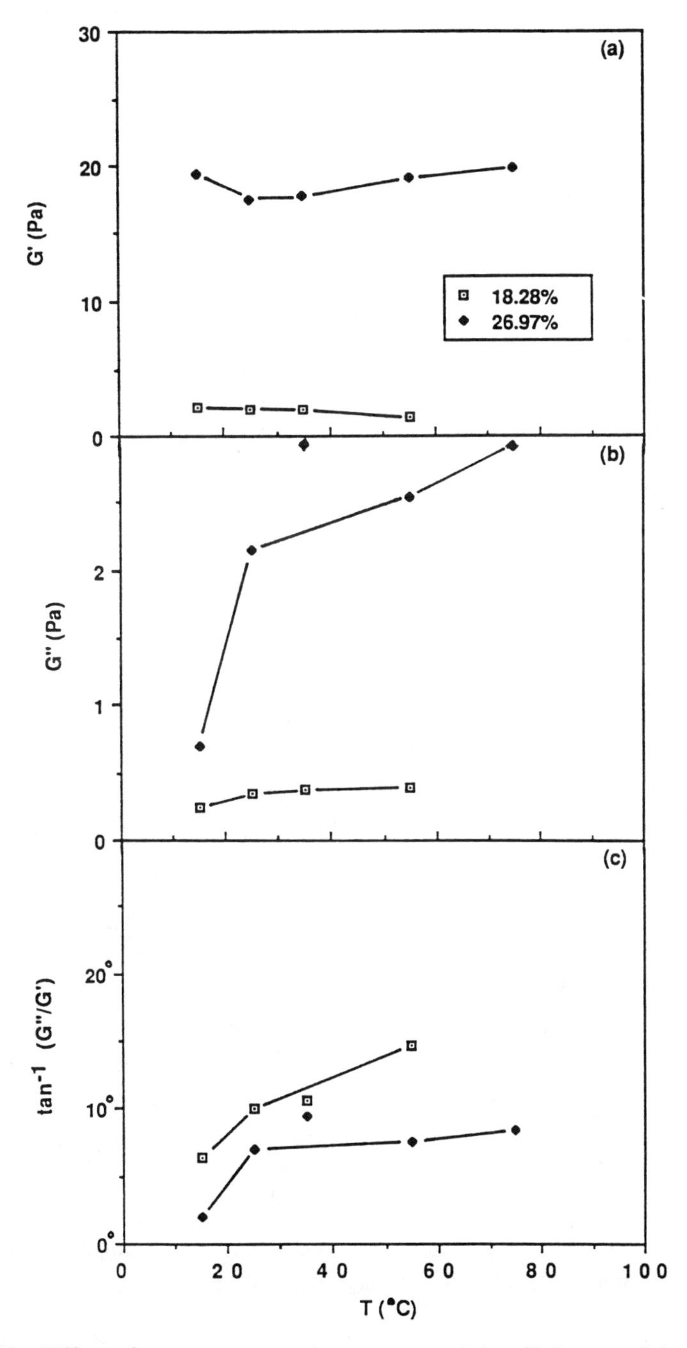

Figure 7. Effect of temperature on the storage modulus G', loss modulus G", and phase shift $\tan^{-1}(G''/G')$ of the cleaned latex A at frequency 2 Hz and various particle concentrations (vol.%).

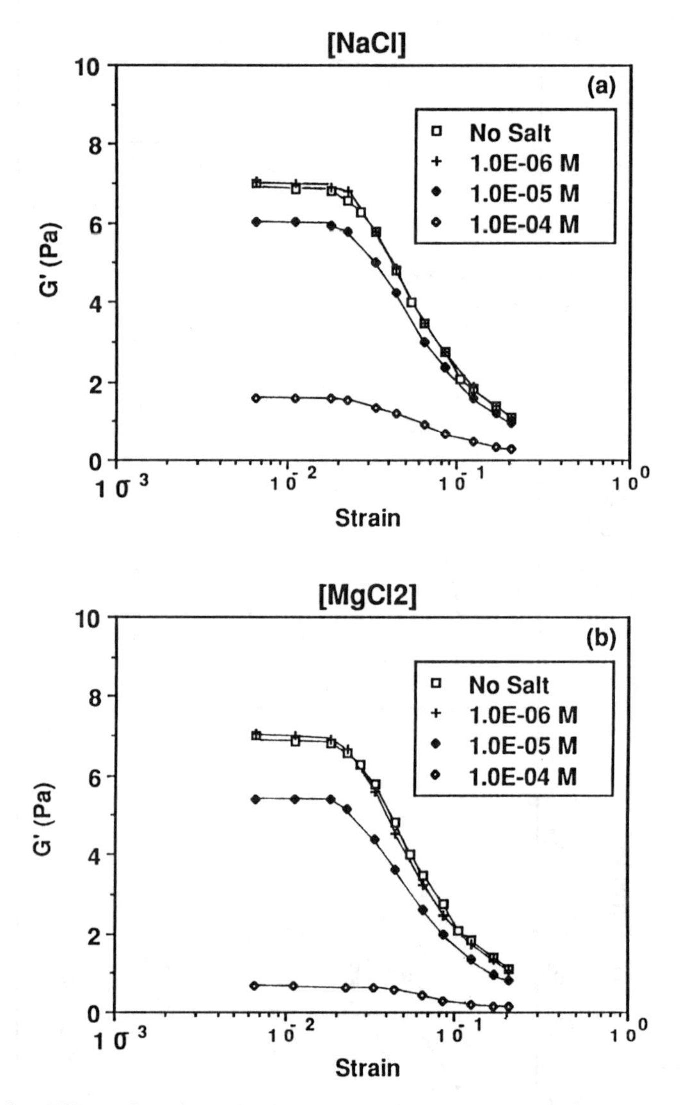

Figure 8. Effect of strain and salt concentration on the storage modulus G' of latex A with $\phi = 20.70\%$ at 25 °C and frequency 2 Hz.

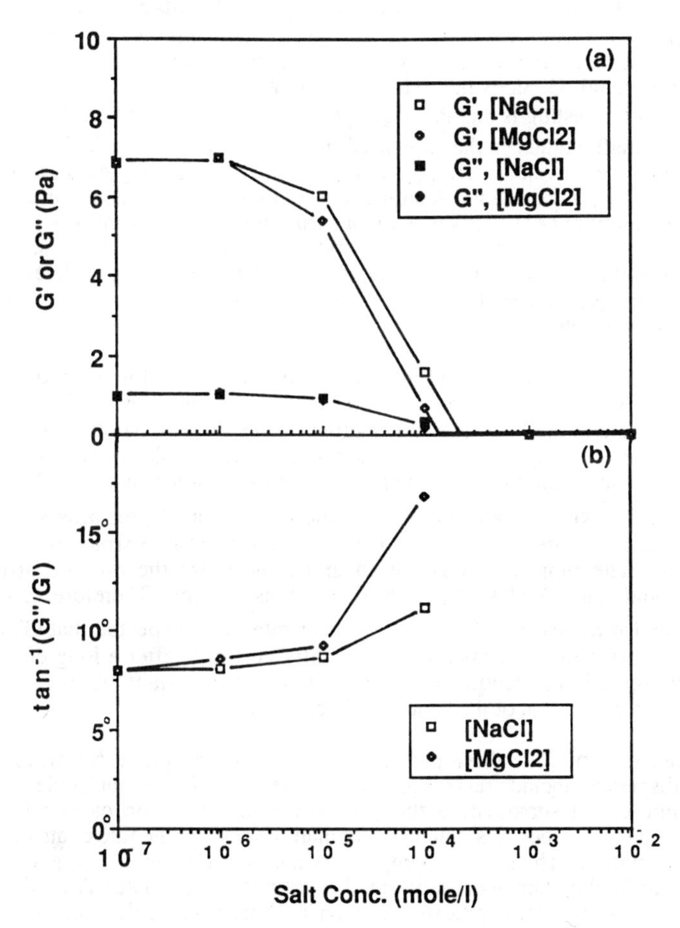

Figure 9. Effect of salt valency and concentration on the storage modulus G', loss modulus G", and phase shift $\tan^{-1}(G''/G')$ of latex A with $\phi = 20.70\%$ at 25 °C and frequency 2 Hz.

increase in the salt concentration, the Debye length is significantly compressed, and hence the elasticity originating from electrostatic interactions between particles vanishes accordingly. While the loss modulus G" or dynamic viscosity η' decreases, the phase shift (i.e., ratio between G" and G') increases with increasing salt concentration. The results also show that the divalent $MgCl_2$ decreases the elasticity and increases the phase shift to a greater extent than does the monovalent NaCl. This is because $MgCl_2$ creates a higher ionic strength and suppresses the Debye length more markedly. Once the range of the electrostatic interaction falls below the interparticle separation, the elasticity disappears and dispersion becomes disordered. On the other hand, G" does not exhibit significant specificity to salt valency type. What is most interesting is that the $MgCl_2$ data are superimposed on the NaCl data at the same normality either in the case of G' and G" or $\tan^{-1}$(G"/G'). This indicates that the secondary electroviscous effect produced by the two electrolytes is identical at the same equivalent concentration. A similar observation has been made for the viscosity data in the literature (*13*). This is really a surprising finding as most of the colloidal properties are sensitive to the valency and nature of the counterions. This also demonstrates that counterions play a bigger role than the coions in determining the colloidal behavior. One mole/l of Mg^{2+} works like two mole/l of Na^+ to screen out part of the particle charge.

Flow Oscillation. Many applications exist where it is crucial to have knowledge of the structure recovery in a dispersion after steady shearing for a finite time period. Such cases include the levelling and sagging problems with paintis, film formation after high speed printing, etc. Figure 10 shows the effect of shear strength on the flow oscillation of the cleaned polystyrene latex A with a volume fraction of 26.97% at 15 °C. G' increases while G" and δ drop as a function of time elapsed after shearing. All these changes indicate a gradual recovery of the crystal-like structure of the latex dispersion. The more rigorous the shear is, the more the ordered structure is perturbed, and hence the less elastic the dispersions become. Therefore, G' decreases while G" and δ increase with increasing shear rate. It is expected that G', G" and δ obtained at different shear rates will eventually converge after a long enough time. This flow oscillation technique has potential to be an important tool for studying the kinetics of crystallization of the structured latex dispersion.

Stress Relaxation. The three possible stress relaxation processes which exist in the latex dispersion include relaxation due to movement of water molecules, relaxation of the ionic cloud surrounding the particles, and relaxation caused by particle diffusion. The last process is much slower than the rest, so the rate controlling step is the particle motion. The effect of temperature and particle concentration on the stress relaxation and relaxation spectra of the cleaned polystyrene latex A is illustrated in Figure 11. Figure 11(a) indicates that on a short time scale (i.e., at very high frequency), the characteristic time of the deformation was much shorter than that of microstructural rearrangement, so the ordered dispersion stored the energy, and the plateau glassy modulus Gg was observed. As a matter of fact, Gg is equal to $G(\infty)$, the storage modulus measured at an extremely high frequency in the oscillation experiment. The above correlation is supported by the following examples. The 17.5 Pa relaxation modulus measured at 0.08 sec (i.e. 2 Hz) for the cleaned latex A with ϕ=26.97% at 35 °C is close to the 17.0 Pa storage modulus at a frequency of 2 Hz obtained from Figure 3. Also, it is noted that the predicted storage modulus of 20.6 Pa at a frequency of 16 Hz by the empirical equation $G'=13.59(2\pi f)^{0.09}$ obtained from Figure 3 agrees satisfactorily with the relaxation modulus of 19.6 Pa measured at 0.01 sec.

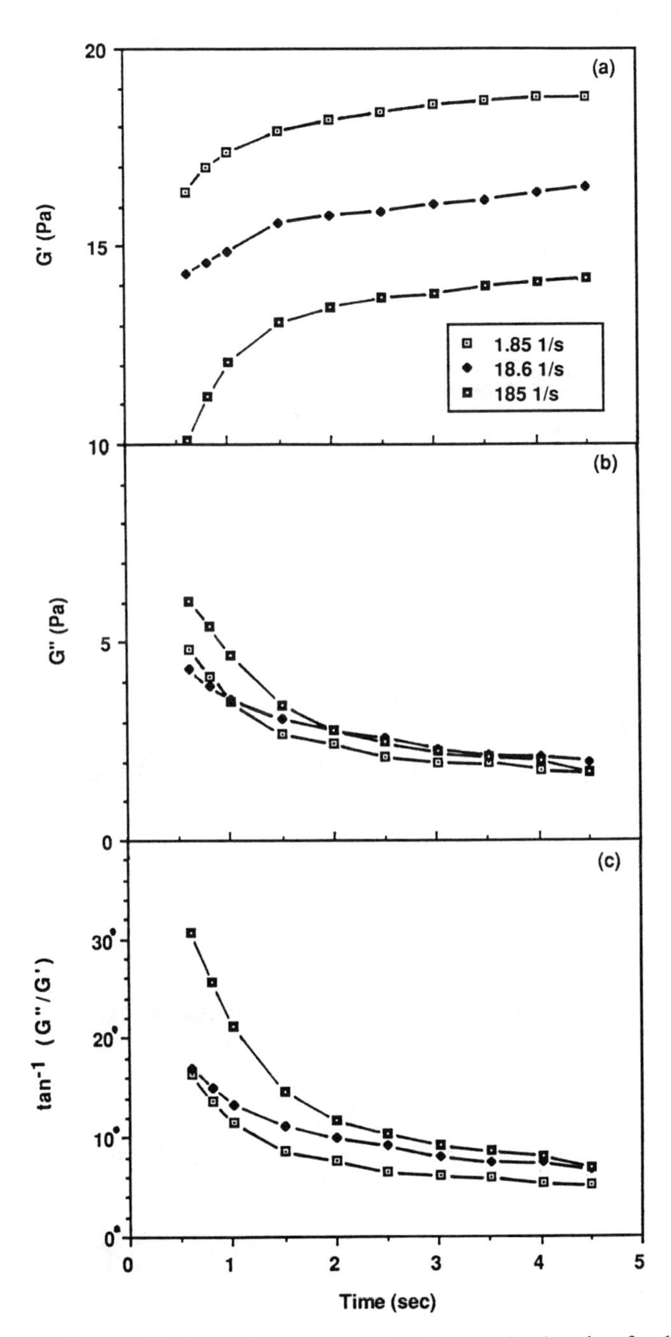

Figure 10. The oscillation at frequency 2 Hz after steady shearing for 1 second of the cleaned latex A with $\phi = 26.97\%$ at 15 ^{o}C.

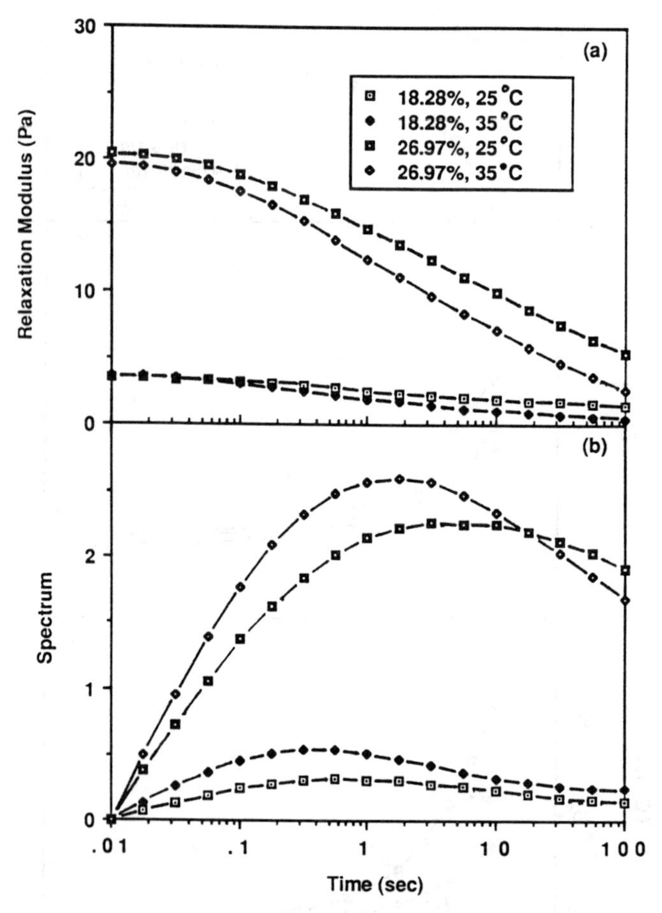

Figure 11. Effect of temperature and particle concentration on the stress relaxation and relaxation spectra of the cleaned latex A.

If one viscoelastic function is known over a wide range of time or frequency, then in principle any other viscoelastic function can be derived using exact transformation equations given by the mathematical theory of linear viscoelasticity (*14*). The exact interrelations among viscoelastic functions include :

$$G'(\omega) = Ge + \omega\int_0^\infty [G(t) - Ge] \sin\omega t \, dt \tag{1}$$

$$G''(\omega) = \omega\int_0^\infty [G(t) - Ge] \cos\omega t \, dt \tag{2}$$

$$G(t) = Ge + \int_{-\infty}^{\infty} H(\lambda) \exp(-t/\lambda) \, d\ln\lambda \tag{3}$$

where Ge is the equilibrium modulus (at long times), ω, the angular frequency (ω=2πf), t, time, λ, relaxation time, G(t), relaxation modulus, and H(λ), relaxation spectrum. An empirical equation to represent the stress relaxation modulus was proposed by Smith (*15*) :

$$G(t) = Ge + \frac{(Gg - Ge)}{[1+(t/t_0)]^{1-m}} \tag{4}$$

where Gg is the glassy modulus (at short times), t_o and m, positive empirical constants. This equation was found to fit the G(t) data in Figure 11(a) very well. For example, the equation with Gg=19.6 Pa, Ge=0, m=0.57 and t_o=1.00 represents well the experimental data for the cleaned latex A with ϕ=26.97% at 35 °C. Upon substituting equation (4) into equation (3) and taking the inverse Laplace transform, H(λ) can thus be obtained (*15*). The integrals in equations (1) and (2) can also be performed numerically or graphically since the starting function G(t) has been expressed in analytical form. G' and G" thus obtained can then be compared with experimental data. The above mathematical exercise enables one to test the applicablity of linear viscoelastic theory on structured latex dispersions. Such an important research area deserves further attention.

Based on the concept that electrostatic interactions dominate the stress relaxation behavior of the ordered structure by slowing the long range diffusive motion of the particles, a broad spectrum of relaxation processes will be expected to occur. Indeed, Figure 11(b) shows the distribution of relaxation times of the ordered dispersion. A shorter characteristic (peak) relaxation time is obtained with a lower particle concentration or higher temperature. This is anticipated since the structured dispersions under the same conditions become less elastic due to the decreased strength and range of the electrostatic repulsion force between particles. The same particle concentration dependence of the most frequently occurring relaxation time has been reported by Goodwin et al. (*16*). Neither the Maxwell Element (*14*) nor Four-Element model (combination of Maxwell and Voigt Element in series) was found to successfully describe the viscoelastic behavior of the ordered dispersions, such as G(t), G', and G". This is not surprising since the limited number of relaxation times attained from these models contradicts what is observed in Figure 11(b). Hence, it is expected that a generalized Maxwell model (Maxwell Elements in parallel) (*14*) will better account for the broad relaxation spectrum.

The order of magnitude of G(∞) or Gg may be written in terms of the magnitude of the thermal fluctuations, δ, of a particle around a crystal lattice point as

$$Gg \equiv \frac{f}{b} \equiv \frac{(kT/<\delta^2>)}{b} \quad (5)$$

where f is the force constant of a particle, b is the unit lattice constant, k is the Boltzmann constant, and T is the absolute temperature. A previous optical study (*10*) has shown that ordered dispersions with high particle concentrations possess the fcc (face-centered cubic) crystal structure. Since four particles per unit cell comprise a fcc crystal, if $<\delta^2>^{1/2}/b$ is replaced by a dimensionless parameter β, we obtain the elastic modulus as a linear function of number density (N) of particles,

$$Gg \equiv \frac{NkT}{4\beta^2} \quad (6)$$

Lindemann's law of crystal melting (*17*) says that $\beta<0.1$ for a stable crystal. Equation (6) gives the β values as 0.09 and 0.05 for the cleaned latex with particle volume fractions of 18.28% (2.76×10^{13} cm^{-3}) and 26.97% (4.07×10^{13} cm^{-3}) at 25 °C, respectively. $\beta>0.1$ is expected for $\phi<15\%$ indicating that ordered latexes with lower particle concentrations are vulnerable to elevated temperature and excess electrolyte, and are easily perturbed to become disordered. β represents the degree of particle fluctuation away from the lattice point and is a measure of the strength of electrostatic interactions between particles. Overlooking the finite number of particles per unit cell results in higher β values as found in Mitaku's (*5*) and Okubo's (*7*) work.

Comparison Between Experimental and Theoretical Shear Modulus. Following the treatment of Goodwin et al. (*9*), the shear modulus of ordered latexes can be derived from the increase in electrostatic energy caused by a small shear strain imposed upon a structured dispersion. In this analysis, high frequency deformation did not allow a measurable relaxation to occur and only nearest neighbor interactions using pairwise additivity were considered. The analytical expression for the high frequency limiting storage modulus $G(\infty)$ has the following forms :

for $\kappa a<3$

$$G(\infty) = 4\pi\alpha D\varepsilon_0 a^2\psi_d^2 \left(\frac{\kappa^2R^2+2\kappa R+2}{R^4}\right) \exp[-\kappa(R-2a)] - \frac{2A_{121}\alpha}{3a^2R}\left\{\left[\frac{3s^2+4}{(s^2-4)^3}\right]+\left(\frac{3}{s^4}\right)-\left[\frac{2(3s^2-4)}{s^2(s^2-4)^2}\right]\right\} \quad (7)$$

for $\kappa a>10$

$$G(\infty) = \frac{2\pi\alpha D\varepsilon_0\psi_d^2 a}{R}\left(\frac{\kappa^2\exp[-\kappa(R-2a)]}{\{1+\exp[-\kappa(R-2a)]\}^2}\right) - \frac{2A_{121}\alpha}{3a^2R}\left\{\left[\frac{3s^2+4}{(s^2-4)^3}\right]+\left(\frac{3}{s^4}\right)-\left[\frac{2(3s^2-4)}{s^2(s^2-4)^2}\right]\right\} \quad (8)$$

$$\alpha = \left(\frac{3}{32}\right)\phi_m\, n \quad (9)$$

where α is a geometric factor, ϕ_m, the maximum close packing volume fraction of the particles, n, the number of nearest neighbor particles, D, the dielectric constant of the

continuous medium (78.54 for water at 25 °C), ε_0, the permittivity of free space (8.854×10^{-12} coul/v m), ψ_d, the Stern potential in mv, a, the particle radius in nm, R, the nearest interparticle distance in nm, $1/\kappa$, the Debye length in nm, A_{121}, the Hamaker constant in Joule, and s, the ratio between R and a.
The second term in the G(∞) equation which accounts for the contribution from the short range van der Waals attraction is negligibly small, whereas the first term which represents the contribution from the long range electrostatic repulsion between particles dominates in the ordered latex system.

Previous studies by reflection spectrophotometry (*10*) have suggested the ordered dispersions with high particle concentrations possess the fcc (face-centered cubic) crystal structure. The nearest interparticle distance R can thus be obtained from $(R/2a)^3 = 0.74/\phi$ by knowing the particle concentration ϕ. The geometric factor α can be calculated as 0.8325 with $\phi_m=0.74$ and n=12.

In dealing with concentrated dispersions, Russel (*18*) made a correction on the Debye length ($1/\kappa_0$) by considering the increased counterion density and the decreased liquid volume due to the presence of particles. The corrected Debye length ($1/\kappa$) is shorter than $1/\kappa_0$ corresponding to the added salt alone. The corrected Debye length can be expressed as :

$$(a\kappa)^2 = \left(\frac{1}{1-\phi}\right)\left[(a\kappa_0)^2 + 3\phi(1+a\kappa_0)\left(\frac{e\psi_d}{kT}\right)\right] \quad (10)$$

The zeta potential ζ obtained from the microelectrophoresis measurements at infinite dilution was used to approximate ψ_d in calculating $1/\kappa$ under given conditions. Shear moduli thus obtained deviate one or two orders of magnitude from the observed Gg. This indicates that the Stern potential used for the calculations might not be realistic. In order to find an appropriate ψ_d, the Gg measured from stress relaxation experiments was plotted against $G(\infty)/\psi_d^2$ calculated from equation (7) in Figure 12. A value of $\psi_d=16.5$ mv could be obtained from the slope of the linear plot by assuming a constant Stern potential.

Comparison of shear moduli between experiment and theory can be made now since every parameter needed for equations (7) and (8) is known. Before doing this, one extra note has to be made. What has been measured experimentally is the storage modulus G' at a frequency of 2 Hz which corresponds to the relaxation modulus obtained on time scale of 0.08 sec. It is obvious from Figure 11 that the value of G' at 2 Hz should be a little below but close to the value of Gg (i.e., G(∞)).

Figure 13 illustrates the effect of sodium chloride concentration on the elastic modulus of latex A with ϕ= 20.70% at 25 °C and compares the experimental G' at 2 Hz with the theoretical G(∞). At low added salt concentrations, a plateau modulus was observed because the total ionic strength, dominated by the counterion density (3rd term in equation (10)), is constant. With further additions of salt, the foreign electrolyte (2nd term in equation (10)) in conjunction with counterions control the ionic strength and compress the Debye length, and hence the elasticity caused by the electrostatic interactions between particles attenuates accordingly. Finally, no elasticity was detected at high salt concentrations. The predicted G(∞), based on the corrected

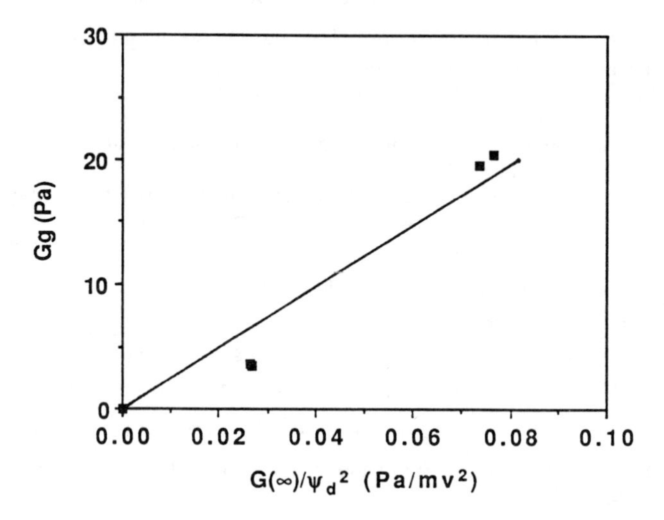

Figure 12. The experimental glassy modulus, Gg, against the calculated $G(\infty)/\psi_d^2$ for various volume fractions and temperatures of the cleaned latex A.

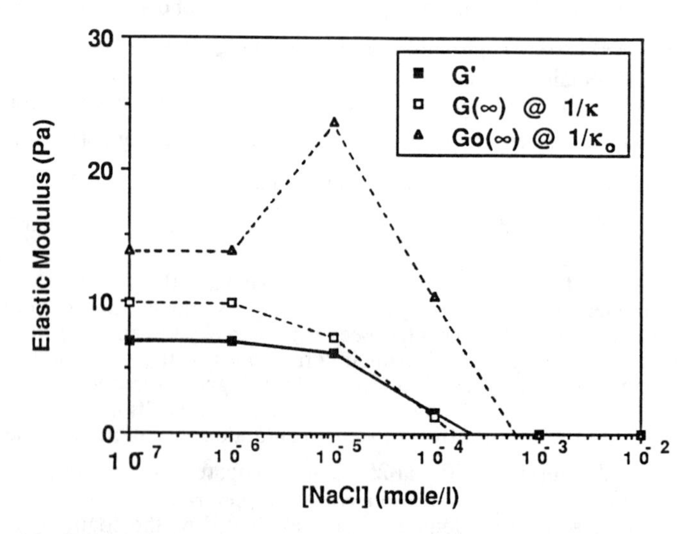

Figure 13. Effect of salt concentration on the elastic modulus of latex A with ϕ = 20.70% at 25 °C; G' at 2 Hz : experimental storage modulus, $G(\infty)$, $G_o(\infty)$: theoretical high frequency limiting modulus calculated by using corrected and uncorrected Debye length, respectively.

Debye length, agrees nicely with the the experimental G'. However, failing to recognize the effect of the counterion on the Debye length leads to an incorrect prediction as shown by the maximum. From now on, the theoretical G(∞) will be calculated by applying the corrected Debye length.

The strength and range of electrostatic repulsion between particles is very sensitive to the counterion valency of the salt. Figure 14 shows the effect of salt valency on the elastic modulus. The theoretical G(∞) agrees not only qualitatively but also quantitatively with the experimental G'.

The effect of temperature on the elastic modulus of the cleaned latex A is shown in Figure 15. Again, the theoretical G(∞) follows satisfactorily the experimental G'.

Figure 16 illustrates the effect of particle concentration on the elastic modulus of the cleaned latex A at 25 °C. Once again, the predicted G(∞) agrees with the experimental data not only in trend but also in magnitude. An increase in particle concentration causes a decrease in particle separation. The interparticle distance can also be adjusted by varying the particle size while keeping the particle concentration constant. From equations (7) and (8), we believe that the shear modulus of ordered latexes will decrease with increasing particle size for a given particle concentration.

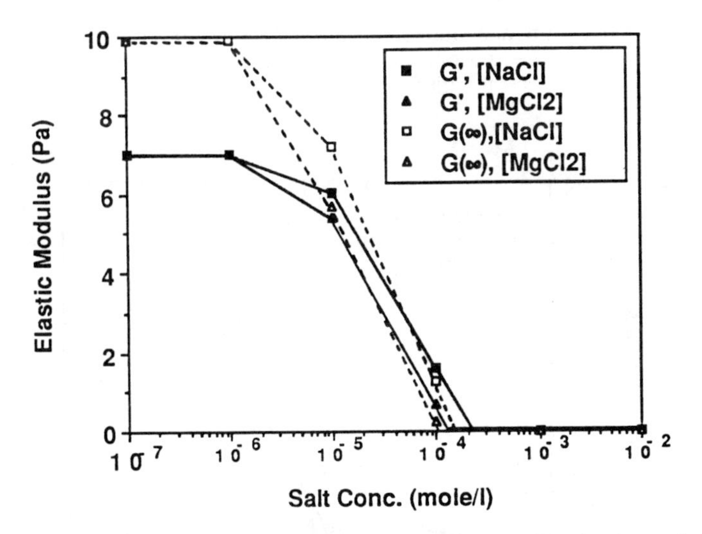

Figure 14. Effect of salt valency and concentration on the elastic modulus of latex A with ϕ = 20.70% at 25 °C; G' at 2 Hz : experimental storage modulus, G(∞) : theoretical high frequency limiting modulus.

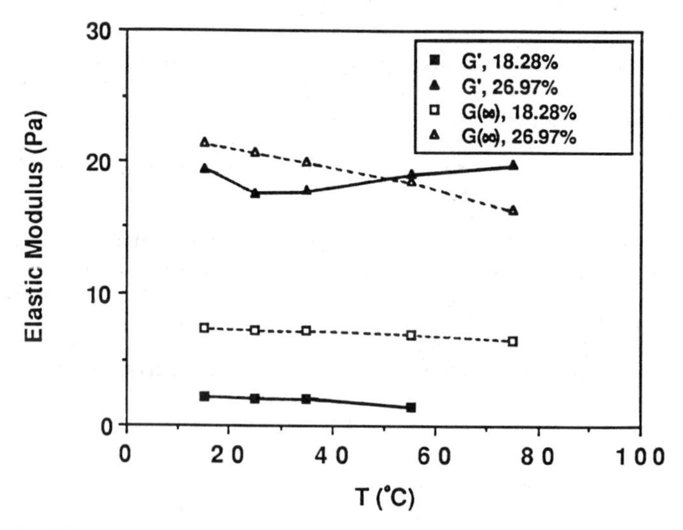

Figure 15. Effect of temperature on the elastic modulus of the cleaned latex A at various particle concentrations (vol.%); G' at 2 Hz : experimental storage modulus, G(∞) : theoretical high frequency limiting modulus.

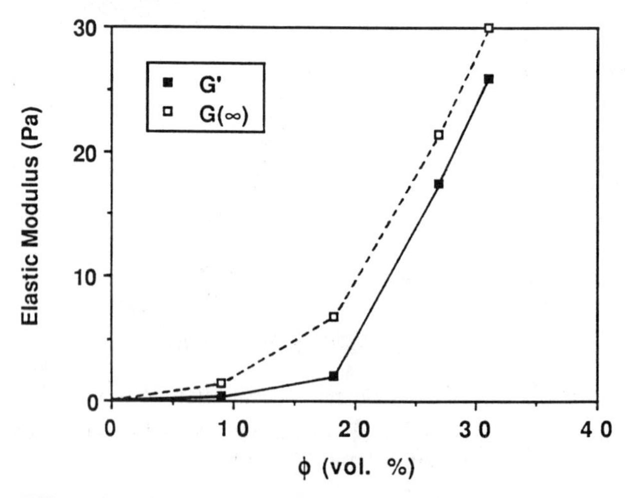

Figure 16. Effect of particle concentration on the elastic modulus of cleaned latex A at 25 °C; G' at 2 Hz : experimental storage modulus, G(∞) : theoretical high frequency limiting modulus.

Literature Cited

1. Hiltner, P.A. and Krieger, I.M., *J. Phys. Chem.* 1969, *73*, 2386.
2. Crandall, R.S. and Williams, R., *Science* 1977, *198*, 293.
3. Buscall, R., Goodwin, J.W., Hawkins, M.W. and Ottewill, R.H., *J. Chem. Soc. Faraday Trans.* 1, 1982, *78*, 2873.
4. Mitaku, S., Ohtsuki,T., Enari, K., Kishimoto, A. and Okano, K., *Japanese J. Appl. Phys.* 1978, *17*, 305.
5. Mitaku, S., Ohtsuki, T. and Okano, K., *Japanese J. Appl. Phys.* 1980, *19*, 439.
6. Dubois-Violette, E., Pieranski, P., Rothen, F. and Strzelecki, L., *J. Physique* 1980, *41*, 369.
7. Okubo, T., *J. Colloid Interface Sci.* 1990, *135*, 259.
8. Benzing, D.W. and Russel, W.B., *J. Colloid Interface Sci.* 1981, *83*, 178.
9. Buscall, R., Goodwin, J.W., Hawkins, M.W., and Ottewill, R.H., *J. Chem. Soc., Faraday Trans.* 1, 1982, *78*, 2889.
10. Lee, Y.P. and Micale, F.J., in *Surface Phenomena and Fine Particles in Water Based Coatings and Printing Technology*, Sharma, M.K. and Micale, F.J., eds., Plenum Press, N.Y., 1991.
11. Ottewill, R.H. and Shaw, J.N., *J. ElectroAnal. Chem.* 1972, *37*, 133.
12. Hiemenz, P.C., *Principles of Colloid and Surface Chemistry*, Marcel Dekker, N.Y., 1977.
13. Krieger, I.M. and Eguiluz, M., *Trans Soc. Rheol.* 1976, *20*, 29.
14. Ferry, J.D., *Viscoelastic Properties of Polymers*, John Wiley & Sons, N.Y., 1980
15. Smith, T.L., *J. Polym. Sci. : Part C* 1971, *35*, 39.
16. Goodwin, J.W., Gregory, T., Miles, J.A. and Warre, B.C.H., *J. Colloid Interface Sci.* 1984, *97*, 488.
17. Lindemann, F.A., *Z. Phys.* 1910, *11*, 609.
18. Russel, W.B., in *Theory of Dispersed Multiphase Flow*, Meyer, R.E., ed., Academic Press, N.Y., 1983, pp 1.
19. Alfrey, J. and Doty, P., *J. Appl. Phys.* 1945, *16*, 700.

RECEIVED December 4, 1991

Chapter 21

Detection and Concentration Measurements of Proteins Adsorbed onto Polystyrene and Poly(styrene–acrolein) Latexes

Stanislaw Slomkowski and Teresa Basinska

Center of Molecular and Macromolecular Studies, Polish Academy of Sciences, Sienkiewicza 112, 90–363 Lodz, Poland

Described is modification of the standard Lowry method, designed for determination of protein concentration in solution, in a manner enabling determination of protein adsorbed onto latex particles. Applicability of this method has been checked for human serum albumin (HSA) and gamma globulin (IgG), and for rabbit gamma globulins (IgGF and IgGFgD, antibodies against human fibrinogen and fragment D of human fibrinogen, respectively) adsorbed onto polystyrene, poly(styrene/acrolein), and poly(styrene/acrolein)·rhodamine 6G latex particles.

The development of a simple and reliable method, enabling detection and measurement of the concentration of proteins immobilized on the surface of latex particles, is important for applications of latexes in the life sciences. Some examples were described recently in monographs edited by Guiot and Couvreur (*1*) and by Rembaum and Tökes (*2*). Papers collected in the mentioned monographs, concentrated mainly on new drug delivery systems, diagnostics, and cell sorting, represent only a small fraction of all papers published within the last decade on applications of latexes in medical, veterinary, and plant disease diagnostics and treatment.

Immobilization of biologically active compounds on latex particles is useful for visualization of reactions involving proteins (e.g. in the simple agglutination tests (*3-8*)) and is helpful for isolation of reaction products, selected macromolecules and cells (*9-19*). Latexes are used also as carriers transporting drugs (often proteins) to organs and as devices regulating the drug release (*20-26*). In biological studies and medical applications, latex particles usually fulfill the auxiliary functions; the main functions are related to the immobilized proteins. Thus, it is essential to have a simple method enabling the easy and fast detection of immobilized proteins and determination of their concentration.

0097–6156/92/0492–0328$06.00/0

Methods Used for Determination of the Concentrations of Immobilized Proteins

Unfortunately, almost all methods described in the literature were developed for the detection and quantification of proteins in solution (*27-28*). The few, which could be used also for immobilized proteins (e.g. radiochemical or Kjeldahl methods) have rather limited applicability.

Radiochemical method can be used only for proteins labelled with radioisotopes (cf examples given in Ref. *29-34*), require special equipment, and access to certified laboratories. The Kjeldahl method, which consists of determination of nitrogen (constituting about 16 wt% of proteins), is not very sensitive (usually more than 5 mg of protein is required in the analyzed sample (*35*)) and can not be used for latexes made from polymers containing nitrogen.

Today, the most common approach is based on determination of the difference between the protein content in the initial solution, used for immobilization, and in the supernatant, obtained after removal of latex particles with attached proteins (*36-42*). The protein content in solution (initial solution or supernatant) could be determined by UV spectroscopy (absorption at 280 nm due to the tryptophan) or by other well known methods (e.g. ninhydrin (*27*), Lowry (*43*), or Bradford (*44*) method). However, determination "by difference" is accompanied with an error that is a combination of the errors of each separate measurement. In cases when immobilization is carried on so that only a small fraction of the initial protein is immobilized, the quantity of adsorbed protein may be within the resulting error. Moreover, one can not always be sure that the whole "missing" protein has been attached to latex, especially when supernatants are additionally purified from the traces of latex particles by filtration.

The difficulties connected with the determination of immobilized proteins "by difference" inspired us to investigate the possibility of adapting the methods designed primarily for measurements in solution to the direct determination of proteins adsorbed onto latex particles. All methods based on dye binding to protein (e.g. the Bradford method) were excluded because measurement of the concentration of dyes, immobilized as complexes with proteins on latex particles, would pose no fewer problems than the direct determination of adsorbed proteins.

The most promising methods were based on reactions of proteins with soluble, low molecular weight compounds yielding soluble products strongly absorbing in the visible region.

Several years ago, Eskamani et al used the ninhydrin method for the determination of proteins immobilized on collagen (*45*). We wanted to check whether the classical Lowry method (*43*), still very often used for determination of proteins in solution and about 100 times more sensitive than the ninhydrin method, could be used also for determination of immobilized proteins.

The Lowry method is based on the reaction of protein (P) with Cu^{2+} cations under basic conditions. At these conditions proteins are present in polyanionic form. To the protein-$(Cu^{2+})_n$ complexes a mixture of phosphomolybdic and phosphotungstic acids is added.

The reactions are schematically illustrated by Equations (1) and (2).

$$P^{2-} + Cu^{2+} \longrightarrow P^{2-} \cdot Cu^{2+} \quad (1)$$

$$3\,P^{2-} \cdot Cu^{2+} + 12MoO_3 \cdot H_3PO_4 + 12WO_3 \cdot H_3PO_4 \longrightarrow$$

$$\longrightarrow 3\,P + Cu_3(PO_4)_2 + 6Mo_2O_5 + 6W_2O_5 \quad (2)$$

where P^{2-} denotes the negatively charged protein segment.

The tungstic and molybdic pentoxides eventually formed in this process strongly absorb in the visible region (maxima at 750 nm). The protein determination by the Lowry method is indirect, depends on the nature of protein, and usually need calibration with standard protein solutions.

The first attempt to use the Lowry method for determination of immobilized proteins was reported by Bunting and Laidler (*46*) for acrylamide supports. However, the results were negative, presumably because of interference by free acrylamide apparently still present in the gel.

We wanted to investigate whether the Lowry method could be used for the direct determination of proteins adsorbed on the surface of polystyrene latex particles, the common polymeric supports for proteins in diagnostic tests, and on the surface of poly(styrene/acrolein) latex particles. It is known that latexes with aldehyde surface groups are convenient carriers because they can immobilize proteins without prior activation (*12-14,47,48*). For immobilization we have chosen proteins which are commonly used in diagnostic tests, namely gamma globulins (human and rabbit), and proteins often used in protein - polymer studies, namely human serum albumins.

Preliminary results of the determination of gamma globulins and albumin at the surface of the polystyrene latexes are reported in a paper (Basinska,T., Slomkowski,S. *J.Biomater.Sci.Polym.Ed.* in press). In this paper we will concentrate on the comparison between detection of proteins immobilized on poly(styrene/acrolein) latex particles, capable of binding proteins covalently and detection of proteins adsorbed onto polystyrene latex particles, for which attachment is only due to hydrophobic interactions. We will also present some preliminary results on the determination of proteins adsorbed on the surface of a latex containing rhodamine G6, a dye used in the synthesis of colored and fluorescent latexes especially suitable for biological studies.

Synthesis of Latexes and their Characterization

The polystyrene latexes were synthesized by emulsifier-free polymerization following the procedure proposed by Okubo et al (*49*). The detailed description of the synthesis and purification of latexes is given in our previous paper (Basinska,T., Slomkowski,S. *J.Biomater.Sci.Polym.Ed.* in press). The average size and size distribution of the latex particles ($\bar{D}_w/\bar{D}_n$, where $\bar{D}_w$ and $\bar{D}_n$ denote the weight and number average diameters) were determined by scanning electron microscopy (JEOL 35C 200 Å apparatus). The concentration of acidic groups on the surface (A^-) was determined by conductometric titration (cf e.g. Ref. *50*). The composition of the polymerization mixture and values of the parameters characterizing the latex particles are summarized in Table I.

Table I. Latexes Used for Protein Adsorption

Reaction mixture[a]	Weight g	$\bar{D}_n$ µm	$\bar{D}_w/\bar{D}_n$	$[A^-]\cdot10^6$ mol/m^2	$[-CHO]\cdot10^6$ mol/m^2
MI		0.52	1.004	3.38	-
Styrene	9.06				
MII		0.36	1.002	2.57	1.76
Styrene	9.06				
Acrolein	0.84				
MIII		0.30	1.30	-	-
Styrene	9.06				
Acrolein	0.50				
Rhodamine 6G	0.017				

[a] 0.044 g $K_2S_2O_8$ and 102 g of H_2O was used in each synthesis.

The poly(styrene/acrolein) latexes were obtained by emulsifier--free radical emulsion-precipitation polymerization. The initial concentrations of monomers and initiator ($K_2S_2O_8$) are given in Table I. The polymerization mixture was prepared in the following way. An emulsion containing the all of the styrene and 92 ml of water was prepared under argon in the homogenizer in a similar manner as in the synthesis of the polystyrene latex. Acrolein (1.0 ml freshly distilled under reduced pressure) was dissolved in 10 ml of deoxygenated water. The polystyrene emulsion and the acrolein solution were mixed in the reaction flask and than initiator ($K_2S_2O_8$) was added. Polymerization was carried on with stirring (300 rpm) under argon at 65°C for 28 hrs. The latexes were isolated and purified in a similar manner as the polystyrene ones.

The average size and size distribution of the poly(styrene/acrolein) latex particles as well as the concentration of negative ions at their surface were determined in the same way as described for the polystyrene latexes.

It is very well known that the synthesis of latexes from mixtures of monomers gives particles with a surface chemical composition different from that in the bulk (*49*). In particular, in the case of aqueous medium polymerization, the fraction of the more hydrophilic monomer is much higher at the surface than, on average, in the bulk polymer. Therefore, we expected that, in the synthesis of latexes from the styrene/acrolein mixtures, the fraction of acrolein monomeric units on the surface would differ from the corresponding fraction in bulk and should be determined in separate experiments.

The determinations of the concentrations of -CHO groups (from acrolein) on the surface of latex particles were performed using 2,4-dinitrophenylhydrazine (DNPH) as an analytical reagent and/or by X-ray photoelectron spectroscopy (XPS). DNPH is very well known as a compound which reacts quantitatively with aldehydes and ketones, yielding the corresponding hydrazones (*51*,*52*).

$$R-\overset{\overset{O}{\|}}{C}-H \; + \; H_2N-NH-C_6H_3(NO_2)_2 \longrightarrow R-CH=N-NH-C_6H_3(NO_2)_2 \; + \; H_2O \qquad (3)$$

The spectra of DNPH and of the hydrazones differ noticeably (λ_{max} equal to 350 and 360 nm for DNPH and hydrazones, respectively). Moreover, for hydrazones, the absorption maximum (λ_{max} = 360) and the extinction coefficient ($\varepsilon(360)$ = 22.000 l/(mol·cm)) are only slightly dependent on the nature of aldehyde (*51*).

The concentration of -CHO groups on the surface of the poly(styrene/acrolein) latex particles was determined according to the following procedure. A known amount of the latex particles in aqueous suspension was added to a saturated solution of DNPH in ethanol. DNHP was used in about 1000 times molal excess with respect to the maximum concentration of aldehyde groups present in the latex. After 5 hrs of mixing, the latex particles were allowed to settle (usually 2 days), and the supernatant containing excess DNPH was separated. The latex was washed several times with fresh portions of ethanol. Washing was repeated until the spectrophotometric measurements indicated that the supernatants are DNPH-free. These latex particles, with DNPH attached to their surface via the aldehyde groups, were isolated and dried under reduced pressure to a constant weight. A known amount of dry latex was dissolved in a known volume of CH_2Cl_2 and the UV spectrum of the solution was registered. In this UV spectrum (Figure 1) the signal due to DNPH immobilized on the latex particles can be clearly seen at 363 nm (spectrum A). For comparison, the spectrum of the polystyrene latex (synthesized without acrolein) treated with DNPH in the same way as the poly(styrene/acrolein) latex is also given in Figure 1 (spectrum B). It is evident that the DNPH is attached covalently only to the latex particles with the aldehyde groups. From the spectrum given in Figure 1 it was possible to calculate the concentration of the aldehyde groups on the surface of latex particles. For the analyzed sample we obtained $2.8 \cdot 10^{-5}$ mol/g (moles of -CHO groups per 1 g of latex) or, since the surface area of the poly(styrene/acrolein) latex particles is equal to 15.87 m^2/g (Table I), $1.76 \cdot 10^{-6}$ mol/m^2.

The XPS spectrum of the poly(styrene/acrolein) latex particles is given in Figure 2. Signals at 285 and at 533 eV correspond to electrons ejected from 1s orbitals of the carbon and oxygen atoms,

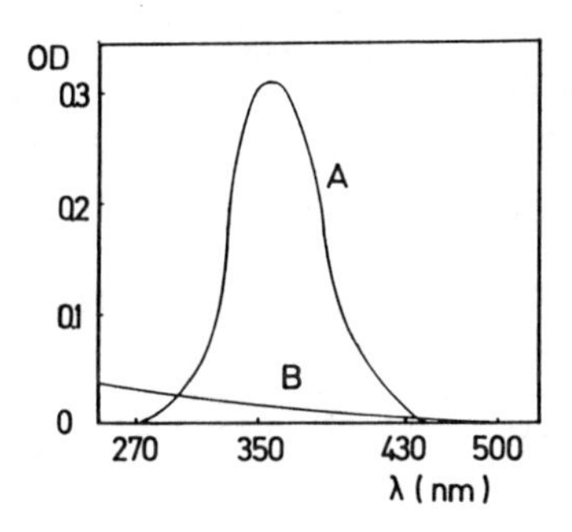

Figure 1. UV spectra of the poly(styrene/acrolein) latex particles with attached DNPH (A) and of polystyrene latexes treated with DNPH (B) and dissolved in CH_2Cl_2. Polymer concentration $5.2 \cdot 10^{-1}$ g/l. Concentration of aldehyde at the surface [-CHO] = $= 2.8 \cdot 10^{-5}$ mol/g ($1.76 \cdot 10^{-6}$ g/m^2).

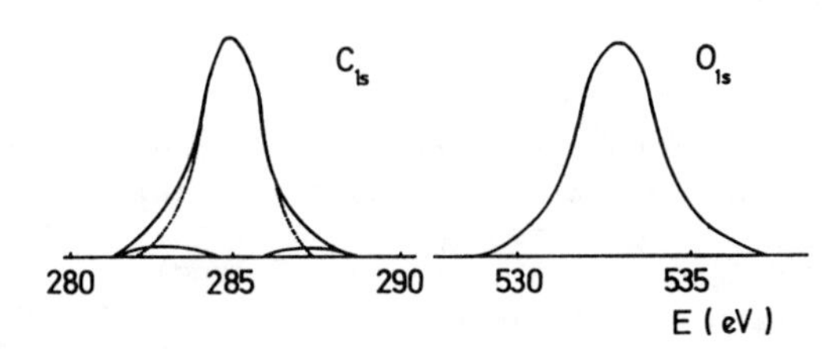

Figure 2. The XPS spectrum of the poly(styrene/acrolein) latex particles.

respectively. The C_{1s} signal is distorted. The high energy shoulder, at 287.5 eV, is due to the C=O carbon atoms (*49*). The low energy shoulder may be due to the differential charge transfer (latex particles are not at the same potential). After resolving this spectrum into components it was possible to calculate the fraction of carbonyl carbon atoms. From the XPS spectrum it follows that the carbonyl carbon atoms constitute 4% of all carbon atoms at the surface. The comparison of the signals due to the oxygen and carbon atoms (the sensitivity ratio of O_{1s}/C_{1s} signals is equal to 3.1) indicates that the ratio of the oxygen atoms to all carbon atoms equals 0.06. The apparent excess of oxygen atoms is presumably due to the oxygen atoms in the sulphate groups introduced from the initiator and/or to traces of water molecules adsorbed onto the surface. The fraction of the carbonyl carbon atoms equal to 0.04 (calculated from the C_{1s} signals) corresponds to 26.7 mol% of acrolein and 73.4 mol% of styrene.

The latex containing rhodamine 6G, a dye strongly fluorescent in the visible region (emission maximum at 598 nm) was obtained in the similar way as the previously described poly(styrene/acrolein) latex. Rhodamine (Aldrich) was dissolved in the acrolein-water solution and 15 min. later was added to the polymerizing mixture. The synthesis, which is very sensitive to pH, has not been optimized and the size distribution of latex particles was broad (Table I). From these latex particles the dye did not migrate into solution even during prolonged storage in water, buffers, and in alcohols.

The concentrations of latex particles in suspensions were determined turbidimetrically. The dependence of OD on the wavelength, measured for the polystyrene latex ($\overline{D}_n$ = 620 nm), is given as an example in Figure 3. The calibrations are shown in Figure 4. For concentrations below $2.5 \cdot 10^{-5}$ g/ml the relationships between OD and concentrations are represented by the straight lines; at higher concentrations, deviations from linearity make the turbidimetric determinations difficult.

Proteins

In our studies we used laboratory grade HSA (Cohn fraction V, Sigma) without further purification and commercial HSA purified from fatty acids and lyophilized (products of the Central Laboratory of Sera and Vaccines (CLSV), Warsaw, Poland).

Chromatograms of HSA from Sigma and from CLSV are given in Figure 5. From a comparison of these chromatograms it appears that the fractions of monomer, dimer, and polymer (quaternary structure) in these albumins are similar.

Human gamma globulin (Cohn fractions II and III, Sigma) was used without purification.

Rabbit gamma globulins were antibodies against human fibrinogen, and against the fragment D of human fibrinogen (IgGF, and IgGFgD, respectively). IgGF and IgGFgD were obtained from Prof. C.Cierniewski, Medical Academy, Lodz, Poland, who isolated these proteins from the blood sera of immunized rabbits and purified them by affinity chromatography on agarose gel with immobilized fibrinogen and/or FgD.

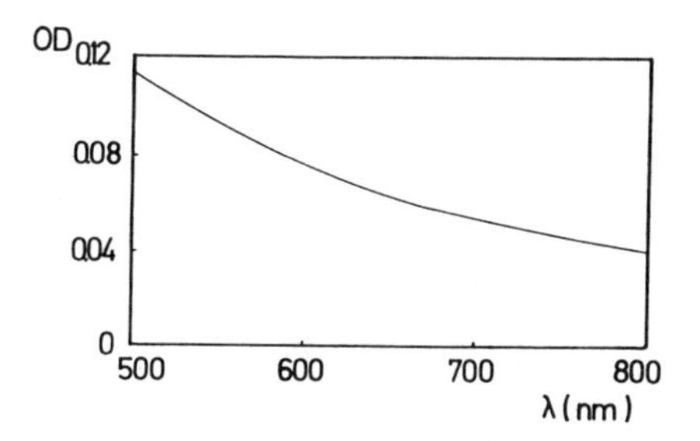

Figure 3. Dependence of OD on the wavelength for the aqueous polystyrene latexes (MI) ($\overline{D}_n$ = 620 nm). Concentration of latex particles $7.74 \cdot 10^{-6}$ g/ml.

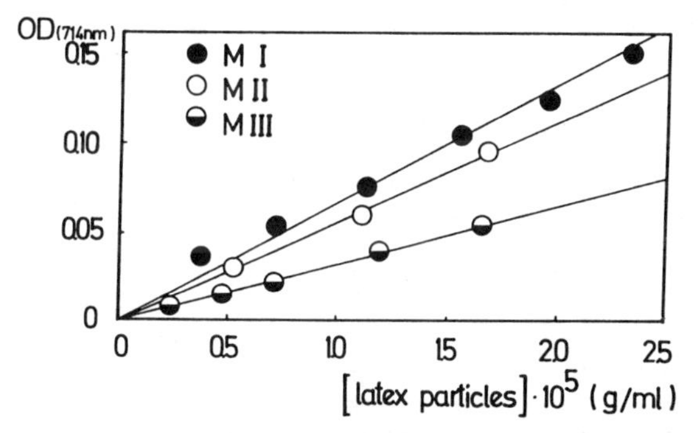

Figure 4. The dependence of OD(714) on the concentration of latexes (slopes in parentheses): ● polystyrene (MI) ($6.46 \cdot 10^3$ ml/g), ○ poly(styrene/acrolein) (MII) ($5.54 \cdot 10^3$ ml/g), ◒ poly(styrene/acrolein)·(rhodamine 6G) (MIII) ($3.38 \cdot 10^3$ ml/g)).

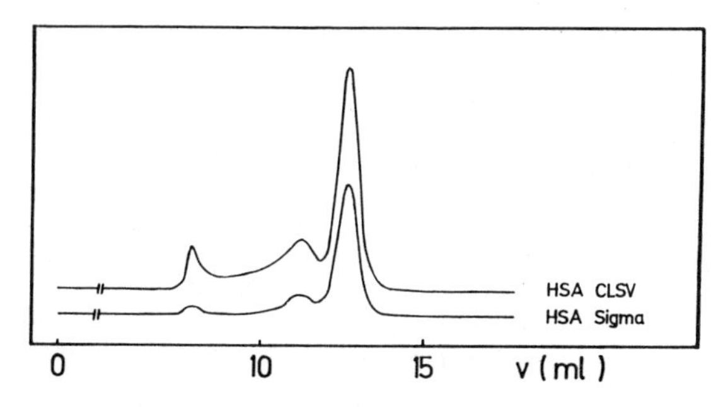

Figure 5. FPLC chromatograms of HSA from Sigma and CLSV. Conditions: column = FPLC Superose 12, eluent = phosphate buffer (pH = 7.0), flow rate = 1 ml/min, UV detection at 280 nm, sensitivity = 0.1.

Adsorption of Proteins

Proteins were adsorbed onto the latex particles by gentle mixing of a protein dissolved in phosphate buffered saline (PBS, pH 7.4) with the aqueous latex. The adsorption was carried out at 20°C for 24 hrs. In a typical experiment 9 ml of the protein solution and 1.0 ml of the latex were used. The concentrations of proteins in the mixtures were varied from $1.0 \cdot 10^{-5}$ to $4.8 \cdot 10^{-4}$ g/ml and the concentrations of latexes were varied from $3.0 \cdot 10^{-3}$ to $1.6 \cdot 10^{-2}$ g/ml.

The reference solution (PI) was obtained by diluting the protein stock solution with an amount of water equal to that of the latex.

Determination of the Proteins Adsorbed onto Latexes

The Lowry reagents used for the protein determination, i.e., the alkaline copper (AC) and phosphotungstic and phosphomolybdic acids (Folin-Ciocalteau reagents) solutions were prepared according to the well known procedures (*53*).

The concentrations of proteins in the initial solutions were determined by the standard Lowry procedure (*43*). An example of the visible spectrum of the tungstic and molybdic pentoxides obtained by treating the HSA solution with the Lowry reagents is shown in Figure 6. Similar spectra were also obtained for the other proteins studied. The dependence of OD(750), determined from these spectra, on the protein concentration in solution (cf Table II) gave the calibration lines shown in Figures 9 and 10. For HSA the slope and intercept were equal to $3.84 \cdot 10^{3}$ and 0.012, respectively; for gamma globulins the slope was equal to $4.731 \cdot 10^{3}$ and intercept to 0.097. In the latter case the intercept was high because corrections for blank absorption were not introduced.

Experiments with adsorbed proteins were performed in the following way. A sample containing latex with adsorbed and free protein (in the cases when protein was used in an excess) was centrifuged (11.000xG) and the first supernatant (SI) was isolated. The pelletted latex particles were washed four times with fresh portions of PBS (each time resuspended and centrifuged) and the supernatants SII, SIII, SIV, and SV were collected. Finally, the latex particles were resuspended in PBS (suspension denoted as SU).

The protein content in the reference PI solution and in each supernatant was determined according to the Lowry procedure. Supernatants SIV and SV were free from protein, i.e. their spectra did not differ from the blank obtained by treating the PBS buffer without protein with the Lowry reagents. The well dispersed SU suspension, free from unbound protein, was treated with the Lowry reagents in a similar manner to the protein solutions. Thirty minutes after addition of the Lowry reagents, latex particles were removed by centrifugation (13.000xG) and the spectrum of the supernatant, shown in Figure 7, was measured. Latexes without adsorbed protein were analyzed in the same manner and these spectra were used as blanks.

Usually it was difficult to isolate 100% of latex particles by centrifugation at 13.000xG. On the other hand, for centrifugal forces over 13.000xG it was difficult to resuspend the pelletted latex particles. Thus, some traces of particles in supernatants interfered with the absorptions of the solution. This problem was solved by the least square method enabling separation of any analyzed spectrum (for the least square calculations absorbances denoted as OD_{AN}) into the "absorbance" and "turbidity" components. The spectra for protein in solution (for calculations denoted as OD_A) and for the latex (denoted as OD_B) were used as references. Examples of these spectra are given in Figures 6 and 3, respectively. For the fitting procedure spectra were converted into the tabular form (λ_i, $OD_{AN}(\lambda_i)$, $OD_A(\lambda_i)$, $OD_B(\lambda_i)$). The best fit is achieved for parameters a and b for which S

$$S = \Sigma(OD_{AN}(\lambda_i) - a\cdot OD_A(\lambda_i) - b\cdot OD_B(\lambda_i))^2 \qquad (4)$$

reaches minimum. This requires $\partial S/\partial a = \partial S/\partial b = 0$. Differentiation and rearrangements give:

$$a\cdot\Sigma OD_A(\lambda_i)^2 + b\cdot\Sigma OD_A(\lambda_i)OD_B(\lambda_i) = \Sigma OD_A(\lambda_i)OD_{AN}(\lambda_i)$$

$$a\cdot\Sigma OD_A(\lambda_i)OD_B(\lambda_i) + b\cdot\Sigma OD_A(\lambda_i)^2 = \Sigma OD_A(\lambda_i)OD_{AN}(\lambda_i) \qquad (5)$$

Solution of the above set provides the required values for a and b. Thus, in the analyzed case, $a\cdot OD_A$ and $b\cdot OD_B$ at each particular wavelength represent the "absorbance" and "turbidity" components, respectively. An example of the experimentally measured complete spectrum, the calculated "turbidity" component, and of the corrected pure "absorbance" spectrum due to the protein is shown in Figure 8.

The described fitting procedure can be easily extended for more than two reference spectra. In this case the new terms should be introduced into Equation (4) (e.g. $c\cdot OD_C(\lambda_i)$, $d\cdot OD_D(\lambda_i)$, ...). The condition $\partial S/\partial a = \partial S/\partial b = \partial S/\partial c = \partial S/\partial d = ... = 0$ (required for the best fit) leads to the set of equations enabling calculations of a, b, c, and d and eventually, to determination of the contribution of each component.

The calibration dependencies, i.e. dependencies of absorptions registered for solutions with known concentrations of proteins treated with the Lowry reagents, were used to determine the protein concentrations in the starting solution and in the supernatants. For the supernatants, the estimate "absorbance" component of the spectra was used. The difference between the protein content of the PI solution and of the first supernatant (SI) was assumed to be equal to the amount of protein adsorbed onto the latex particles (determination "by difference"). Parallel, the "absorbance" components from the spectra for supernatants, obtained after centrifugation of the reaction mixture containing SU (suspension of latex particles free from unbound protein) with added Lowry reagents, were also determined. The plots of OD(750), due to W_2O_5 and Mo_2O_5 formed in SU, versus the corresponding concentrations of adsorbed protein determined "by difference" are shown in Figures 9 and 10. Values of protein concentration and OD(750) are collected in Table II.

We performed several adsorption experiments with the ratio of latex particles to protein concentrations higher than 50 g/g,

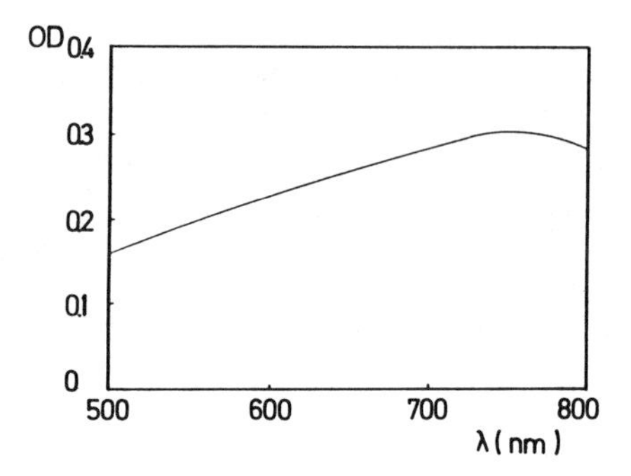

Figure 6. UV spectrum of HSA (Sigma) solution treated with the Lowry reagents; [HSA] = $8.93 \cdot 10^{-5}$ g/ml.

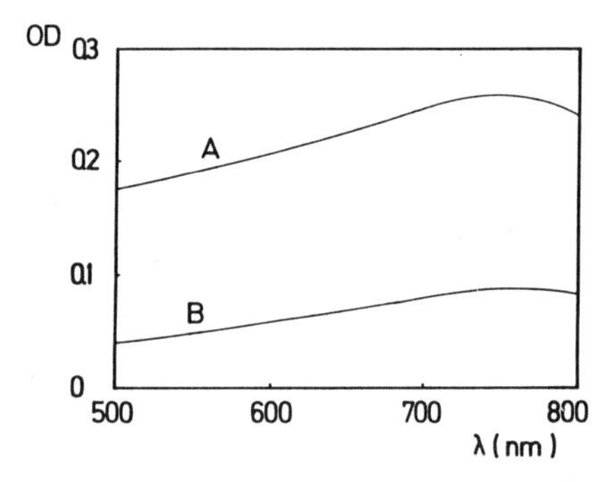

Figure 7. UV spectra of supernatants obtained after treating (A) polystyrene latexes (MI) with adsorbed human gamma globulin (Sigma) and (B) MI microspheres without protein with the Lowry reagents. Concentrations of latex particles = $3.43 \cdot 10^{-3}$ g/mol. Concentration of gamma globulin used for adsorption = $2.84 \cdot 10^{-5}$ g/ml.

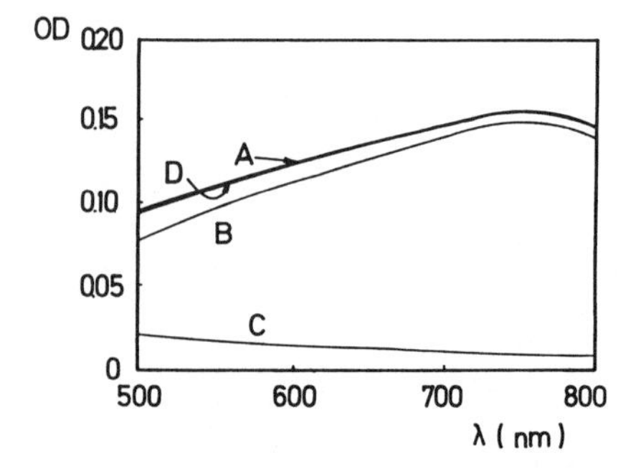

Figure 8. Adsorption of HSA on polystyrene latex particles (MI): (A) UV spectrum of supernatant obtained after treating latex with adsorbed HSA (Sigma) with the Lowry reagents ([latex particles] = $4.5 \cdot 10^{-3}$ g/ml, [HSA] = $9.96 \cdot 10^{-3}$ g/g); (B) calculated contribution of absorption and (C) of the light scattering, (D)-- sum of the spectra (B) and (C).

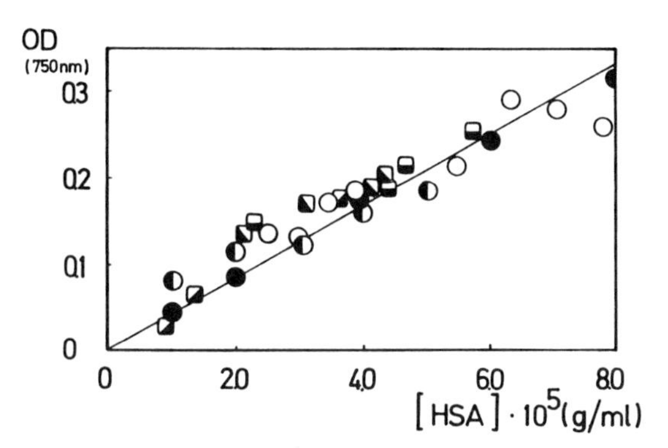

Figure 9. Plot of OD(750) versus concentration of HSA. Points for the calibration line - ● HSA(Sigma) in PBS. Experiments with partial protein adsorption, protein concentration determined "by difference": ○ HSA(CLSV) on latex MI. Experiments with complete protein adsorption, protein concentration equal to the concentration of protein added: ◪ HSA(Sigma) on latex MI; ◐ HSA(CLSV) on latex MI; ⬓ HSA(Sigma) on latex MII; ◩ HSA(Sigma) on latex MIII.

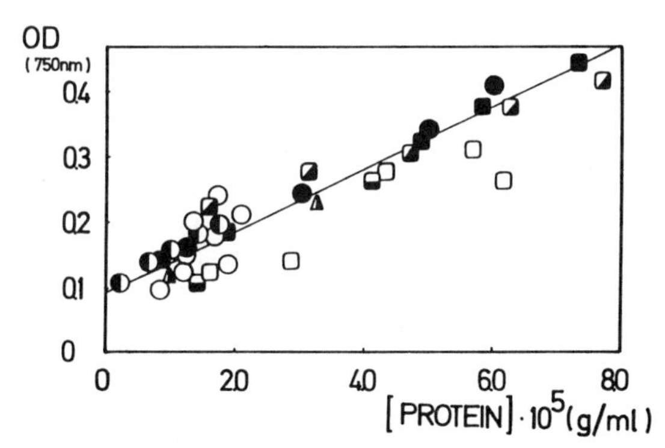

Figure 10. Plot of OD(750) versus concentration of gamma globulins. Points for the calibration line - ■ IgGF, ● IgGFgD in PBS. Experiments with partial protein adsorption, protein concentration determined "by difference": □ IgGF on latex MI; ○ IgGFgD on latex MI. Experiments with complete protein adsorption, protein concentration equal to the concentration of protein added: ◩ IgGF on latex MI; ◐ IgGFgD on latex MI; ⬓ IgG(Sigma) on latex MII; ▲ IgG(Sigma) on latex MIII.

Table II. Protein Concentration and OD(750) for Spectra Obtained after Reaction the Lowry Reagents with Protein in Solution and/or Adsorbed onto Latex Particles

Type of latex	Protein	Protein concentration·10^5 g/ml	OD(750)[a] (1 cm cell)
Human serum albumins			
Protein in solution (for calibration)			
	HSA(Sigma)	1.00	0.045
		2.00	0.085
		3.00	0.130
		4.00	0.170
		6.00	0.247
		8.00	0.320
		10.00	0.390
Protein on latex particles (protein concentration determined "by difference")			
MI	HSA(CLSV)	2.55	0.135
		3.28	0.167
		3.80	0.163
		3.92	0.183
		5.44	0.216
		6.43	0.290
		7.08	0.176
		7.85	0.253
Protein on latex particles Experiments with complete protein adsorption (protein concentration equal to the concentration of protein introduced)			
MI	HSA(CLSV)	1.00	0.081
		2.00	0.120
		3.00	0.127
		4.00	0.158
		5.00	0.184
MI	HSA(Sigma)	2.46	0.138
		3.15	0.167
		4.19	0.196
		4.20	0.202
MII	HSA(Sigma)	2.37	0.140
		4.31	0.194
		4.84	0.214
		5.82	0.256

Table II. (continued)

Type of latex	Protein	Protein concentration·10^5 g/ml	OD(750)[a] (1 cm cell)
MIII	HSA(Sigma)	0.97	0.123
		1.26	0.148
		3.55	0.176
	Gamma globulins		
	Protein in solution (for calibration)		
	IgGF	0.98	0.130
		1.96	0.182
		4.90	0.326
		5.88	0.370
		7.35	0.430
		9.80	0.531
	IgGFgD	1.20	0.153
		3.00	0.249
		4.80	0.342
		6.00	0.405
		9.00	0.540
	Protein on latex particles (protein concentration determined "by difference")		
MI	IgGF	0.75	0.075
		1.52	0.122
		2.93	0.139
		4.34	0.289
		5.69	0.303
		6.14	0.268
	IgGFgD	0.82	0.098
		1.02	0.140
		1.24	0.147
		1.38	0.200
		1.59	0.227
		1.72	0.165
		1.98	0.123
		2.05	0.202

Continued on next page

Table II. (continued)

Type of latex	Protein	Protein concentration·10^5 g/ml	OD(750)[a] (1 cm cell)
	Protein on latex particles Experiments with complete protein adsorption (protein concentration equal to the concentration of protein introduced)		
MI	IgGF	1.54	0.220
		3.08	0.280
		4.62	0.309
		6.16	0.360
		7.70	0.411
	IgGFgD	0.36	0.102
		0.74	0.140
		1.08	0.149
		1.44	0.172
		1.80	0.188
MII	IgG(Sigma)	1.41	0.119
		4.13	0.257
MIII	IgG(Sigma)	1.00	0.135
		3.25	0.235

[a] For experiments with protein on latex particles OD(750nm) was determined from the "absorbance" component of the spectra obtained after treating suspension with the Lowry reagents and after subsequent removal of the main part of latex particles by centrifugation. For HSA absorptions were corrected for blank. For gamma globulins absorptions were not corrected.

ensuring that all of the protein initially in solution was attached to latex particles. The completeness of protein adsorption was proven by absence of the protein in supernatant obtained after centrifugation of the sample of such suspension. Because the mentioned suspensions did not contain unbound protein they did not require washing prior to reaction with the Lowry reagents. Analysis of the spectra registered after treatment of the suspensions with the Lowry reagents (similar as for SU) gave the "absorbance" components. Values of OD(750) of these "absorbance" components are given in Table II and are plotted in Figures 9 and 10 versus the corresponding concentrations of proteins adsorbed onto latex particles. In the discussed experiments, with the complete protein adsorption, concentration of adsorbed protein was assumed to be equal to the concentration of protein introduced to suspension.

Figures 9 and 10 indicate that absorbances for adsorbed proteins are close to the corresponding lines obtained with unbound ones. As a measure characterizing deviation of experimental points from the calibration line we used parameter s (the average residual):

$$s^2 = \sum_{i=1}^{n} (OD_i(calc) - OD_i)^2/n \qquad (6)$$

where n is equal to the number of experimental points, $OD_i(calc)$ and OD_i denote the optical densities calculated from the calibration dependence and measured, respectively for a given protein concentration.

For experiments with concentration of adsorbed protein determined "by difference" s = 0.048 for HSA and 0.056 for gamma globulins. For these experiments the correlation coefficient between OD(750 nm) and protein concentration was equal to 0.738 and 0.813 for HSA and gamma globulins, respectively.

For experiments with complete protein adsorption, i.e. for which concentration of adsorbed protein was equal to the well known concentration of protein introduced to latex, s = 0.037 for HSA and 0.028 for gamma globulins. The correlation coefficient (OD(750 nm) versus protein concentration) was equal to 0.896 and 0.964 for HSA and gamma globulins, respectively. Thus, the direct treatment of protein adsorbed onto latex particles with the Lowry reagents and the measurement of OD(750 nm) for resulting supernatant can be used for determination of the adsorbed protein concentration. This approach, in distinction from the method "by difference", we call the "direct" determination. The lower limit of "direct" determination (corresponding to OD(750 nm) = 0.025 in 1 cm cell) is equal to 5 μg of adsorbed protein in 1 ml of the latex.

Conclusions

Results indicate that the reagents used in the Lowry method react in the same way with proteins in solution as with the proteins adsorbed onto the surface of the polystyrene, poly(styrene/acrolein), and poly(styrene/acrolein)·(rhodamine 6G) latexes. Apparently, any changes in the protein conformation accompanying the hydrophobic adsorption (on the polystyrene latex particles) and/or immobilization (on the poly(styrene/acrolein) latex particles) via the Schiff base linkages do not decrease the access of the Lowry reagents to the polypeptide segments and do not change the reactivity of these segments with these reagents. Calibrations obtained for proteins in solution can be used for the direct determination of the concentration of proteins adsorbed onto either of these surfaces by hydrophobic interactions and/or (for poly(styrene/acrolein) latexes), additionally immobilized via the Schiff base linkages.

It is important to note that the "direct" determination is more accurate than the still commonly used determination "by difference". This can be concluded from the following reasoning. The correlation of OD(750 nm), which can be used for the "direct" determination of the concentration of adsorbed protein, with well defined protein concentration (experiments with complete protein adsorption) is higher than for protein concentration determined "by difference". Moreover, the average deviation of experimental points from calibration line is lower in the former case.

The "direct" determination enables determination of adsorbed proteins also in the case of commercial diagnostic formulations when often information on the protein concentration is not given by the producer.

The incorporation of rhodamine 6G into the latex does not affect the direct determination of proteins adsorbed onto the surface of the latex particles.

Authors express their gratitude to prof. C.Delamar from the Paris VII University for the XPS analysis.

The work was supported by the M.Sklodowska-Curie Fund, Grant PAN/NIST-90-42.

Literature Cited

1. *Polymeric Nanoparticles and Microspheres*; Guiot,P.; Couvreur,P., Eds.; CRC Press: Boca Raton, Fl, 1986.
2. *Microspheres: Medical and Biological Applications* ; Rembaum,A.; Tökes,Z., Eds.; CRC Press: Boca Raton, Fl, 1988.
3. Rembaum,A.; Yen,S.-P.S. *J.Macromol.Sci.-Chem.* **1979**, *A13* , 603.
4. Gerber,M.A.; Spadaccini,L.J.; Wright,L.L.; Deutsch,L. *J.Pediatr.* **1984**, *105*, 702.
5. Krambovitis,E.; Illmurray,M.B.; Loch,P.E.; Hendricksen,W.; Holzel,H., *Lancet*, **1984**, 1229.
6. Anderson,J.; Ehrnst,A.; Larsson,P.H.; Hedlung,K.O.; Norrby,E. *J.Infect.Dis.* **1987**, *156*, 928.
7. Riggin,C.H.; Beltz,G.A.; Hung,C.H.; Thorn,R.M.; Marciani,D.J. *J.Clin.Microbiol.* **1987**, *25* , 1772.
8. Johnson,J.; Duffy,K.; New,L.; Holliman,R.E.; Chesum,B.S.; Fleck, D.G. *J.Clin.Pathol.* **1989**, *42* , 536.
9. Lim,R.W.; Molday,R.S.; Huang,H.; Yen,S.-P.S. *Biochem.Biophys.Acta* **1975**, *349* , 377.
10. Terman,D.S.; Tavel,T.; Petty,D.; Tavel,A.; Harbeck,R.; Buffaloe, G.; Carr,R. *J.Immunol.* **1976**, *116* , 1337.
11. Rembaum,A.; Dreyer,W.J. *Science* , **1980**, *208* , 364.

12. Margel,S.; Rembaum,A. *Macromolecules*, **1980**, *13*, 19.
13. Marens,L.; Offarim,M.; Margel,S. *Biomat., Mar.Dev.,Art.Org.* **1982**, *10*, 157.
14. Margel,S.; Beitler,U.; Offarim,M. *J.Cell.Sci.* **1982**, *56*, 157.
15. Vartdal,F.; Kvalkheim,G.; Lea,T.E.; Bosnes,V.; Gaudernack,G.; Ugelstad,J.; Albrechtsen,D. *Transplantation*, **1987**, *43*, 366.
16. Burg,K.; Mauz,O.; Noetzel,S.; Sauber,K. *Angew.Makromol.Chem.* **1988**, *157*, 105.
17. Nustad,K.; Danielsen,H.; Reith,A.; Funderus,S.; Lea,T.; Vartdal, F.; Ugelstad,J., p.53 in Ref.2.
18. Platsoucas,C.D.; Chae,F.H.; Collins,N.; Kernan,N.; Rembaum,A.; Good,R.A.; Reilly,R.O.; Ugelstad,J. p.89 in Ref.2.
19. Anderson,I.C.; Shpall,E.J.; Leslie,D.S.; Nuystad,K.; Ugelstad, J.; Peters,W.P.; Bast, Jr., R.C. *Cancer Res.* **1989**, *49*, 4659.
20. Ichihara,I.; Sakamoto,K.; Mori,K.; Akagi,M. *Cancer Res.* **1989**, *49*, 4357.
21. Oppenheim,R.C., p.1 in Ref.1.
22. Couvreur,P.; Grislain,L.; Lenaerts,V.; Brasseur,F.; Guiot,P., p.27 in Ref.1.
23. Gurny,R., p.127 in Ref.1.
24. Benoit,J.-P.; Puisieux,F., p.137 in Ref.1.
25. Tice,T.R.; Labrie,F.; McRae-Degueurce,A.; Dillon,D.L.; Mason, D.W.; Gilley,R.M. *ACS Polym.Prep.* **1990**, *31* (*2*), 185.
26. Fitsch,R.M.; Scholsky,K.M.; p.101 in Ref.2.
27. Plummer,D.H. *An Introduction to Practical Biochemistry;* McGraw-Hill: London, 1978.
28. Mikes,O. *High-Performance Liquid Chromatography of Biopolymers and Bioligomers*; Elsevier, Amsterdam, 1988; part B, chapter 7.5.
29. Yalow,R.S. *Science*, **1978**, *200*, 1239.
30. Bohnert,J.L.; Horbet,T.A. *J.Colloid Interface Sci*. **1986**, *111*, 363.
31. vanDulm,P.; Norde,W. *J.Colloid Interface Sci*. **1983**, *91*, 248.
32. Jönsson,U.; Lundström,I.; Rönberg,I. *J.Colloid Interface Sci.* **1987**, *117*, 127.
33. Lahav,J. *J.Colloid Interface Sci*. 1987, *119*, 162.
34. Jenissen,H. *J.Colloid Interface Sci*. **1986**, *111*, 570.
35. Clarke,H.T.; Haynes,B. *The Handbook of Organic Analysis;* Edward Arnold: London, 1975; p.233.
36. Gabel,D.; Axen,R. *Methods in Enzymology*; Mosbach,K., Ed.; Academic Press: New York, N.Y., 1976, vol.44; p.383.
37. Soderquist,M.E.; Walton,A.G. *J.Colloid Interface Sci.* **1980**, *75*, 386.
38. Tamai,H.; Fujii,A.; Suzawa,T. *J.Colloid Interface Sci.* **1987**, *118*, 176.
39. Bagchi,P.; Birnbaum,S.M. *J.Colloid Interface Sci*. 1981, *83*, 460.
40. Beddows,C.G.; Gothrie,J.T. *J.Appl.Polym.Sci*. **1988**, *35*, 135.
41. Norde,W.; McRitchie,F.; Nowicka,G.; Lyklema,J.; *J.Colloid Interface Sci*. **1986**, *112*, 447.
42. Sandwick,R.K.; Schray,K.J. *J.Colloid Interface Sci.* **1988**, *121*, 1.

43. Lowry,D.H.; Rosenbrough,N.J.; Farr,A.L.; Randall,R.J. *J.Biochem.* **1951**, *193*, 265.
44. Bradford,M.M. *Anal.Biochem*. **1976**, *72*, 248.
45. Eskamani,A.; Chase,T.; Freudenberger,J.; Gilbert,S.G. *Anal. Biochem*. **1974**, *57*, 412.
46. Bunting,P.S.; Laidler,K.J. *Biochemistry*, **1972**, *11*, 4477.
47. Margel,S. *J.Polym.Sci.,Polym.Chem. Ed*. **1984**, *22*, 3521.
48. Chang,M.; Colvin,M.; Rembaum,A. *J.Polym.Sci.,Polym.Lett. Ed.* **1986**, *24*, 603.
49. Okubo,M.; Yamamoto,Y.; Kamei,S. *Colloid Polym.Sci.* **1986**, *267*, 603.
50. Vanderhoff,J.W. *Pure Appl.Chem*. **1980**, *52*, 1263.
51. Lohman,F.H. *Anal.Chem*. **1958**, *30*, 972.
52. Cheronis,N.D.; Ma,T.S. *Organic Functional Group Analysis by Micro and Semimicro Methods*; Interscience: New York, N.Y. 1964; p.145.
53. Folin,O.; Ciocalteu,V. *J.Biol.Chem*. 1927, *73*, 627.

RECEIVED December 4, 1991

Chapter 22

Synthesis, Analysis, and Immunodiagnostic Applications of Polypyrrole Latex and Its Derivatives

P. J. Tarcha, D. Misun, D. Finley, M. Wong, and J. J. Donovan

Abbott Laboratories Diagnostics Division, North Chicago, IL 60064–3500

Poly(pyrrole) latex particles possess several unique properties in regard to their use as solid phase supports for immunoassays. Firstly, they are intensely colored black presumably due to their free radical nature, which is delocalized throughout the extensively conjugated chain. The particles, with immobilized protein on their surfaces, serve not only as a support, but as an easily visualized protein label. Secondly, the surface can be modified with reactive groups, which provide for covalent linkage of the appropriate biomolecules.

Chromatography-based immunoassays were demonstrated in the useful clinical ranges for hepatitis B surface antigen, AIDS antibody, and the pregnancy marker human chorionic gonadotropin.

Visual colloidal indicators have been used in immunodiagnostics for decades, the most well known being the slide agglutination test. In this application, the particles are coated with antibodies, antigens, or other ligand-binding molecules which bind the analyte of interest. The presence of the analyte causes multiple inter-particle associations to form, causing the suspension, originally smooth in appearance, to become grainy or clumped when agglutination occurs. Gold colloid coated with antibodies or antigens have also been used to prepare agglutination tests, where analyte-induced agglutination causes the sol to change in color (*1*).

Visual immunoassay tests in various other configurations have been devised employing latex particles which contain absorbed dye (*2*) or functional groups which react with dyes (*3*) to enhance their detectability. A typical test involves coating a strip of nitrocellulose or cellulose acetate with a ligand that specifically binds to the analyte of interest. The strip can be overcoated with other components to prevent non-specific binding of protein to the strip. A sample which may contain the analyte of interest is added. After some time the strip is washed and the colored or colorable latex, which contain surface immobilized ligand specific for the analyte, is added. An appropriate time is given to allow the indicator latex to bind to the strip. Colorable latexes require an additional step in that a component is added which reacts with the latex to render it colored. In both cases the presence or absence of analyte is

0097–6156/92/0492–0347$06.25/0

determined by observing the presence or absence of color visually or with the aid of a microscope.

Poly(pyrrole) latex is uniquely suited for use in visual immunoassay applications (*4*) because of its intense, intrinsic black color, presumably due to electronic transitions within the conjugated chain structure. This optical absorbance is very strong ($e_{max} \sim 10^5$ l mol^{-1}cm^{-1}). It has several advantages over dyed latexes in that this level of optical absorbance cannot be approached by absorbing dyes into a pre-existing latex particles. Furthermore, the latex can be surface-functionalized to permit covalent linkage of appropriate biomolecules to the particle surface.

The purpose of this study was to develop methods for preparing, purifying, and characterizing poly(pyrrole) latex suitable for use as a visual label in immunoassays. Methods for producing functional groups on the particles' surfaces were developed and both passive adsorption and covalent linkage of model ligand binding molecules to the particles' surfaces were studied. Several model immunoassays were demonstrated.

Materials and Methods

Ferric chloride hexahydrate, polyvinyl alcohol (124,000 - 186,000, 89% hydrolyzed), bromoacetyl bromide, and triethylene tetramine were obtained from Aldrich and were used without further purification. Pyrrole was also obtained from Aldrich and was distilled under reduced pressure and stored at < 4 C^o under nitrogen until used. Thioglycolic acid was obtained from Sigma and used as received. Highly purified N-methyl pyrrolidone was obtained from Burdick and Jackson and used as received. Dialysis of the polypyrrole latex was done using 25 mm diameter Spectra/Por 4 dialysis membrane, 12,000 - 14,000 molecular weight cut-off. Ion-exchange was done using a mixed bed prepared from equal amounts with respect to exchange capacity of Dowex 1X8 anion (1.5 mEq/ml) and Dowex 50WX8 cation (2.2 mEq/ml). Each individual resin was washed by the method of Vanderhoff et al. (*5*). Antibodies and antigens were produced by Abbott Laboratories Diagnostics Division (ADD) for use in commercial immunodiagnostic kits. Rabbit antibody to digoxin was affinity purified with an acetylouabain-trisacryl column (*6*). An aliquot was radio-iodinated by ADD by standard methods employing chloramine T (*7*), purified by Sephadex G-50 column chromatography, and used within 30 days of preparation. Tracer amounts of iodinated antibody were mixed with unlabeled antibody for use in coupling studies. Nitrocellulose strips (5 mm pore size) were obtained from Schleicher and Schuell, Keene, NH.

Analysis of particle diameter was done on a Phillips CM 12 transmission electron microscope using formvar-coated grids and carbon shadowing. The mean average diameters were determined from the measurement of at least 100 particles from each sample. Pre-weighed aliquots of purified, surface-modified latexes were titrated with standard 0.01 N NaOH or HCl using a Metrohm 655 Dosimat digital buret and a YSI model 32 conductivity meter with a 0.1 cell constant dip-type cell. The meter was attached to a strip chart recorder. Zeta potential values were obtained on a DELSA Model 440 zeta meter manufactured by Coulter, Hialeah, FL. Centrifugations and washings were done on a Sorvall refrigerated centrifuge using Corex glass tubes and a SS-34 rotor. Radioactivity was measured on an Abbott Laboratories Auto-Logic gamma counter. Infrared spectra were obtained with a Nicolet 5SXC FT-IR spectrometer using standard KBr pellet techniques.

Synthesis of Poly(pyrrole) Latex . Poly(pyrrole) latex was prepared by a method similar to that of Armes, et al. (*8*). A 0.60 g sample of polyvinyl alcohol (PVOH) was dissolved in 100 ml of distilled water by stirring overnight with a magnetic stirrer. The solution was then filtered into a 500 ml RB flask equipped with a magnetic stirrer. Then 17.6 g of ferric chloride hexahydrate was dissolved in 100 ml of distilled water, filtered, and combined with the PVOH solution. With stirring, 2.0 ml of distilled pyrrole was added all at once and and solution turned black within a few seconds, signifying that polymerization had started. The reaction was allowed to proceed for four hours. It was then centrifuged at 10,000 rpm (~ 15,000 x G) for about 30 minutes, which pelleted most of the particles. The black supernatant was carefully decanted and was replaced with approximately half its volume with fresh distilled water. The particles were dispersed by vortexing or with the aid of a sonic bath and the suspension was dialyzed against 3 - 14 liter volumes of distilled water for about 4 days total. If the latex was to be used without further functionalization, it was treated with a mixed bed of ion-exchange resin in the H+ and OH- forms by shaking in batch form for 1 hour. A 1 ml aliquot of exchanged latex was spun down rapidly on a bench top high speed centrifuge and the conductivity of the supernatant was measured. The ion-exchange process was repeated until the conductivity of the supernatant was the same as distilled water. If the latex was to be further derivatized, it was not ion-exchanged, but solvent exchanged into N-methyl pyrrolidone (NMP).

Solvent Exchange of Aqueous Poly(pyrrole) Latex into N-methyl Pyrrolidone The dialyzed poly(pyrrole) latex was spun down once again on the Sorvall centrifuge under the same conditions as described above. The supernatant was decanted and replaced with approximately 1/2 the volume of N-methyl pyrrolidone. At this point, the latex was approximately 4 % solids. It redispersed easily into NMP and was colloidally stable. It did not become a solution, since a much greater dilution into NMP exhibited a strong Tyndall beam, when a focussed light beam was passed through it.

The wet latex in NMP was further freed of water before the bromoacetyl derivative was made. For a 100 ml batch of wet poly(pyrrole) latex, 20 ml of toluene was added and the latex was refluxed under reduced pressure using a Dean-Stark trap. After all the water was removed via the azeotrope, the toluene was distilled off. Karl Fischer analysis indicated that this procedure was adequate for drying to an acceptable level, since all analyses gave values less that 0.1% water.

Synthesis of Bromoacetylated Poly(pyrrole) by N-Acylation (Figure 1). Anhydrous poly(pyrrole) latex suspension in NMP (20 g, 3.51% solids, 10 mEq) was placed in a 50 ml RB flask fitted with a magnetic stirrer and closed with a stopper containing a petcock, used to relieve pressure. Between 0.90 ml and 0.07 ml (depending on the theoretical conversion desired) of bromoacetyl bromide was added dropwise, the flask was stoppered, and placed in a thermostatically controlled water bath at 80 ^{o}C. The slight pressure which developed was relieved and the flask was heated for 4 hours. The reaction was allowed to cool to room temperature and an an equal volume of 50/50 NMP/distilled water was added. This caused the latex to aggregate and the solids were spun down on a slow speed clinical centrifuge. The supernatant was removed and replaced with 5% sodium carbonate containing 0.1% Brij 35 (polyoxyethylene-23-lauryl ether), if the material was to be further functionalized, or 0.1% Brij 35 only, if the material was to be used for binding protein. The aggregated latex was sonicated for about three minutes at 150 watts using a Branson ultrasonic probe, which resuspended the particles and formed a stable colloid.

Synthesis of Bromoacetylated Poly(pyrrole) by Friedel-Crafts Acylation (Figure 1). Poly(pyrrole) latex suspended in dry NMP (35.4 g, 5.1% solids, 1.8 g polymer, (27 mEq)) was placed in a RB flask equipped with a magnetic stirrer and chilled in an ice bath. Then 4.3 g of $AlCl_3$ (32 mEq) was added to the latex followed by the dropwise addition of 2.55 ml of bromoacetyl bromide (29 mEq). The flask was removed from the ice, warmed to 40 °C, and allowed to react at this temperature for 2 hours. The latex was then filtered through filter paper to remove residues. It was colloidally stable in NMP and could not be easily centrifuged. It was precipitated by the addition of ethyl ether and the precipitate was washed 2X with acetone and dried. The material could be redispersed into water with the aid of a sonic probe at 125 Watts energy. The latex, bromoacetylated by the Friedel-Crafts chemistry, was not used for further derivatization, antibody binding, or immunoassays in this study.

Synthesis of Carboxyl Surface-Functionalized Latex (Figure 2A). To a 10 g sample of bromoacetylated poly(pyrrole) latex (0.70 g polymer, 54 % bromoacetylated based on Br analysis, 29 mEq bromoacetyl groups) 10 ml of 1 : 1 water: NMP was added. The resulting aggregates were spun down on a centrifuge and redispersed into 20 ml of 5% sodium carbonate with the aid of an ultrasonic probe. Added 0.72 ml (10 mEq) of thioglycolic acid and allowed to react for 6 hours at room temperature with stirring. The reaction mixture was spun down on a centrifuge, the black-colored supernatant was removed, and the remaining plug of material was redispersed into 0.1% Brij 35 with the aid of sonication. The dispersed latex was dialyzed for three days against 2 - 14 liter volumes of distilled water. The latex was then ion-exchanged and analyzed for carboxylic acid by conductometric titration.

Synthesis of Amino Surface-Functionalized Poly(pyrrole) Latex (Figure 2B). Bromoacetylated poly(pyrrole) latex in NMP was chilled in ice (30 g, 2.45 g polymer, 3.3 mEq bromoacetyl groups), Added 2.1 ml of triethylene tetramine (14 mEq) and the ice bath was removed. The reaction was allowed to go overnight during which aggregation occurred. The aggregated latex was spun down and redispersed into 0.1% Brij 35. The pH was made acidic (pH = 5.9) with HCl and the latex was dispersed with the aid of a probe sonicator (125 Watts, 2 min.), after which it remained colloidally stable. It was dialyzed against 3 - 14 liter volumes of distilled water over three days. The latex was then finally cleaned by ion-exchange.

Binding of ^{125}I Antibody to Poly(pyrrole) Latexes (Figure 3A - 3C). Affinity purified rabbit anti-digoxin antibody was mixed with a small amount of the same specimen which had been ^{125}I labeled. This was used in the binding studies described below:

Carboxyl-modified Latex. Serial dilutions of labeled antibody were mixed with carboxyl modified latex in 20 mM MES buffer , pH 4.6. Fresh 1-ethyl-3-(3-dimethylaminopropyl) carbodiimide (EDAC) was added to a final concentration of 14 mM.

Amino-modified Latex. Serial dilutions of labeled antibody were mixed with amino-modified latex in 50 mM MOPS buffer, pH 7.3. Glutaraldehyde was added to a final concentration of 10 mM.

Bromoacetylated Latex. Serial dilutions of labeled antibody were mixed with bromoacetylated latex in 0.5 M sodium phosphate buffer, pH 9.2.

Adsorption control. Serial dilutions of labeled antibody were mixed with unmodified polypyrrole latex in 20 mM MES buffer, pH 4.6. Fresh EDAC was added to a final concentration of 14 mM.

Adsorption control 2. Serial dilutions of labeled antibody were mixed with unmodified polypyrrole latex in 50 mM MOPS, pH 7.3.

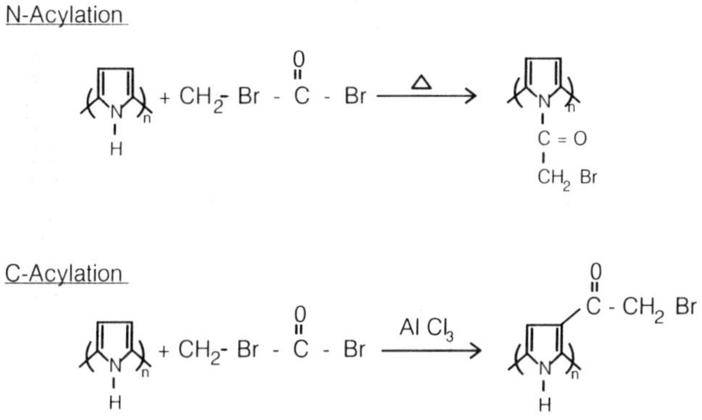

N-Acylation

$$\text{poly(pyrrole)(N-H)} + CH_2\text{-Br-}\overset{O}{\overset{\|}{C}}\text{-Br} \xrightarrow{\Delta} \text{poly(pyrrole)-N-C(=O)-}CH_2Br$$

C-Acylation

$$\text{poly(pyrrole)(N-H)} + CH_2\text{-Br-}\overset{O}{\overset{\|}{C}}\text{-Br} \xrightarrow{AlCl_3} \text{poly(pyrrole)(N-H)-C(=O)-}CH_2Br$$

Fig. 1 N-acylation and C-acylation of poly(pyrrole) latex with bromoacetyl bromide

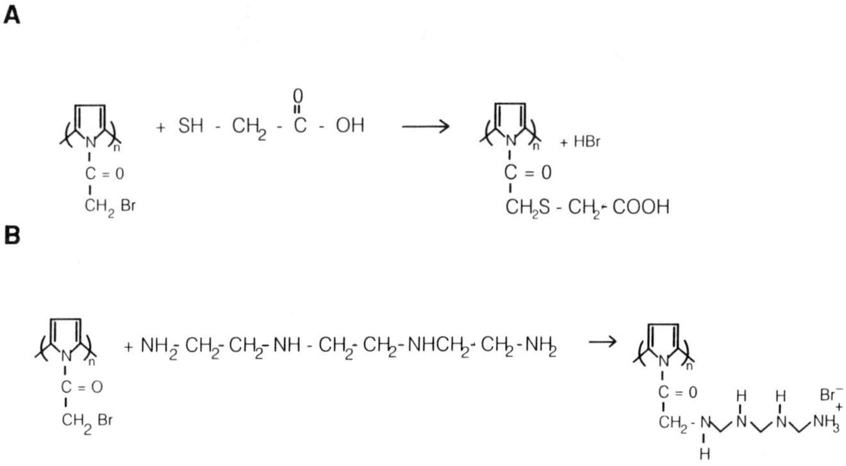

A

$$\text{poly(pyrrole)-N-C(=O)-}CH_2Br + SH\text{-}CH_2\text{-}\overset{O}{\overset{\|}{C}}\text{-OH} \longrightarrow \text{poly(pyrrole)-N-C(=O)-}CH_2S\text{-}CH_2\text{-COOH} + HBr$$

B

$$\text{poly(pyrrole)-N-C(=O)-}CH_2Br + NH_2\text{-}CH_2\text{-}CH_2\text{-NH-}CH_2\text{-}CH_2\text{-NH}CH_2\text{-}CH_2\text{-}NH_2 \longrightarrow \text{poly(pyrrole)-N-C(=O)-}CH_2\text{-NH-}CH_2CH_2\text{-NH-}CH_2CH_2\text{-NH-}CH_2CH_2\text{-}NH_3^+\ Br^-$$

Fig. 2 A) Conversion of bromoacetylated poly(pyrrole) latex to a carboxylated latex, B) to an aminated latex.

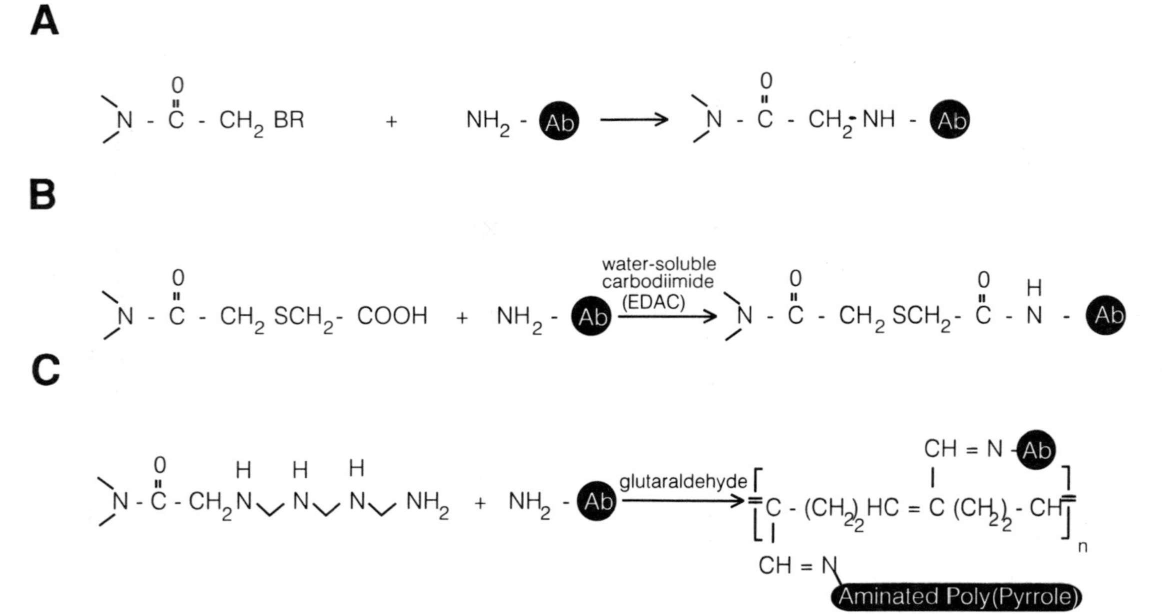

Fig. 3 Binding of antibodies to: A) bromoacetylated poly(pyrrole) latex, B) carboxylated latex, and C) aminated latex.

Incubation and clean-up. The final concentration of the latexes was 0.5% solids. The solutions were mixed end-over-end overnight at room temperature. The latex was separated from unbound antibody by centrifugation and removal of the supernatant. The latex was resuspended with fresh buffer and sonicated to disperse the pellet. The centrifugation and washing step was repeated a total of 3 times.

Measurements. Measured aliquots of latex were transferred to test tubes and diluted with a solution of 25 mM MOPS, 0.05% Brij, pH 7.3, and gamma counted. Data was plotted in terms of μg antibody bound/mg of latex as a function of antibody added.

Coating of Anti-Biotin Antibody on Poly(pyrrole) Latex for Subsequent Use in a Hepatitis Immunoassay as an Indicator Reagent. A 10 milliliter volume of unmodified poly(pyrrole) latex (0.1% solids in distilled water) was placed in a vial. Tris(hydroxymethyl)aminomethane (Tris) buffer (1 ml, 100 mM, pH 7.5) was added with stirring, followed by 50 microliters of bovine serum albumin and 500 microliters of 1 mg/ml anti-biotin antibody. The mixture was incubated at room temperature for 1 h., after which the particles were separated from the unbound antibody by centrifugation and washing 1 time. The particles were resuspended in Tris buffer containing BSA.

Preparation of a HIV-p24 Antigen/Poly(pyrrole) Indicator Reagent for Use in an HIV Antibody Immunoassay. One ml of unmodified poly(pyrrole) latex (0.1% solids in distilled water) was placed in a vial. A volume of 100 microliters of 100 mM of borate buffer, pH 8.0 was added with stirring followed by 20 ml of 0.75 μg/ml HIV-p-24 antigen (the major core protein of the virus responsible for AIDS). The mixture was incubated at room temperature for one hour, after which BSA was added to the resultant indicator reagent latex.

Results and Discussion

Bromoacetylation of Poly(pyrrole). Conversion to the bromoacetylated derivative of both N-acylated and ring aceylated derivatives was determined by elemental analysis for bromine. Final yields can be affected by the presence of moisture and amine impurities in the NMP and loss of derivatized material due to solubilization.. Generally, if a 1 : 1 ratio of bromoacetyl bromide relative to pyrrole units is used, greater than 50 % recoverable conversion to the N-acylated derivative is possible. Analysis of the supernatant from the bromoacetylation reaction via visible spectroscopy and gravimetric analysis indicates that as much as 50% of the derivatized poly(pyrrole) can be lost due to solubilization. Under Friedel-Crafts conditions, a conversion to 25% ring acetylation was achieved.

Carboxylated and Aminated Poly(pyrrole). Conversion to the carboxylated and aminated derivatives once again appeared to produce solubilized material which became detached from the particle and was removed during purification. The work-up involved centrifugation and washing at least 2 times with 0.1 % Brij 35, followed by dialysis against distilled water, followed by ion-exchange with a mixed bed in the H+ and OH- forms. After ion-exchange, an aliquot of latex was pelleted on a centrifuge, and the conductivity of the supernatant was measured. Ion-exchange was repeated if the conductivity of the supernatant was greater than 2 times that of distilled water. After the first titration, the titrated latex was re-ion-exchanged using 100 -150 fold excess exchange capacity of resin based on the endpoint determined. The percent solids of the latex was re-determined and the latex was re-titrated. Table I shows the results of conductometric titration of four surface-modified latexes. The data provides a further indication that the individual modified poly(pyrrole) chains are somewhat solubilizable, and may be detached during the ion-exchange process. All latexes were ion-exchanged to the level that no polyelectrolyte remains free in solution; however, re-exchange of previously titrated aliquots showed that samples 19-carboxyl and 52-amino resulted in a loss of surface charge density.

Infrared spectroscopy provided further evidence that the carboxyl derivative was indeed formed. Figure 4 compares unmodified poly(pyrrole) to a carboxylated derivative. The absorption at 1723 cm^{-1} is indicative of the surface bound carbonyls from probably both the bromoacetylation and carboxylation.

Residual bromine was determined to measure the amount of bromoacetyl groups that were not converted to carboxyl derivatives and remained with the clean particles. This data is shown in Table II.

Antibody Immobilization by Adsorption. The proposed mechanisms of polymerization of aromatic molecules has been covered by several authors for both chemical and electropolymerizations (*9,10,11,12*). The approximate stoichiometric reaction of the chemical polymerization with $FeCl_3$ as the oxidant is thought to be (*8*):

3x [pyrrole, N–H] + 7 x Fe (III) Cl_3 → [terpyrrole unit, N–H, N–H, N–H, Cl^-]$^+$ + 7 x Fe (II) Cl_2 +6 x HCl

The polymer obtains a net positive charge. Zeta potential measurements on ion-exchanged, unmodified polypyrrole latex in distilled water yielded values of approximately +7 millivolts. In addition, radical defects have been observed by ESR spectroscopy in neutral and doped poly(pyrrole)(*13*). The role, if any, these radicals may play in the binding of proteins to the surface of unmodified particles is not clear.

If the surface of the particles are saturated with antibody, rinsed several times with buffer, and subjected to overnight incubations with sodium dodecyl sulfate, only a small amount (15%) of the bound protein can be detached. In contrast, subjecting antibodies bound to polystyrene, presumably by simple adsorption, to similar conditions, resulted in approximately 70% detachment and removal (*14*). At the pH that antibodies were coated onto the unmodified latex, they are either neutral or negatively charged. Perhaps the charge attraction of the multivalent protein to the oppositely charged particle surface contributes to the tenacious binding, similar to the difficulty one experiences in trying to displace Fe^{+3} ions from a cation exchange resin.

Alternatively, if the resonance stabilized radicals have sufficient reactivity, one might expect that exposing the particle to mercaptoethanol may result in proton abstraction and a change in the protein binding properties of the surface. In fact, treatment of unmodified latexes with 160 mM thiol resulted in an 80% reduction in the ability of the particles to bind antibodies. Confirmation of the chemistry occurring as a result of the thiol exposure remains to be determined.

TABLE I

PHYSICAL CHARACTERISTICS OF CARBOXYL AND AMINO POLY(PYRROLE) LATEXES

LOT #	Titratable Groups (theoretical) meq./g polymer	Actual Groups 1st ion-exc., meq./g polymer	Actual Groups 2nd. ion-exc. meq./g polymer	Particle size (nm)	Surf. Charge Density. 2nd ion-exc. (μeq./cm^2) X 10^4
35-carboxyl	15	0.76	0.74	114	19.1
51-carboxyl	3.8	0.63	0.58	114	15.0
19-carboxyl	0.75	0.22	0.12	112	3.0
52-amino	11	0.69	0.33	119	8.9

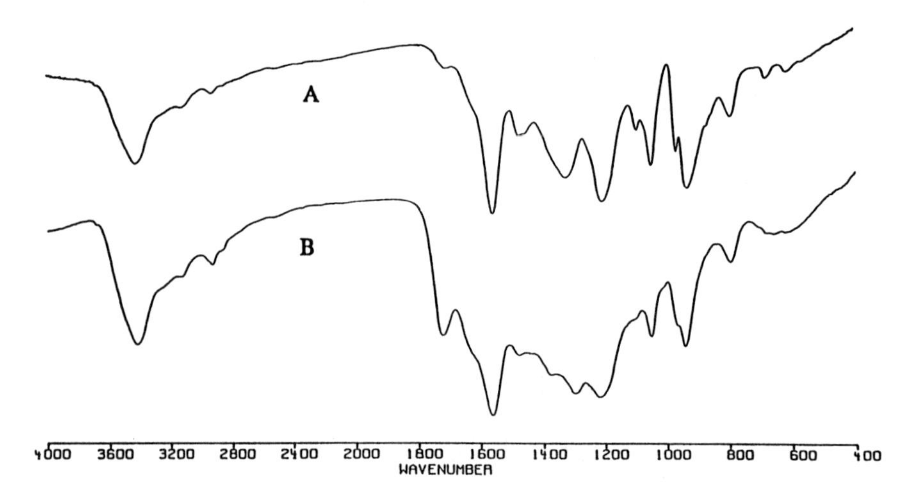

Fig. 4 Infrared spectra of A) unmodified poly(pyrrole) (lot #28), and B) carboxylated poly(pyrrole), (lot# 35-carboxyl).

TABLE II

PERCENT BROMOACETYL GROUPS BEFORE AND AFTER REACTION WITH THIOGLYCOLIC ACID IN NMP

lot #	before carboxylation	after carboxylation
carboxyl-25	23	5.8
carboxyl-35	9.0	6.5

Covalent Immobilization of Antibody. The covalent coupling of rabbit antibody to carboxyl-modified poly(pyrrole) using EDAC (14 mM, Figure 5) resulted in approximately 9.0 μg IgG bound per mg of latex at an antibody concentration of 50 μg/ml. In comparison, the Adsorption Control 1 (unmodified poly(pyrrole) latex), also in the presence of EDAC, bound approximately 3.5 μg IgG per mg of latex or about 60% less antibody under the same conditions.

In addition to binding 2.5 times as much antibody as the adsorption control, the EDAC coupling procedure resulted in a greater fraction of tightly bound antibody. This was demonstrated by the removal of only approximately 5% antibody under conditions of incubation in 1% sodium dodecyl sulfate at room temperature followed by a 60 minute incubation at 45 °C.

Figure 6 shows that even higher binding was obtained upon covalently coupling rabbit antibody to amino-modified poly(pyrrole) using 10 mm glutaraldehyde at pH 7.3, or by reacting the antibody directly to bromoacetylated poly(pyrrole) at pH 9.2. At an antibody concentration of 50 μg/ml, 12.5 μg and 15.5 μg IgG per mg of latex became immobilized, for amino-modified and bromoacetylated latex, respectively. In contrast, adsorption at pH 7.3 to unmodified latex, in the absence of any coupling agents, resulted in 6.5 μg of antibody bound per mg of latex, using the same starting concentration of antibody.

In the three covalently coupled and the two adsorption examples of antibody attachment to carboxyl- amino-, bromoacetylated-, or unmodified latex, the amount of antibody bound increased nearly linearly with increasing antibody concentration as seen in Figs. 5 and 6. If we assume the aminated latex particles have a diameter of 119 nm and a density of 1.36 (*15*), and the diameter of an antibody is 14 nm, a monolayer of antibody would be equivalent to about 251 μg IgG per mg of latex. The highest level of binding is 17 μg IgG per mg of latex and represents about 7% of a monolayer of antibody bound to the surface, or about 20 antibodies bound per particle. At this fraction of a monolayer binding, little steric hindrance is expected during chemical coupling, and the nearly linear coupling curves in Figs. 5 and 6 are consistent with this calculation.

In order to get a measure of the antibody that was loosely bound to the carboxylated latex, aliquots of the EDAC coupled antibody-latex conjugate were incubated overnight in 1% SDS followed by a 60 minute incubation at 45 °C. The latex was washed by centrifugation and resuspension four times, then counted to determine the fraction of Ab remaining bound to the latex. It was found that 94.5% of the antibody remained bound to the latex. In the case of the Adsorption 1 control, 84.8% of the antibody remained bound to the latex, demonstrating that antibody binds almost as tenaciously to unmodified poly(pyrrole) as antibody bound presumably by covalent linkages.

Visible Absorption Spectrum of Poly(pyrrole). Poly(pyrrole) polymer absorbs strongly in the visible region depending on allowed optical transitions shown in the energy level diagram (*11*). It is this property that allows it to be a useful visual label for immunoassays. Figure 7 shows the visible spectrum of a typical unmodified poly(pyrrole) latex. The regions of the spectrum where allowed transitions occur are indicated, where region 1 is a valence band (VB) to conduction band (CB) transition, region 2, a valence band to antibonding cation level transition and region 3, a bonding cation level to an antibonding cation level transition. The absorbance properties of the surface-modified latexes were not practically changed from an immunoassay applications standpoint.

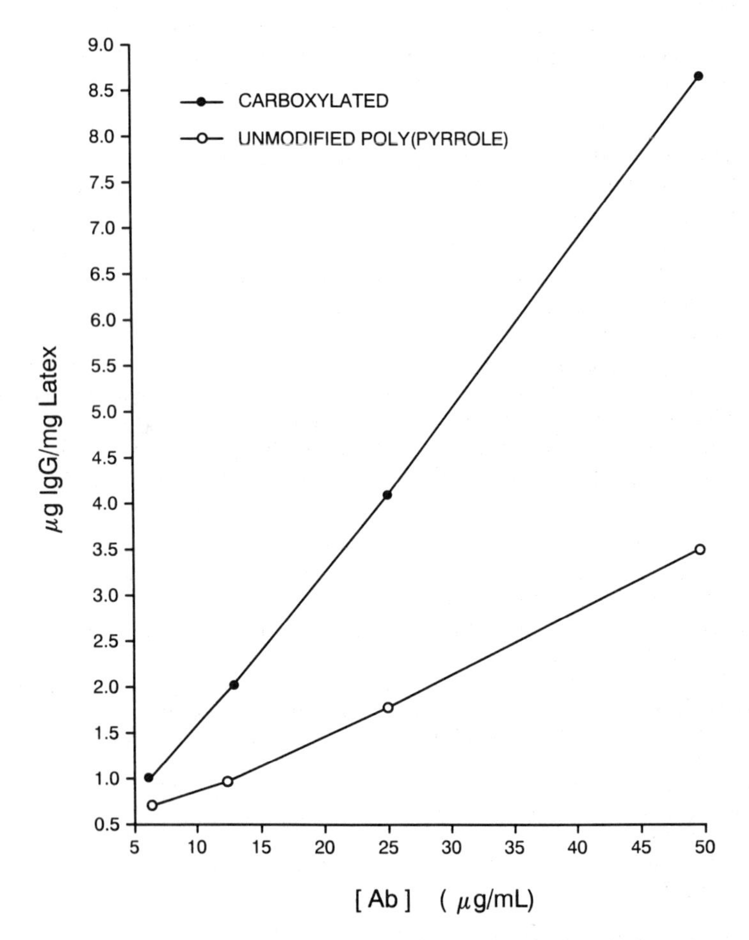

Fig. 5 Covalent coupling of rabbit anti-digoxin antibody to carboxylated poly(pyrrole) latex using EDAC, (lot# 35-carboxyl), and adsorption under the same coupling conditions using EDAC, to unmodified poly(pyrrole) latex (lot# 28, adsorption control 1).

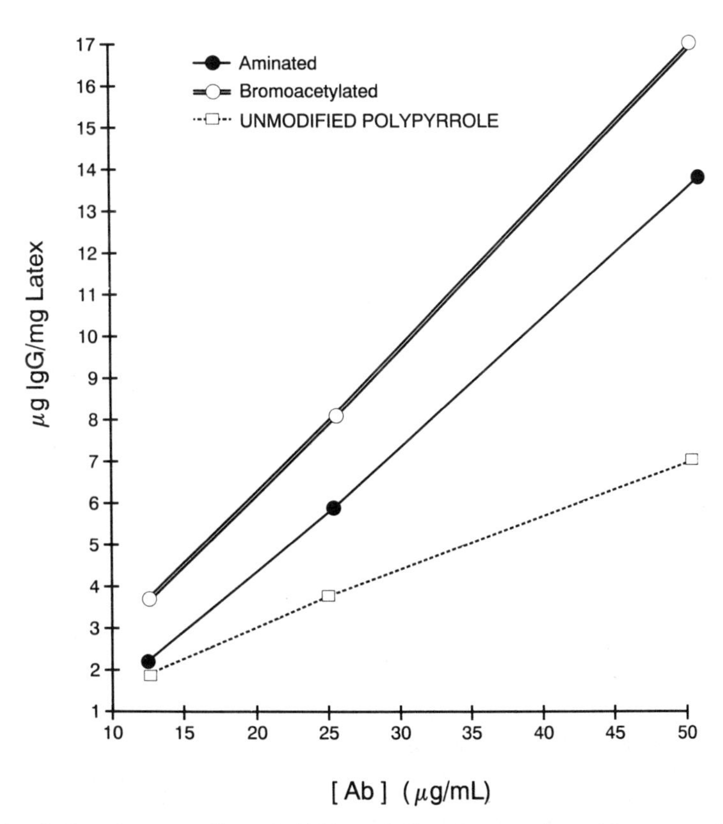

Fig. 6 Covalent coupling of rabbit anti-digoxin to aminated latexes using glutaraldehyde (lot# 52-amino), bromoacetylated latex (lot# 25-bromoacetyl), and unmodified latex (lot# 28, adsorption control 2).

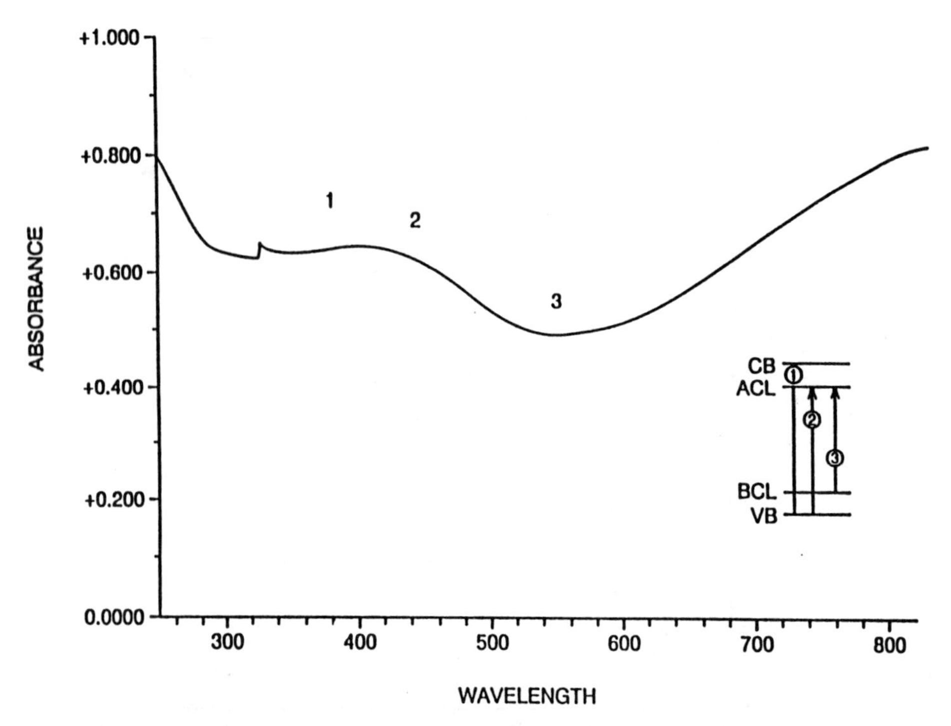

Fig. 7 Visible absorption spectrum of unmodified poly(pyrrole) latex in water showing the valence band to conduction band transition (1), valence band to antibonding cation level transition (2), and the bonding cation to antibonding cation level transition (3) regions.

Chromatography-Based Immunoassays. In the assay formats described below, the basic principle involves the use of a nitrocellulose paper chromatography strip, containing immobilized antibody or antigen, localized in a specific region of the strip. Addition to the bottom, of sample and/or poly(pyrrole) immunoreagent latex, which travel along the strip by capillary action, results in the immobilization of the latex in that region through an immunoreaction, if the analyte of interest is present.

Hepatitis B Surface Antigen (HBsAg) Indirect Sandwich Assay. A capture reagent was prepared by immobilizing anti-HBsAg antibody in a spot midway along on a nitrocellulose test strip. A blotter was fixed to the top end of the strip. The strip was then contacted at the bottom to a sample consisting of 120 μl of human plasma containing a defined amount of HBsAg. The sample was drawn up the strip by capillary action past the immobilized capture antibody, so that HBsAg in the sample would be captured by the immobilized anti-HBsAg antibody. This was immediately followed by 10 μl of a 2 μg/ml biotinylated anti-HBsAg antibody solution which was contacted to the bottom of the strip and became bound upon migration to the immobilized HBsAg. Finally, the strip was put in contact with the indicator reagent, 15 μl of poly(pyrrole) latex containing surface immobilized anti-biotin antibody, which also migrated along the strip.

In samples containing a threshold amount of HBsAg, the antigen reacted with the capture reagent and was immobilized on the nitrocellulose test strip. The biotinylated anti-HBsAg antibody then reacted with the antigen, and the indicator reagent completed the immunocomplex by binding to the biotinylated anti-HBsAg antibody. The immobilized indicator reagent was detectable as a dark area on the strip, and the intensity of the color was directly related to the quantity of HBsAg in the sample. A schematic representation of this assay format is shown in Figure 8, and Figure 9 presents the intensity of the color obtained for various concentrations of HBsAg analyte. The results indicate that as the analyte concentration increased, the indicator reagent and thus the detectable color also increased on the test strip. Antigen concentrations of 10 ng/ml and less were readily detected. This level of sensitivity meets the requirement for a third generation assay suitable for clinical use in the United States (*16*).

Anti-HIV Antibody Immunoassay. A capture reagent was prepared by immobilizing HIV-p24 antigen in a spot midway along a nitrocellulose strip. The test sample, human plasma containing various dilutions of antibody specific to HIV-p24 was mixed with an equal volume of the HIV-p24 antigen/poly(pyrrole) indicator reagent described earlier. Forty microliters of the indicator reagent/test sample mixture was then applied to the nitrocellulose strip, and the fluid was drawn up the strip past the immobilized antigen. In samples containing antibody to HIV-p24, that antibody reacted with the antigen previously deposited on the strip, as well as with the antigen of the indicator reagent to form an antigen/antibody/antigen sandwich immunocomplex on the nitrocellulose strip. A schematic of this type of assay is shown in Figure 10. The immobilized indicator was detected as a dark area on the strip, and the intensity of the color was directly related to the quantity of anti-HIV-p24 antibody in the test sample as demonstrated by the results shown in Figure 11. The highest dilution tested (1:1000) approached the limit of sensitivity of the commercial Abbott HIV immunoassay.

Self-Performing Qualitative Assays. Qualitative assays for analytes of interest may be preformed on strip configurations which require only the addition of sample

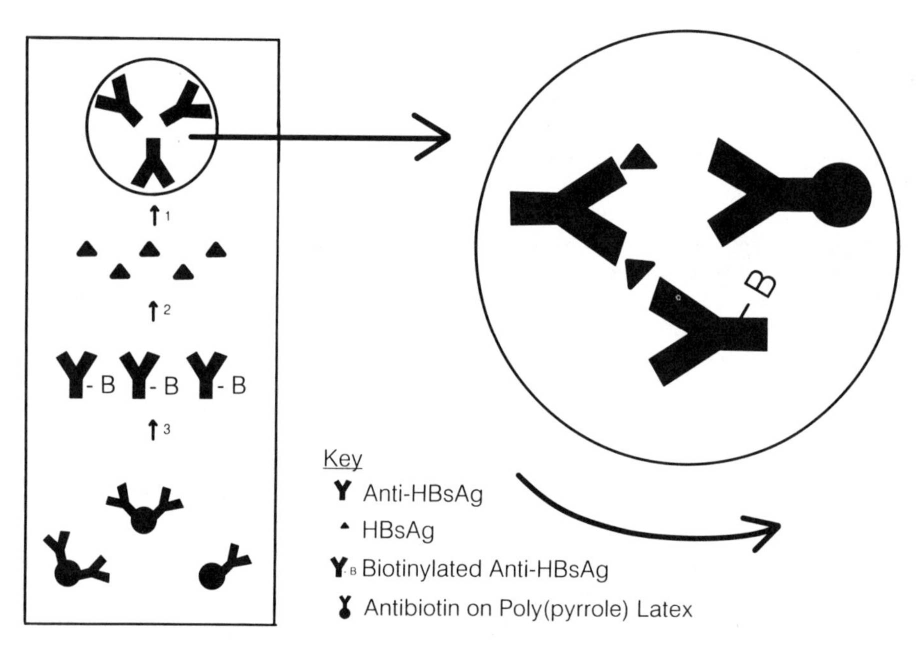

Fig. 8 Schematic representation of a chromatography-based immunoassay for hepatitis B surface antigen (HBsAg) using a poly(pyrrole) latex as a labeled immunoreagent.

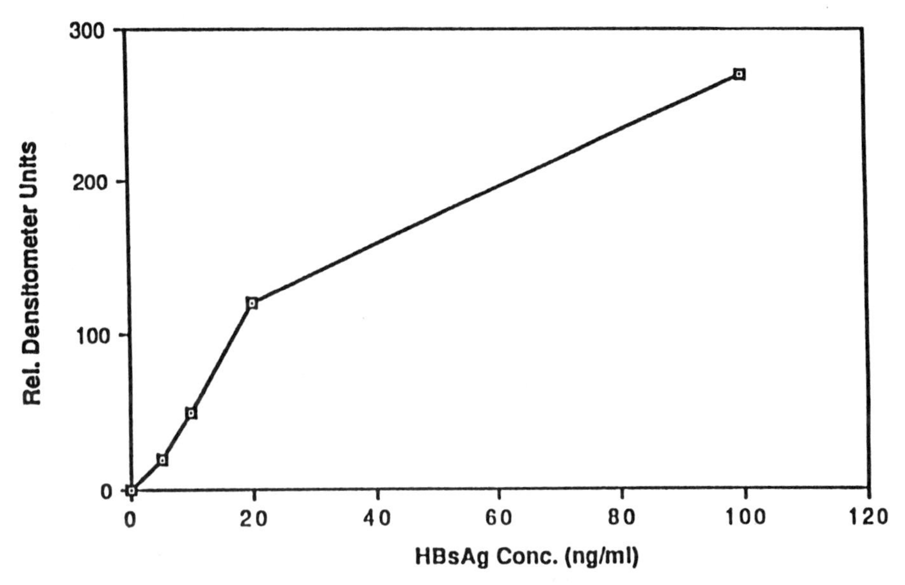

Fig. 9 Relative densitometer reading vs. HBsAg concentration for an assay of the format illustrated in Fig. 8.

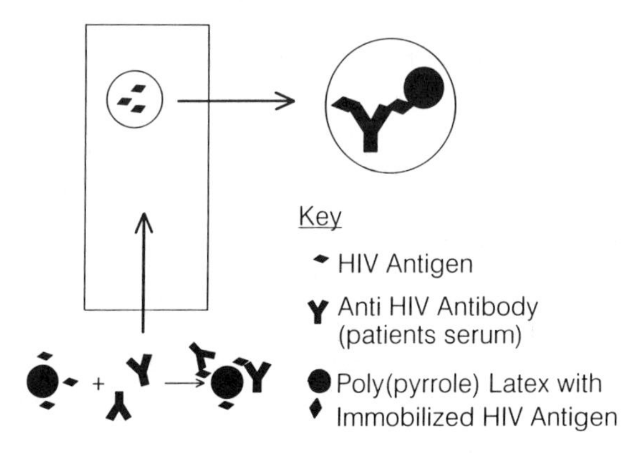

Fig. 10 Schematic representation of a chromatography-based immunoassay for anti-HIV antibody using a poly(pyrrole) latex as a labeled immunoreagent.

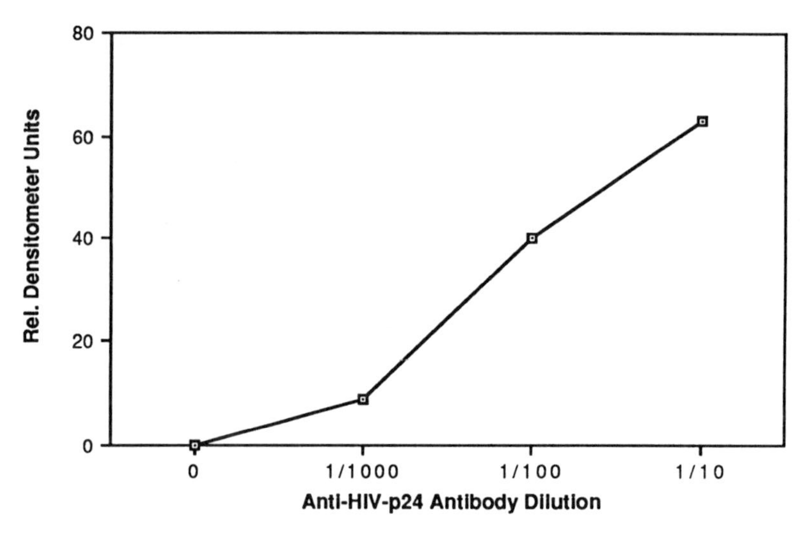

Fig. 11 Relative densitometer reading vs antibody dilution for an assay of the format illustrated in Fig. 10.

to obtain the result. Rather than give detailed procedures for the coating of the poly(pyrrole) latex with antibodies and/or antigens, the recipes of which are similar to those described earlier, a description of the configuration of the assay will be given with an example of the results. The example which will be described will be a self-performing assay for the protein hCG, a marker for pregnancy. Figure 12 shows a nitrocellulose strip, where dried poly(pyrrole) latex containing surface-immobilized antibody against the hCG antigen is deposited at the bottom of the strip. Midway along the strip, hCG antigen was deposited in a horizontal bar. On a vertical bar, which intersects the horizontal bar, was deposited antibody against hCG. When patient's sample, which may or may not contain antigen, is added to the dried latex, the particles become hydrated and redispersed and migrate along the strip. If there is no antigen in the sample, the particles, which contain antibody, bind to the horizontal bar which contains pre-deposited antigen, forming a minus sign. If the sample contains antigen, it binds to the antibody on the particles and the antigen-antibody-particle complex is captured by the vertical bar which contains antibody. The particles are always in excess, hence those particles which were unsuccessful in competing for the limited antigen, become captured by the horizontal bar completing the plus sign. This is further illustrated by Figure 13 in schematic form and an actual assay for a positive and a negative sample is shown in Figure 14. The positive sample shown was 250 mIU.ml, a sensitivity sufficient to detect pregnancy during the first missed menses.

Summary

Poly(pyrrole) latex is a superior polymer colloid label for visual immunoassay formats. Its large apparent visible extinction coefficient is much greater than that attainable by impregnating or dying "white" latex particles with strongly absorbing chromophores. The particles can be surface-functionalized by bromoacetyl bromide resulting in pendant bromoacetyl groups. These functionalities will couple with

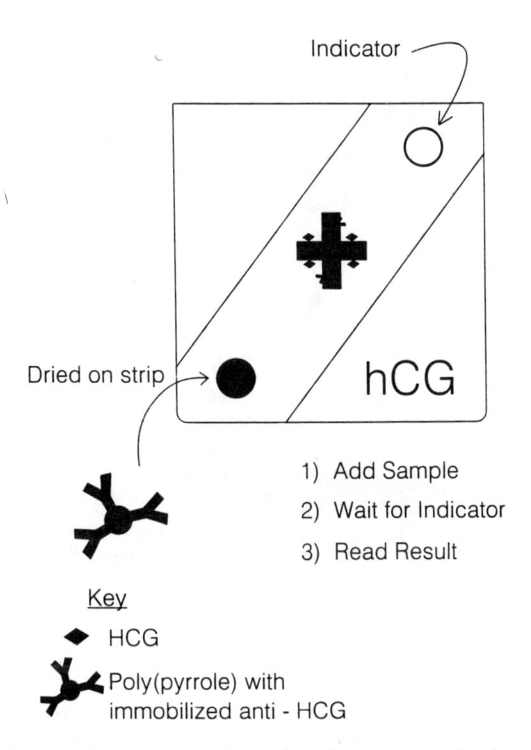

Fig. 12 Schematic representation of a chromatography-based self-performing assay for the pregnancy marker hCG, employing a dried poly(pyrrole) latex immunoreagent.

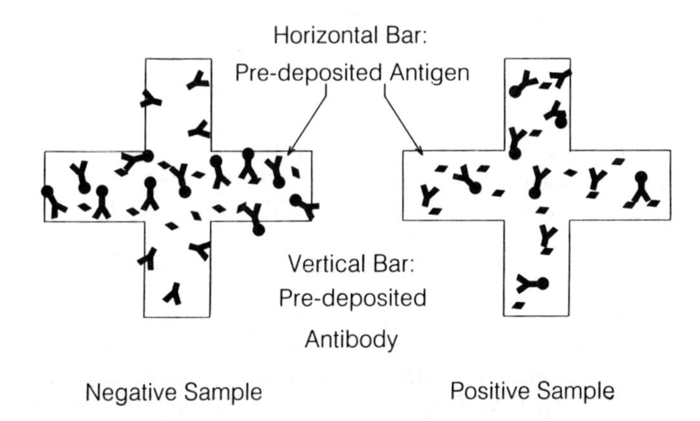

Fig. 13 Detailed schematic representation of the indicator bars on the self-performing assay shown in Fig. 12.

Fig. 14 Actual results of a self-performing assay for the presence of the pregnancy marker hCG.

proteins directly, or the groups can be further reacted with thioglycolic acid or multifunctional amines to form carboxylated and aminated surfaces, respectively. Biological ligands can be coupled to these surfaces with the aid of the appropriate coupling agents. Such methods result in significantly higher immobilization levels than attainable through adsorption to unmodified poly(pyrrole).

Several model chromatography-based immunoassay models were demonstrated, including a self-performing assay. In this particular format, the operator simply adds sample to the strip containing the poly(pyrrole) immunoreagent, and can read an unambiguous yes/no result in a matter of minutes.

Literature Cited

1. Leuvering, J.H.W. *U.S. Patent* 4,313,734, 1982.
2. Dorman, L.C.; Bangs, L.B. *U.S. Patent* 4,419,453, 1983.
3. de Jaeger, N.; Monbaliu, M.J.; Noppe, M.J.M.; Konigs, F.J. *U.S. Patent* 4,837,168, 1989.
4. Tarcha, P.J.; Wong, M.; Donovan J.J. *Eur. Pat. Appl.* 360,088, March 28, 1990.
5. Vanderhoff, J.W.; van den Hul, H.J.; Tausk, R.J.M.; Overbeek. J. Th. G. *Clean Surfaces: Their Preparation and Characterization for Interfacial Studies* G. Goldfinger, Ed., Marcel Dekker, New York, 1970, pp. 15.
6. Grenier, F.C.; Pry, T.A.; Kolaczkowski, L. *U.S. Patent* 4788136, 1988.
7. Jeffcoate, S.L.; Gilby, E.D.; Edwards, R. *Clin. Chim Acta,* 1973, 43, 343-349.
8. Armes, S.P.; Miller, J.F.; Vincent, B. *J. Colloid Interface Sci.,* 1987, 118, 410-416.
9. Kovacic, P.; Jones, M.B. *Chem. Rev.,* 1987, 87, 357-379.
10. Beck, F.; Oberst, M.; Jansen, R. *Electrochimica Acta,* 1990, 35, 1841-1848.

11. Genies, E.M.; Pernaut, J.M. *J. Electroanal. Chem.*, 1985, 191, 111-126.
12. Hsing, C.F.; Kovacic, P; Khoury, I.A *J. Polym. Sci., Polym. Chem. Ed.*, 1983, 21, 457-466.
13. Scott, J.C.; Pfluger, P.; Krounbi, M.T.; Street, G.B. *Physical Rev. B,* 1983, 28, 2140-2145.
14. Chu, V.P.; Tarcha, P.J. *J. Appl. Polym. Sci.,* 1987, 34, 1917-1924.
15. Diaz, A.F.; Hall, B. *IBM J. Res. Develop.*, 1983, 27, 342-347.
16. Lostia, O.; Lorenzini, P.; Maroccia, E. *Boll. Inst. Sieroterapico*, 1983, 62, 381-392.

RECEIVED December 26, 1991

Chapter 23

Conducting Polymer Composites

Polypyrrole–Metal Oxide Latexes

R. E. Partch, S. G. Gangolli, D. Owen, C. Ljungqvist, and E. Matijević

Department of Chemistry and Center for Advanced Materials Processing, Clarkson University, Potsdam, NY 13699

Electrically conducting latexes of polypyrrole coated on inorganic cores were obtained by deposition of pyrrole on catalytically active particles. The finely dispersed core materials were hematite (polyhedral and spindle-type), silica coated with hematite, magnetite, and cerium(IV) oxide. The degree of polymer coverage could be controlled by varying the aging time and the properties of the carrier particles. Also, yttrium basic carbonate was coated on polypyrrole latex. The so prepared coated powders were examined by transmission electron microscopy, and further characterized by elemental and thermogravimetric analyses, as well as by electrophoresis, X-ray diffraction, and conductivity measurements. The d.c. conductivities of all but the magnetite particles coated with polymer were comparable to those reported for pure polypyrrole and they increased with pressure.

The synthesis, characterization, and processing of organic conducting polymers show potential use in batteries, molecular electronic devices, conductive paints, anti-corrosion coatings, controlled-release composites, and copy machines (*1-4*). Highly conjugated polyacetylene, polyaniline, and polyheterocycle varieties (*5-7*) have conductivities ranging from 10^{-3} to 10^{5} S cm^{-1}, depending on the method of preparation and whether they are doped with an oxidizing or reducing additive (*8-10*). Polypyrrole, one of the polyheterocycles, is intensely colored (black) in the oxidized form while its reduced, nonconducting product is transparent green, and as such it can undergo electrochromic switching (*10-12*). Nonconjugated polymers derived from several butadiene derivatives exhibit lower levels of conductivity, the mechanism of which is the focus of considerable debate (*13-16*).

The potential uses of conducting polymers are often diminished due to their sensitivity to oxygen or poor mechanical properties. Polypyrrole has the advantage of being air stable but has poor malleability and is, therefore, difficult to process into

NOTE: This chapter is Part II of a series.

0097–6156/92/0492–0368$06.00/0

a device. This problem has been partially overcome by using preparative chemistry that incorporates specific counterions or dispersants into the polymer as it is synthesized. For example, p-toluenesulfonate ions are included in electrochemically generated free-standing films of polypyrrole (*17-19*). Stable dispersions of powders of the polymer can be prepared when steric stabilizers are mixed with pyrrole in the solution containing the ferric ion initiator (*20,21*).

Other techniques that have been employed to improve the processing capabilities of conducting polymers are to coat *inert* core particles such as latex, SiO_2, and WO_3, or to impregnate films with metal oxide powders (*22-25*). Of special relevance to the present work are reports of polymerization of pyrrole intercalated in solid FeOCl (*26*) and in Fe(III)-zeolites (*27*).

Here, a new approach is employed, in which particles of hematite, silica doped with hematite, magnetite, and cerium oxide having reactive sites on their surfaces can be coated with polypyrrole of controlled thickness without the use of electrochemical or soluble initiators (*28*).

Organic material can also serve as a core onto which layers of different chemical composition may be deposited. Such composites having an inorganic coating are useful as less dense substitutes for purely inorganic particles and, if heated, may act as precursors to hollow particles (*29*). The experimental parameters for coating polypyrrole latex with yttrium basic carbonate have now been determined.

Experimental

Materials. Submicron polyhedral and spindle-type hematite (α-Fe_2O_3), and spherical magnetite, cerium oxide, and polypyrrole particles were prepared following previously published procedures (*20,28,30-33*).

Polyhedral Hematite. A solution 0.018 mol dm^{-3} in $FeCl_3$ and 1×10^{-3} mol dm^{-3} in HCl was placed in a preheated oven at 100°C and aged for 24 h. The so obtained dispersion was repeatedly washed with doubly distilled water and the particles were separated by centrifugation. The final suspension contained 6.5 mg cm^{-3} α-Fe_2O_3 particles of 0.065 μm in the mean diameter. The isoelectric point (i.e.p.) of the original particles was at pH 4.3. Repeated washings of the powder with water raised the i.e.p. to pH 7.5.

Spindle-type Hematite. Forced hydrolysis of a 2×10^{-2} mol dm^{-3} ferric chloride solution containing 2×10^{-4} mol dm^{-3} NaH_2PO_4 was carried out by heating the mixture for 2 days in a preheated oven at 96°C. The dispersion was either filtered and the solids purified by rinsing with water, or used as obtained after the completion of aging. These hematite particles were ~ 0.3 μm long and the working suspension contained 10.2 mg cm^{-3} solids. The i.e.p. of the particles in the original mother liquid was at pH 4.0. After rinsing dispersions twice with distilled water the i.e.p. was shifted to pH 8.0.

In both preparations of α-Fe_2O_3 particles, the separated supernatant solutions, after sufficient washings, gave a negative test for ferric ions when treated with 1,10-phenanthroline (*34*).

Hematite dispersions used in some of the coating experiments were modified from their "as prepared" state as follows. Both kinds of α-Fe_2O_3 particles with the i.e.p. at pH > 7 were treated with 0.04 mol dm^{-3} HCl for 24 h, thoroughly washed and, finally, redispersed in distilled water. The sample of polyhedral α-Fe_2O_3, having the i.e.p. at pH 4.3, was placed in 2×10^{-4} mol dm^{-3} NaH_2PO_4 solution for 24 h, and then treated as above. Another sample of the same α-Fe_2O_3 particles, with the i.e.p. at pH 8, was admixed into a dilute NaCl solution for 24 h, washed, and redispersed in distilled water.

Silica Coated with Hematite. A dispersion of uniform spherical SiO_2 particles (0.8 μm in diameter), prepared by hydrolysis of tetraethyl orthosilicate (TEOS) (*35*), was coated with tiny α-Fe_2O_3 particles by adapting a method for deposition of hematite on colloidal TiO_2 (*36*). One milligram of this silica was dispersed in 1.0 cm^3 of 0.03 mol dm^{-3} HCl solution and equilibrated for 24 h. A stock ferric chloride solution was then added and the dispersion diluted to 30 cm^3 with doubly distilled water to give a final concentration of 1×10^{-3} mol dm^{-3} HCl and 2×10^{-3} mol dm^{-3} $FeCl_3$. This dispersion was aged in a preheated oven at 100°C for 24 h, the resulting coated particles separated by decanting the supernatant liquid, and the precipitate washed a few times with water. The final dispersion contained a small amount of free hematite particles which were then separated by centrifugation; the heavier coated silica particles settled, while hematite which remained suspended was readily removed by decantation.

The final suspension was made by admixing 30 mg of coated silica particles into 1 cm^3 of distilled water.

Cerium(IV) Oxide. An 8×10^{-2} mol dm^{-3} cerium sulfate solution, containing 1.2×10^{-3} mol dm^{-3} H_2SO_4, was hydrolyzed at 90°C for 48 h. The resulting particles were rinsed several times with doubly distilled water, the solids separated by centrifugation, and the supernatant solution discarded. Further washings were done with 2×10^{-3} mol dm^{-3} NaOH solution, followed by water. The so obtained cerium oxide was either dispersed in water or calcined at 600°C for 2 h before further use. For coating studies 18.7 mg of calcined powder was dispersed in 1 cm^3 of distilled water.

Magnetite. Polydisperse, low density particles were prepared by adding premixed and deoxygenated solutions of KOH and KNO_3 to freshly prepared $FeSO_4$ solution under nitrogen. The final concentration of the reagents was 0.125, 0.2, and 0.125 mol dm^{-3}, respectively. The gel obtained was aged at 90°C for 4 h after which the black magnetite was separated from the liquid by sedimentation in a magnetic field and washed several times with distilled water.

The surface of the magnetite particles was activated by treatment with 0.03 mol dm^{-3} HCl or 15% H_2O_2, each for 1 or 5 days, followed by thorough washing with distilled water before exposure to pyrrole. The solids content in stock suspensions after 1 and 5 day treatment with HCl was 29.0 and 26.4 mg cm^{-3}, and with H_2O_2 was 31.0 and 31.6 mg cm^{-3}, respectively.

Magnetite particles having a narrow size distribution and higher density were formed by heating a sample of the previously described polyhedral hematite in hydrogen at 400°C for 3 h (*32*). The black powder was then treated with HCl or H_2O_2 as above, washed with distilled water, filtered and air dried.

Polypyrrole. The monomer was polymerized in the presence of poly(vinyl alcohol) using $FeCl_3$ as oxidant (*20*).

Analyses. Cores and coated particles were inspected in transmission and scanning electron microscopes. Thermogravimetry data were obtained with a Perkin-Elmer TGA 7 analyzer, interfaced with model 3700 Data Station. Electrophoretic mobility measurements were carried out at a constant ionic strength of 0.001 mol dm^{-3} $NaClO_4$ with the DELSA 440 (Coulter Electronics) instrument. X-ray diffraction patterns were obtained with the Siemens D500 apparatus using Cu Kα radiation. Conductivities were measured by a four-point probe technique on pressed pellets of the particles inside a Teflon holder.

Techniques. The general procedure for coating metal oxide particles with polypyrrole with no soluble ferric ion present involved mixing 0.5 cm^3 of the particle dispersion (or 24 mg of dry polyhedral magnetite) with 0.5 cm^3 of a solution 50% by volume in each pyrrole and ethanol, diluting to 5.0 cm^3 with distilled water, and stirring the suspension in a tightly closed vessel at 100°C. Purging with N_2 before sealing the container had no effect on the results. At the end of the desired period of time, the product was cooled, acidified with 5 cm^3 of 0.4 mol dm^{-3} HCl, and then mixed with 10 cm^3 diethyl ether. More than 98% of the dispersed matter was in the aqueous phase. After removing the solid the supernatant liquid was clear. In a few experiments, a trace amount of a less dense solid was located at the water-ether interface, and could be easily withdrawn with a pipette. The solids were separated by centrifugation, filtration or magnetic field, washed with distilled water, and dried at room temperature.

A typical procedure for coating polypyrrole latex with yttrium basic carbonate involved making a dispersion containing 2.4 mol dm^{-3} urea, 1.2 weight percent poly(vinylpyrrolidone), 0.05 g dm^{-3} polypyrrole particles and 2 x 10^{-3} mol dm^{-3} $Y(NO_2)_3$ in water. Samples of these suspensions were aged at temperatures ranging from 70-150°C for various lengths of time. Product isolation was achieved by centrifugation or filtration, washing with distilled water, and drying.

Results

Table I summarizes the essential results of coating metal oxides with polypyrrole which are described in detail below.

Polymerization on Polyhedral α-Fe_2O_3. The electron micrograph in Figure 1a illustrates the morphology of α-Fe_2O_3 particles which had the i.e.p. at pH 4.3. Dispersions of these rust-colored polyhedra turned almost black within 5 min when mixed with pyrrole, ethanol, and water at 100°C. There was no evidence that particulate polypyrrole had formed in the solution and subsequently adsorbed onto the iron oxide surface. The thickness of the polypyrrole coating could be controlled by adjusting the time of contact of the metal oxide with the polymerization medium. The electron micrograph in Figure 1b is of particles coated over a period of 5 days.

Table I. Coating of Metal Oxides with Polypyrrole

Sample No.	Oxides Treated with Pyrrole	Coating	C/H	Net TGA wt. loss (%)	i.e.p. (after coating)
1	Polyhedral α-Fe_2O_3	Yes	11.3	37.0	3.9
2	Spindle-type purified α-Fe_2O_3	No	---	---	8.0
3	Spindle-type α-Fe_2O_3 treated with HCl	Yes	11.6	27.0	4.0
4	Spindle-type α-Fe_2O_3 in the mother liquor	Yes	12.0	48.6	3.8
5	α-Fe_2O_3 coated SiO_2	Yes	7.5	11.2	4.2
6	Calcined CeO_2 (1 day)	Yes	10.2	9.0	---
7	Calcined CeO_2 (5 days)	Yes	11.2	30.0	---
8	Fe_3O_4	No	---	---	---
9	Fe_3O_4 treated with HCl	Yes	---	32.7	---
10	Fe_3O_4 treated with H_2O_2	Yes	---	23.0	---

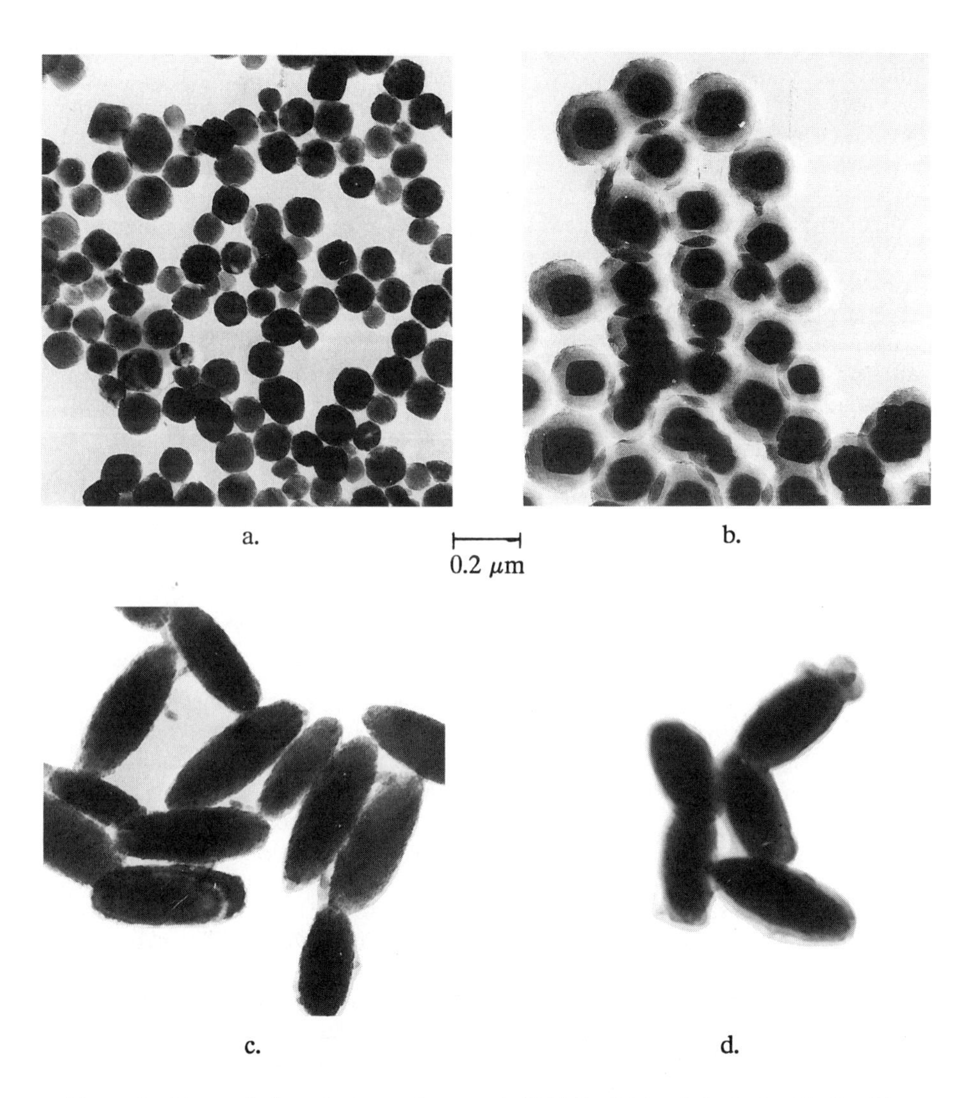

Figure 1. Transmission electron micrograph (TEM) of (a) polyhedral α-Fe_2O_3, (b) polypyrrole coated particles of the same cores, (c) spindle-type α-Fe_2O_3, and (d) coated spindle-type α-Fe_2O_3 pretreated with HCl. (reprinted with permission from J. Colloid Interface Sci., ref. 28)

The ability of these cores to initiate the polymerization of pyrrole is dependent both on the method of preparation of the hematite dispersion and on the solvents used during polymerization. For example, if the α-Fe_2O_3 particles are repeatedly washed with distilled water until their i.e.p. is at pH 7.5 before mixing with pyrrole, no reaction will take place. Likewise, α-Fe_2O_3 with an i.e.p. at pH 4.3 reacted with dilute NaH_2PO_4, then after being thoroughly washed, proved to be much less active and only a thin coating of polypyrrole was formed, even after 5 days. The same particles can be reactivated by stirring in 10^{-4} mol dm^{-3} HCl and then washing until the supernatant solution is neutral.

Polymerization on Spindle-shaped α-Fe_2O_3. The electron micrograph in Figure 1c is of coated spindle-shaped hematite particles. The cores were used as prepared in the original mother liquor (i.e.p. at pH 4.0). The same hematite, after washing several times with distilled water, does not initiate polypyrrole formation on the particle surfaces or in solution following the described coating procedure. However, after stirring in 10^{-4} mol dm^{-3} HCl for several hours and washing until the supernatant solution was neutral, these solids could be uniformly coated with polypyrrole (Figure 1d) at about the same rate as polyhedral α-Fe_2O_3.

Polymerization on SiO_2 doped with α-Fe_2O_3. Silica particles coated with hematite proved effective in initiating polypyrrole formation, while SiO_2 alone was inert. Figure 2a illustrates such composite particles; the dark spots appearing in the polymer coating are due to the nodules of hematite. Entry 5 in Table I shows that the SiO_2/α-Fe_2O_3 core particles are covered to a lesser degree with polypyrrole than pure α-Fe_2O_3 cores and that the product contains a lower C/H ratio.

Polymerization on CeO_2. Both original and calcined CeO_2 particles initiate the formation of polypyrrole on their surfaces at a faster rate than that induced by α-Fe_2O_3. This result is in direct contrast to a report that CeO_2 particles could not be coated, even when the reaction medium contained aqueous $FeCl_3$ (*25*). The transmission electron micrograph showed that the outer portion of each uncalcined particle was quite porous. This property allows pyrrole monomer to impregnate into the surface and makes the polymer coating difficult to be observed by electron microscopy. As expected, calcination of the CeO_2 powder before coating densifies the surfaces and subsequently deposited polymer is clearly visible (Figure 2b).

Polymerization on Fe_3O_4. Magnetite powder does not initiate polypyrrole formation under the conditions of the experiments reported unless the particles are first treated with an oxidizing agent. Figures 3a and 3b are electron micrographs of Fe_3O_4 particles formed by forced hydrolysis of $FeSO_4$, followed by immersion in either HCl or H_2O_2, before and after coating with polypyrrole, respectively.

Magnetite particles formed by reduction of hematite at elevated temperature are more uniform in size and also require activation in order to initiate polymer coating. Entries 9 and 10 in Table I and Figures 3c and 3d characterize those coated with polypyrrole after surface activation with HCl or H_2O_2, followed by mixing with the monomer at 100°C for 5 days.

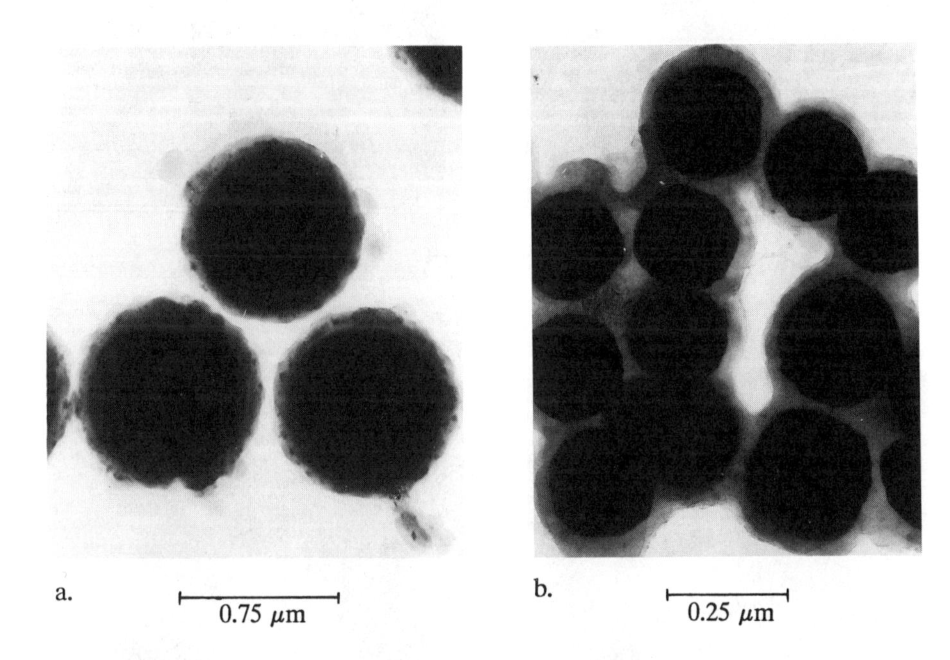

Figure 2. TEM of polypyrrole coated particles of (a) $SiO_2/\alpha\text{-}Fe_2O_3$, and (b) calcined CeO_2. (reprinted with permission from J. Colloid Interface Sci., ref. 28)

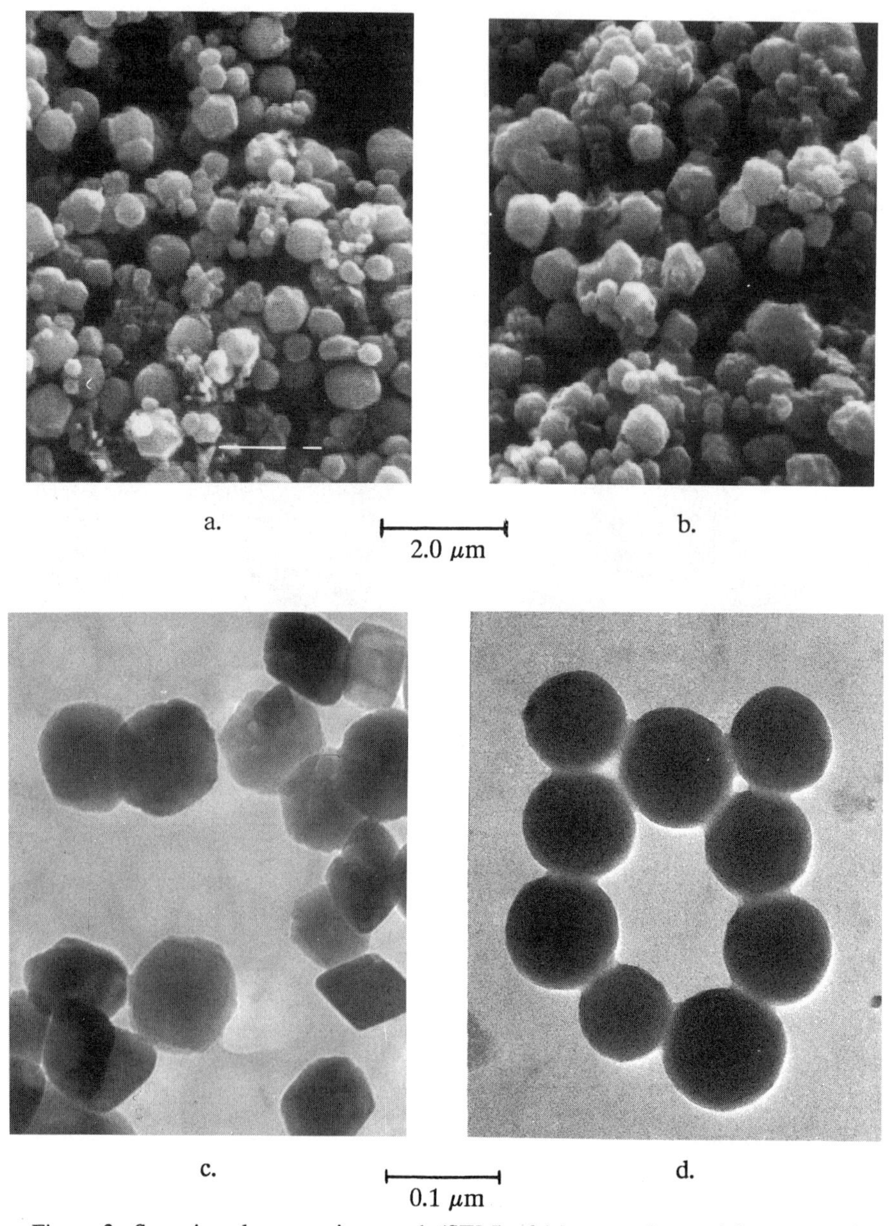

Figure 3. Scanning electron micrograph (SEM) of (a) magnetite particles prepared by hydrolysis of $FeSO_4$, (b) and the same particles coated with polypyrrole; TEM of magnetite obtained from hematite (c) treated with HCl and coated with polypyrrole, and (d) treated with H_2O_2 and coated with polypyrrole.

Coating $Y(OH)CO_3$ on Polypyrrole Latex. Forced hydrolysis of yttrium ions in aqueous urea solutions containing suspended polypyrrole latex at elevated temperature results in the formation of coated particles. Their morphology varies widely as a function of reactant concentrations, reaction temperature, and aging time (Figure 4).

Particle Characterizations. Figure 5 displays the electrophoretic mobility data of pure polypyrrole and of the two kinds of hematite cores coated with this polymer. In the case of polyhedral particles essentially the same dependence on the pH is observed, indicating that the particles were fully covered with the polymer. The mobilities of the negatively charged coated spindle-type α-Fe_2O_3 are much higher than of the cores alone, over the basic pH range.

The electrophoretic mobility data for polypyrrole particles coated with $Y(OH)CO_3$ places the i.e.p. at pH 8.2, well above that for pure polypyrrole. The coated particles shown in Figure 4a have sharply decreasing mobility in the pH range 8.2-9.0 compared to core polymer, while those shown in Figure 4c have a curve superimposable on that of polypyrrole.

Figure 6 displays the TGA data for the polyhedral hematite coated with polypyrrole (sample shown in Figure 1b). It is typical of all samples listed in Table I. There is a rather smooth weight loss commencing at ~ 140°C and ending at ~ 450°C. The inflection in the curve suggests that either two different polymer species are evolved, or that the mechanism of degradation changes as the temperature increases, which was not evident in the TGA of the other types of composites. The non-crosslinked polypyrrole coatings described here degrade and volatilize, which is in contrast to the conversion of crosslinked polystyrene latex to higher surface area carbon powders (*37*).

The TGA measurements with magnetite particles prepared from hematite and treated with H_2O_2 exhibited a small weight loss commencing at 225°C, a broad peak between 350°C, and 450°C, and no weight change above 500°C. Overall, the sample gains approximately 10% weight. In contrast, the same magnetite treated with HCl shows only a 3% weight loss at 300°C. The unexplained peak associated with the former core material also appears on the TGA curve of the particles coated with polypyrrole and the total weight loss reported in entry 10 of Table I takes this into account.

Thermogravimetric analysis of polypyrrole latex coated with $Y(OH)CO_3$ indicates that the particles typically contain more than 60% inorganic material. Major weight loss occurs smoothly from 200° to 500°C, and then more slowly up to 700°C. No attempt was made to verify that the powder remaining was Y_2O_3.

X-ray diffraction spectra have shown that pure polypyrrole is amorphous with a broad peak centered around $2\theta \approx 22$ (*18,38*). Figure 7 compares the X-ray data of the spindle-type particles and for the same material coated with polypyrrole (Table I, entry 4, and Figure 1d). In both cases, the spectrum is characteristic of hematite.

Conductance. Figure 8 shows the d.c conductivities of pure polypyrrole, and polypyrrole coated on either α-Fe_2O_3 or SiO_2/α-Fe_2O_3. While the values for all systems depend on pressure as expected, the coated particles are less conductive than pure polymer. Furthermore, the solids with the pure hematite cores are more efficient than the silica particles partially covered with hematite.

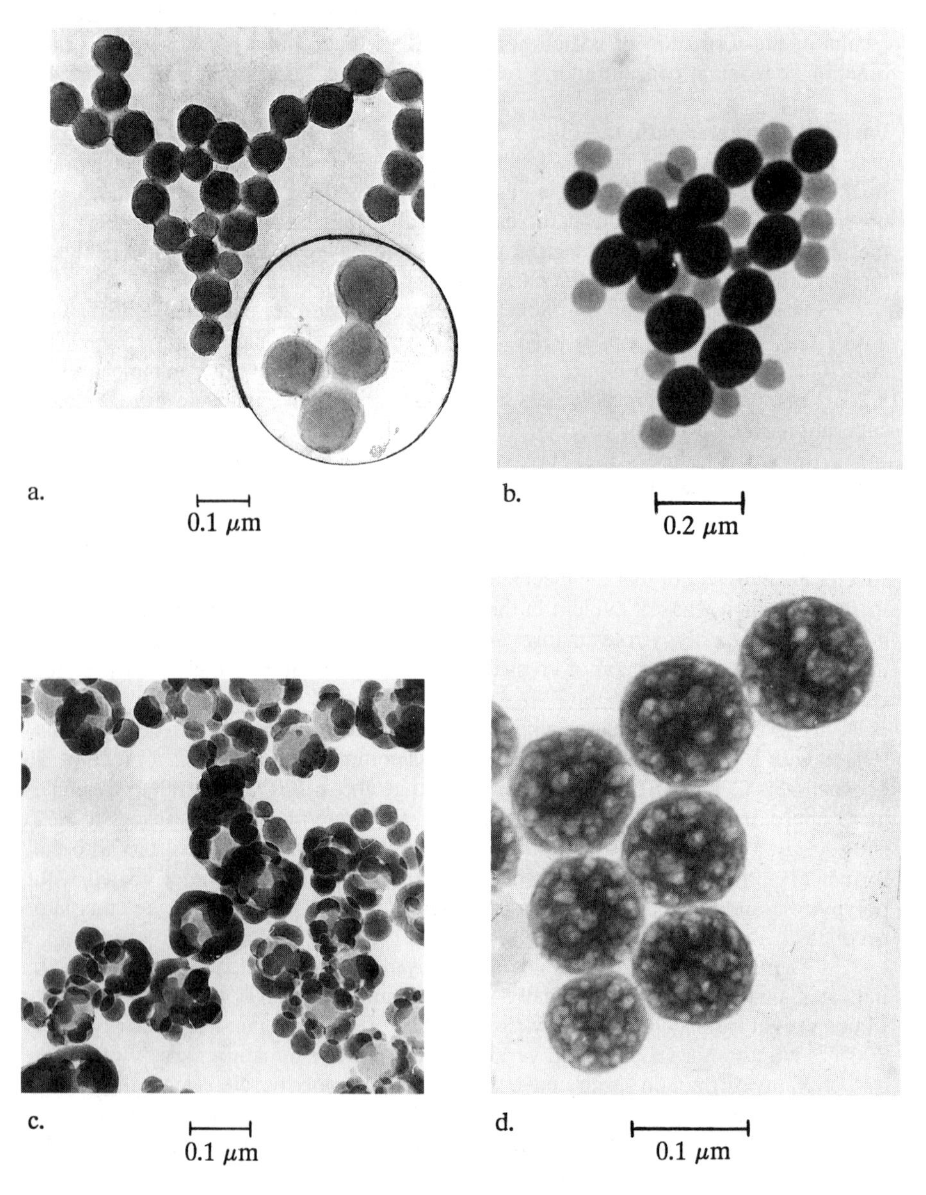

Figure 4. TEM of a polypyrrole latex (100 mg dm^{-3}) coated with $Y(OH)CO_3$ under conditions (a) 10^{-3} mol dm^{-3} $Y(NO_3)_3$, 3.6 mol dm^{-3} urea, 1.2 weight percent PVP, 87°C, 40 h; (b) 10^{-2} mol dm^{-3} $Y(NO_3)_3$, 1.8 mol dm^{-3} urea, 1.2 weight percent PVP, 90°C, 6 h; (c) 1.2 x 10^{-3} mol dm^{-3} $Y(NO_3)_3$, 1.8 mol dm^{-3} urea, 0.5 weight percent PVP, 86°C, 24 hr; and (d) 10^{-2} mol dm^{-3} $Y(NO_3)_3$, 1.8 mol dm^{-3} urea, 1.2 weight percent PVP, 100°C, 24 h.

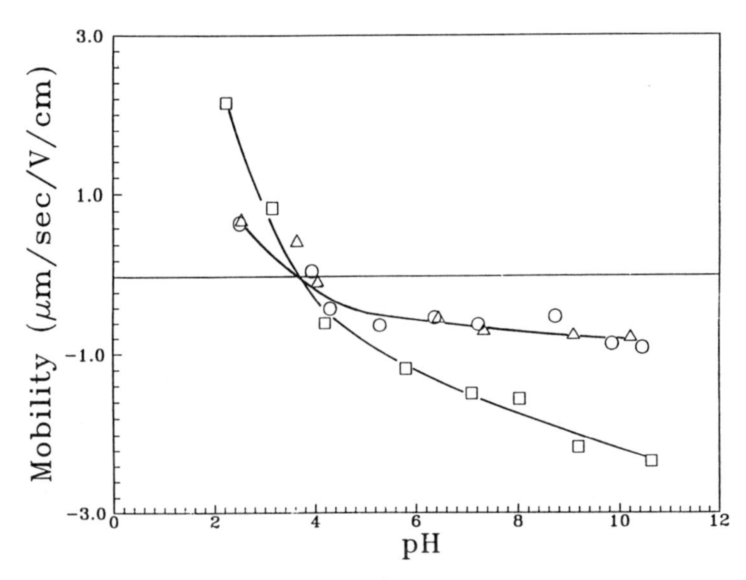

Figure 5. Electrokinetic mobilities of polypyrrole (○), and polypyrrole coated $SiO_2/\alpha\text{-}Fe_2O_3$ (Δ) and of spindle-type $\alpha\text{-}Fe_2O_3$ particles (□), as a function of the pH. (reprinted with permission from J. Colloid Interface Sci., ref. 28)

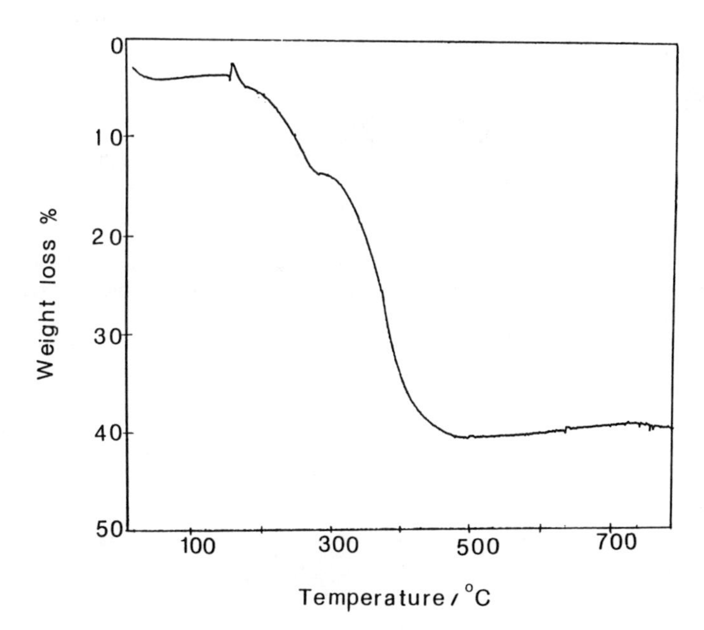

Figure 6. Thermogravimetric analysis of polyhedral $\alpha\text{-}Fe_2O_3$ particles coated with polypyrrole in N_2 atmosphere. (reprinted with permission from J. Colloid Interface Sci., ref. 28)

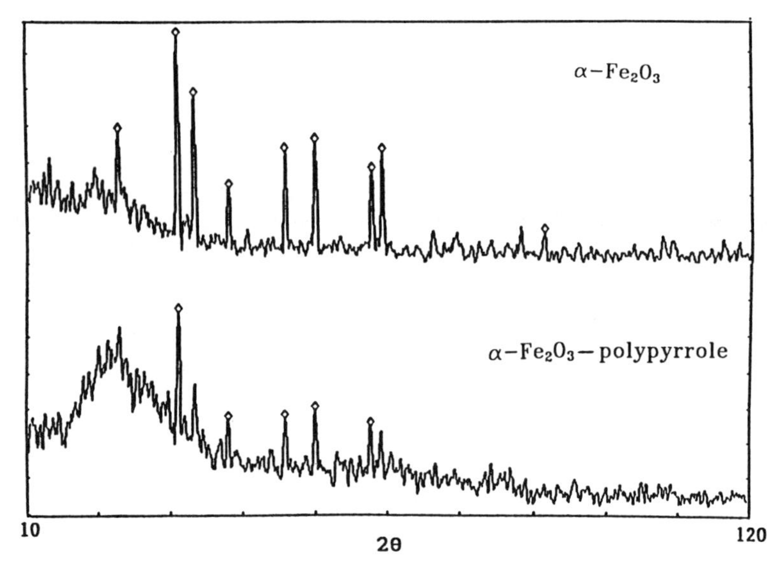

Figure 7. X-ray diffraction pattern for spindle-type α-Fe_2O_3 and of the same particles coated with polypyrrole. (reprinted with permission from J. Colloid Interface Sci., ref. 28)

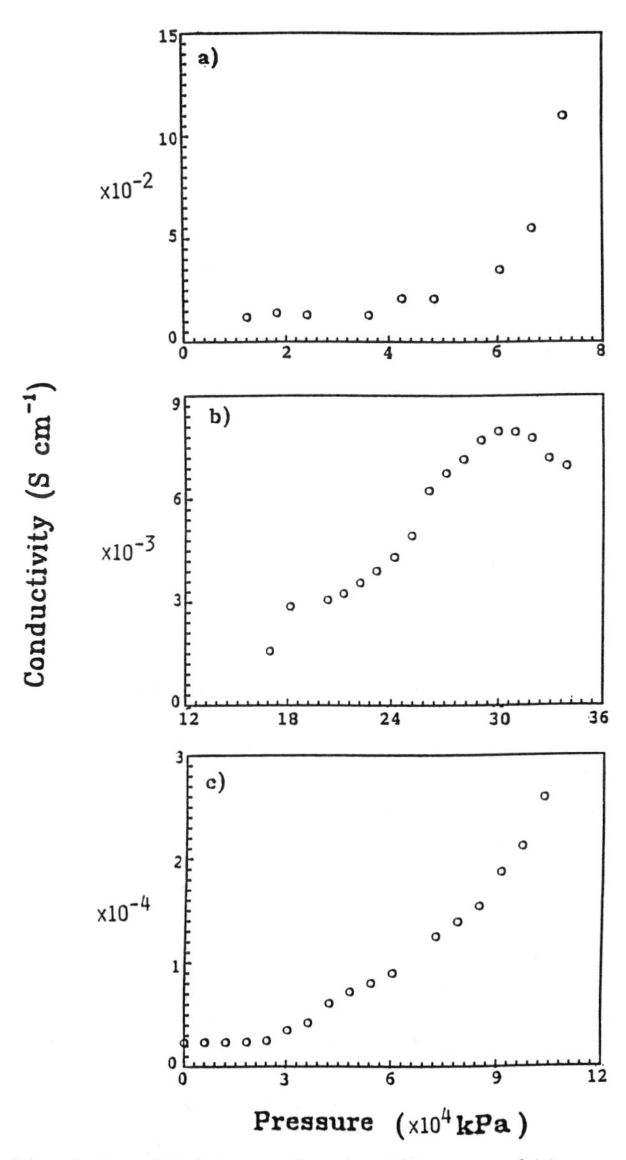

Figure 8. Electrical conductivity as a function of pressure of (a) pure polypyrrole, (b) spindle-type α-Fe_2O_3 coated with polypyrrole, and (c) SiO_2/α-Fe_2O_3 coated with polypyrrole. (reprinted with permission from J. Colloid Interface Sci., ref. 28)

The conductivity of pressed powder of magnetite coated with polypyrrole is significantly less than that of the composites containing hematite. Values of 10^{-10}-10^{-8} S cm^{-1} were obtained for both entries 9 and 10 in Table I.

Discussion

This study appears to be the first strong evidence that the polymerization of pyrrole can be initiated at a site on the surface of metal oxide particles without the need to include zeolite support structures, to add solute oxidants, or to carry out electrolysis oxidation. All of these latter conditions have heretofore been widely used for the synthesis of pure polypyrrole, in particulate or film form, as well as for coating polypyrrole on the surface of suspended solids. In contrast, no free ferric ions could be detected in the solutions which catalyzed pyrrole polymerization in this work.

The ability of polyhedral-shaped α-Fe_2O_3 particles to initiate the polymerization of pyrrole is dependent on both the method of preparation of the hematite dispersion and on the solvents used during polymerization. For example, if α-Fe_2O_3 particles are repeatedly washed with distilled water until their i.e.p. is at pH 7.5 before mixing with pyrrole, the dispersion does not induce polymerization. Likewise, treatment of α-Fe_2O_3 with dilute NaH_2PO_4 reduces activity and only a thin coating of polypyrrole is formed after a lengthy reaction. The finding that inactive particles can cause polymerization of pyrrole after they have been treated with HCl indicates that the presence of Cl^- in hematite enhances its catalytic activity. A low i.e.p. of α-Fe_2O_3 has been associated with chloride ion contamination (*39*). Indeed, if a thoroughly purified dispersion of hematite is aged for extended periods of time, the i.e.p. decreases due to diffusion of Cl^- from the particle interior into the surface layer (*30,39*). These anions can be again removed by extensive washing. In the present experiments, the Cl^- ion was reintroduced by equilibrating purified hematite particles with HCl. The effect of the phosphate ion is also readily understood, if one considers that this anion is more strongly complexed than Cl^- and, therefore, it cannot be replaced by treating the dispersion with NaCl.

In a previously reported work, it was shown that the adsorption of polypyrrole synthesized in a $FeCl_3$ solution required the core particles to have an i.e.p. at pH < 5, which is probably due to the charge difference between the substrate and the polymer (*25*).

In one experiment, in which methyl acetate was replaced for a portion of the water in the polymerization reaction, it was qualitatively observed that a heavier polypyrrole coating formed than in the water-ethanol medium. This result suggests that the choice of solvent may be a useful variable in controlling the rate at which polymer chain growth takes place on surfaces. Results from other laboratories have shown only that the conductivity of pure polypyrrole depended on the kind of ester solvent used with $FeCl_3$ (*40*).

Systems with spindle-type α-Fe_2O_3 exhibit in general a similar behavior as the polyhedral hematite. Coating of the former particles, activated with HCl, is regulated by the history of the core particles; when filtered after treatment by 10^{-4} mol dm^{-3} HCl and thoroughly washed (Table I, entry 3), the cores are covered by about half the amount of polypyrrole than when used in the mother liquor (Table I, entry 4).

On the other hand, regardless of the type of hematite used, the C/H ratio in the coated product is the same (Table I, entries 1,3,4).

Silica coated with iron oxide proved effective in initiating polypyrrole formation, although to a lesser degree than hematite alone. In contrast, silica alone was inert. An earlier study had shown that SiO_2 could be coated with the polymer only when the particles were suspended in an aqueous solution of pyrrole and $FeCl_3$ (*25*).

The polymerization on CeO_2 particles, as used in this work, proceeded faster than on hematite. This example again indicates the sensitivity of the substrate in terms of its catalytic activity.

Magnetite particles from either preparative route are capable of initiating the polymerization of pyrrole only after treatment with HCl or H_2O_2. This may be due to the ability of the latter reactants to convert surface ferrous into ferric ions. It is not yet known what the influence of repeated washing and i.e.p. value, or the nature of anion has on Fe_3O_4 activity towards the organic monomer. TGA data show that over 30% w/w polymer coating is formed on these particles in only 24 h while it takes 5 days to achieve the same loading on hematite cores. Comparison of Figures 3c and 3d indicates that the method of surface activation influences the morphology of the Fe_3O_4-polypyrrole latex formed.

Yttrium basic carbonate deposition on cationic polystyrene occurs when latex is suspended in a heated solution containing ions of the metal, urea, and polyvinylpyrrolidone (PVP) as a dispersant (*29*). It has now been determined that polypyrrole can also serve as core material but the reaction conditions that result in smooth versus rough coating are different than for polystyrene. This is presumably due mainly to the surface charge on particles of polypyrrole altering the ability of nuclei of $Y(OH)CO_3$ to heterocoagulate on the polymer surface. The polypyrrole latex was sterically stabilized by poly(vinyl alcohol) (PVA) during its preparation; however, when PVA replaced PVP in the coating reactions the solid isolated was a mixture of uncoated polypyrrole and $Y(OH)CO_3$ particles (Figure 4b). The present results suggest that for a smooth coating of the inorganic material to be formed the concentration of urea should be higher and the reaction carried out at a lower temperature and for longer periods of time than when the core particles are composed of polystyrene (similar to Figure 4a).

The electron micrographs and TGA data prove that the surfaces of various core particles have been covered with an organic substance, but it is only by measuring their conductivity that the coating can be compared to pure polypyrrole. When a sample of the latter is pressed into a pellet the value obtained (Figure 8, top) is due to the continuum of the single chemical in the cavity of the cell. Depending on the method of polymer synthesis, conductivities ranging from 10^{-3}-10 S cm^{-1} have been reported for polypyrrole and theories have been developed to explain how the property varies with pressure and temperature (*9,17,41,42*). The situation is more complicated when the powder of the coated particles is placed in the four-point probe apparatus, because the pressure exerted may cause the harder core material to squeeze the softer polymer coating into isolated pockets in the matrix and thereby yield erroneous conductivity data. Fortunately it is possible to use literature data on the core oxides as an indicator. Hematite has a very low conductivity (less than 10^{-12} S cm^{-1}) and becomes an n-type semiconductor when irradiated with light of

$\lambda < 570$ nm (*43,44*). The data in Figure 8 clearly imply that polypyrrole, and not hematite or silica, is responsible for the conductivity of the two types of composites evaluated. Our preliminary measurements of the conductivity of the magnetite-polypyrrole composite suggest that in this case the inorganic material may be playing a significant role. Since a magnet was used to isolate the particles after completion of the suspension polymerization reactions there is no question that they contain magnetite. However, pyrrole did not react with Fe_3O_4 cores unless they were first treated with HCl or H_2O_2, which are reagents capable of converting a surface layer into nonmagnetic but more reactive Fe_2O_3 (*45*). Magnetite has a conductivity in excess of 100 S cm^{-1} at room temperature (*46*) but would have a much lower value if coated with hematite. Even though the magnetite-polypyrrole composites contain from 23-32 weight percent polymer, similar to that in the hematite-containing systems, their conductivity (10^{-10}-10^{-8} S cm^{-1} at 0-6 x 10^4 kPa) seems to be controlled by the small amount of surface hematite needed to initiate the polymerization leading to coating.

Conclusion

It has been demonstrated that active sites on the surfaces of particles of α-Fe_2O_3, α-Fe_2O_3/SiO_2, Fe_3O_4 and CeO_2 are capable of initiating the polymerization of pyrrole. This is the first report of such a successful reaction without aqueous oxidant or an electric current present in the reaction medium. The coating thickness may be easily controlled by selecting reaction conditions. Preliminary conductivity measurements verify that the polypyrrole is similar to that described in the literature.

Acknowledgments

Supported in part by U.S. Air Force Contract F49620-85-C-0142 and by the New York Science and Technology Foundation. Correspondence should be addressed to Richard Partch. The authors are grateful to Professor S. Arajs and W. Cai and M. Zhang, Physics Department, Clarkson University for helpful discussions and conductivity measurements.

Literature Cited

1. *Handbook of Conducting Polymers*; Skotheim, T., Ed.; Marcel Dekker: New York, 1986; Vols. 1-2.
2. Patil, A.; Heeger, A.; Wudl, F. *Chem. Rev.* **1988**, *88*, 183.
3. Kaner, R.; MacDiarmid, A. *Scientific American*, February 1988; p. 106.
4. Yassar, A.; Roncali, J.; Garnier, F.; Michel, J.-M.; Bonnebat, C. *Fr. Demande* **1988**, 2616790.
5. Reynolds, J. *CHEMTECH.* **1988**, 441.
6. Keller, T. *CHEMTECH.* **1988**, 635.
7. MacDiarmid, A.; Epstein, A. *Faraday Discuss. Chem. Soc.* **1989**, *88*, 317.
8. Duke, C.; Gibson, H. In *Kirk-Othmer's Encyclopedia of Chemical Technology*; Mark, H.; Othmer, D.; Overberger, C.; Seaborg, G., Eds.; Wiley Interscience: New York, 1982, Vol. 18; p. 755.

9. Reynolds, J. *J. Molec. Electronics* **1986**, *2*, 1.
10. Patil, A.; Ikenone, Y.; Wudl, F.; Heeger, A. *J. Am. Chem. Soc.* **1987**, *109*, 1858.
11. Foot, P.; Simon, R. *J. Phys. D: Appl. Phys.* **1989**, *22*, 1598.
12. Armour, M.; Davies, A.; Upadhyay, J.; Wassermann, A. *J. Polym. Sci., A-1* **1967**, *5*, 1527.
13. Wegner, G.; Rühe, J. *Faraday Discuss. Chem. Soc.* **1989**, *88*, 333.
14. Borman, S. *Chemical & Engineering News*, May 7, 1990; p. 53.
15. Cholli, A. *Chemical & Engineering News*, June 4, 1990; p. 2.
16. Mao, H.; Ochmanska, J.; Paulse, C.; Pickup, P. *Faraday Discuss. Chem. Soc.* **1989**, *88*, 165; Heeger, A.; *ibid.*, 203.
17. Glatzhofer, D.; Ulanski, J.; Wegner, G. *Polymer* **1987**, *28*, 449.
18. Zhong, C.; Doblhofer, K.; Weinberg, G. *Faraday Discuss. Chem. Soc.* **1989**, *88*, 307.
19. Gibson, H.; Bailey, F.; Epstein, A.; Rommelmann, H.; Pochan, J. *J. Chem. Soc. Chem. Comm.* **1989**, 426.
20. Armes, S.; Aldissi, M.; Agnew, S. *Synthetic Metals* **1989**, *28*, C837.
21. Iyoda, T.; Ando, M.; Kaneko, T.; Ohtani, A.; Shimidzu, T.; Honda, K. *Langmuir* **1987**, *3*, 1170.
22. Lyons, M.; McCormack, D.; Smyth, O.; Bartlett, P. *Faraday Discuss. Chem. Soc.* **1989**, *88*, 139, 177 and references therein.
23. Cooper, E.; Vincent, B. *J. Phys. D: Appl. Phys.* **1989**, *22*, 1580.
24. Yassar, A.; Roncali, J.; Garnier, F. *Polym. Commun.* **1987**, *28*, 103.
25. Yoneyama, H.; Shoji, Y.; Kawai, K. *Chem. Lett.* **1989**, 1067.
26. Kanatzidis, M.; Tonge, L.; Marks, T.; Marcy, H.; Kannewurf, C. *J. Am. Chem. Soc.* **1987**, *109*, 3797.
27. Bein, T.; Enzel, P. *Angew. Chem.* **1989**, *101*, 1737.
28. Partch, R.; Gangolli, S. G.; Matijević, E.; Cai, W.; Arajs, S. *J. Colloid Interface Sci.* **1991**, *144*, 27.
29. Kawahashi, N.; Matijević, E. *J. Colloid Interface Sci.* **1990**, *138*, 534.
30. Matijević, E.; Scheiner, P. *J. Colloid Interface Sci.* **1978**, *63*, 509.
31. Ishikawa, T.; Matijević, E. *Langmuir* **1988**, *4*, 26.
32. Amin, N.; Arajs, S.; Matijević, E. *Phys. Status Solidi A* **1987**, *101*, 233.
33. Hsu, W. P.; Rönnquist, L.; Matijević, E. *Langmuir* **1988**, *4*, 31.
34. Bassett, J.; Denney, R.; Jeffrey, G.; Mendham, J. In *Vogel's Quantitative Inorganic Analysis*; Longman: England, 1982, Fourth Ed.
35. Stöber, W.; Fink, A.; Bohn, E. *J. Colloid Interface Sci.* **1968**, *26*, 62.
36. Gherardi, P.; Matijević, E. *J. Colloid Interface Sci.* **1986**, *109*, 57.
37. Gangolli, S.; Partch, R.; Matijević, E. *Colloids Surf.* **1989**, *41*, 339.
38. Cvetko, G.; Brungs, M.; Burford, R.; Skyllas-Kazacos, M. *J. Mater. Sci.* **1988**, *23*, 2102.
39. Hesleitner, P.; Babić, D.; Kallay, N.; Matijević, E. *Langmuir* **1987**, *3*, 815.
40. Armes, S.; Aldissi, M. In *Proceedings 21st Europhysics Conference on Macromolecular Physics*, Lodz, Poland, September 1989.
41. Lundberg, B.; Sundqvist, B.; Inganas, O.; Lundström, I.; Saloneck, W. *Mol. Cryst. Liq. Cryst.* **1985**, *118*, 155.

42. Ezquerra, T.; Mohammadi, M.; Kremer, F.; Vilgis, T.; Wegner, G. *J. Phys. C: Solid State Phys.* **1988**, *21*, 927.
43. Balbashow, A.; Lebedev, A.; Pavlova, S.; Bakhetuzov, V. *Acta Phys. Pol. A* **1985**, *A68*, 457.
44. Grätzel, M.; Kiwi, J.; Morrison, C. *J. Chem. Soc. Faraday Trans. I* **1985**, *81*, 1883.
45. Gribanov, N.; Bibik, E.; Buzunov, O.; Naumov, V. *J. Magn. Magn. Mater.* **1990**, *85*, 7.
46. Tannhauser, D. *J. Phys. Chem. Solids* **1962**, *23*, 25.

RECEIVED December 4, 1991

Chapter 24

Nucleation of Emulsion Polymerization in the Presence of Small Silica Particles

Philippe Espiard[1], André Revillon[1], Alain Guyot[1], and James E. Mark[2]

[1]**Laboratoire des Matériaux Organiques, Centre National de la Recherche Scientifique, B.P. 24, 69390 Vernaison, France**
[2]**Polymer Research Center and Department of Chemistry, University of Cincinnati, Cincinnati, OH 45221–0172**

Several silica were used : either non functionalized silica from commercial origin or prepared according to the Stöber method, or functionalized silica prepared upon surface coverage of non functionalized silica with functional coupling agents such as $X(CH_2)_3Si(OR)_3$ where X is either a methacryloyl group or the precursor amino group for an azo compound. Some organophobic silica can be prepared via a sol-gel process from co-hydrolysis and co-condensation of tetraethoxysilane (TEOS) and the functional coupling agent with water in inverse microemulsion in the presence of a suitable surfactant. The average diameter of the silica particles is in the range 20-90 nm. When the silica is hydrophilic, encapsulation with polymer upon emulsion copolymerization of highly hydrophobic monomers (styrene, butyl acrylate) is not observed : a regular latex is produced independently from the silica, even if coverage of silica with surfactant has been previously carried out. Partial success has been obtained using limited amounts of methyl methacrylate. Very good results have been obtained for ethyl acrylate polymerization with a semi-continuous feed process, provided a suitable nonionic surfactant is used. The dispersability of functionalized silica is poor, but may be improved after a suitable treatment with ammonia, and a combination of an alcohol and a nonionic surfactant. Then, each latex particle is nucleated by the silica, again in the case of ethyl acrylate.

Dispersion of small mineral particles in polymer matrices has received an increasing amount of interest these last few years. A popular approach is to prepare these particles in situ, and this has been carried out extensively by one of us (J.E.M) using various elastomer matrices *(1)*. An extension of this approach leads to the preparation of new materials called ceramers *(2)*. On the other hand, some years ago a Japanese team gave the first reports of emulsion

0097–6156/92/0492–0387$06.00/0

polymerization of methyl methacrylate (MMA) in the presence of various inorganic materials such as calcium carbonate *(3,4)*, graphite *(5)*, calcium sulfite *(6)*, or barium sulfate *(7,8)*. Most of these experiments led to encapsulation of the particles, even if polymerizations have been carried out without emulsifier. A patent from Union Carbide *(9)* claims a general method for polymer encapsulation of dispersed solids. Basic research involving encapsulation of silica has been reported recently. For instance Hergerth et al *(10-14)* have covered quartz powder with a diameter of 26 nm, with methyl methacrylate or vinyl acetate. In some cases previous treatments have been carried out in order to anchor organic materials onto the silica *(15,16)* or onto TiO_2 *(17,18)*. Surface active monomer has been used in one case *(19)* and finally there is one report about vinyl modified silica covered with polyvinylpyrrolidone *(20)*.

The present paper describes our first work along this line, in order to try to understand the possibilities and the limitations of this approach, and to determine the necessary conditions to reach the final goal which is to obtain latex particles containing one and, hopefully, only one silica particle in its core. Therefore, we have used a variety of silica, functionalized or not, as well as various monomers which were more or less hydrophilic.

Experimental

The Ludox AS40 (Dupont) and Degussa Aerosil A 200 V are commercial products. The preparation of silica via a sol-gel process in inverse microemulsion has been previously described *(23)*. Trimethoxypropylmethacrylate silane (MPS), γ-aminopropyl triethoxy silane (APS) and azobiscyanovaleric acid (ACVA) are commercial products from Fluka.

Functionalization of Commercial Silica. A first technique to prepare these functionalized silicas is to react a typical coupling agent for composite materials with the silanol groups of a commercial silica, (Degussa A 200 V) according to scheme I.

The typical procedure for reaction Ia or Ib (scheme I) is as follows : Dried Degussa A 200 V silica is suspended in dry toluene, i.e. toluene distilled over molecular sieve under nitrogen atmosphere. After addition of an excess of MPS or APS (8 $\mu mol/m^2$), the mixture is refluxed for 16 hours under a N_2 atmosphere, the weight ratio of toluene/silica being 30. The support was washed 3 times with toluene, separated by centrifugation and dried in vacuum.

The grafting of ACVA onto APS modified silica (scheme Ic) is carried out by reaction of 4.5 g of the modified silica with 1 g of ACVA in 250 ml of dry tétrahydrofuran (THF), to which 0.8 ml of triethylamine and 0.5 g of chloroformiate are added. Temperature is kept at -78°C for the first half hour and then allowed to reach -10°C, where the reaction was run for 17 hours. Grafted silica is separated from THF by centrifugation.

Emulsion Polymerizations with Silica

Ludox Silica and Modified Degussa Silica. Emulsion polymerizations are performed at 60°C in a 1.5 L jacketed flask under a nitrogen atmosphere. Prior to polymerization, silica particles are dispersed in water which is then flushed with nitrogen in order to remove oxygen. These silica particles are emulsified with different surfactants : sodium dodecyl sulfate (SDS, Prolabo), or dodecyl trimethylammonium bromide (Sigma), or polyoxyethylene lauryl ether (E4C12, Aldrich), or nonylphenyl polyoxyethylene (NP30). All experiments were carried out with deionized water. Different kinds of monomers were involved : styrene

(Prolabo), butyl acrylate (Norsolor), methyl methacrylate (Aldrich) and ethyl acrylate (Kodak). Each of them was distilled at reduced pressure to remove the inhibitor.

The monomer/water weight ratio was kept constant at 10 % whereas the silica content was variable. Silica/monomer weight ratios from 1/1 to 1/64 were investigated. When experiments were carried out using a semi-continuous feed process, an addition rate of 0.34 ml/min was chosen in order to prevent the formation of polymer particles without silica inside them. Potassium persulfate ($K_2S_2O_8$, Fisher) was employed as initiator at 0.7 g/L. The emulsion was stirred at 250 rpm during polymerization.

Monomer conversion was determined by taking samples during the entire course of the polymerization and by measuring their solids content. Hydroquinone was added to each sample in order to stop the reaction. Verification of silica encapsulation was undertaken by different techniques on the final latex :

- by measuring the particle size (transmission electron microscopy, photon correlation spectroscopy, hydrodynamic chromatography)
- by centrifugation at 18000 rpm and analyzing the different observed layers (if any) by elemental analysis and Fourier Transform Infra-red spectroscopy.
- by comparison of the number of initial silica particles with the number of final latex particles

Silica from the Water-in-Oil (W/O) Emulsion. Polymerization with this last kind of silica was undertaken by a three step process :

1st step : w/o polymerization of 0.8 g of acrylic acid (Aldrich) at 60°C in a 100 ml round bottom flask containing 50 ml of the emulsified silica obtained by the sol-gel process from hydrolysis and condensation of TEOS and MPS in a water-in-toluene process *(23)* ; the average particle diameter being 64 nm. 0.017 g of 2,2 azoisobutyronitrile (AIBN, Prolabo) was used as initiator. The resulting emulsion had some characteristic bluish effects.

2nd step : the w/o emulsion was inverted in an o/w emulsion and the remaining toluene evaporated with an evaporator.

3rd step : emulsion polymerization of ethyl acrylate took place in this emulsion at 60°C using semi-continuous feed process with 0.85 g/L of $K_2S_2O_8$ as initiator.

Results and Discussion

Silica Particles

Hydrophilic Silica. We have first used non-functionalized commercial silica, Ludox AS40, i.e., silica containing only hydroxyl groups on its surface, which consists of non-porous uniform particles with an average diameter of 22 nm, and a surface area of 140 m^2/g ; the particles are stabilized by ammonium ions and carry a negative charge, the pH of a 40 weight % suspension being 9.2.

Hydrophobic silica. Then we have been working with a second kind of silica ; functionalized silica, i.e silica carrying functional, alkyl groups on the surface.

SCHEME I. FUNCTIONALIZATION OF SILICA WITH COUPLING AGENTS
a) grafting of a methacryloyl comonomer (MPS)
b) grafting of the aminopropyltriethoxysilane (APS)
c) reaction of the azobiscyanovaleric acid (ACVA) with the APS modified silica.

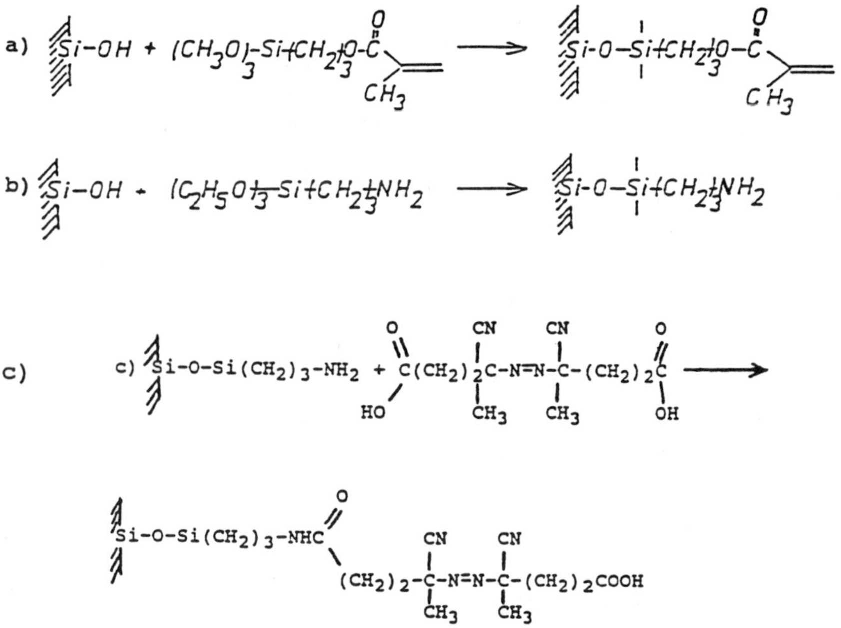

a) $\equiv Si{-}OH + (CH_3O)_3{-}Si{-}(CH_2)_3{-}O{-}C(=O){-}C(CH_3){=}CH_2 \longrightarrow \equiv Si{-}O{-}Si{-}(CH_2)_3{-}O{-}C(=O){-}C(CH_3){=}CH_2$

b) $\equiv Si{-}OH + (C_2H_5O)_3{-}Si{-}(CH_2)_3{-}NH_2 \longrightarrow \equiv Si{-}O{-}Si{-}(CH_2)_3{-}NH_2$

c) $\equiv Si{-}O{-}Si(CH_2)_3{-}NH_2 + HO{-}C(=O)(CH_2)_2C(CN)(CH_3){-}N{=}N{-}C(CN)(CH_3){-}(CH_2)_2C(=O){-}OH \longrightarrow$

$\equiv Si{-}O{-}Si(CH_2)_3{-}NHC(=O){-}(CH_2)_2{-}C(CN)(CH_3){-}N{=}N{-}C(CN)(CH_3){-}(CH_2)_2COOH$

Grafting of methacryloylpropyl trimethoxy silane (MPS) gives a grafted reactive monomer (scheme Ia) while successive reactions of γ-aminopropyltriethoxy silane (APS) with surface silanol groups and of azobiscyanovaleric acid (ACVA) with the amino groups of the grafted APS silica gives a radical generator (schemes I b and c). These reactions were already well documented in the literature *(20-22)*.

These grafted silicas were characterized from elemental analysis (C for MPS grafted and N for APS and ACVA grafted materials) from which it can be estimated according to the Berendsen formula *(24)* that 90 % of the silanol groups have been reacted with MPS while the yield was 72 % in the case of APS, the conversion from APS to ACVA being 61 %.

Infrared spectroscopy (Figures 1-3) and ^{29}Si CPMAS NMR spectra (Figures 4 and 5) have been used for further characterization. The grafting of MPS is evidenced by the infrared bands at 1708-1719 cm^{-1} $\nu(C{=}O)$, 1637 cm^{-1} $\nu(C{=}C)$ 2737 cm^{-1} $\nu(CH_2)$ and 2922-2944 cm^{-1} $\nu(CH_3)$ (Figure 1). For APS, the two bands at 3352 and 3301 cm^{-1} are characteristic for non-hydrogen bonded and hydrogen bonded NH_2 groups, respectively in addition to the $\nu(CH_2)$ at 2922 cm^{-1} and $\nu(CH_3)$ at 2973 cm^{-1} (Figure 2). For the ACVA-APS grafted initiator there is a broad band from 2500 to 3500 cm^{-1} due to OH groups of the

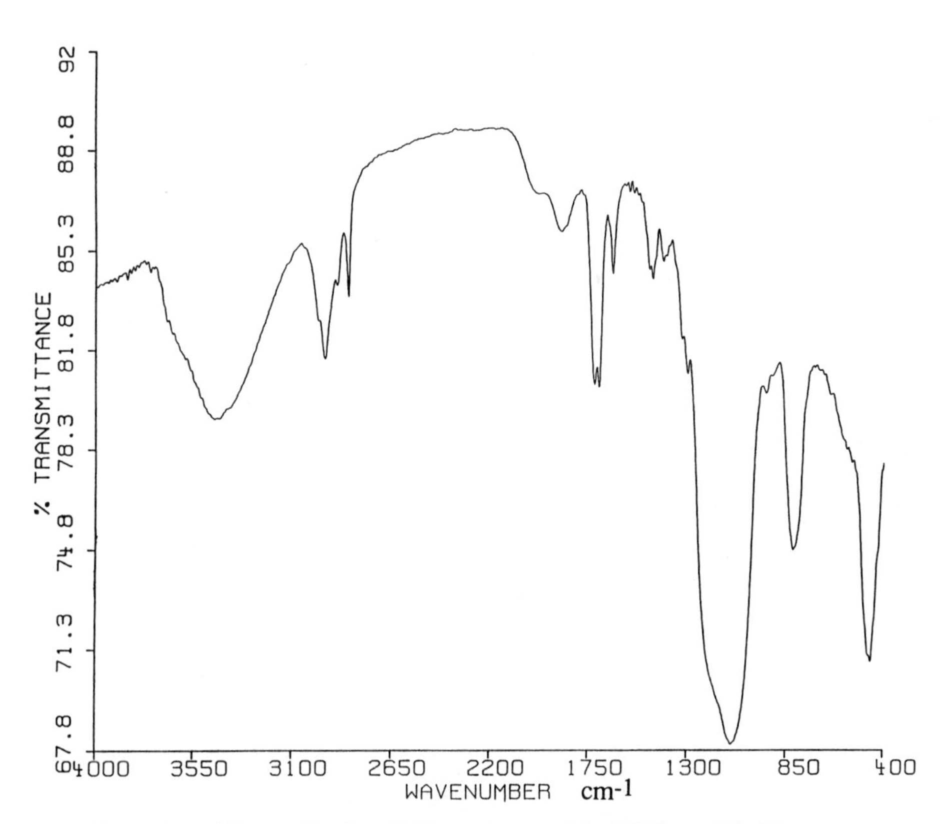

Figure 1. Diffuse reflection FTIR spectrum of the MPS modified Degussa silica.

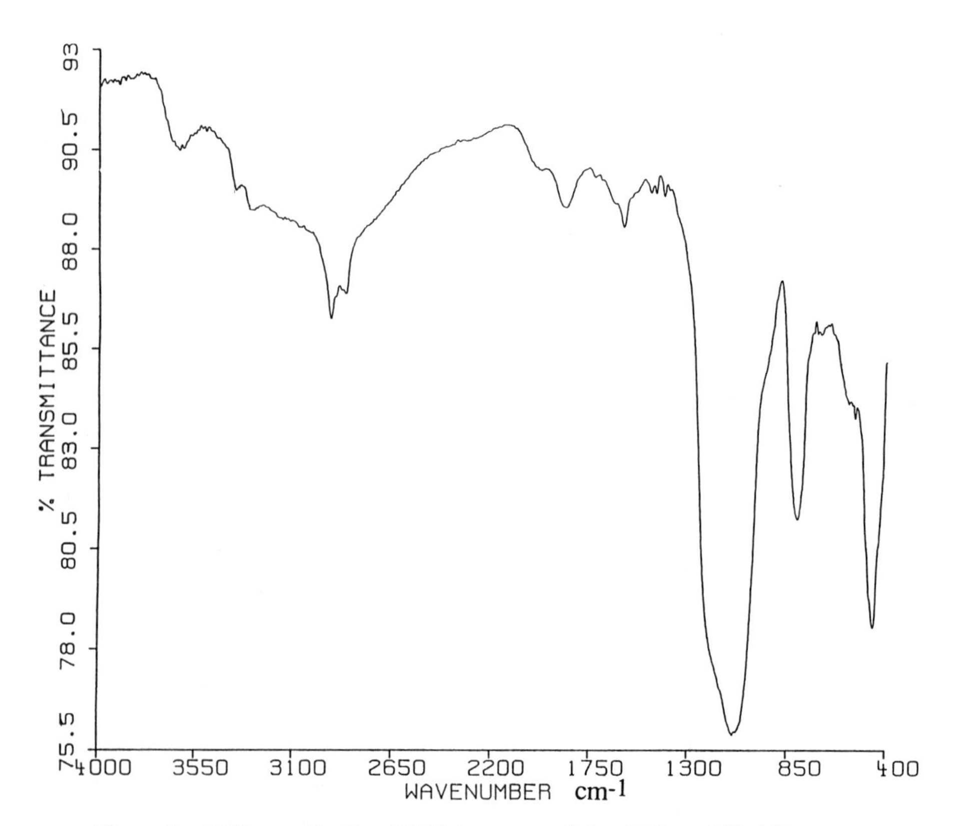

Figure 2. Diffuse reflection FTIR spectrum of the APS modified Degussa silica.

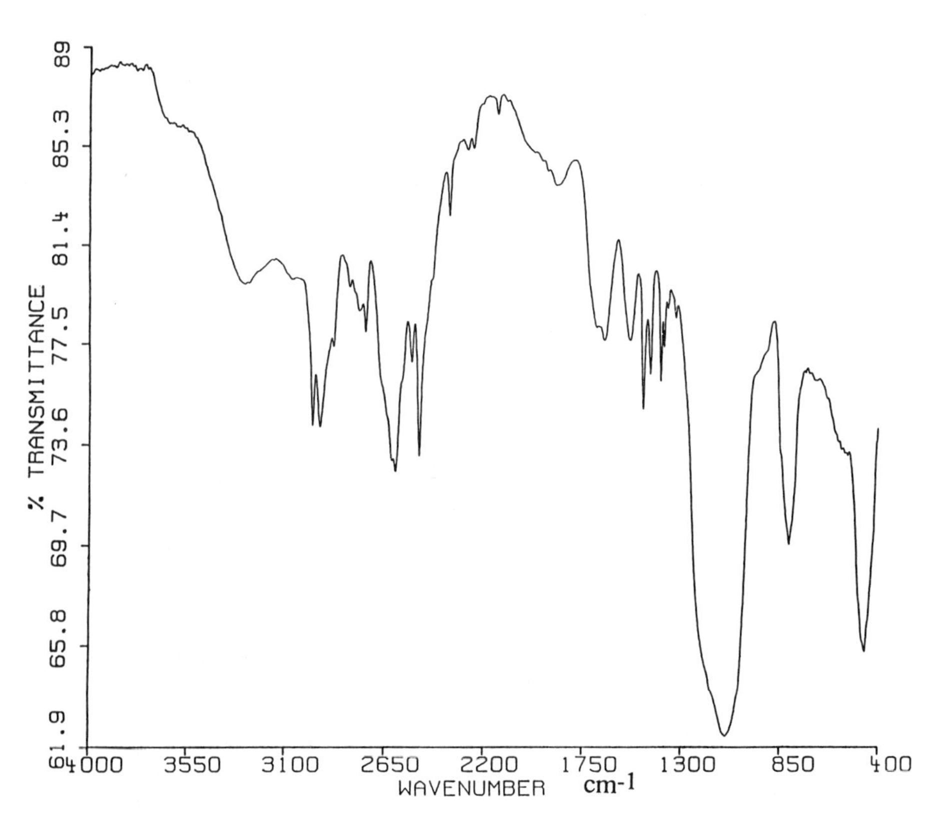

Figure 3. Diffuse reflection FTIR of the ACVA modified APS modified Degussa silica.

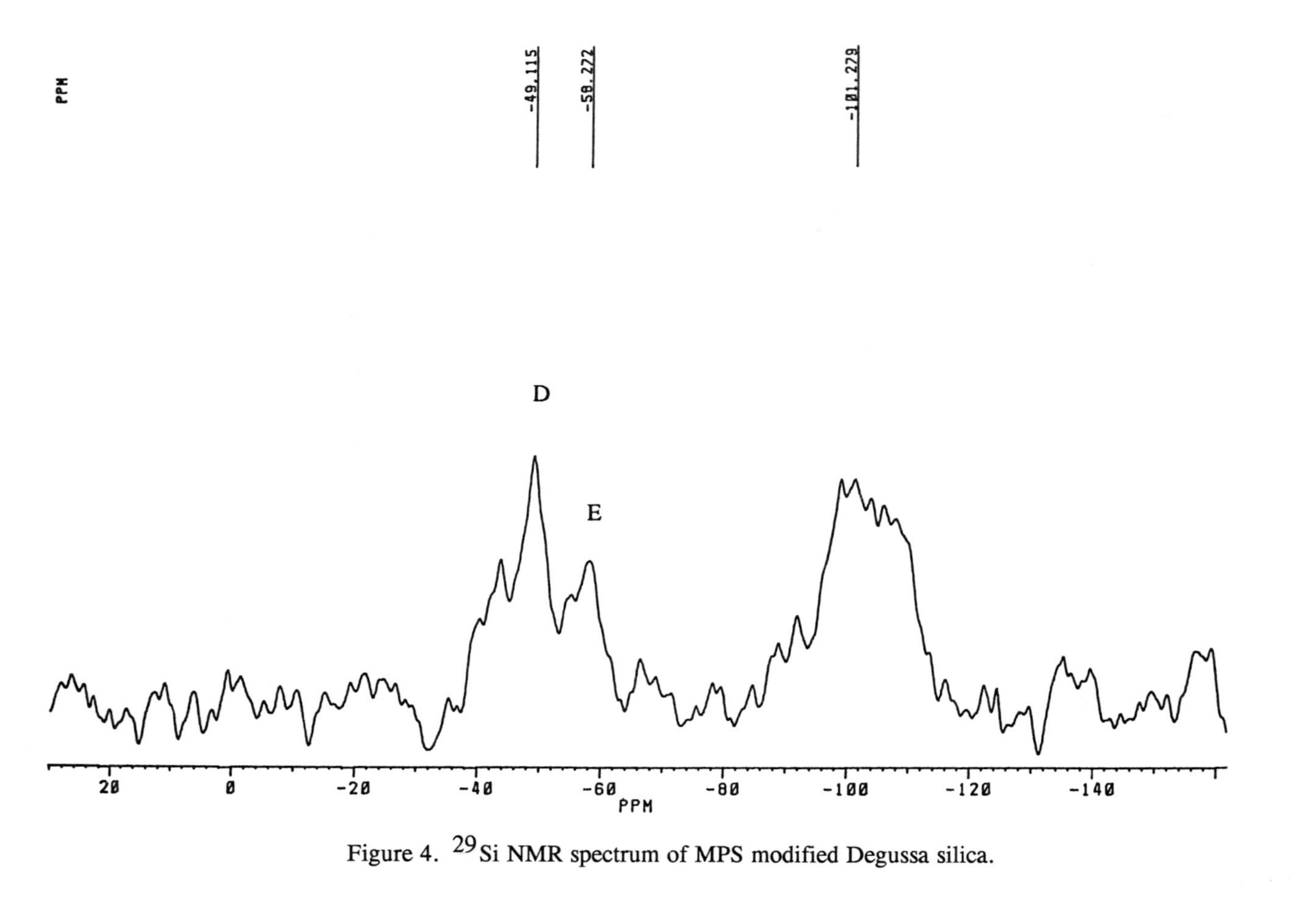

Figure 4. ^{29}Si NMR spectrum of MPS modified Degussa silica.

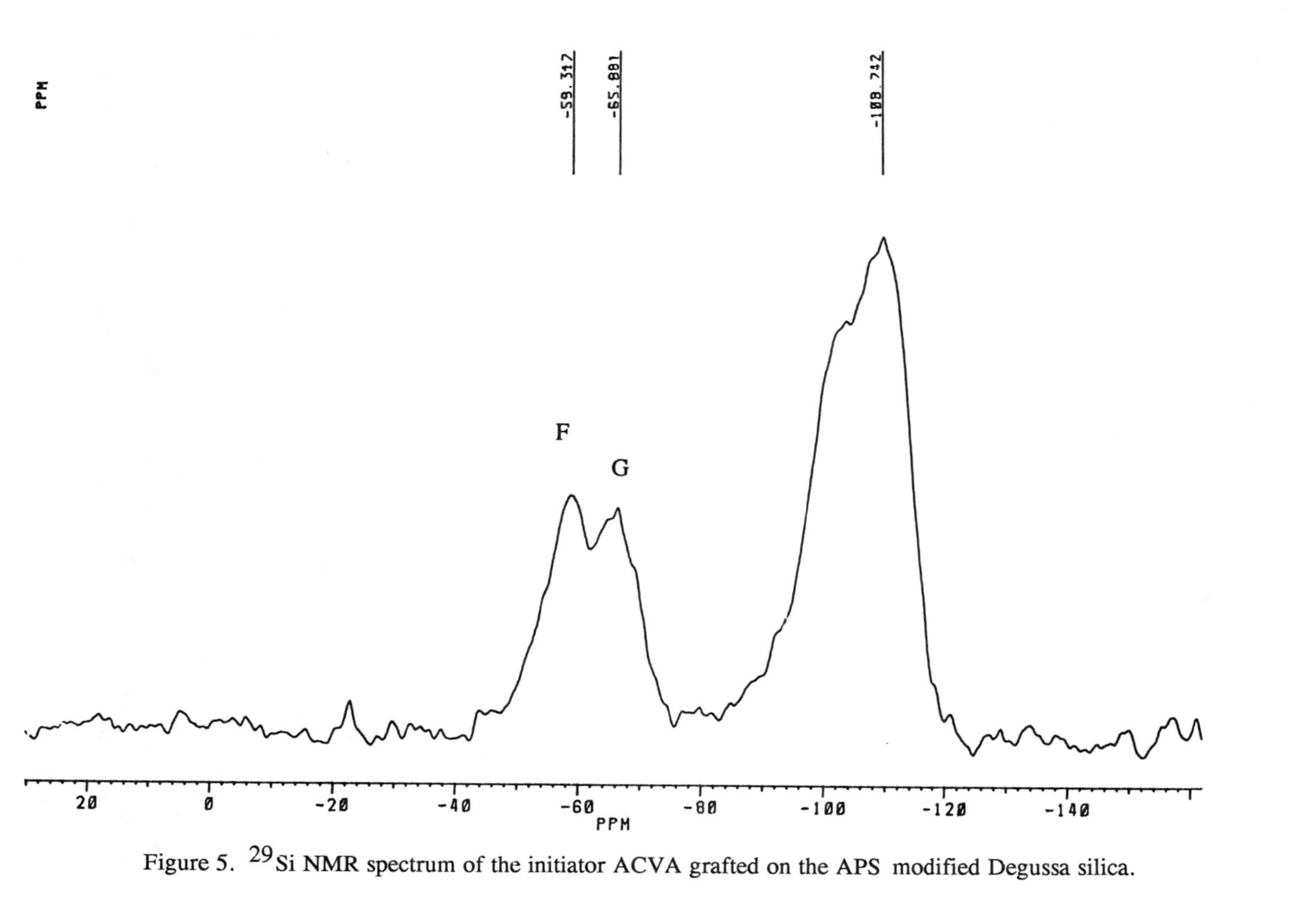

Figure 5. ^{29}Si NMR spectrum of the initiator ACVA grafted on the APS modified Degussa silica.

carboxylic bands, as well as many poorly resolved bands in the range 1500-1800 cm^{-1} due to ν(C=O) of amide and acid groups, ν(N=N) of the azo group or ν(C-N) of the amide ; the most characteristic band is the narrow one at 2300 cm^{-1} for ν(C=N) (Figure 3).In Figure 4, in addition to the bands between -80 and -120 ppm due to the surface Si atoms of the silica (Si-O-Si), Si-OH and $Si(OH)_2$, one can observe the bands D and E respectively, at -49 and -58 ppm which can be assigned to the following structures in scheme II. Figure 5 shows similar structures F and G associated with the bands at -58 and -65 ppm, respectively.

SCHEME II: Structure of ^{29}Si CPMAS NMR observable Si atoms ; R being for the propyl chains of MPS or ACVA-APS grafted groups

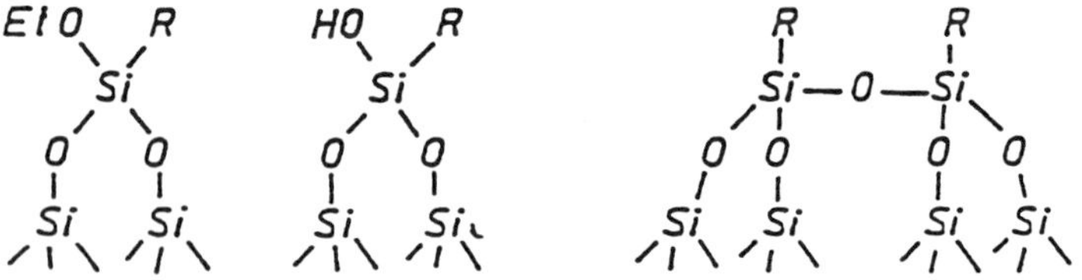

(F) (G)

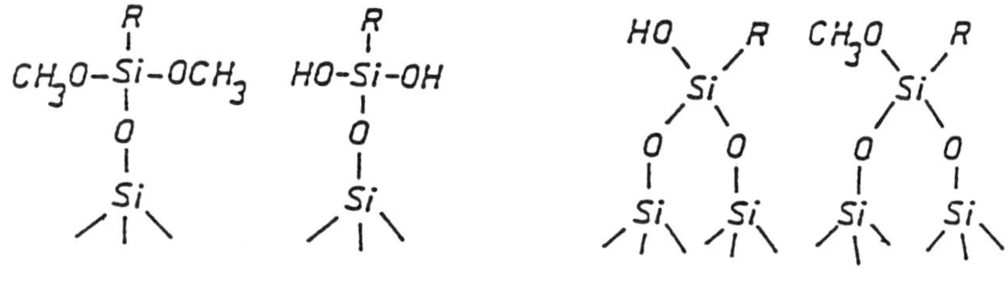

(D) (E)

These modified commercial silicas were hardly dispersable in water and lead to aggregates, unless suitably treated with ammonia at pH 9.5 to make them less hydrophobic ; in addition a small amount of isobutanol must be added to the water. Basic treatment regenerates hydroxyl groups on silica particles which were chemically and physically aggregated. Element analysis of the ammonia treated silica indicated that none of the surface MPS was hydrolyzed whereas thermal gravimetric analysis showed an increase of the OH groups.

A second technique used to prepare functionalized silica was described in a previous paper *(23)* ; a sol-gel technique in inverse microemulsion was used to prepare directly functionalized silica upon co-hydrolysis and co-condensation of TEOS and MPS with an inverse microemulsion of water-in-toluene in the presence of diethylhexyl sodium sulfosuccinate (Aerosol AOT) as an emulsifier.

The main characteristics of the silica used for emulsion polymerization are summarized in Table I.

Table 1. Hydrophilicity and size of the silica

SILICA	FUNCTIONALIZATION	CHARACTER	PRIMARY PARTICLE DIAMETRE/nm
I Ludox	no	Hydrophilic	22
II Degussa A200V	MPS	Hydrophobic	Aggregates
II' Degussa A200V treated at pH 9.5 + Isobutanol	MPS	Hydrophobic	37
III Degussa A200V	APS-ACVA	Hydrophobic	Aggregates
IV W/O emulsion	MPS	Hydrophobic	64

In a first series of polymerization experiments, emulsion copolymerization of the hydrophobic monomers, styrene and butyl acrylate (in the mole ratio 1/1) was carried out in the presence of silica samples I, II and III, and sodium dodecyl sulfate (SDS) as an emulsifier. The SDS amount was limited so as not to form micelles of SDS but only have SDS either solubilized in water or adsorbed onto the surface of silica. The required amount of SDS (2 g/L) was deduced from an independent study of the adsorption of SDS onto silica by conductimetric titration, showing that the formation of micelles may be observed if the amount of SDS is higher than 2.4 g/L. (i.e the critical micelle concentration CMC).

When hydrophilic silica I was used, no encapsulation was observed : the emulsion polymerization initiated with potassium persulfate produces a latex which can be separated from silica upon centrifugation, and non-modified silica can be recovered ; this is confirmed by FTIR analysis of both phases (latex and silica). The same negative results are obtained, whatever the ratio between monomers and silica, which varied from 10 to 140. Most probably, the emulsifier which is initially either adsorbed onto the silica particles or dissolved in the water phase is essentially used to stabilize the latex which is formed ; actually, the size of the latex particles remains practically constant around 75 nm in diameter, whatever the amount of silica ; in these experiments, the concentration of SDS was fixed at 2 g/l. However, when silica was covered with the nonionic emulsifier NP30, encapsulation by the copolymer styrene-butylacrylate occured but a polydisperse latex was the result.

Using sample II modified by MPS, one gets two kinds of materials : again a latex corresponding to 96.5 % of the monomers and a powder of silica encapsulated with polymer. Identical results were obtained using sample III modified with APS and then ACVA. Attempts to vary the mode of introduction of the monomer (semi-continuous process instead of batch), the nature of the emulsifier (cationic instead of anionic) and its amount (less or more than the CMC) did not lead to improvements of the extent of encapsulation.

At this point, it was decided to work with more hydrophilic monomers, such as methyl methacrylate (MMA) or ethyl acrylate (EA), so as to get a better affinity between monomers and silica. In addition, a better dispersion of the MPS modified Degussa silica was obtained after a suitable treatment with ammonia at pH 9.5. Taking into account that 80 % of the MPS is located on the silica surface, this treatment makes the silica less hydrophobic. The initial dispersion is further improved if a small amount of isobutanol is added. It was also decided to limit the amount of monomer employed, the ratio of monomer to silica being limited to 4 in batch experiments ; otherwise semi-continuous process under monomer starvation conditions was used. The results of a first series of experiments is reported in Table II, where a non ionic emulsifier $C_{12}H_{25}(OCH_2CH_2)_4OH$ ($C_{12}EO_4$) was used. According to the literature (25-27) non-ionic emulsifiers containing ethylene oxide groups are strongly adsorbed on silica surfaces. In addition, their adsorption is higher and stronger than ionic emulsifiers *(32)* such as SDS. Thus, silica dispersion is expected to be improved when non-ionic emulsifiers are used instead of anionic surfactants.

Table II. Emulsion polymerization in the presence of silica particles using $C_{12}H_{25}$-$(OCH_2CH_2)_4OH$ as an emulsifier

SILICA SAMPLE	MONOMER	MONOMER SILICA	CONVERSION %	d /nm b)	LATEX/SILICA PARTICLE NUMBER RATIO a)
I	MMA	1/1	50	35	0.24
I	EA	1/1	38	32	1.13
	EA	2/1	41	37	0.67
	EA	4/1	56	38	1.09
	EA	8/1	39	40	1.27
II'	MMA	3/1	-	30 1700	-
II'	EA	3/1	37	-	-
	EA	4/1	40	36	2.53
	EA	16/1	39	45	10.6

a) Ratio of the final number of latex particles to the initial number of silica particles.
b) Particle diameter number from transmission electron microscopy

Two main remarks : when MMA is used as monomer the silica is not perfectly dispersed and some aggregates of silica particles may be observed, mostly with the functionalized silica (II'). Also, in all the experiments, the conversion remains limited. Attempts to improve these results through a more careful purification of the monomers or addition of more initiator and the choice of the initiator were not successful. When the monomer is EA, the dispersion of the silica is much better and, if the amount of monomer remains limited, the number of latex particles remains close to that of the silica ; this feature is well obeyed with the non-functionalized silica ; however when the functionalized silica is used, there is a definite trend to the formation of pure latex when the amount of monomer is increased.

One core/shell experiment has been carried out using hydrophilic silica I. The purpose of such a core/shell reaction is to encapsulate the hydrophilic Ludox silica I, by the hydrophobic copolymer styrene/butyl acrylate. The latter was shown previously to have no affinity with silica I. Then a layer of PMMA is first created which has enough hydrophilic affinity with silica and enough hydrophobic affinity with the copolymer which is formed later. The first step of MMA polymerization is actually a pretreatment of the hydrophilic silica. In that first step MMA is polymerized with a silica to monomer ratio of 4. The conversion was limited to 33 %. The encapsulated material has a silicon content of 39 % and a diameter of 30 nm. In a second step an equimolar mixture of styrene and butyl acrylate is added so that the ratio of monomers to silica is 1/1. The conversion of that second step was again limited to 37 % and the diameter was increased to 38 nm. The silicon content was still 35 %, so that attempts to produce a film lead to a very brittle material.

A next series of experiments was carried out replacing the C12EO4 emulsifier by another one with a higher HLB, a nonylphenol oxyethylated material (NP30) with an average of 30 ethylene oxide units. The NP30 adsorption on the hydrophilic silica I and hydrophilic silica II' was proved by measuring the CMC through surface tension measurements. The higher the silica content, the higher was the CMC. This result is in agreement with other studies *(25-28)* showing that adsorption of non-ionic surfactants takes place through the ethylene oxide chains on hydrophilic silica and through nonylphenyl chain on hydrophilic silica. Much better results were then obtained with both non-functionalized or functionalized silica. These data are reported in Tables III and IV, respectively. Higher monomer conversions were obtained ; in both cases the number of latex particles remains close to that of silica. In both cases, elemental analysis of the latices separated after centrifugation and washed with toluene confirms the encapsulation of silica. It follows that the size of the particles was decreasing as the amount of silica was increasing. Figure 6 shows transmission electron microscopy of different latexes

A kinetic study has been carried out in both cases. It is clear from the curves shown in Figures 7 and 8 that in both cases the nucleation period of the emulsion polymerization is longer if more silica is used. This is a strong argument to support the idea that the nucleation actually takes place through capture of the growing radicals by silica particles.

The silica prepared upon microemulsion was stabilized by the rather large amount of aerosol AOT emulsifier used for its preparation. However it is rather hydrophobic and tends to agglomerate unless enough non-ionic emulsifier is added. Another hydrophobic silica was prepared using the sol-gel procedure described by Stober *(29)* and then modified by Vrij *(30)* and by Philipse *(31)*. When these silica were used in the polymerization, aggregation was observed whatever the monomer. So, starting from the sol-gel silica IV, which was initially in a water-in-oil emulsion. It was decided to make it hydrophilic through a first step of inverse emulsion polymerization of acrylic acid. Then the emulsion is inverted in water and finally a semi continuous process of ethyl acrylate emulsion polymerization was carried out. The data reported in Table V show that the encapsulation process took place.

Conclusion

The encapsulation of very small silica particles through emulsion polymerization involves a particular set of conditions to be successful. First, the silica must be hydrophilic enough i.e, containing some hydroxyl groups, to be well dispersed in water. Second the monomer must have enough water solubility to encapsulate hydrophilic silica. Methyl methacrylate, ethyl acrylate and as shown by the literature *(12)*, vinyl acetate are convenient. However with styrene or butyl acrylate suitable adsorption of a nonionic surfactant onto hydrophobic silica is required in order to give a hydrophobic character to its surface. A core/shell process seems possible after a first encapsulation step. Concerning MPS modified Degussa silica hydrophilic and hydrophilic monomers are appropriate as this silica carries both character. Third, the amount of monomer involved at the same time must be limited, for instance using a semi-continuous process under starvation conditions. Fourth, an emulsifier with a high HLB is preferred. It seems that the major feature of these conditions is to provide a high capture efficiency by the silica particles of the growing radicals born in the water phase.

Table III. Emulsion polylmerization of ethyl acrylate in the presence of non-functionalized silica I

WEIGHT RATIO MONOMER/SILICA	THEORETICAL SiO_2 CONTENT IN THE POLYMER a)	CONVERSION %	LATEX DIAMETER nm b)	RATIO nb OF PARTICLES LATEX/SILICA	PARTICLES C %	Si %	EXPERIMENTAL SiO_2 % CONTENT IN THE POLYMER
-	0 %	95	76	-	-	-	-
64/1	1.5 %	93.6	60	2.53	43.9	11.7	1.6
32/1	3.0 %	92.8	56	1.15	42.7	12.0	3.3
16/1	5.9 %	92.8	48	0.92	42.2	12.5	6.3
8/1	11.0 %	92.6	43	0.62	41.5	13.7	12
4/1	20.0 %	91.5	37	0.50	39.0	15.8	21.5
2/1	33.0 %	92	28	0.60	34.9	19.8	35.2

a) SiO_2 % calculated suuming full conversion of the monomer
b) Particle diameter number from transmission electron microscopy
c) SiO_2 % calculated from monomer conversion

Table IV. Emulsion polymerization of ethyl acrylate in the presence of functionalized silica II'

WEIGHT RATIO MONOMER/SILICA	THEORETICAL SiO % IN THE POLYMER	CONVERSION %	LATEX DIAMETRE nm b)	RATIO nb OF PARTICLES LATEX/SILICA	C %	Si %	EXPERIMENTAL SiO_2 CONTENT IN THE POLYMER c)
-	0	95	76	-	-	-	-
64/1	1.5	91.6	74	1.46	47.7	12.2	1.7
32/1	3.0	89.5	54	1.54	43.9	12.9	3.4
16/1	6.0	86.3	46	1.3	41.5	13.8	6.7
8/1	11.0	83.5	41	0.88	40.2	15.6	13.0
2/1	33.3	80.9	38	2.5	32.3	20.7	38.2

a) SiO_2 % calculated assuming full conversion of the monomer
b) Particle diameter number from transmission electron microscopy
c) SiO_2 % calculated from monomer conversion

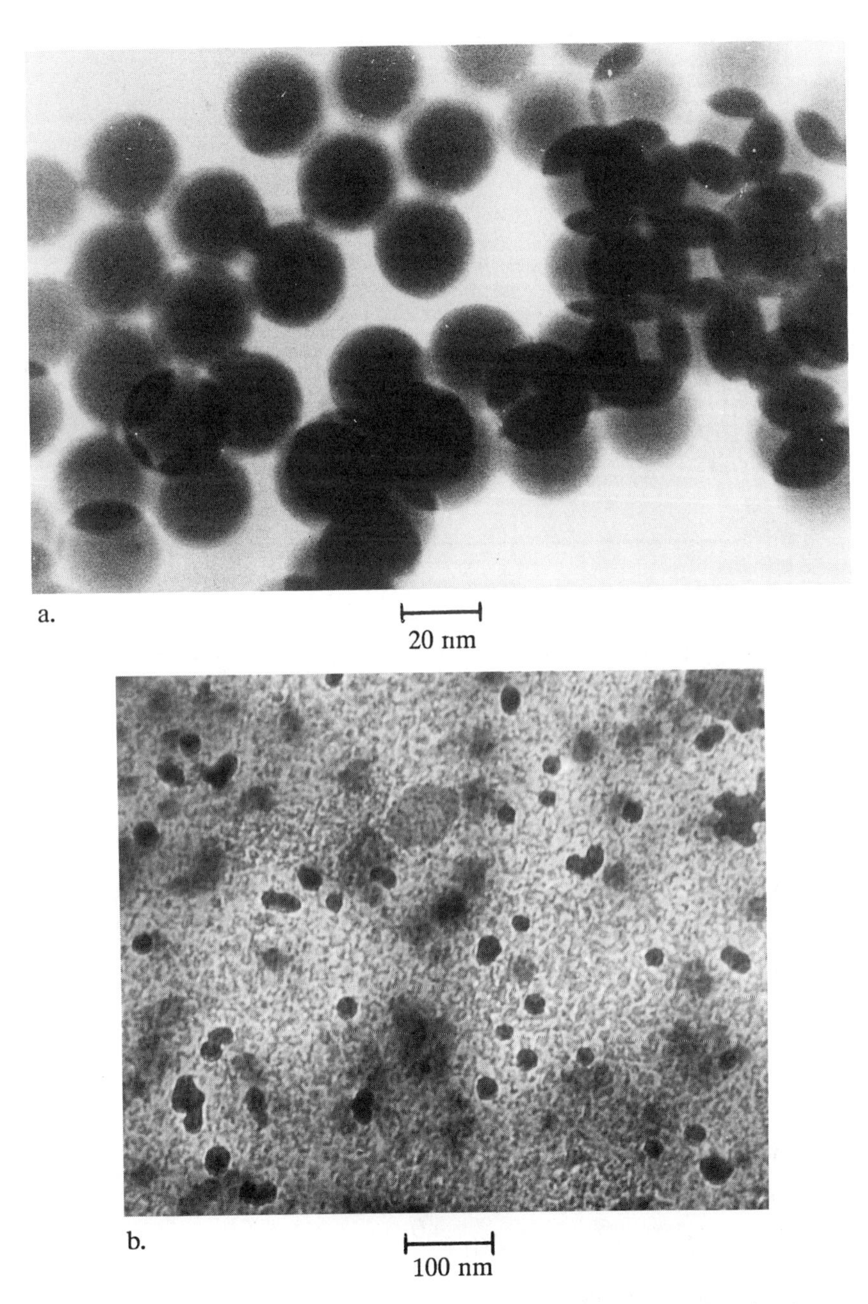

Figure 6. Transmission Electron Microscopy of polyethylacrylate in the presence of a) non functionalized silica I b) functionalized silica II'.

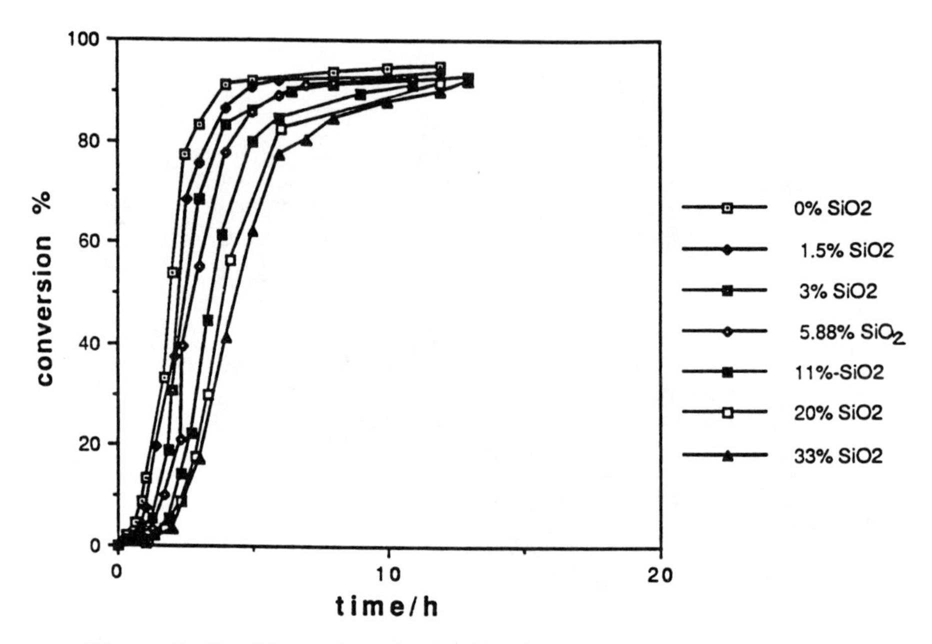

Figure 7. Emulsion polymerization kinetic curves of ethylacrylate with Ludox silica.

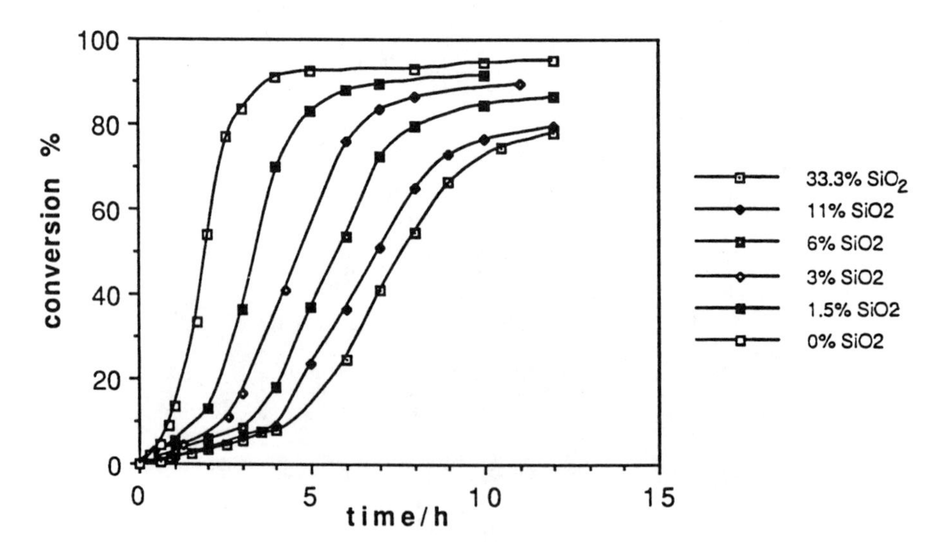

Figure 8. Emulsion polymerization of ethylacrylate with MPS modified Degussa silica. Kinetic curves.

Table V. Two steps core shell emulsion polymerization of ethylacrylate from functionalized silica prepared in water/oil microemulsion

1st step : inverse emulsion polymerization of acrylic acid onto sol-gel silica IV		
	D /nm	Number of particles x 10^{-17}
SILICA	64	1.10
POLYMER	71	0.78

2nd step : emulsion polymerization of ethyl acrylate onto acrylic acid grafted silica IV				
Weight ratio EA/silica	Conversion %	SiO % in the polymer	D /nm	Ratio nb of particles latex/silica
1/10	90.2	10	79	2.19
1/14	91.4	7	75	1.09

Acknowledgments

The authors are indebted to Dr. M.F. Llauro-Darricades and Petiaud for the NMRG experiments ; they want to thank Rhône-Poulenc for support.

Literature cited

1 Mark J.E and Schaeffer D.W. *Mat. Res. Soc. Symp. Proc.*1990 *171*, 51
2 Wilkes G.L., Huang H.H.and Glaser R.H.A.C.S. *Adv. Chem. Ser.* 1990 *224*, 207
3 Yamaguchi T., Ono T. and Ito H. *Angew. Makromol. Chem.* 1973 *32*, 177
4 Hasegawa M , Arai K and Saito S. *J. Chem. Eng.* Japan 1988 *21*, 30
5 Yamaguchi. T., Ono T. and Saito S. *Angew. Makromol. Chem* 1976 *53*, 65
6 Arai M. , Arai K. and Saito S. *J. Polym. Sci. A, Polym. Chem* 1982 *20*, 1021
7 Konno M., Arai K. and Saito S. *J. Polym. Sci. A, Polym. Sci* 1987 *25*, 223
8 Hasegawa M., Arai K. and Saito S. *J. Polym.Sci.A, Polym. Chem.*1987 *25*, 3117
9 Union Carbide Co. Europ. Pat. EP 010498A2 (1983)
10 Hergerth W.D. and Schmutzler K. *Acta Polym.* 1985 *36*, 472
11 Hergerth W.D., Peller M. and Hauptmann P. *Acta Polym.* 1986 *37*, 468
12 Hergerth W.D., Starre P., Schmutzler K. and Wartewig S.*Polymer* 1988 *29*, 1323
13 Hergerth W.D., Steinau U.J., Bittrich H.J., Simon G. and Schmutzler K. *Polymer* 1989 *30*, 254

14 Hergerth W.D., Schmutzler K. and Wartewig S. *Macromol. Symp.* 1990 *31*, 123
15 Furuzawa K., Kimura Y. and Tagawa T. *J. Colloid Interf. Sci.* 1986 *109*, 69
16 Cooper E.C. Vincent B. *J. Colloid Interf. Sci.* 1989 *132*,592
17 Caris C.H.M., Van Elven L.P.M., Van Herk A.M. and German A.L. *British. Polym. J.* 1989 *21*, 133
18 Caris C.H.M., Van Herk A.M. and German A.L. *Makromol. Chem. Macromol. Symp.* 1990 *35-36*, 535
19 Nagai K., Onishi Y., Ishiyama K. and Kuramoto N.J. *Applied Polym. Sci.* 1989 *38*, 2183
20 Wullf G., Oberkobush D. and Minarik M. *Reactive Polym.* 1965 *3*, 261
21 Nakajima M., Shono T. and Kimura K. *Chem. Lett.* 1965 *747*
22 Okamoto Y and Katada K. *J. Liquid Chrom.* 1986 *9*, 369
23 Espiard P., Mark J.E., and Guyot A. *Polymer Bull*, 1990 *24*, 173
24 Berendsen G.E., De Galan L., *J. of liquid chromatography* library, 1978 *1*, 561
25 Somasundaran P., Snell E.D., Fu E., Xu Q.200th *ACS* Meeting, Washington 1990 (to be published in ACS serie)
26 Van Den Boomgaard TH., Tadros TH.F., Lyklema *J.,J. Colloïd Inter. Sci.* Vol *116,* (1), 8-16 (1987)
27 Rapprecht Van H., Liebl H., *Kolloid-Z.U.Z. Polymer* 1972 *250*, 719-723
28 Stober W., Fink A., *J. Colloïd Inter. Sci.* 1968 *26*, 62
29 Van Helden A.K., Jansen J.W., Vrij A. *J. Colloïd Inter. Sci.* 1981 *81*, 354
30 Philipse A.P. *Colloïd* and *Polymer Science* 1988 *266*,11741180
31 Gao Y., Yue C., Lu S., Gu W., Gu T. *J. Colloïd Inter. Sci.* 1984 *100*, 581

RECEIVED December 26, 1991

Chapter 25

Pigmented Polymer Particles with Controlled Morphologies

Wei-Hsin Hou, Thomas B. Lloyd, and Frederick M. Fowkes

Center for Polymer Science and Engineering, Department of Chemistry, Lehigh University, Bethlehem, PA 18015

In this study, mechanisms of phase separation as well as microencapsulation of pigments from polymer solutions are presented. Two types of pigmented polymer particles with different morphologies have been prepared by the microencapsulation process : 1) particles with pigment enriched on the surface and 2) particles with more fully encapsulated pigment. The morphologies as well as the surface properties of the encapsulated particles are controlled by the interactions between pigment, polymer and solvent which are dependent upon the functional groups on the pigment surface, the chemical structure of the polymer and the solvency. Photon correlation spectroscopy and scanning electron microscopy showed that the pigmented polymer particles have narrow size distribution and the size varied from 0.5μm to 2μm depending upon the pigment content and the solvency. Different techniques such as BET adsorption, densitometry, and microelectrophoresis are used to characterize the surface properties of the pigmented polymer particles. These results support a proposed mechanism for the microencapsulation and delineate morphologies of the encapsulated particles.

The process of microencapsulation started with nature's creation of the first living cell. However, no technique had been successfully developed to copy this ingenious process of nature until about fifty years ago. The first process of microencapsulation was developed by Barry Green at National Cash Register Company in the 1930s. Since then, more and more scientists have entered this fascinating field and are producing new processes and new applications which encompass fields from graphic arts to agriculture to medicine. Microencapsulation is the development of small solid particulates,

0097–6156/92/0492–0405$06.00/0

liquid droplets, or gas bubbles with a coating. Basically three different microcapsule structures; namely mononuclear capsules, microcapsule aggregates and double-wall structures have been prepared for different applications*(1)*. There are many approaches to carry out microencapsulation. In general, these approaches can be classified into five categories : polymerization*(2,3)*, phase separation*(4)*, emulsification*(5,6)*, grinding*(7,8)* and spray drying*(9)*. Each technique has its advantages and disadvantages for preparing encapsulated particles; however, in all of these techniques it is still difficult to the control morphologies as well as the surface properties of the encapsulated particles. Therefore, the development of simple techniques to produce uniform particles with controlled morphologies is a challenge and is very important industrially. In this study, attention will be paid to thc phase separation process because it has potential for simplicity in terms of both materials and time.

Formation of colloids by phase separation of polymer solutions by varying the solvency has been reported by several authors*(10,11,12)*. It has been found that under certain conditions, aqueous mixtures of polyvinyl alcohol and colloidal silica formed a coacervate which contains 40% silica*(13)*. Similar studies have also been carried out in polymer solutions containing ceramic powder*(14)*. However, the mechanisms of phase separation of polymer solutions in the presence of particulates are still not well understood. As advanced materials for paints, ceramic composites, magnetic materials and liquid toners, well controlled properties of the materials are required. Therefore, it is important to understand the mechanism of microencapsulation during phase separation and to control the morphology of the encapsulated particles.

Experimental

Materials. Two different carbon black pigments, Monarch 1000 (MO) and Regal L 330 (RL) (Cabot Corporation) were used in this study because of their fairly different surface characteristics. The carbon black pigments were heated at 110°C for 8 hours in a vacuum oven to get rid of adsorbed water, and stored in a desiccator for the further studies.

The polymer used in this study was Griltex Nylon (GN), a nylon 6/nylon 12/nylon 6,12 random copolymer, which was received from EMS-American Grilon/Emser Industries. The polymer was made by ring opening polymerization, and has a molecular weight of approximately 50,000. It was chosen by our sponsor for its quality in liquid toners.

The solvent used for these experiments was a 200 proof ethyl alcohol (Midwest Grain Products, Pekin, Illinois) which is a good solvent for the polymer at higher temperatures but a poor solvent at lower temperatures. Non-solvent used in this study was Isopar G, an isoparaffinic hydrocarbon (Exxon).

Surface Titration of Carbon Blacks. The characterization of surface acidity and basicity of powders by flow microcalorimetry is a versatile technique pioneered by Fowkes and co-workers*(15)*. Briefly, the flow of pure solvent and solution is controlled by twin gear-driven Perfuser VI syringe pumps. An HPLC switching valve allows either pure solvent or solution to flow to the bed in which carbon black is located. Heats of adsorption or desorption are detected while solution or pure solvent are alternately passing through carbon black. The liquid that flowed through the bed was conveyed directly into a Perkin-Elmer LC-75 UV flow detector so that the UV absorbance (or concentration) of solutes could be determined, thereby allowing quantitative measurement of solute adsorption.

Cyclohexane (distilled in glass, Burdick & Jackson Laboratories Inc., Michigan) dried with molecular sieves was used as a carrier solvent. Two solutes, phenol (Aldrich 99%), 20 mM, and pyridine (Fisher, Certified), 20 mM, were used as an acidic and a basic probe, respectively. Carbon blacks were carefully dried before use.

The number and strength of the acidic and basic sites on the RL and MO surfaces relative to the pyridine and phenol probes are summarized in Table I. The results show that RL is more basic than MO due to over twice as many basic sites on the surface, although the molar heat is a little bit smaller than MO. On the other hand, RL is less acidic due to the 40% weaker molar heat of pyridine adsorption than MO, although it has the same number of acidic sites as MO.

Table I : The number and strength of the acidic and basic sites on the Regal L 330 and Monarch 1000 pigment surfaces measured by flow microcalorimetry by using cyclohexane solutions of pyridine as a basic probe and of phenol as an acidic probe at 30°C

Carbon Black	Basic Character (titrated by phenol)			Acidic Character (titrated by pyridine)		
	mcal/m^2	μmole/m^2	Kcal/mole	mcal/m^2	μmole/m^2	Kcal/mole
RL	1.07	0.54	2.0	1.56	0.43	3.62
MO	0.57	0.23	2.4	2.56	0.43	5.95

Encapsulation of Carbon Blacks from Polymer Solutions. Griltex nylon (as received, 1.56 grams) and carbon black pigment (0.0156, 0.078, 0.156, 0.312 and 0.78 grams, respectively) were added to 200 ml ethanol in a 500 ml flask. The mixture was sonified by ultrasonic water bath for 20 minutes to break up the carbon black agglomerates and then heated in a 70°C water bath for one hour to form a clear polymer solution in which the carbon black pigment was dispersed. The flask was connected with a condenser to avoid loss of ethanol. The sonification/heating cycle was then repeated and the suspension finally sonified and poured into 20 ml vials which were put in a stirred ice-water bath very quickly to cool. Cooling proceeded for half an hour to assure completion of the process. The cooling rate is shown in Figure 1. The temperature decreased from 70°C to 0°C at a initial rate of 1°C/sec., and the phase separation of the solution was observed visually after 150 seconds.

The precipitated suspension was poured into a flask containing Isopar G (400 ml) and shaken by hand to mix the ethanol and Isopar G fully. Centrifuging at 1000rpm for 5 minutes was sufficient to separate the precipitate and the solvent mixture. The supernatant was removed and the precipitate was washed with Isopar G and dried by spreading on a watch glass under vacuum at room temperature overnight. The dried powders were stored in a desiccator for further studies.

Heats of Competitive Interaction. Measurements of heats of interactions, as stated below, were carried out in a differential C80 Setaram Calorimeter. The system to be studied is located in a sample vessel with the two interactants separated by a membrane, and this is placed in a calorimetric block, together with a reference vessel. The temperature of the block is well controlled. Rupture of the membrane allows the interaction to generate a certain heat which causes the temperature to change. Compared to the reference cell, the temperature change is converted to a net heat of interaction. Heats of immersion of carbon blacks in ethanol and heats of adsorption of Griltex nylon onto the carbon black surfaces from ethanol solutions were measured by the calorimeter and are listed in Table II.

Table II : The heats of immersion and adsorption measured by Setaram Calorimeter

Carbon Black	Heat of Immersion in ethanol at 30°C (cal/m^2)	Heat of Adsorption of Griltex Nylon 1wt% (cal/m^2)
RL	-0.0383	-0.0163
MO	-0.0193	-0.0361

Characterization of Surface Properties. Two series of samples, varying in carbon black and loading, were made by the above procedures. They are designated as follows : Griltex Regal - parts carbon black per 100 parts (wt.) nylon. Thus, we have GR-1, -5, -10, -20 and -50 and GM-1, -5, -10, -20, -50.

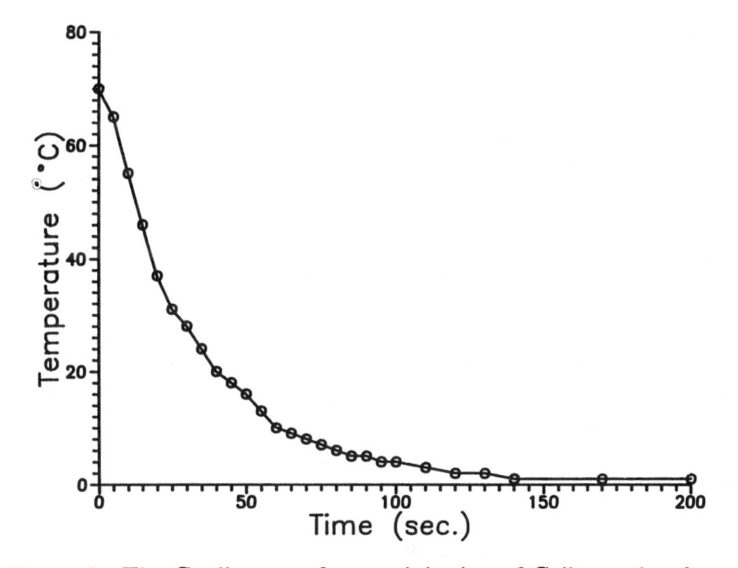

Figure 1 : The Cooling rate for precipitation of Griltex nylon from ethanol solutions in the presence of carbon black pigment.

Particle Size. The dried pigmented polymer particles were redispersed in Isopar G by ultrasonic water bath using basic barium petronate as a stabilizing agent. Particle size was measured by photon correlation spectroscopy at 25°C using Coulter N4 MD, a dynamic light scattering instrument.

Surface Area. The surface area of the pigments and pigmented polymer particles were measured by the MONOSORB surface area analyzer, a single-point dynamic BET procedure. Nitrogen with a fractional pressure of 228 mm-Hg ($p/p_0=0.3$) was used as the adsorbate and helium was used as the carrier. The samples were carefully dried before each measurement.

Blackness Index. The same amount (by weight) of the pigments and pigmented polymer particles were coated on white substrates of constant area. This resulted in a coating which is thick enough to completely opacify the substrate. The blackness measurements were carried out using the Macbeth RD514P Densitometer.

Zeta Potential. The particles were dispersed in Isopar G containing different concentrations of cupric naphthenate, a positive charge control agent, to form dispersions having 0.025 wt% of particles. The dispersions were sonified for five minutes and tumbled on a roller overnight. The electrophoretic mobility of the particles were measured by the Doppler Effect Laser Scattering Analyzer (Coulter DELSA 440) at 25°C. Flow profiles were made and further converted to van Gils profiles to determine the mobility of particles at the theoretical stationary layers. More detail about the measurement and analysis of microelectrophoretic mobility are described elsewhere*(16)*. Electrophoretic mobility was then converted to zeta potential according to the Hückel equation.

Results and Discussion

Microencapsulation Mechanism. In a polymer solution with the presence of pigment, three interactions are considered : 1) between solvent and polymer, 2) between solvent and pigment, and 3) between pigment and polymer. Strengths of these interactions are strongly dependent upon the chemical nature of each component. At higher temperatures, the three interactions reach equilibrium in which solvent molecules and polymer chains are in one phase. When the solvency is reduced by lowering temperature, the interaction between solvent and polymer decreases, causing polymer chains to collapse and form a second phase. In this condition, polymer chains will either precipitate by themselves to form a solid polymer phase, or precipitate around the pigment to form capsules. Therefore, during the precipitation, there is a competition between polymer chains and solvent molecules to interact with pigment particles. If the interaction between pigment and solvent is stronger than that between pigment and polymer, the pigment surface will be more efficiently wetted by the solvent molecules. Thus, during phase separation, the pigment will tend to contact with the solvent rather than to be associated with the polymer. So, after phase separation, the pigment will be either separate from the polymer or be partially encapsulated.

On the other hand, if the interaction between pigment and polymer is stronger than that between pigment and solvent, the polymer chains will preferentially adsorb on the pigment surface. So during phase separation, those adsorbed polymer chains will precipitate onto the pigment surface and form a thick polymer layer fully encapsulating the pigment. Accordingly, the morphology of the final products of the phase separation are predictable if the interactions between polymer, pigment and solvent are known. That is, if we can control the interaction between polymer, solvent and pigment by changing the surface properties of pigment, polymer composition or solvency, we are able to control the morphologies of the encapsulated particles.

In this study we designed experiments to measure the heat of immersion of pigment in solvent using a batch calorimeter to judge the interaction between ethanol and the two different carbon blacks. Results of the measurements are listed in Table II.

Heats of immersion have long been used to characterize various aspects of powder surfaces*(17,18)*. It has been reported that nonpolar pigments produce small heats even with polar liquids and that polar pigments produce large heats with liquids of opposite polarity and smaller ones with nonpolar liquids. The value of the heat of immersion is indicative of the strength of the interaction between solvent and pigment.

Table II shows that the heat of immersion of RL carbon black in acidic*(19)* ethanol is about twice that of the MO. This result is consistent with the surface characteristics of the pigments titrated by flow microcalorimetry. Table I shows that RL is much more basic than MO. The heat of interaction between solvent and pigment is also reflected in the stability of the pigments in ethanol. RL is stable over a month in ethanol without sedimentation at $25^{o}C$, but MO flocculated and settled in ten minutes under the same conditions. The better stability of Regal L 330 in ethanol is, therefore, due to the stronger interaction between solvent and pigment.

In addition, Table II shows that the heat of adsorption of Griltex Nylon polymer onto MO is larger than onto RL. Compared to RL as shown in Table I, the more acidic surface of MO is thought to be responsible for the larger heat of adsorption of the polymer (the nylon is primarily basic due to the amine and carbonyl group).

From measured heats in each system, it is found that the interaction of RL with ethanol is stronger than with Griltex nylon. During phase separation, when the temperature of the solution is lowered, the polymer chains come out of the solution and form embryo capsules, a condensed phase, in which the solvent molecules are being squeezed out while the polymer chains are condensing. Finally most of the solvent molecules are squeezed out of the condensed phase and solid polymer/carbon black particles are formed. Because the pigment has a stronger interaction with the solvent than with the polymer, when the solvent is squeezed out of the polymer particles, the pigment particles tend to migrate with the solvent to the surface of the polymer particles in which the pigment has most of its surface in contact with the solvent. From the heat of adsorption, we have evidence that there are still some polymer chains associated with Regal L 330 pigment, although there are not much. Photon correlation spectroscopy (PCS) and scanning electron microscopy (SEM) show that the final encapsulated particles have very uniform size and microelectrophoresis shows that the particles have only a single mobility, both of which indicate no separate phases (polymer and pigment). Therefore, we believed that the RL was held largely on or just within the surface of the capsules, but not separated from the capsules. The mechanism of the encapsulation and the morphology of the final encapsulated particles are schematically illustrated in Figure 2a. Six carbon black particles are used in the sketch because GR-50 were calculated to have that number per nylon particle.

In contrast, the interaction of Monarch 1000 with Griltex nylon is stronger than with ethanol. Thus the pigment particles are preferentially covered by the polymer chains. During phase separation, the polymer chains adsorbed on the pigment surface will act like nuclei for precipitation of the polymer. Since the pigment particles are more strongly associated with the polymer than with the solvent, they tend to stay with the polymer when the solvent is being squeezed out of the polymer particles. This favors complete encapsulation of the pigment by the precipitated polymer. The mechanism and the final encapsulated particles are shown schematically in Figure 2b.

Characterization of Surface Properties of Pigmented Polymer Particles.

Particle Size. Photon correlation spectroscopy shows, in Table III, that the size of the pigmented polymer particles decreased with an increase of the pigment content, and that the relationship is not linear but roughly logarithmic. At lower pigment contents, particle size is strongly dependent on pigment content; however, at higher pigment contents the dependence decreases, perhaps due to greater flocculation of the pigment. Indeed GM-25 and GM-50 could not be kept dispersed well enough to encapsulate all of the Monarch pigment.

In polymer solutions, if each primary carbon black floc forms one nucleus and grows independently, the capsule size should decrease linearly with the pigment content. Since we found that the capsule size is not linearly dependent upon the pigment content, we do not think that this mechanism is fully correct. In another model, if the polymer chains not only grow on each carbon black floc, but also undergo self nucleation, the final particles should end up with a broad size distribution because the rate of polymer chain growth with or without pigment are different. Since the scanning electron microscopy (SEM) results, as shown in Figure 3, tell us that the final precipitated particles have very uniform size, we do not think that this is the mechanism. It should be noted that evidence from experiments with unpigmented phase separation of uniform nylon spheres indicates that the cooling rate of $1^{o}C$/sec is so fast that little fractionation occurs*(20)*.

(a) GN/RL/EtOH

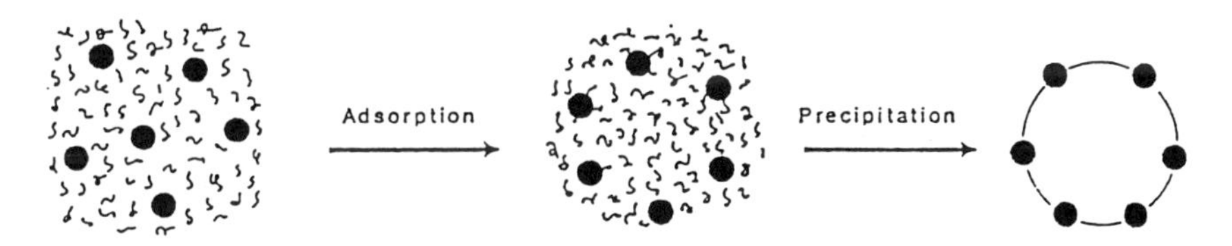

(b) GN/MO/EtOH

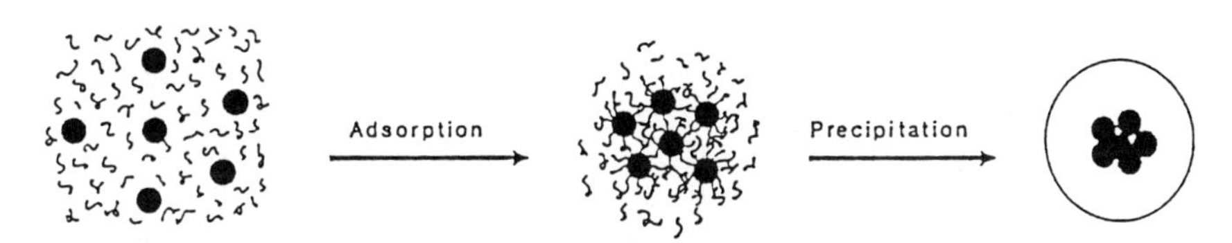

Figure 2 : Mechanisms of the microencapsulation of (a) Regal L 330 and (b) Monarch 1000 carbon black from Griltex/ethanol solutions at an initial cooling rate of 1°C/sec.

We suspect that each primary carbon black floc acts as a nucleus during phase separation, but that these flocs may differ in size as their concentration was increased. However, flocculation of embryonic particles at an increasing rate with increasing concentration is also probable. Accordingly, in this model the final capsule size would not only depend on the nucleation due to the pigment but also on flocculation of nuclei or embryonic particles (wherein the RL particles would be relatively free to migrate to the surface).

Surface Area. Table III shows that the surface area of the RL encapsulated particles increases significantly with increasing pigment content. However, the surface area of the MO encapsulated particles demonstrate a somewhat lower increase. Table IV shows that the MO pigment has a surface area about three times larger than the RL. Therefore, if Monarch 1000 is on the surface, GM particles should have larger area than equally loaded particles with RL on the surface. However, we found the reverse to be true, i.e., the surface area of the Regal L 330 pigmented encapsulates is larger than those pigmented with Monarch 1000 at each level of pigment content. The PCS and SEM results show us that the particle size of GR and GM encapsulates are almost the same, and therefore, the higher area of the GR particles over the GM is due to more RL roughening the particle surface.

Table III : Particle size and surface area of different pigmented polymer particles which were made by the precipitation technique

	Pigment Content (wt%)	Particle Diameter By PCS (nm)	Surface Area By BET (m^2/g)	Blackness Density
GR-1	1.0	1600	17.3	0.37
GR-5	4.8	960	28.4	0.81
GR-10	9.1	840	32.5	1.16
GR-25	16.7	670	36.3	1.44
GR-50	33.3	580	47.4	1.71
GM-1	1.0	1560	17.4	0.24
GM-5	4.8	930	27.1	0.60
GM-10	9.1	770	28.1	0.76

* 0 = total reflection of incident light; 2 = total adsorption of light

If we take the particle diameter of these pigmented polymer particles measured by the PCS and convert them to surface area by assuming that the particles are spherical and have smooth surfaces and that the density of the pigments and polymer are 1.83 and 0.8 g/cm^3, respectively, then we find that

Figure 3 : SEM photographs of (a) GM-10 and (b) GR-10 pigmented polymer particles made by the phase separation process.

the ratio of the surface areas measured by the BET adsorption to that calculated from the PCS diameter is revealing. The ratio, an indication of roughness, increases for the RL system but is almost constant or slightly decreases with an increase of pigment content for the MO system (see Table V). So, for the RL system, the higher relative roughness is an indication of more pigment on the particle surface.

Table IV : Surface area of carbon black pigments measured by single point BET adsorption and blackness

Pigment	Surface Area (m^2/g) Reported by Cabot Corporation	Surface Area (m^2/g) Measured by BET Adsorption	Blackness Density Measured by Densitometer
RL	94	71.7	1.65
MO	343	211.4	1.87

Table V : Relative surface roughness and amount of pigment on the particle surface

	Pigment Content (wt%)	Surface Area calculated from PCS (m^2/g)	Relative Roughness S_{BET}/S_{PCS}	Pigment on Surface (%)
GR-1	1.0	4.63	3.74	
GR-5	4.8	7.35	3.86	100
GR-10	9.1	7.99	4.07	87
GR-25	16.7	9.23	3.93	62
GR-50	33.3	9.05	5.24	75
GM-1	1.0	4.75	3.66	
GM-5	4.8	7.59	3.57	25
GM-10	9.1	8.72	3.22	23

Moreover, using the particle size from PCS results and the surface area measured by the BET adsorption (from Table III), the amount of pigment on the surface of the encapsulated particles was calculated based on the following assumptions : 1) The density of the pigments and Griltex nylon are 1.83 g/cm^3 and 0.8 g/cm^3, respectively, 2) In a given sample, each particle has the same size and is spherical, 3) For samples, GR-1 and GM-1, which contain 1 wt%

of pigment, the amount of pigment on the particle surface is negligible, 4) The surface roughness is independent of the particle size, and 5) Nitrogen can not diffuse into the particles. The calculations are summarized in Table V which shows that more than 60% of Regal L but only 25% of Monarch 1000 pigments are on the surface of the capsules. This is another piece of evidence to support the proposed morphologies of the two different encapsulated particles.

Blackness Index. It is known that, for carbon blacks, light adsorption is more important than light scattering in developing tinting strength, although the later is not negligible. The blackness of carbon black is related to its particle size as well as its aggregate size. Small particle size and/or low degree of aggregation of carbon blacks favor blackness. For a pigmented polymer particle, the blackness of the particles is dependent upon the amount of carbon black in the particles, the morphology of carbon black in the particles, the refractive index of the polymer and carbon black, the path of incident light passed before or after being reflected/scattered or adsorbed by the polymer and the size of the particles.

The blackness of the pigments and pigmented polymer particles were measured with a Macbeth RD514P Densitometer and the results are listed in Table III and IV. They show that the blackness of both carbon black pigmented polymer particles increases with the increase of pigment content. Since the MO exhibits higher blackness density than the RL pigment, the MO pigmented particles should have higher blackness density than the RL counterpart given the same morphology. However, the blackness density of the RL pigmented polymer particles is higher and increased more significantly than the MO pigmented polymer particles as the pigment content was increased. This provides two bits of information. Firstly, in the GM series, the pigment was coated with a thick polymer layer, so incident light was reflected/scattered by the polymer layer before being adsorbed by the pigment. This caused the MO pigmented particles to have lower blackness density than the RL pigmented particles which have relatively more pigment located on the particle surface. Secondly, the MO pigment particles are aggregated in the center of particles reducing their adsorption efficiency relative to the RL pigmented particles whose pigment particles are less aggregated. Accordingly, the proposed morphologies of the different encapsulated particles are further supported by the blackness measurements.

Zeta potential. According to Fowkes' proton charging mechanism for non-aqueous colloids*(21)*, the charge formation is directly related to the acid-base interactions between the charging species and the functional sites on the particle surfaces. An acidic particle in a basic medium receives a negative charge, and a basic particle in an acidic medium receives a positive charge. For example, rutile*(22)* and anatase*(23)*, which have both acidic and basic surfaces, are positive in n-butanol and n-pentanol, which are acidic liquids. However, rutile become negative in more basic liquids such as butylamine*(22)* and nitrobenzene*(24)*.

Lloyd et al.*(25)* titrated rutile pigments with different surface treatments by flow microcalorimetry and found that silica-coated titanium oxide has a more acidic surface and alumina-coated titanium oxide has a more basic surface. They further found that the silica-coated sample had a larger negative zeta potential than the alumina-coated sample in xylene solutions containing lecithin which is considered to be a basic dispersant. The alumina-coated titanium oxide, however, had a larger positive zeta potential than the silica-coated titanium oxide in xylene solutions containing alkyd resin, a primarily acidic dispersant.

From zeta potential measurements, surface characteristics of particles can be qualitatively evaluated. Microelectrophoretic mobility results (see Table VI)

Table VI : Zeta potential (mV) of RL and MO carbon blacks in different solvents and Isopar G solutions containing 0.025wt% of basic barium petronate (BaPB) and cupric naphthenate measured by Coulter DELSA at 25°C

Pigment	THF	$CHCl_3$	Isopar G/ BaPB	Isopar G/ Cu Naphthenate
Regal L 330	-165	+90	-143	+45
Monarch 1000	-150	+30	-150	+15

show that RL and MO carbon black pigments have very similar negative zeta potential values in basic tetrahydrofuran and basic barium petronate/Isopar G solution. However, in acidic chloroform and cupric naphthenate/Isopar G solution, RL pigment has positive zeta potentials about three times larger than MO pigment. These results indicate that both pigments have similar acidic characteristics, but RL is considerably more basic than MO, which is consistent with the flow microcalorimetry data as shown in Table I. Accordingly, the dramatic difference of the zeta potential for RL and MO pigments in acidic liquids and solutions containing acidic charge control agents is a good approach to differentiate the two pigmented polymer particles. Figure 4a shows that the MO pigmented particles have positive zeta potential values very close to the polymer, but very different from the Monarch 1000 itself. However, Figure 4b shows that the RL pigmented particles have positive zeta potential values almost identical to Regal L 330 itself but different from the polymer. The electrophoretic results give strong evidence that the RL pigmented polymer particles have a large amount of the pigment on the surface, and therefore, have surface characteristics similar to the pigment. However, the MO pigmented polymer particles have the pigment mostly fully

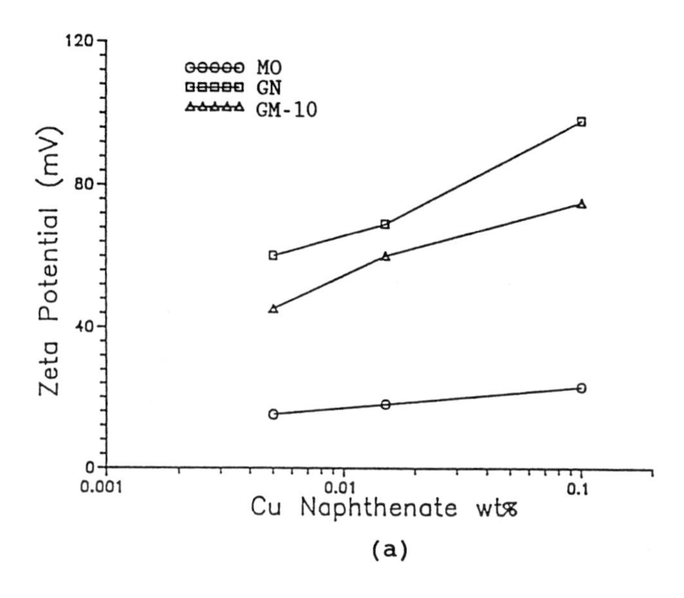

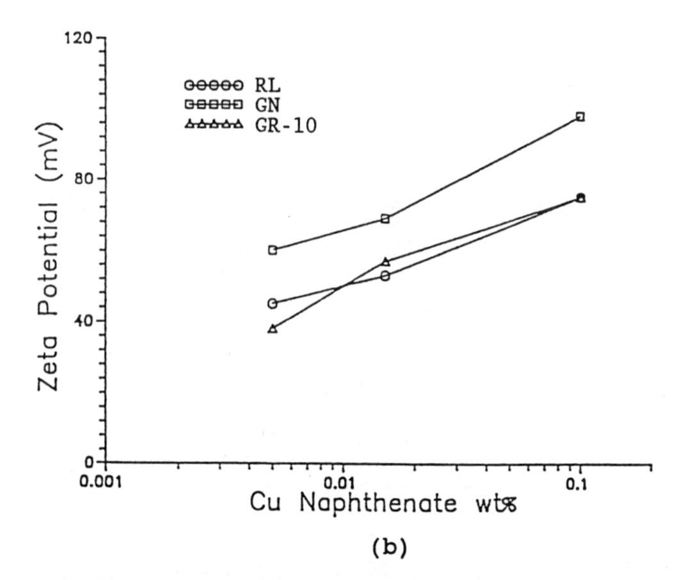

Figure 4 : The zeta potential of polymer, pigments and pigmented polymer particles in Isopar G solutions containing different concentrations of cupric naphthenate measured by Coulter DELSA at 25°C.

encapsulated, and therefore the surface characteristics are similar to the polymer itself. Once again, the proposed encapsulation mechanisms and encapsulated particle morphologies are strongly supported.

Conclusions

The interactions between polymer, pigment and solvent play an important role in controlling the morphologies of pigmented polymer particles during phase separation from polymer solutions. Two types of pigmented polymer particles with different morphologies were prepared : a) particles with pigment largely on the surface and b) particles with more fully encapsulated pigment. Pigmented polymer particles with different surface characteristics have different physical properties such as surface area, optical density and surface potential.

Postscript

Professor Frederick M. Fowkes passed away on October 17, 1990; this work was done with his advice.

Acknowledgement

This research project was sponsored by Coulter Systems Corporation, Bedford, MA. The encouragement of Dr. Kenneth A. Lindblom is highly appreciated.

Literature Cited

1. Thies, C.; *Encyclopedia of Polymer Science and Engineering*; Wiley-Interscience: New York, NY, 1987; Vol. 9, pp. 724.
2. Huang, T. C. Ph.D. Dissertation; Lehigh University, 1986.
3. Luzzi, L. A.; Zoglio, M. A.; Maulding, H. V. *J. Pharm. Sci., 59*, 338(1970)
4. Green, B. K.; Schleicher, L. U.S. Patent 2,800,457 (1957)
5. Micale, F. J. European Patent 238,035 (1987)
6. Micale, F. J. U.S. Patent 4,665,107 (1987)
7. Landa, B.; Hall, J.; Gibson, G. A. U.S. Patent 4,842,974 (1989)
8. Vollmann, H.; Soden, B.; Herrmann, H. U.S. Patent 4,594,305 (1986)
9. Brenner, *J. Perfum. Flavour, 8*, 40(1983)
10. Siegel, B. M.; Johnson D. H.; Mark H. *J. Polym. Sci. 5*, 111(1950)
11. Kumaki, J. *Macromolecules, 21*, 749(1988)
12. Brown, H. R.; Wignall, G. *Macromolecules, 23*, 683(1990)
13. Iler, R. K. *J. Colloid Interface Sci., 51*, 388(1975)
14. Moffatt, W. C.; Bowen, H. K. *J. Material Sci. Letters, 6*, 383(1987)
15. Joslin, S. T.; Fowkes, F. M.; *Ind. Eng. Chem. Prod. Dev., 24*(30), 69(1985)

16. Fowkes, F. M.; Chen, W. J.; Fluck, D. J.; Hou, W. H. *Particulate Science and Technology* (1990 in press)
17. Chessick, J. J.; Young, G. J.; Zettlemoyer, A. C. *Trans. Faraday Soc., 50*, 587(1954)
18. Harkins, W. D.; Jura, D. J. *Amer. Chem. Soc., 66*, 919(1944)
19. Gupta, A.; Patel, M. *J. Materials Sci. Letters, 7*, 1021(1988)
20. Hou, W. H.; Lloyd, T. B.; *J. Appl. Polym. Sci.* (in press 1991)
21. Fowkes, F. M. *Discuss. Faraday Soc., 42*, 243(1966)
22. Romo, L. A. *J. Phys. Chem., 67*, 386(1963)
23. Griot, O. *Trans. Faraday Soc., 62*, 2904(1966)
24. Micale, F. J.; Lui, Y. K.; Zettlemoyer, A. C. *Discuss. Faraday Soc., 42*, 238(1966)
25. Lloyd, T. B.; Li, J.; Fowkes, F. M.; Brand, J. R.; Dizikes, L. J. *Coatings Technology* (1991 in press)

RECEIVED December 4, 1991

Chapter 26

Polymer Colloids as Catalyst Supports

Warren T. Ford, Rickey D. Badley[1], Rama S. Chandran[2], S. Hari Babu, M. Hassanein[3], Sanjay Srinivasan[4], Hayrettin Turk[5], Hui Yu, and Weiming Zhu

Department of Chemistry, Oklahoma State University, Stillwater, OK 74078

We have designed and used functional colloidal particles as catalyst supports. The colloidal catalysts are prepared by emulsion copolymerization as cationic or anionic latexes. Subsequent reactions of the latexes can produce particles with up to 80 mol percent of charged repeat units. Anionic transition metal complex catalysts and anionic reagents bind to cationic particles, which serve as sites of locally high concentration of reactant and catalyst in aqueous dispersions. Cationic catalysts bind similarly to anionic particles. Surface modification of colloidal silica with silane coupling agents also can produce colloidal catalysts. The colloidal supports increase the activity of most catalysts tested for oxidations and hydrolyses of organic compounds.

Heterogeneous catalysis relies on high surface area supports to expose the active catalysts to reactants in the surrounding fluid. In the high activity limit the rates of heterogeneous reactions depend only on the rates of mass transfer of reactants to and products away from the active sites on the surface *(1)*. High surface area is normally achieved by use of macroscopic particles of porous materials such as silica, alumina, and active carbon as supports. High porosity may lead to catalytic reactions limited by rates pore diffusion. The active catalysts are often colloidal particles or clusters of metals, oxides, or sulfides that are not inherently very active, but capable of fast chemical reactions at high temperature when finely divided and adsorbed on a high surface area support. Under high temperature fast reaction conditions, mass transport and pore diffusion are often the rate-limiting steps. The oxide supports and the catalysts themselves usually are chemically stable for use at temperatures as high as synthetic transformations of organic compounds can be performed.

Polymers too can be obtained in porous, high surface area forms, but most polymers lack the high temperature stability of oxide materials. Over the last two decades polymer supports have been investigated for catalysis of a variety of chemical reactions, mostly in 20-100 ^{o}C temperature range in organic solvents *(2-4)*. The

[1]Current address: Phillips Petroleum Company, Bartlesville, OK 74004
[2]Current address: National Starch and Chemical Company, Bridgewater, NJ 08807
[3]Current address: Tanta University, Tanta, Egypt
[4]Current address: Ethyl Corporation, St. Louis, MO 63104
[5]Current address: Anadolu University, Eskisehir, Turkey

0097–6156/92/0492–0422$06.00/0

near-ambient use temperatures allow less stable but much more active homogeneous molecular organic and organometallic catalysts to be employed. The solvent adds further important considerations to the design and understanding of catalytic processes. Catalysts that otherwise would be soluble must be fixed to the support by ionic or covalent bonds. Since solvent may dissolve polymers, most polymer supports are cross-linked but still swellable by solvent and used in gel form. Swelling means that active sites can be located in the gel phase as well as on the surface, and that reactant intraparticle diffusion may be still another physical process limiting the rates of chemical reactions.

Supports for conventional heterogeneous catalysts are macroscopic particles used in batch, packed bed, or fluidized bed reactors. The catalytic species itself may be colloidal particles of a metal or oxide bound to the macroscopic support. Molecular, homogeneous catalysts bind to colloidal as well as to macroscopic polymer particles. Colloidal particles provide high surface area and can be prepared in a variety of sizes and compositions. Functional monomers can create specific binding sites for catalysts in the polymer. We review in this paper progress in design and use of polymer latexes and colloidal sol-gel silica as catalyst supports.

In 1984 we discovered naively that cationic latexes used as phase transfer catalysts coagulated in the presence of concentrated aqueous sodium cyanide *(5)*. We resumed the study of colloidal particle supports about two years later with emphasis on oxidation reactions that do not require high concentrations of electrolytes. In addition to our research, colloidal polymeric sulfonic acids *(6)* and imidazoles *(7-9)* have been used as hydrolysis catalysts, poly(sodium acrylate) and poly(sodium styrenesulfonate) latexes were used as catalysts for reaction of $Co(NH_3)_5Br^+$ with hydroxide ion *(10)*, and enzymes have been immobilized as catalysts on latexes *(11)*.

Stable colloidal particles are either charge stabilized or steric stabilized. To retain high surface area a colloidal catalyst must not coagulate and precipitate during use. Electrolytes which screen charged particle-particle repulsions, and organic components which cause collapse of the polymer loops and tails of a steric stabilizer on the surface may cause coagulation.

Most of our investigations of colloidal catalyst supports have used charge stabilized particles dispersed in water. The latexes produced by emulsion polymerization have been cross-linked to prevent dissolution in water or any organic solvent. Ionic active catalysts bind electrostatically to the charged particles. These colloidal catalysts in some ways resemble other types of catalysts, such as micelles, polyelectrolytes, and ion exchange resins. Like micelles they may have charged surfaces and nonpolar interiors. A colloidal catalyst typically has a diameter of 100 nm and is polymeric inside, whereas a micelle has a diameter of 5 nm and consists of aliphatic hydrocarbon inside. Like polyelectrolytes, colloidal catalysts may have highly expanded charged polymer chains in water, but those chains are anchored to the cross-linked polymer support. Ion exchange latexes with charged sites both on the surface and inside the particles may have the same primary structures and compositions as ion exchange resins. However, typical ion exchange resins have diameters of 10^4 to 10^6 nm, whereas a typical latex diameter is 10^2 nm. Frequently with ion exchange resin catalysts intraparticle diffusion of reactants as well as intrinsic chemical reactivity limit the reaction rates. Since surface area per unit mass of catalyst is inversely related to particle diameter and to intraparticle diffusion times, colloidal particles offer an average diffusion pathlength at least 100 times shorter for reactants to reach active sites, eliminating the intraparticle diffusional limitation to reactions. Thus high surface area does not necessarily mean that reactions occur only on the surface of the particles.

The goals of our colloidal catalysis research are to understand where and how chemical reactions proceed in the colloidal environment and to create highly active catalysts. The fundamental understanding should aid the practical design of catalysts.

Colloidal polymer supports are likely to provide more active catalysts than larger polymer particles for two reasons: High surface area can overcome mass transport and diffusional limitations to reaction rates, and solubility of the organic reactants in the polymer concentrates reactants in the same phase with the active catalytic sites. These reasons to expect high activity do not consider intrinsic reactivity at the active sites, which might be higher or lower than in larger polymeric catalysts. Colloidal catalytic reactions may occur at the particle surface, which must be charged and highly hydrated to stabilize the colloid, or inside the particles in environments that could vary widely, from that of a lipophilic hydrocarbon polymer to that of a hydrated ion exchange resin. The interior of the particle may be tailored to suit the hydrophile/lipophile balance needed for high intrinsic chemical reactivity. The groups on the particle surface can be varied, but they must be charged to stabilize the colloid.

Colloidal Supports

Anionic oxidizing agents, anionic catalysts, or both have been employed in oxidation and hydrolysis reactions with colloidal catalysts, so we will emphasize positively charged supports.

Emulsion polymerization of nonpolar monomers such as styrene promoted by cationic surfactants such as hexadecyltrimethylammonium chloride (CTACl) produces cationic latexes having all of the positive sites derived from the surfactant. If persulfate is used as the initiator, the polymer surface will also have a small number of negatively charged sulfate sites which will serve as counterions to the surfactant. Dissociation of part of the chloride counterions provides the net positive charge on the particle.

Much larger numbers of cationic sites can be created in two steps by copolymerization with a monomer such as chloromethylstyrene followed by reaction with an amine to form ammonium ions. One can synthesize a colloidal particle having a quaternary ammonium ion in almost every repeat unit by use of only chloromethylstyrene and a small amount of a cross-linking monomer, which prevents the highly ionic quaternized product from dissolving in water. Since cationic polyelectrolytes also can be highly active catalysts for reactions of anions *(12)*, we have studied only cross-linked latexes. Cationic latexes can even be produced using an anionic surfactant and chloromethylstyrene followed by amine treatment. The CMS latex made with sodium dodecyl sulfate early during treatment with trimethylamine coagulates at its composition of zero charge and then quickly redisperses as positively charged particles after further quaternization *(13)*.

Since our goal is to understand latexes as catalysts, and cationic micelles can also be active catalysts, we have prepared some latexes with polymerizable, micelle-forming surfactants such as monomers **1-4** (Scheme 1) *(14,15)*. Similar 2-tailed surfactants have been used to stabilize bilayer vesicles by polymerization *(16-20)*. A surfactant covalently bound to the particles cannot act independently as a micellar catalyst, but it might be possible for excess surfactant such as CTACl to dissociate and act as an aqueous phase catalyst. After copolymerizations of 2 mol percent of the monomers **1-3** with 1% DVB (divinylbenzene) and 97% styrene, we proved that the charged monomers were covalently bound by ultrafiltration of the latexes to remove soluble electrolytes and analysis of both the filtrate and the particles for bromide ion. The particles contained 95-97% of the amount of bromide ion originally charged as monomer. Ultrafiltration was used to purify most of the latex catalysts by removal of soluble by-products of the synthesis. After ultrafiltration and extensive washing of the particles with water the conductivity of the filtrate is a factor of 5 or 10 less than the conductivity of the initial filtrate from the reaction mixture. However, even the final filtrate usually has conductivity at least ten times that of the wash water. The latexes from monomers **1-3** were slightly polydisperse 60 nm diameter spheres *(14)*.

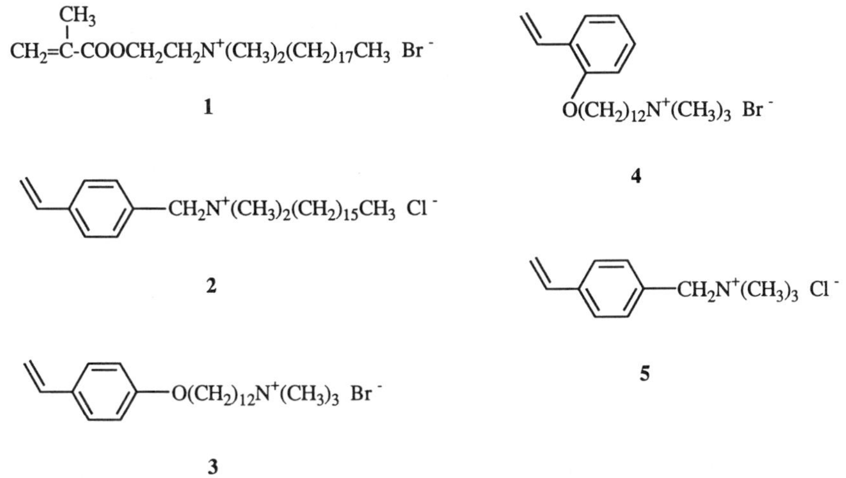

$CH_2=C(CH_3)-COOCH_2CH_2N^+(CH_3)_2(CH_2)_{17}CH_3\ Br^-$

1

$CH_2=CH-C_6H_4-CH_2N^+(CH_3)_2(CH_2)_{15}CH_3\ Cl^-$

2

$CH_2=CH-C_6H_4-O(CH_2)_{12}N^+(CH_3)_3\ Br^-$

3

$CH_2=CH-C_6H_4-O(CH_2)_{12}N^+(CH_3)_3\ Br^-$

4

$CH_2=CH-C_6H_4-CH_2N^+(CH_3)_3\ Cl^-$

5

Scheme 1. Quaternary ammonium ion monomers

The monomer **4**, first reported by Tsaur and Fitch *(21)*, was used to explore the effects of monomer and divinylbenzene levels on the particle sizes and stabilities of latexes. The particle size decreased and the polydispersity increased as the amount of surfactant **4** increased from 0.5 to 5.0 mol percent and as the amount of DVB in the monomer mixture increased from 1 to 5 mol percent. Within experimental error all of surfactant **4** was incorporated into particles. Transmission electron microscopy showed 22-95 nm diameter particles with nonspherical shapes from the various polymerizations. Particle aggregation increased with the amount of DVB. The morphology suggested that the final particles were formed by aggregation of smaller primary particles.

Since all of the latexes produced from polymerizable monomers **1**-**4** were prone to coagulation in the presence of the organic substrates in catalysis experiments, we turned to more highly ionic latexes prepared with vinylbenzyltrimethylammonium chloride (**5**), the non-micelle-forming salt formed by quaternization of chloromethylstyrene with trimethylamine *(22)*. Early during emulsion polymerization copolymer containing the quaternary ammonium ions serves as the charge stabilizing surface of the growing colloidal particles. Up to 5 mol percent of sodium styrenesulfonate was incorporated into cross-linked polystyrene latexes by the shot-growth technique of Kim, El-Aasser, and Vanderhoff *(23)*. Their technique is similar to but not identical with seed growth emulsion polymerization. In our syntheses of cationic latexes an initial batch of monomers containing 1 mol percent of **5** was polymerized to 90-95% conversion, and then a second shot of monomers containing a larger amount of **5** was added, and the polymerization was completed. In the presence of 90-95% polymerized particles, all of the second shot of monomer **5** was incorporated into existing particles instead of forming new ones. The products of shot growth polymerization are highly monodisperse, as in seed growth polymerizations. Use of the shot growth technique with monomer mixtures containing 1% DVB and varied amounts of styrene and chloromethylstyrene followed by quaternization of the chloromethyl groups with trimethylamine produced the family of latexes in Table I which contain from 0.6 to 60 mol percent of quaternary ammonium ion repeat units. The particle number in shot growth emulsion polymerization is established during the first stage, and by use of 1.05 weight percent of charged monomer **5** in every polymerization along with varied mixtures of styrene and chloromethylstyrene, approximately the same size of particles was produced over the entire range of copolymer composition.

Table I. Ion Exchange Latex Particles Using Monomer 5[a]

radii in nm by		TEM			DLS[d]		
sample	N^+, mol %	d_w[b]	d_n[c]	d_w/d_n	no buffer	pH 8.0 TAPS	V_{wet}/V_{dry}[e]
1N+	0.63	151.01	49.4	1.01	167.4	165.6	1.32
2N+	0.55	157.1	147.3	1.07	159.2	158.8	1.03
5N+	1.27	153.9	152.7	1.01	171.6	171.2	1.37
10N+	6.2	156.5	155.2	1.01	182.8	179.0	1.49
25N+	16.9	163.5	161.8	1.01	218	218.0	2.4
50N+	34.1	147.1	146.5	1.00	274	250	4.9
75N+	60.4	154.3	143.4	1.01	380	336	10.3

[a]Ref. 22. [b]Weight average diameter. [c]Number average diameter. [d]Hydrodynamic equivalent radius at 90° scattering angle in the single scattering limit. [e]Ratio of particle volume in 5.0 mM TAPS [N-tris(hydroxymethyl)methyl-3-aminopropanesulfonic acid] buffer to dry volume.

From the sizes of the particles in Table I measured by dynamic light scattering it is clear that the samples with large amounts of quaternary ammonium ion repeat units are swollen in water, up to 10 times in a 5 mM buffer compared with the dry volume. Buffer of the same concentration used in kinetic experiments was chosen for particle size measurements. The particles in Table I all have about the same dry size and cover a wide range of interior hydrophile/lipophile balance.

The shot growth technique was demonstrated initially for polystyrene latexes with sodium styrenesulfonate (NaSS) as the charged monomer *(23)*. Latexes with amine ligands to activate Co^{2+} for autoxidation catalysis were prepared with up to 20 mol percent 4-vinylpyridine, 3.3 mol percent NaSS, and 1% DVB *(24)*. Thus the shot growth technique works well with a basic, more hydrophilic comonomer such as 4-vinylpyridine.

Supports for Co^{2+} were produced also by more conventional emulsion copolymerizations of styrene and 1% DVB with 20-76% acrylic acid and with 20% methacrylic acid *(24)*. Attempts to bind Co^{2+} directly to the latexes in carboxylic acid form resulted in coagulation, the normal fate of negatively charged colloids in the presence of divalent cations. However, conversion to the potassium carboxylate form first with KOH followed by addition of Co^{2+} gave colloidally stable pink latexes. The K^+ form of the latex is more hydrated than the free acid form, allowing fast ion exchange of K^+ for Co^{2+}. Apparently kinetically slow ion exchange allowed time for particle coagulation when Co^{2+} was added directly to the free acid form of the latex.

Silica can also be made in the form of charged colloidal particles by the sol-gel method. In our laboratory hydrolysis of tetraethyl orthosilicate in aqueous alcoholic ammonia formed 60 nm diameter particles which were surface modified with thiol, sulfonic acid, amine, and ammonium ion sites using silane coupling agents *(25)*. In neutral and weakly basic solutions the silica particles are negatively charged due to partial ionization of the surface silanol groups. The parent particles are colloidally stable in aqueous ethanol but not in water alone. Silica modified with thiopropyl groups slowly coagulates over a period of weeks in ethanol but is stable indefinitely in N,N-dimethylformamide (DMF). Oxidation of the thiol groups to sulfonic acid with *tert*-butyl hydroperoxide gave a strongly acidic colloidal catalyst *(26)*. The surface of colloidal silica can be modified with any silane coupling agent if a good solvent for the modified surface is used. Thus C_{18}-modified silica is colloidally stable in hydrocarbons, methacryloxypropyl-modified silica is stable in ethanol-toluene mixtures, and silica with adsorbed triblock poly(ethylene oxide-propylene oxide) is stable in water.

Attempts to convert negatively charged silica to positively charged particles with quaternary ammonium groups on the surface have resulted in very low coverage of the cationic sites.

Catalysis

We have tested most of the latexes and silicas described above as catalyst supports for oxidation or hydrolysis reactions of organic compounds summarized in Tables II and III. Details are in the original papers. Only a few comments about the catalytic processes are included here.

Table II. Oxidations with Colloidal Catalysts

reaction	catalyst[a]	support[b]	ref.
$CH_3(CH_2)_9SH + O_2$ ---> $[CH_3(CH_2)_9S]_2$	CoPcTS	SCL	14
$HOCH_2CH_2SH + O_2$ ---> $(HOCH_2CH_2S)_2$	CoPcTS	SCL, IEL, silica	c
2,6-di-*tert*-butylphenol + O_2 ---> 2,6,2',6'-tetra-*tert*-butyl-1,4-diphenoquinone	CoPcTS	IEL	30
styrene + NaOCl ---> styrene oxide	$Mn^{III}(Cl)TPPCl_8TS$	IEL	31
alkenes + NaOCl ---> epoxides and diols	$Mn^{III}(Cl)TPPCl_8TS$	IEL	d
alkenes + $KHSO_5$ ---> epoxides and diols	$Mn^{III}(Cl)TPPCl_8TS$	IEL	d
cyclooctene + H_2O_2 ---> cyclooctene oxide	$(NH_4)_6Mo_7O_{24}\cdot 4H_2O$	IEL	32
tetrahydronaphthalene (tetralin) + O_2 ---> α-tetrol + α-tetralone	$Co^{II}(pyr)_6$	ASL, ACL	24

[a]CoPcTs = cobalt phthalocyaninetetrasulfonate. $Mn^{III}(Cl)TPPCl_8TS$ = chloromanganese(III) meso-tetrakis(2,6-dichlorophenyl)porphyrintetrasulfonate. $(NH_4)_6Mo_7O_{24}\cdot 4H_2O$ forms unidentified oxomolybdate anions in hydrogen peroxide. $Co^{II}(pyr)_6$ = Co^{II} in the presence of at least 6 molar equivalents of pyridine. [b]SCL = cationic surface charged latex. IEL = cationic ion exchange latex. ASL = anionic sulfonate latex. ACL = anionic carboxylate latex. [c]Hari Babu, S., Ford, W. T., *J Polym Sci. Part A: Polym. Chem.*, in press. [d]Zhu, W., Ford, W. T., *J. Polym. Sci. Part A: Polym. Chem.*, in press.

Table III. Hydrolyses with Colloidal Catalysts

reaction	catalyst	support	ref.
$(C_6H_5O)_2P(O)OC_6H_5$-*p*-NO_2 ---> $(C_6H_5O)_2PO_2^- + ^-OC_6H_5$-*p*-$NO_2$	*o*-iodosobenzoate	IEL	22
diazinon ---> diethyl thiophosphoric acid	$-SO_3H$	silica	26
$(C_6H_5)_3CF$ ---> $(C_6H_5)_3COH$	$-SO_3H$	silica	26

We have employed only readily available primary oxidants, such as dioxygen, hydrogen peroxide, potassium monoperoxysulfate, and sodium hypochlorite, with the aim of demonstrating processes that could be used on an industrial scale. Most of the oxidations have been carried out with water-immiscible substrates in the absence of added organic solvent in triphase dispersions of water, substrate, and colloidal particles. In every case one or more of the colloidal supports increased rates of reaction by two to thirty times compared with rates using the same catalyst in aqueous dispersions lacking the colloidal supports, but we have found no new spectacularly large rate increases. In kinetic studies the method of agitation of the triphase mixtures and the relative amounts of oxidant, substrate, catalyst particles, binding sites in the particles, and binding sites occupied by catalyst all affect the rates of oxidation, and most of these parameters are interdependent. The overall kinetics of most of the reactions depend on mass transport of the water-immiscible substrate as well as intrinsic chemical reactivity. Although none of the mechanisms of these heterogeneous oxidations is well understood at present, they have sufficient potential for large scale applications that further investigation to understand the kinetics is warranted.

The autoxidation of mercaptans to disulfides catalyzed by cobalt phthalocyaninetetrasulfonate (CoPcTS, Scheme 2) is known as "sweetening" in the petroleum industry *(27)*. Schuit, German, and their coworkers discovered very high co-catalytic activity of cationic polyelectrolytes in aqueous solutions for autoxidation of mercaptoethanol *(28,29)*. In our oxidations of water-soluble mercaptoethanol (Table II) at 25 °C, surface modified silicas, surface charged latexes made from monomer **1**, and ion exchange latexes with bound CoPcTS are slightly more active than CoPcTS in the absence of cationic colloids and CoPcTS bound to commercial anion exchange resins, but much less active than the cationic polyelectrolyte [3,6]ionene $\{[(CH_2)_3N^+(Me)_2(CH_2)_6N^+(Me)_2]_n\ 2Br^-\}$ (Hari Babu, S., Ford, W. T., *J. Polym. Sci. Part A: Polym Chem.,* in press.) 1-Decanethiol was autoxidized efficiently at 35 °C with CoPcTS bound to latexes prepared from monomers **1**-**3**. The rate enhancements due to latexes were about the same as those due to CTABr micelles *(14)*.

2,6-Di-*tert*-butylphenol is converted to only a quinone dimer by dioxygen and CoPcTS on ion exchange latexes at 70 °C, whereas the monomeric quinone is formed also in water alone *(30)*. There must be a higher concentration of the active species that dimerizes in the latex phase than in a solution lacking latex. Although CoPcTS is reputed to be stable to oxidation, its visible spectrum changed and catalytic activity decreased during the oxidations.

Epoxidation is one of the most important oxidation reactions in synthetic chemistry, both for industrial production of ethylene oxide and propylene oxide and for exploratory syntheses. The epoxidations in Table II show useful selectivities but not high overall activity for oxidations of alkenes. At 25 °C the hindered Mn^{III} porphyrin $Mn^{III}(Cl)TPPCl_8TS$ (Scheme 2) bound to an ion exchange latex in strongly basic 0.38 M NaOH catalyzes the NaOCl (laundry bleach) epoxidation of styrenes but not aliphatic alkenes *(31)*. At 40 °C, cyclooctene, cyclohexene, α-methylstyrene, and β-methylstyrenes are oxidized to mixtures of epoxides and diols, and 1-alkenes fail to react (Zhu, W., Ford, W. T., *J. Polym. Sci., Part A: Polym. Chem*, in press) The relative reactivities depend upon both the nucleophilicity of the alkene toward the electrophilic oxidizing species, presumed to be a Mn^V=O porphyrin, and the solubility of the alkene in the latex, which decreases in the order aromatic alkenes > cycloalkenes > acyclic alkenes. Although the $Mn^{III}(Cl)TPPCl_8TS$, fully substituted with chlorine atoms at the 2,6-positions of the phenyl rings, is one of the more oxidatively stable Mn porphyrins known, it is deactivated by oxidation of the porphyrin ring during the NaOCl reactions. An oxomolybdate (Table II), generated from ammonium molybdate and hydrogen peroxide and bound to ion exchange latexes, also selectively catalyzes

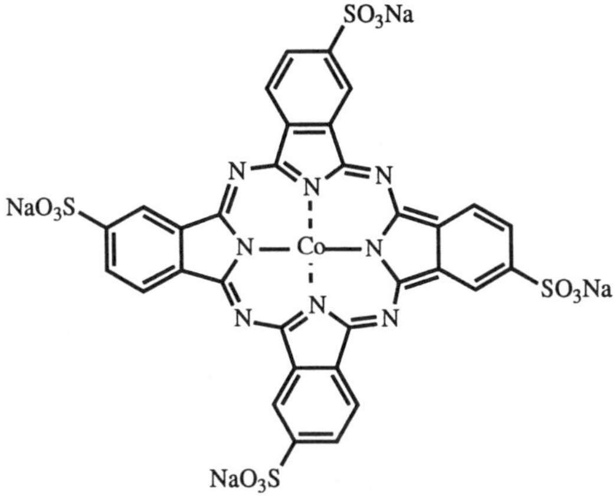

CoPcTS

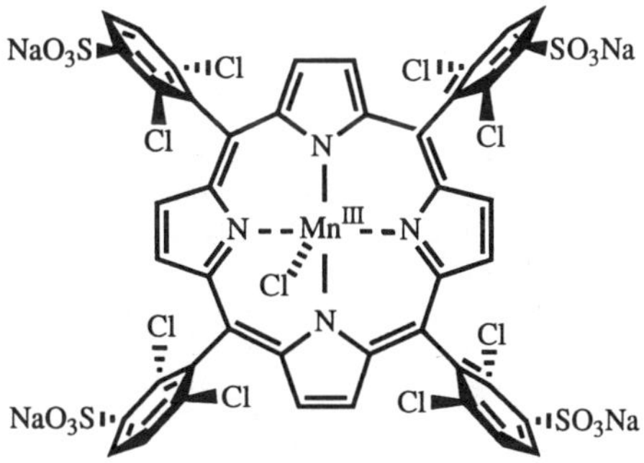

$Mn^{III}(Cl)TPPCl_8TS$

Scheme 2. Water soluble phthalocyanine and porphyrin oxidation catalysts

epoxidation of cyclooctene but not 1-octene at 40 °C *(32)*. The latex-bound oxomolybdate catalyst gradually lost half of its activity over five catalytic cycles. Potassium hydrogen monoperoxy sulfate, a commercial mixture of $2KHSO_5 \cdot KHSO_4 \cdot K_2SO_4$, is a powerful oxidant that reacts with alkenes in the absence of the Mn porphyrin catalyst and latexes *(33)*. The presence of the Mn porphyrin and latex has little effect on $KHSO_5$ activity.

The autoxidation of tetralin to tetralone (Table II) is used in the production of methyl N-(1-naphthyl)carbamate, a widely used insecticide, and cobalt ion catalyzed autoxidation is used generally for large scale oxidation of alkyl aromatic hydrocarbons to carboxylic acids, such as the conversion of *p*-xylene to terephthalic acid for production of polyesters. The Co^{II}-pyridine complex in an acrylic acid/styrene copolymer latex is only two times more active than in aqueous solution lacking latex *(24)*. The oxidation proceeds via tetralin hydroperoxide at 50 °C, a low temperature for such processes.

o-Iodosobenzoate ions are highly active catalysts for the hydrolysis of phosphate esters (Table III). Moss, Alwis, and Bizzigotti *(34)* discovered that CTACl micelles greatly increased the activity and achieved a half-life of hydrolysis of *p*-nitrophenyl diphenyl phosphate of about 10 seconds at pH 8 and 25 °C. We found that 0.2 mg/mL of ion exchange latex particles are as active as CTACl micelles, and that the latexes $25N^+$ and $50N^+$ (25% and 50% of quaternary ammonium ion repeat units) have the highest activities of the latexes in Table I.

The colloidal silica-sulfonic acid produced by surface modification of 55 nm diameter sol-gel silica has almost as high activity for hydrolysis of the insecticide diazinon (diethyl 2-isopropyl-6-methyl-4-pyrimidinyl phosphorothioate) as does aqueous HCl at 62 °C, and much higher activity than polystyrene-based gel and macroporous sulfonic acid ion exchange resins (Table III) *(26)*.

Conclusions

Latex and colloidal silica supports promote the oxidations and hydrolyses of organic compounds in aqueous dispersions at or near ambient temperature with readily available oxidants and catalysts. The new catalytic processes are attractive for their ability to work in the absence of organic solvent even with water-immiscible reactants. They might be used for the destruction of specific components of industrial wastewater or for chemical manufacturing processes. Ideally we would like to have a catalyst capable of autoxidizing all organic compounds in water to carbon dioxide, but there is not and may never be such a catalyst.

Although the use of water-immiscible substrates in our research shows the utility of colloidal catalysts, it makes the heterogeneous reactions difficult to understand. It adds complications of solubilities and mass transfer to heterogeneous catalysis problems that are complex even with water-soluble reactants. Progress toward understanding these heterogeneous reactions will require water-soluble reactants and complete analysis of the distributions of every component of the reaction mixture between the aqueous and colloidal particle phases. Those distributions might be determined both by analysis of suitable kinetic data, as in the pseudo-phase model of micellar catalysis *(35)*, and by extensive chemical analysis of the particle and aqueous phases. With known local concentrations the intrinsic rates of reaction in the particle phase can be determined quantitatively. This will enable analysis of the contributions of increased local concentrations in the particle phase and the intrinsic catalytic activity to the observed rates of reaction. Then systematic study of composition and structure of gel and surface-charged particles will reveal how the environments created via variations in colloidal particle synthesis contribute to the catalytic activity.

Acknowledgment. We thank the U.S. Army Research Office for support of our colloidal catalyst research.

Literature Cited

1. Satterfield, C. N. *Mass Transfer in Heterogeneous Catalysis*, MIT Press: Cambridge, MA, 1970.
2. Chauvin, Y.; Commereuc, D.; Dawans, F. *Progr. Polym. Sci.* **1977**, *5*, 95.
3. Pittman, C. U., Jr., In: *Comprehensive Organometallic Chemistry*; Wilkinson, G.; Stone, F. G. A.; Abel, E. W., Eds.; Pergamon: Oxford, **1983**,vol. 8, pp 553-611.
4. Garrou, P. E.; Gates, B. C. In: *Syntheses and Separations Using Functional Polymers*, Sherrington, D. C.; Hodge, P., Eds.; Wiley: Chichester, 1988, pp 123-147.

5. Bernard, M.; Ford, W. T.; Taylor, T. W. *Macromolecules* **1984**, *17*, 1812.
6. Fitch, R. M. In: *Macromolecules*; Benoit, H.; Rempp, P., Eds.; Pergamon: Oxford, 1982, p 39.
7. Hopkins, A.; Williams, A. *J. Chem. Soc., Perkin Trans. II* **1983**, 891.
8. Kitano, H.; Sun, Z.-H.; Ise, N. *Macromolecules* **1983**, *16*, 1306.
9. Sun, Z.; Yan, C.; Kitano, H. *Macromolecules* **1986**, *19*, 984.
10. Ishiwatari, T.; Maruno, T; Okubo, M.; Okubo, T.; Ise, N. *J. Phys.Chem.* **1981**, *85*, 47.
11. Kitano, H.; Nakamura, K.; Ise, N. *J. Appl. Biochem.* **1982**, *4*, 34.
12. Ise, N. *Acc. Chem. Res.* **1982**, *15*, 171.
13. Campbell, G. A.; Upson, D. A. *Macromol. Syn.* **1990**, *10*, 1.
14. Hassanein, M.; Ford, W. T. *Macromolecules* **1988**, *21*, 525. Hassanein, M.; Ford, W. T. *J. Org. Chem.* **1989**, *54*, 3106.
15. Choubal, M.; Ford, W. T. *J. Polym. Sci. Part A: Polym. Chem.* **1989**, *27*, 1873.
16. Bader, H.; Dorn, K. I.; Hupfer, B.; Ringsdorf, H. *Adv. Polym. Sci.* **1985**, *64*, 1.
17. Fendler, J. H.; Tundo, P. *Acc. Chem. Res.* **1985**, *14*, 45.
18. Hayward, J. A.; Johnston, D. S.; Chapman, D. *Ann. N. Y. Acad. Sci.* **1985**, *446*, 267.
19. O'Brien, D. F.; Klingbiel, R. T.; Specht, D. P. Tyminski, P. N. *Ann. N. Y. Acad. Sci.* **1985**, *446*, 282.
20. Sadownik, A.; Stefely, J.; Regen, S. L. *J. Am. Chem. Soc.* **1986**, *108*, 7789.
21. Tsaur, S.-L.; Fitch, R. M. *J. Colloid Interface Sci.* **1987**, *115*, 450.
22. Ford, W. T.; Yu, H. *Langmuir* **1991**, *7*, 615.
23. Kim, J. H.; Chainey, M.; El-Aasser, M. S.; Vanderhoff, J. W. *J. Polym Sci. Part A: Polym. Chem.* **1989**, *27*, 3187.
24. Chandran, R. S.; Srinivasan, S.; Ford, W. T. *Langmuir* **1989**, *5*, 1061.
25. Badley, R. D.; Ford, W. T.; McEnroe, F. J.; Assink, R. *Langmuir* **1990**, *6*, 792.
26. Badley, R. D.; Ford, W. T. *J. Org. Chem.* **1989**, *54*, 5437.
27. Frame, R. R. U.S. Patent 4,298,463, **1981**.
28. Zwart, J.; van der Weide, H. C.; Bröker, N.; Rummens, C.; Schuit, G. C. A.; German, A. L. *J. Mol. Catal.* **1977-78**, *3*, 151.
29. van Welzen, J.; van Herk, A. M.; Kramer, H.; German, A. L. *J. Mol. Catal.* **1990,** *59*, 291, 311.
30. Turk, H.; Ford, W. T. *J. Org. Chem.* **1988**, *53*, 460.
31. Turk, H.; Ford, W. T. *J. Org. Chem.* **1991**, *56*, 1253.
32. Srinivasan, S.; Ford, W. T. *New J. Chem.* **1991**, *15*, 693.
33. Zhu, W.; Ford, W. T. *J. Org. Chem.* **1991**, *56*, 7022.
34. Moss, R. A.; Alwis, K. W.; Bizzigotti, G. O. *J. Am. Chem. Soc.* **1983**, *105*, 681.
35. Bunton, C. A.; Savelli, G. *Adv. Phys. Org. Chem.* **1986**, *22*, 213.

RECEIVED December 4, 1991

Chapter 27

Nonaqueous Polymer Colloids

Stabilization of Poly(vinyl acetate) Particles with the Diblock Copolymer Poly(styrene-*b*-[ethylene-*co*-propylene])

J. V. Dawkins and S. A. Shakir

Department of Chemistry, Loughborough University of Technology, Loughborough, Leicestershire LE11 3TU, United Kingdom

Nonaqueous dispersions of poly(vinyl acetate) in n-alkanes have been prepared in the presence of the diblock copolymer poly(styrene-b-[ethylene-co-propylene]) (S-EP) as steric stabilizer. The particle size over the range 0.13 to 0.31μm may be varied by performing seeded radical dispersion polymerizations of vinyl acetate as a function of monomer content and concentration of S-EP diblock copolymer. Estimates of surface coverage of the stabilizing EP copolymer on particles have been obtained. Controlled studies of flocculation with particles in a dispersion medium of a binary liquid mixture of n-heptane and n-propanol as a function of temperature suggest that the steric stabilization mechanism operates for particles stabilized with surface layers of EP copolymer. The thickness of the surface layer was determined from viscosity studies of the dispersions at 298, 308 and 318 K. The dimensions were observed to be somewhat larger than the dimensions of free EP copolymer chains in solution.

The term polymer colloids to describe polymer dispersions with particle diameters below 10 μm has been used for twenty years following an ACS Symposium organised by Fitch *(1,2)*. A later Symposium Volume illustrated the substantial interest in dispersions of polymer particles in water *(3)*. Nonaqueous polymer colloids are prevented from flocculation by steric stabilization in which each particle is surrounded by a surface layer of adsorbed polymeric stabilizer *(4)*. Initial studies of nonaqueous polymer colloids involved dispersion polymerization in the presence of a graft copolymer of a monomer dissolved in a diluent which is a precipitant for the polymer *(5)*. The stabilizing copolymer contains A blocks which are insoluble in the dispersion medium and act as anchors for the B blocks which are swollen by the diluent and which extend away from the particle surface. Much of this early work was concerned with the preparation and properties of nonaqueous poly(methyl methacrylate) (PMMA) dispersions *(5)*, and Fitch and Kamath reported on the interfacial behaviour of graft copolymers stabilizing PMMA particles *(6)*, Our work has involved the preparation of non-aqueous PMMA dispersions in aliphatic hydrocarbons stabilized with well-defined AB diblock copolymers based on

0097–6156/92/0492–0432$06.00/0

polystyrene anchor blocks *(7,8)*. Extensive studies of the properties of these PMMA dispersions stabilized by AB poly(styrene-b-dimethyl siloxane), abbreviated to PS-PDMS, and by AB poly(styrene-b-[ethylene-co-propylene]), abbreviated to S-EP, have been reported *(9-12)*.

The preparation of non-aqueous polymer colloids other than PMMA in the presence of stabiliser containing A blocks of PS is of interest because effective anchoring of the copolymer may be influenced by the degree of compatibility between the PS anchor blocks and the polymer chains in the particle core. Interpretation of scattering intensities from small angle neutron scattering experiments on PMMA particles stabilized in n-alkanes with PS-PDMS indicated segregation of the PS blocks into domains at or near the surfaces of the particles *(13)*. It would appear that effective anchoring arises because the PS blocks which are incompatible with PMMA chains become trapped within a hard polymer matrix. The present study on poly(vinyl acetate) (PVAC) particles was therefore undertaken in order to examine stabilization for a core polymer having a glass transition temperature below the dispersion polymerization temperature. Whilst there have been studies of PVAC particles stabilized in hydrocarbon media by grafted chains *(14-17)*, the only work involving block copolymers appears to be preliminary studies involving PS-PDMS *(18,19)*. These initial experiments on dispersion polymerisation of VAC with PS-PDMS in aliphatic hydrocarbons had indicated that effective anchoring for soft particles might require covalent grafting of the PS blocks to some core PVAC chains *(18,19)*. Here, we report on the dispersion properties of non-aqueous polymer colloids of PVAC stabilized by S-EP.

Experimental

Block copolymer and dispersions. Samples of the S-EP diblock copolymer were kindly provided by Dr B. Wright, Shell Research Centre, Thornton, Chester, Dr A. Bull, Shell Research BV, Amsterdam and Ms K.F. Churchley, Shell Centre, London. Characterization data for S-EP provided by Dr Wright were number average molar mass M_n = 104,000 g mol^{-1}, weight average molar mass M_w = 118,000 g mol^{-1}, and a styrene content of 38.5% by weight, indicating a molar mass M_{EP} = 64,000 g mol^{-1} for the EP copolymer block. Our characterization experiments by gel permeation chromatography (GPC) indicated that S-EP had a polydispersity M_w/M_n ~1.1. This S-EP diblock copolymer has, therefore, a well-defined structure and had been produced by hydrogenating the polyisoprene block in a diblock copolymer of polystyrene-polyisoprene synthesized by anionic polymerization *(20)*.

Vinyl acetate (VAC) monomer (Aldrich Chemical Co. Ltd., stabilized with 4 ppm hydroquinone and 300 ppm diphenylamine) was degassed, and distilled under vacuum when required for use. The initiator azobisisobutyronitrile (AIBN) was double recrystallised from ethanol. The S-EP sample was dispersed in an n-alkane, typically n-heptane (which had previously been dried over molecular sieve, degassed, and distilled under vacuum), by first leaving the mixture overnight at room temperature and then raising the temperature of the stirred mixture to 343 K for 30 min. The apparatus consisting of a round-bottomed flask with a side arm and equipped with stirrer, stopper and condenser contained nitrogen gas throughout. A seeding technique was incorporated into the dispersion polymerization because single stage ("one-shot") polymerizations provided somewhat broader particle size distributions. A typical seed stage involved adding monomer (20% by weight of the total VAC with the equivalent proportion of AIBN) very quickly to the dispersion medium containing the S-EP stabilizer at the polymerization temperature 343 K. After this addition, the seed dispersion was allowed to form for 13 hours. When the seed stage had been accomplished, the growth stage was started by adding the remaining VAC (with AIBN) incrementally as a feed over a period of 2 hours. A typical small-scale dispersion polymerization consisted of S-EP (1g), heptane (20g),

VAC (3.3 g) and AIBN (0.06 g), with a monomer conversion to PVAC of about 70 per cent. Dispersions obtained with polymerization times up to 50 hours were stored at ambient temperature.

Dispersions were washed by repeated centrifuge/diluent exchange cycles to remove unadsorbed S-EP and unconverted monomer. Analysis of the supernatant by infrared spectroscopy demonstrated that six such redispersion cycles were usually sufficient to reduce the excess stabilizer to negligible proportions. Redispersion also provided a way of exchanging the dispersion medium for a different one, and products prepared were redispersed in hexane and heptane. Determinations of the critical flocculation temperature (CFT) on cooling were performed with dispersions in a cell designed and constructed to be accommodated in a Unicam SP600 UV-visible spectrophotometer operating at 600 nm. The sample compartment containing the dispersion was surrounded by a jacket containing water circulated from an external thermostated bath. A rotating magnet beneath the cell rotated a bar stirrer in the sample compartment to ensure constant and efficient mixing of the dispersion. A dispersion in n-heptane (polymer content $2x10^{-3}$ g cm^{-3} in 10 cm^3) was added to the cell at 298 K. Addition of n-propanol (99.9% Aristar grade from B.D.H.) dropwise to the stirred dilute dispersion provided a dispersion medium of n-heptane/n-propanol (70:30, v/v). The stirred contents of the cell were heated to at least 5°C above the CFT and then allowed to cool at the rate 1°C per 360 s, and the temperature at which a significant change in transmittance was observed was recorded as the CFT. Typically, a significant change in transmittance occurred over a temperature range as the dispersion was cooled, and so the estimate of CFT has an experimental error $< \pm 0.4$ K. Flocculation was noticed to be reversible, and addition of further n-heptane or an increase in temperature produced deflocculation. Stopping the stirrer at the flocculation point caused a remarkable decrease in the turbidity as the flocs settled.

Values of mean particle diameter D were estimated from transmission electron micrographs (TEM) by the method described previously *(7)*. The polydispersity of the particles was estimated to be less than 1.03. The size of a particle core was assumed constant on heating from 298 to 318 K. The surface coverage of a particle by S-EP was estimated from the PS content determined by UV spectrophotometry (Kontron UVIKON810) on dry particles at 272 nm. During the preparation of dispersion samples for TEM examination, the swollen surface layer of EP chains collapses onto the surface of a particle when the dispersion medium is removed. From the surface coverage data, the thickness of this interfacial layer of bulk EP copolymer was estimated to be about 2 nm which represents 1.5% increase in the diameter of the smallest particles. Thus, the thickness of the collapsed layer could be neglected, and the TEM diameter was taken as the core diameter.

Viscometry. The relative viscosity η_r of dispersions with a volume fraction ϕ of dispersed phase in the range 0.02-0.12 was measured with a Cannon-Fenske type of capillary viscometer having a capillary diameter 0.55 mm. This diameter was large compared to the diameters of the dispersion particles; thus, corrections for wall-effects could be neglected *(10)*. Flow times for the pure dispersion medium were typically of the order of 200 s and reproducibility was better than $\pm$ 0.1%. Relative viscosities were determined for dispersion particles in n-heptane at three different temperatures, 298, 308 and 318 $\pm$ 0.02 K. The relative viscosity was converted to an absolute viscosity with literature data for the viscosity of the dispersion medium *(21)*. Cumulative errors arising from dilution procedures were avoided by gravimetrically determining the polymer content of samples at each dilution. The viscometer was washed with filtered heptane and filtered chloroform and dried between each determination. Possible deposition of dispersion particles on the walls of the viscometer was avoided by silylating the viscometer and all glassware with a solution of chlorotrimethylsilane (10% w/v) in chloroform *(10)*. Glassware was baked for several hours at 373 K before cooling, then treated with the silylating agent for 24 h, and finally washed thoroughly with filtered chloroform before drying. The silylation of the viscometer in such a manner remained effective for at least six months.

Results and Discussion

Dispersion Polymerization. The progress of polymerization for the growth stage was followed for a range of polymerizations with the AIBN concentration fixed at 1 weight per cent based on monomer and VAC concentration in the range 10-30 weight per cent based on diluent in the presence of S-EP having a concentration in the range 1-6 weight per cent based on diluent. From these polymerizations, a major aim was a controlled particle size coupled with a narrow particle size distribution. Seeded polymerizations were employed to produce PVAC particles for studies of flocculation and for characterization of dispersions by viscometry. The concentration of S-EP stabilizer is an important factor controlling nucleation in the seed stage. Figure 1 demonstrates the variation of particle diameter of dispersions prepared in the presence of different concentrations of S-EP for [VAC] held constant at 20 weight per cent. As the stabilizer concentration increased, smaller particles were produced, as predicted by the theories of particle formation *(5)*. The data in Figure 1 plotted on logarithmic axes gave a straight line which obeyed the relationship

$$D \propto c^{-0.62} \qquad (1)$$

in which D is the average particle diameter and c is the concentration of the S-EP stabilizer in solution. A similar relationship for dispersions of PMMA stabilized by graft copolymers has been reported by Barrett *(5)* who found that the exponent was in the range -0.5 to -0.6. Dawkins and Taylor *(7)* reported an exponent of -0.77 for dispersions of PMMA stabilized with PS-PDMS. Dispersion polymerizations of VAC were usually performed in the presence of 5 weight per cent S-EP in solution. It should be noted that despite this relatively high concentration, only up to 50% of the S-EP stabilizer was actually incorporated onto PVAC particles. It appeared that higher concentrations were required to prepare stable dispersions since the adsorption of the S-EP block copolymer onto the particle surface might not be readily achieved owing to the low miscibility of PS anchor blocks with the PVAC chains in the particle core.

Figure 2 shows the variation of mean particle diameter prepared with an increasing proportion of monomer in the seed stage with [S-EP] held constant at 5 weight per cent. As the monomer content of the seed stage increased, larger particles were produced, until in the limit all the monomer in the seed stage corresponds to a one-shot polymerization when PVAC particles with a mean diameter of 0.48 μm were produced. The process of particle formation begins when a polymer chain (oligomer) grows in solution until it reaches a threshold molar mass at which it will precipitate and contribute to nucleus formation. The results in Figure 2 for a total overall monomer concentration fixed at 20% for each dispersion are consistent with the proposal that a lower monomer content in the seed stage produces low solvency of the dispersion medium for the propagating chains which precipitate forming more initial nuclei. Since the total overall [VAC] concentration is fixed, smaller particles must result. Our seed-feed method was to employ 1/5 of the total overall [VAC] in the seed stage. The variation of mean particle diameter with the total overall monomer concentration for these variable seed (1/5 of VAC) experiments is shown in Figure 3. These results demonstrate that somewhat higher concentration of VAC during the seed stage together with an increase in VAC concentration at the feed stage produced larger particles.

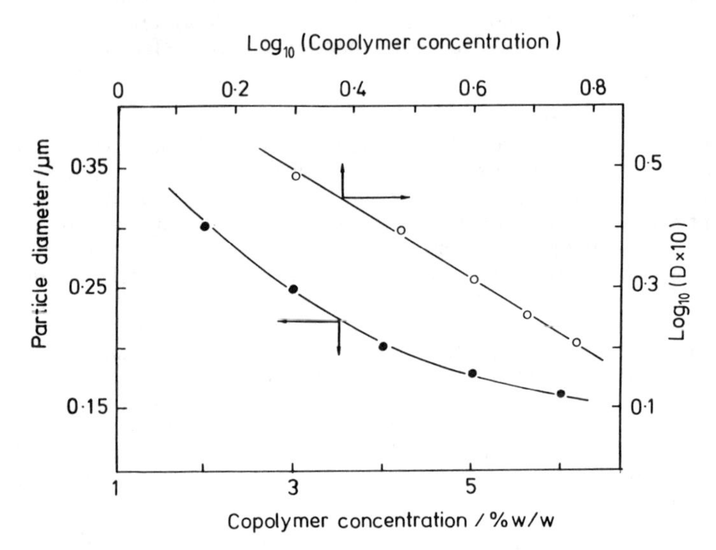

Figure 1. Dependence of mean particle diameter on the concentration of S-EP diblock copolymer in dispersion polymerizations of vinyl acetate.

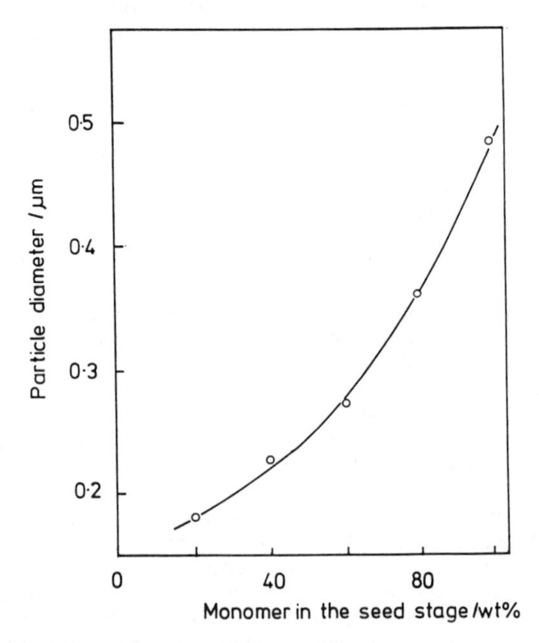

Figure 2. Dependence of mean particle diameter on monomer content in the seed stage in seeded dispersion polymerizations of vinyl acetate.

Particle Characterization and Stability

The surface coverage of PVAC particles by EP copolymer chains can be estimated from the copolymer content and particle diameter. In Figure 4 we display the percentage of block copolymer stabilizer for PVAC and PMMA dispersions which had been freed of excess non-adsorbed S-EP and then dried. It follows that the amount of S-EP surrounding PVAC particles is higher than for PMMA particles and that the block copolymer concentration increased as the particle size decreased. The content of EP copolymer per polymer particle can be calculated from Figure 4 since the block copolymer composition of S-EP was known.

Surface coverage results can be interpreted in terms of the surface area A occupied or stabilized by each EP copolymer chain and the mean separation distance d between adjacent EP copolymer chains, assuming that the PS block in the S-EP stabilizer does not extend significantly into the dispersion medium, that each EP copolymer chain is terminally anchored at the particle surface, and that an EP copolymer chain is anchored at the centre of a regular hexagon of area A. The data for A in Table I indicate that there is no obvious dependence of the area stabilized per chain on particle size, so it can be assumed that total surface coverage is attained. The value of $A = 30\ nm^2$ for an EP copolymer chain is lower for PVAC particles than $A = 44\ nm^2$ previously reported for PMMA particles *(11)* which follows from the block copolymer contents displayed in Figure 4. It may be concluded that the EP stabilizing chains are packed more closely around the surface of a PVAC particle than a PMMA particle, and we calculate that d = 6 nm for PVAC particles compared with the result d = 7 nm previously reported for PMMA particles *(12)*. For a free EP copolymer chain it is calculated that the root-mean-square radius of gyration $<s^2>^{0.5}$ is at least 9 nm *(12)*. With $d < 2 <s^2>^{0.5}$ it is indicated that close-packing of the stabilizing chains promotes overlaps between neighbouring EP copolymer chains in order to attain total surface coverage of particles.

PVAC dispersions free of excess S-EP exhibited excellent long-term stability (over many months). Values of CFT for PVAC dispersions are shown in Table I. It was demonstrated that there was no variation in CFT with particle concentration up to $2 \times 10^{-2}\ g\ cm^{-3}$, in agreement with previous observations for PMMA particles *(9,11)*. As with the surface coverage data in Table I, there is little or no dependence of the CFT on D over the particle diameter range 0.13 to 0.31 μm. In determinations of the CFT with the liquid mixture having a composition n-heptane/n-propanol (70:30, v/v), PVAC dispersions retained stability at experimental conditions when PMMA dispersions flocculated with CFT = 315.4K *(11)*. Theta conditions for this same liquid mixture have been determined for EP copolymer chains, and θ was found to be 317K *(11)*. These results indicate that these dispersions retained stability at theta conditions, with flocculation occurring when the dispersion medium was just worse than a theta system for the EP copolymer chains.

Further work is required to explain the different flocculation behaviour of PVAC and PMMA particles. One possible contribution might be preferential sorption of one liquid component by the stabilizing chains or by the particle core which may change the solvency of the dispersion medium around the stabilizing chains. This has been considered previously in connection with the proposal that preferential sorption of n-propanol on polymer particles may occur *(11)*, but the significant quantity of PS blocks per particle (see Figure 4 and Table I) could provide a barrier to uptake of liquid components from the dispersion medium into the particle core. Another explanation may follow from how S-EP adsorbs onto the two types of polymer particles. We have observed that d is somewhat higher for EP copolymer chains on PMMA particles than on PVAC particles, and the surface layer thickness is higher for EP copolymer chains on PVAC particles than on PMMA particles (see Table II). Figure 4 indicates that the segment density of EP copolymer chains is higher for PVAC particles. Preliminary studies *(22)* indicate that the mode of anchoring of S-EP through PS blocks is different for the two types of particles which

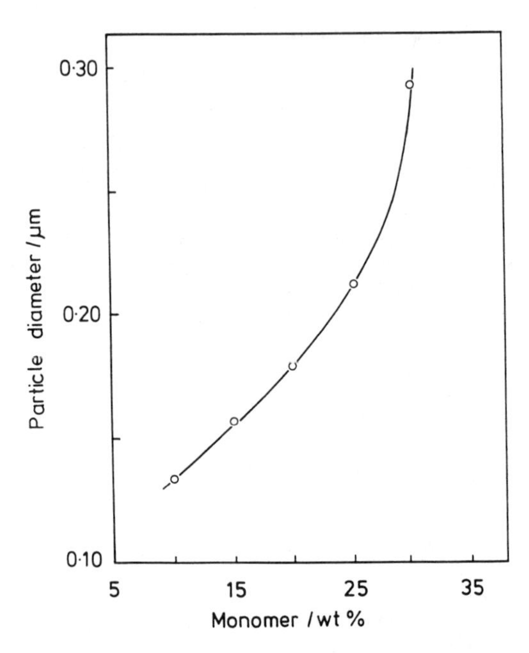

Figure 3. Dependence of mean particle diameter on total monomer concentration in seeded dispersion polymerizations of vinyl acetate.

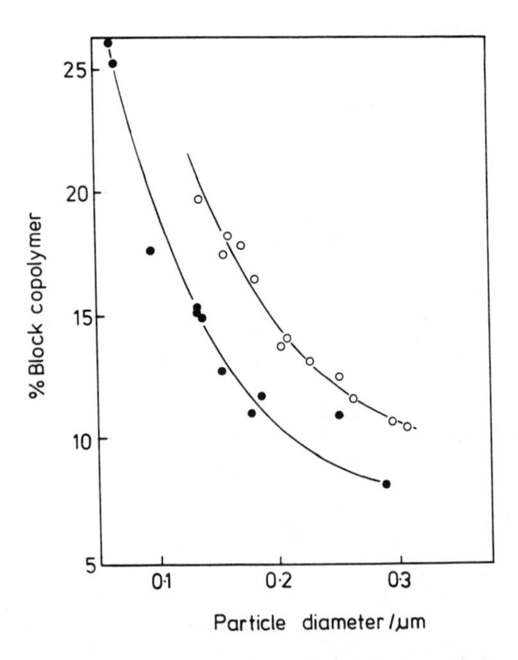

Figure 4. Dependence of concentration of stabilizing S-EP copolymer (%, w/w) on mean particle diameter. o, PVAC particles; ●, PMMA particles from references 11 and 12.

Table I. Dependence of flocculation behaviour on particle size and surface coverage

D/nm	S-EP/ % (w/w)	A/nm^2	CFT/K[a]
136	19.8	30.7	312.9
158	18.3	29.1	313.1
180	16.5	29.0	313.1
203	13.6	32.3	313.0
250	12.5	28.9	312.9
309	10.6	28.1	313.0

where D is mean particle diameter, A is the surface area stabilized by each EP copolymer chain, and CFT is the critical flocculation temperature.

[a]particles in n-heptane/n-propanol (70:30, v/v)

Table II. Chain dimensions of EP copolymer in n-heptane

Temperature/K	δ(PVAC)/nm	$\langle r^2 \rangle^{0.5}$/nm[a]	δ(PMMA)/nm[a]
298	36	25	26
308	41	25	28
318	45	24	30

where δ is the hydrodynamic thickness for the EP surface layer for PVAC and PMMA particles and $\langle r^2 \rangle^{0.5}$ is the root-mean-square end-to-end distance for an EP copolymer chain.

[a]reference 12

may contribute to these observations, and a more detailed study of block copolymer attachment will be considered in a subsequent paper.

Viscosity

Plots of the dependence of the absolute viscosity of PVAC particles in n-heptane at 298 K against the volume fraction of the polymer particle core ϕ_o are shown in Figure 5. The value of ϕ_o was calculated from the total polymer content in the dispersion (obtained gravimetrically), the EP content (obtained from surface coverage data for dry particles), and the density of the particle core (assumed to be 1.19 g cm^{-3} for bulk PVAC *(23)*). The value of the hydrodynamic thickness δ for the EP surface layer may be calculated from the limiting slope of the curves at $\phi_o = 0$ in Figure 5 according to the method employed by Walbridge and Waters *(24)*. However, this method may not be accurate because of errors in locating the curves in Figure 5 at low values of ϕ_o, and so we have preferred to use the procedure involving the relative viscosity η_r reported by Barsted et al. *(25)* The dependence of η_r on ϕ_o is given by Equation (2) which was proposed by Saunders *(26)* who modified the relation suggested by Mooney *(27)*

$$\phi_o/\ln \eta_r = (1/\alpha_o f)-(k\phi_o/\alpha_o). \quad (2)$$

In Equation (2) α_o is the Einstein coefficient (equal to 2.5 for a dilute suspension of rigid non-interacting spheres), k is a crowding factor and the factor f, as defined by Maron et al. *(28)* is given by

$$f = \phi/\phi_o \quad (3)$$

where ϕ is the total volume fraction of the dispersed phase. Viscosity data plotted according to Equation (2) are shown in Figures 6 and 7 for PVAC particles stabilized in n-heptane with S-EP at 298, 308 and 318 K. The quantity $\alpha_o f$ in Equation (2) may be thought of as an effective Einstein coefficient which may be determined from the intercepts in Figures 6 and 7. For dispersions of particles having a value of δ which is significant compared with D, Goodwin *(29)* proposed that the magnitude of f was correctly given by

$$f = [1+(2\delta/D)]^3. \quad (4)$$

and so the effective Einstein coefficient is given by

$$\alpha_o f = \alpha_o [1+(2\delta/D)]^3. \quad (5)$$

Values of $(\alpha_o f)^{1/3}$ determined from the intercepts in Figures 6 and 7 were plotted against the reciprocal of D, as shown in Figure 8 which confirms the linear behaviour predicted by Equation (5) for the range of particle core diameters (130-300 nm) examined. As D^{-1} tends to zero, the surface layer becomes negligible relative to the core diameter and the effective Einstein coefficient approaches the true Einstein coefficient of 2.5. From the intercept in Figure 8 the value of α_o was found to be in good agreement with the true Einstein value, suggesting that the PMMA particles were spherical and free from aggregation. The sphericity of the particles was indicated from TEM observations. The linearity of the plots in Figure 8 implies that δ at each temperature was essentially constant over the particle size considered. Therefore, from Equation (5) values of δ at each temperature may be found from the ratio of the slope to the intercept for each plot in Figure 8. The results are shown in Table II, where the error in δ considering the experimental data for η_r and D is about 1 nm.

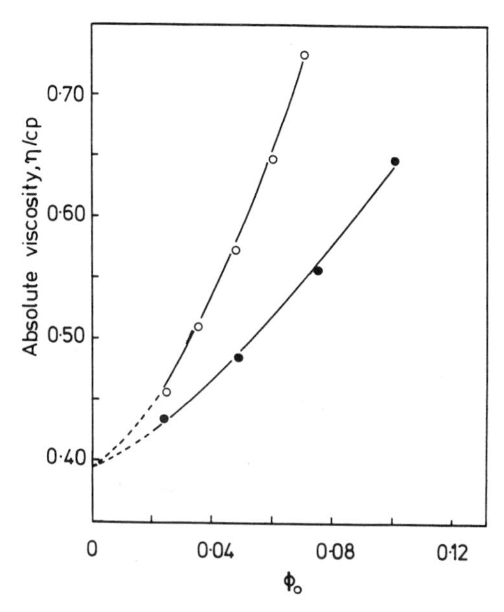

Figure 5. Variation of viscosity of PVAC particles in n-heptane at 298 K with the volume fraction ϕ_o of core polymer. o, D = 137 nm; ●, D = 292 nm.

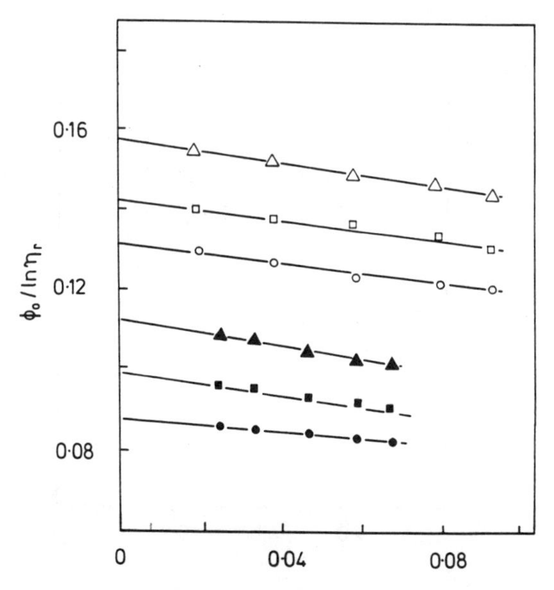

Figure 6. Plots of viscosity data according to Equation (2) for PVAC particles in n-heptane at 298 K (Δ,▲), 308 K (□,■), and 318 K (o,●). Open symbols, D = 202 nm; filled symbols, D = 137 nm

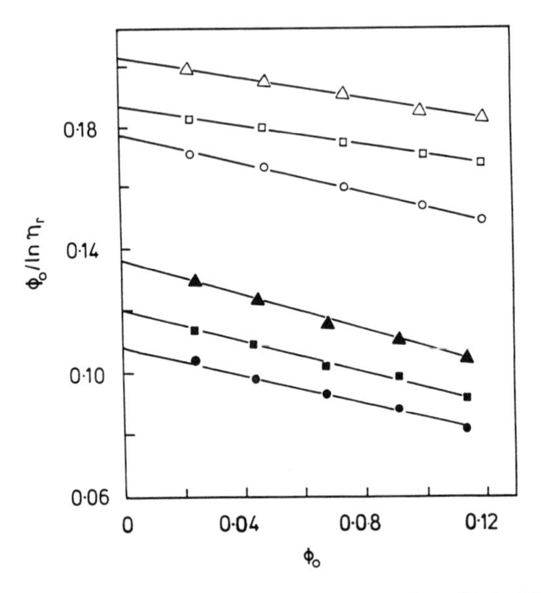

Figure 7. Plots of viscosity data according to Equation (2) for PVAC particles in n-heptane at 298 K (Δ,▲), 308 K (□,■), and 318 K (o,●). Open symbols, D = 292 nm; filled symbols, D = 170 nm.

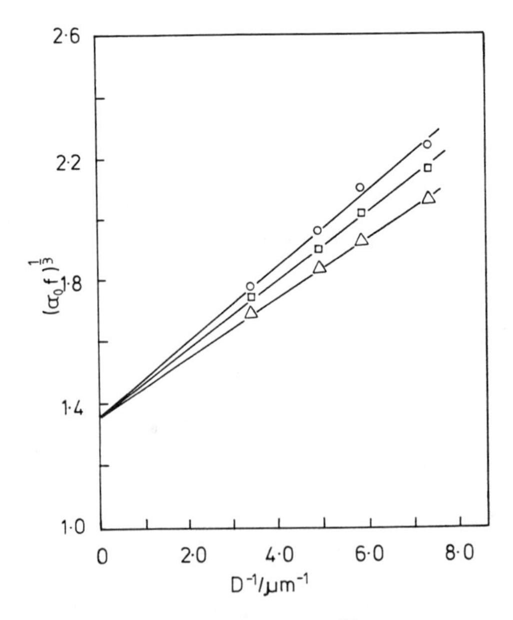

Figure 8. Dependence of the function $(\alpha_o f)^{1/3}$ on the reciprocal of mean particle diameter for PVAC particles in Figures 6 and 7. Δ, 298 K; □, 308 K; o, 318 K.

Predicted values for the root-mean-square end-to-end distance $<r^2>^{0.5}$ have been made elsewhere *(12)* and are shown in Table II. This comparison of chain dimensions indicates that EP copolymer chains terminally anchored at the interface between PVAC and n-heptane appear to be somewhat extended over random coil dimensions for free chains in solution. The surface layer thickness for EP copolymer chains on PVAC particles is larger than for PMMA particles. The dimensions of EP copolymer chains are not expected to change markedly with temperature because n-heptane may be considered to be a good solvent for free copolymer chains. The increase in δ with temperature could be overestimated because D has been estimated by TEM on dried particles. If the particles remained glassy (e.g. PMMA), then the volume change for thermal expansion on heating from 298 to 318 K would be extremely small. If the particles were at or near the glass transition temperature (e.g PVAC), then a larger volume change might be expected together with possible diffusion of n-heptane into the particle core. Consequently, a possible explanation of the data at 318 K is that δ is overestimated for PVAC particles because expansion of the PVAC core has not been incorporated into the values of D. Overall, it does appear that the surface layer is thicker than expected on the basis of the dimensions of free EP copolymer chains, which is consistent with results reported for PDMS steric stabilizers *(10)*. Furthermore, values of δ in Table II exceed 10 nm and so attractive forces between particle cores will not influence stabilization/flocculation behaviour. Consequently, the results of our controlled flocculation experiments are consistent with Napper's view *(30)* on steric stabilization of polymer colloids with long chains in a thick interfacial layer.

Conclusions

Our results demonstrate that S-EP diblock copolymer stabilizes PVAC particles during non-aqueous dispersion polymerization. Flocculation/stabilization behaviour can be interpreted in terms of surface layers of EP copolymer chains around particles. It was observed that data for surface coverage and surface layer thickness of the stabilizing EP copolymer chains were somewhat different for PVAC and PMMA particles.

Literature Cited

1. Amer. Chem. Soc. Symposium, *"Polymer Colloids"*, **1970.**
2. Fitch, R.M., Ed. *"Polymer Colloids";* Plenum: New York, **1971.**
3. Fitch, R.M., Ed. *"Polymer Colloids II";* Plenum: New York, **1980.**
4. Vincent, B. *Adv. Colloid Interface Sci.*, **1974,** *4*, 193.
5. Barrett, K.E.J., Ed. *"Dispersion Polymerization in Organic Media";* Wiley: New York, **1975.**
6. Fitch, R.M.; Kamath, Y.K. *J. Indian Chem. Soc.*, **1972,** *49*, 1209.
7. Dawkins, J.V.; Taylor, G. *Polymer*, **1979**, *20*, 599.
8. Dawkins, J.V.; Maghami, G.G.; Shakir, S.A.; Higgins, J.S. *Colloid Polym. Sci.*, **1986,** *264*, 616.
9. Dawkins, J.V.; Taylor, G. *Colloid Polym. Sci.*, **1980,** *79*, 258.
10. Dawkins, J.V.; Taylor, G. *J.C.S. Faraday I*, **1980**, *76*, 1263.
11. Dawkins, J.V.; Shakir, S.A. *Colloid Polym. Sci.*, **1987,** *265*, 329.
12. Dawkins, J.V.; Shakir, S.A. *Colloids Surfaces*, **1988,** *32*, 345.
13. Higgins, J.S.; Dawkins, J.V.; Taylor, G. *Polymer*, **1980,** *21*, 627.
14. Napper, D.H. *Trans. Faraday Soc.*, **1968,** *64*, 1701.
15. Croucher, M.D.; Hair, M.L. *Colloids Surfaces*, **1980,** *1*, 349.
16. Croucher, M.D.; Hair, M.L. *J. Colloid Interface Sci.*, **1981,** *81*, 257.
17. Croucher, M.D.; Lok, K.P.; Wong, R.W.; Drappel, S.; Duff, J.M.; Pundsack, A.L.; Hair, M.L. *J. Appl. Polym. Sci.*, **1985,** *30*, 593.
18. Dawkins, J.V.; Taylor, G.; Baker, S.P.; Collett, R.W.R.; Higgins, J.S. *Amer. Chem. Soc. Symp. Ser.*, **1981,** *165*, 189.

19. Dawkins, J.V.; Taylor, G.; Ghaem-Maghami G.; Higgins, J.S. *Amer. Chem. Soc. Symp. Ser.*, **1984,** *240*, 267.
20. Noshay, A.; McGrath, J.E. *"Block Copolymers: Overview and Critical Survey",* Academic Press, New York, **1977.**
21. Weast, R.C. *"Handbook of Chemistry and Physics",* 53rd edition, CRC Press, Cleveland, Ohio, **1972.**
22. Shakir, S.A. Ph.D. thesis, Loughborough University of Technology, **1987**
23. Broens, O.; Muller, F.H. *Kolloid Z.*, **1955**, *140*, 121.
24. Walbridge, D.J.; Waters, J.A. *Discuss. Faraday Soc.*, **1966,** *42*, 294,
25. Barsted, S.J.; Nowakowska, L.J.; Wagstaff, I.; Walbridge, D.J. *Trans. Faraday Soc.*, **1971,** *67*, 3598.
26. Saunders, F.L., *J. Colloid Sci.*, **1961,** *16*, 13.
27. Mooney, M., *J. Colloid Sci.*, **1951,** *6*, 162.
28. Maron, S.H.; Madow, B.P.; Krieger, I.M. *J. Colloid Sci.*, **1951,** *6*, 584.
29. Goodwin, J.W., In *"Colloid Science";* Everett, D.H., Ed.; Specialist Periodical Report, The Chemical Society, London, **1975;** *Vol. 2,* Chapter 7.
30. Napper, D.H. *"Polymeric Stabilization of Colloidal Dispersions",* Academic Press, New York, **1983.**

RECEIVED December 4, 1991

INDEXES

Author Index

Affiliation Index

Subject Index

A

K

L

M

O

P

T

W

Z

Production: Betsy Kulamer and Paula M. Bérard
Indexing: Deborah H. Steiner
Acquisition: A. Maureen Rouhi and Anne Wilson
Cover design: Peggy Corrigan

Printed and bound by Maple Press, York, PA